MICROCOMPUTER ORGANIZATION
Hardware and Software

Virginia Polytechnic Institute and State University

MICROCOMPUTER ORGANIZATION

Hardware and Software

Macmillan Publishing Company

New York

Collier Macmillan Publishers

London

Printed in the United States of America

Macmillan Publishing Company
866 Third Avenue, New York, New York 10022

Collier Macmillan Canada, Inc.

Library of Congress Cataloging in Publication Data

Holt, Charles A. (Charles Asbury)
Microcomputer organization.

Includes index.
1. Microcomputers. 2. Computer architecture.
I. Title.
QA76.5.H636 1985 001.64 84-17124
ISBN 0-02-356350-8

Printing: 2 3 4 5 6 7 8 Year: 6 7 8 9 0 1 2

ISBN 0-02-356350-8

PREFACE

This book was prepared for use *in a first course treating digital fundamentals, principles of computer organization, and programming concepts.* Development of system aspects is focused on the personal computer (PC), and the numerous examples are based largely on the IBM PC, which is introduced in Chapter 1 and referred to throughout the text. This is an excellent machine for illustrating basic concepts. Although not as sophisticated as the large mainframe data processors that serve multiple users concurrently, it is a powerful single-user microcomputer with advanced features.

There are various other types of PCs that are similar to the IBM PC, containing essentially the same operating system, the same processor, the same BASIC, and compatible hardware. In fact, the MS Disk Operating System (DOS) of the PC has become somewhat of a standard among PCs with 16-bit processors, and there is little difference in the various versions of BASIC that are used. Hence the assembly language and BASIC programs developed here apply also to these computers. Actually, the intent is not to describe a particular device but *to develop in the student a deep appreciation of the principles of computer ogranization in general, with the PC serving as a tool to illustrate the fundamentals.* While student access to a PC or a compatible machine is helpful, it is certainly not essential.

The text is intended primarily to serve the needs of college students in engineering and computer science. These students are the ones who will design the computers of the future. However, the material should be helpful to those who wish simply to understand the hardware and software that control the PC and similar machines. A thorough understanding of the computer and its operating system, including peripherals and their control, leads to greater proficiency in its use. Prerequisites are *an alert mind, a willingness to study with diligence, and an aptitude for working with numbers*.

This study of basic concepts has a span extending from elementary logic circuits to a complete hardware system, which is interconnected with many input/output components. Emphasized is hardware control by an operator who sends commands through a complex software interface system. The software is based on both assembly language and BASIC. All symbols and logic diagrams conform to IEEE standards.

Bit storage in registers and memory, timing with clocks, coding systems, and an introduction to BASIC programming are all found in Chapter 1, and elementary arithmetic operations, pseudo code, and flowcharts are described in Chapter 2. Immediately thereafter comes the organization of a central processor unit based on the architecture of the 8088 microprocessor, along with assembly language programming. Throughout the book, programming concepts are developed, often with programs that accomplish the same task performed in both assembly language and BASIC. *This parallel treatment of a low-level and a high-level language serves to reinforce basic programming concepts,* such as stacks, loops, and subroutines, and shows clearly the relative advantages and disadvantages of the two methods.

As new principles of computer organization are introduced, related programs are designed. To illustrate, the logic functions described in Chapter 4 are accompanied by logic instructions in both assembly language and BASIC. Input and output instructions and methods accompany the discussion of the programmable peripheral interface (PPI) chip and the timer/counter chip in Chapter 5. Software interrupts are examined along with hardware interrupts in Chapter 6. *This procedure allows the various instructions to be presented gradually, with new ones often related directly to hardware.* A summary of the complete assembly language instruction set is given in the appendixes, and all instructions of both languages are listed in the index for easy reference.

Program creation with an editor, followed by assembly, linking, and debugging, are described in detail in Chapter 7. Reference here is to the line editor, the linker, and the debugger programs of the PC, and the assembler is the IBM MACRO Assembler used for the 8088 processor. Design of both execution (EXE) and command (COM) programs is studied. Also included is debugging of BASIC programs, along with treatment of error and event trapping.

The material that follows Chapter 7 includes direct memory access, the floppy disk system, communications, operating systems, and file management, with considerable emphasis on related assembly language programs. In particular, the diskette read and write programs that are presented *relate the activities of five large-scale integrated circuits, showing how the components fit together and function as a unified system*. These are the interrupt controller, the microprocessor, the direct memory access controller, the floppy-disk controller, and the timer/counter.

The era of data processing is here, and computer usage is not only commonplace but growing rapidly. Many of us are fortunate enough to own PCs or to have access to one, and most engineers, scientists, and other professional men and women work frequently with large computing systems. Often the interface is through a PC. Perhaps the user may never design a digital system. However, *familiarity with computer organization and logic networks provides confidence, increased satisfaction, and greater skill.*

Preparation of the text was fun, even though tedious at times. Many readers will undoubtedly find this extremely important subject to be exciting. It is hoped that the student will be stimulated so that the study will continue. The topics are presented in sufficient depth to provide a firm foundation for investigation of computer systems at a more advanced level.

C. A. H.

Blacksburg, VA

CONTENTS

MICROCOMPUTER ORGANIZATION
Hardware and Software

CHAPTER 1

Computers, Codes, and Numbers

1-1

BASIC COMPUTER CONCEPTS

Digital integrated circuits (ICs) are the building blocks of stored-program general-purpose computers, often referred to as *data processing systems*. Such computers are governed by a *program* consisting of instructions that are executed in sequence. The program and associated data are stored in a read/write memory, which is commonly called *random access memory,* or *RAM*. Random access means that each storage location of the RAM can be reached in approximately the same amount of time. Digital circuits are also used in other systems, such as calculators, instruments, and teletypewriters.

Binary States

In the usual digital system, the electric voltages are restricted to two discrete levels, such as 0 V and 5 V, which are called *binary states*. Because it is impossible to fix a voltage level precisely, each state actually has a range of allowed values, but neither of the two states is defined by a value that lies between the specified ranges. For example, the low state might be defined as a voltage from 0 to 1 V, with the high state having a voltage between 3 and 5 V. In this case, the region from 1 to

3 V must be avoided, as well as values less than 0 and greater than 5 V. The gap between the states is a measure of the amount of noise that can be tolerated.

It is not essential that the signal variable be a voltage. An electric current can be used, or perhaps light or pressure, provided that circuits are designed to provide the discrete states. However, most digital circuits give discrete voltage levels, as shown in Fig. 1-1. As indicated in the figure, the high voltage is called the *high-level H*, or *state H*, and the low voltage is called the *low-level L*, or *state L*.

A logic variable of a digital circuit is a signal that assumes one of two states, the 1 state or the 0 state. These states are represented by the two voltage levels. In *positive logic systems* the 1 state is defined as the high voltage state H and the logical 0 state as the low-voltage state L. The optional modifier ''logical'' distinguishes 0 and 1 from numerical digits. In *negative logic systems* the 1 state is the low level and the 0 state is the high level.

Different systems operate with different voltage levels. Levels of 5 V (high) and 0 V (low) are common. The levels might be -0.8 V (high) and -1.6 V (low). However, all signals can be described by either the 1 and 0 or the H and L designations. We shall do so when analyzing and designing logic systems, thereby avoiding the different voltage levels of the various logic families. A signal is simply 1 or 0. Furthermore, all instructions and data stored in the memory of a computer are viewed as strings of 0s and 1s. For comparison, Fig. 1-2 shows logic levels of both positive logic and negative logic systems in relation to approximate logic level voltages of a system that operates with a supply of 5 V relative to ground (0 V).

Data sheets describing ICs are normally based on the assumption of positive logic. However, many systems utilize a mixture of positive and negative logic, and there is no need to specify state H as always 1 (positive logic) or as always 0 (negative logic). A more flexible system is one using *mixed logic*, which is examined in Chapter 4. In this book, either mixed logic or positive logic is used. Positive logic is chosen in preference to negative logic because it is more natural and common.

A binary digit 0 or 1 is called a *bit*, and a group of bits representing a number, an alphanumeric character, an instruction, or another quantity of significance is called a *word*. A word of 8 bits is a *byte*, and one of 4 bits is a *nibble*. Digital words of processing systems are transferred from one unit to another over a group of wires called a *bus*. In general, any set of conducting lines grouped together because of similar signals constitutes a bus.

Parallel transfer of 16-bit words requires a bus of 16 lines. There are external data buses between an IC and input/output (I/O) devices that supply and receive data. Data transfer between IC packages occurs on a bus. Address buses feed the address words to the memory units. There are also internal buses that transfer data within an IC, or *chip*.

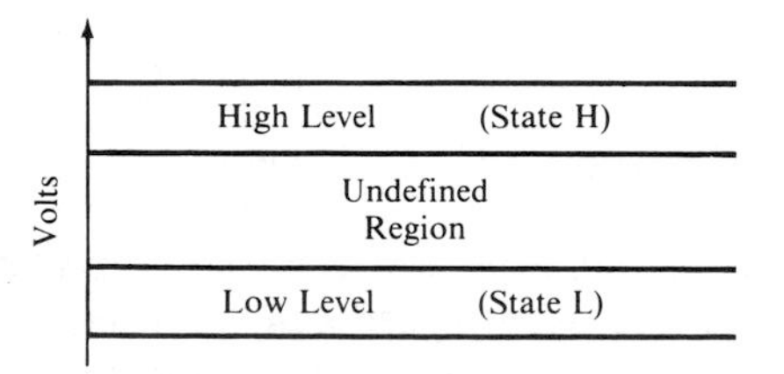

FIGURE 1-1. The voltage levels of a digital system.

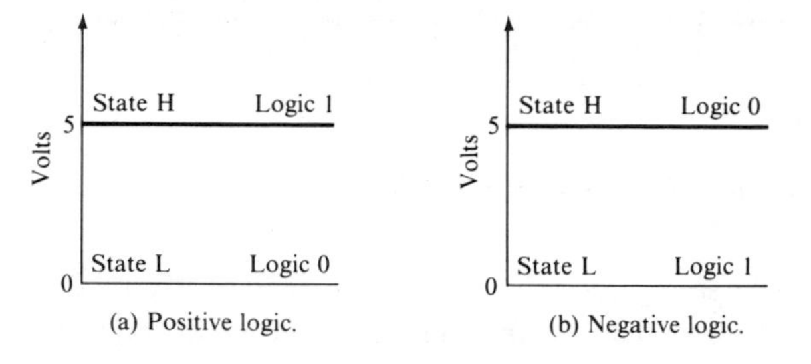

FIGURE 1-2. Comparison of positive and negative logic.

The bus lines of a chip are fabricated in parallel on a thin layer covering the silicon surface. Contacts with elements of the silicon chip are made through suitable openings in the insulating layer. Figure 1-3 shows a bidirectional 8-bit data bus connecting two ICs. Eight bits are transferred simultaneously along the eight conductors from either chip to the other.

The binary system has greater reliability than one with many states. If the difference between the two voltage levels is substantial, introduction of small noise signals has no effect on the logic states. On the other hand, for a specified voltage supply, a system with many voltage levels has a smaller voltage difference between adjacent states, which increases the probability of errors from noise voltages. The widespread use of binary systems is a direct consequence of their *very high reliability*, along with the ease of designing two-state electronic circuits.

The manufacturer of an IC publishes a data sheet describing the operation of the circuit. It is normal practice to represent the input and output logic states by the letters *L* for low and *H* for high. This system has the advantage of independence from the choice of positive or negative logic. With positive logic, *L* denotes state 0 and *H* denotes state 1.

Storage

A binary *storage cell* is an element that contains a bit. Usually the stored bit is present on an output line in the form of either a high or low voltage, and the bit is held until a write operation changes its value. With positive logic, an operation that changes the stored bit from 1 to 0 causes the voltage of the output line to switch from high to low. A group of binary storage cells with associated circuits for loading and perhaps manipulating the bits of the cells constitutes a *register*.

A 16-bit register contains a 16-bit word, an 8-bit register contains a byte, and a

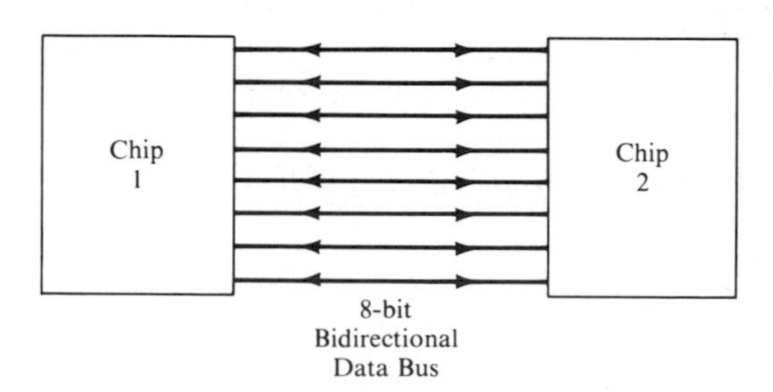

FIGURE 1-3. A data bus connecting two ICs.

0	0	0	0	0	0	0	0

(a) Clear (reset).

1	1	1	1	1	1	1	1

(b) Set (preset).

FIGURE 1-4. Register storage after CLEAR and SET.

1-bit register is a single storage cell. The stored word normally represents some form of encoded information. To *encode* means to represent an entity by means of a code, and the binary code of a register might denote an instruction, an alphanumeric character, a binary or decimal number, or some other information. A word can be written into a register or read from it.

If a cell of a register contains a logical 1, it is said to be *set*, or *preset*. On the other hand, if a 0 is present, the cell is *clear*, or *reset*. Most registers of a computer have a number of bits equal to the width of the data bus. The storage of a byte register after a CLEAR, or RESET, operation is illustrated in Fig. 1-4a, and the contents after a SET, or PRESET, operation are shown in Fig. 1-4b.

The rightmost bit of the word of a register is designated bit 0 and is referred to as the *least significant bit* (*LSB*). The leftmost bit with the highest bit number is the *most significant bit* (*MSB*). For an 8-bit register the MSB is bit 7, as shown in Fig. 1-5. Not all computers use this number system. Sometimes LSB and MSB are used to differentiate between the least and most significant bytes of a 16-bit register.

Combinational and Sequential Circuits. Digital circuits are classified as either combinational or sequential. A *combinational circuit* has outputs that depend on the binary inputs that are present at the time under consideration. For each combination of states of the input word, there is one and only one corresponding state of the output word.

On the other hand, the output word of a *sequential circuit* has more than one state for at least one combination of states of the inputs. Thus, the outputs are functions of variables such as time and the previous internal states of the element, in addition to the inputs. There is *dependence on the past history, or time sequence, of the inputs*. Accordingly, sequential circuits have memory capability. A digital circuit with at least one storage cell is sequential even though there may be numerous combinational components. An example is the register.

To illustrate, two elementary circuits are shown in Fig. 1-6. The circuit of (a) has its output Y always equal to its input X; therefore, it is a combinational circuit. On the other hand, the circuit of (b) is designed so that the state of output Y switches to the other state whenever input X goes from 0 to 1, but when X changes from 1 to 0, the state of Y is not affected. Suppose both X and Y are initially 0. If X changes to 1, output Y goes to 1 and does not return to 0 until X makes a second transition from 0 to 1. This is a sequential circuit because Y can be 1 (or 0) for both values of X, depending on the previous history.

Memory. An array of cells organized into words that can be addressed for

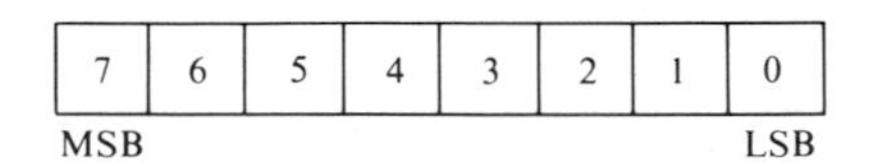

FIGURE 1-5. Bit numbers of a byte.

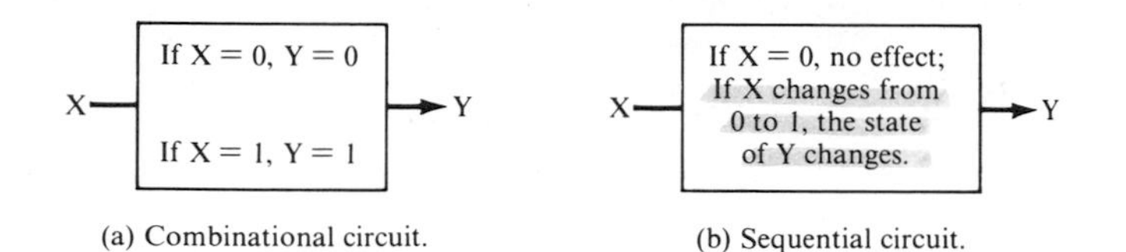

(a) Combinational circuit. (b) Sequential circuit.

FIGURE 1-6. Elementary examples of combinational and sequential circuits.

reading or writing is a *memory*. An *address* is a binary word, with each combination of the bits identifying a particular location in memory. A location may contain one or more bits, but words of 8, 16, 32, and 64 bits are common. The bits of the address are supplied to the memory in parallel on conductors of the address bus, and the stored bits of the cells are written to or read from the memory by means of the wires of the data bus. Transfer operations are controlled by signals of the *control bus*.

Memory storage is usually expressed in terms of *K*, with the letter *K* denoting 2 raised to the tenth power, or 1,024. Although *K* denotes *kilo*, it is not 1,000. To illustrate, a memory of 512K words stores 524,288 words. Frequently, the smallest addressable unit is a byte, and capacity is then expressed in *kilobytes* (*KB*). Larger memories are specified in *megabytes* (*MB*), with *M* for *mega* defined as 2 raised to the twentieth power, or 1,024 squared. Storage of a 64 KB memory is compared with that of a 1 MB memory in Fig. 1-7.

The *read-only memory* (*ROM*) is not a true memory. It is a combinational circuit, whereas a circuit with memory elements is sequential. A ROM consists of a two-dimensional matrix array of transistors, with selected transistors disconnected during the fabrication process. The presence of an active transistor in the array can be defined as a logical 1, with a disconnected transistor being a 0. Each transistor can be regarded as a cell that contains a bit. However, the state of the bit is fixed and cannot be changed by a program instruction. Thus, writing is not possible. For each word fed into the input of the combinational circuit, there is a corresponding output word, which never changes. The input word is the address, and the output is the stored word that is read by activating the address.

Both ROM and RAM are accessed by words of the address bus, and their outputs feed to the data bus. The storage of a ROM is established during fabrication. Some ROMs are designed for programming and reprogramming by the user, but the process requires hardware.

When power is removed, the storage of a RAM is lost, but not that of a ROM. Consequently, semiconductor read/write memories are said to be *volatile*, whereas ROMs are *nonvolatile*, with their 0 and 1 bits wired into the matrix. To avoid loss of programs and data of RAM when power is interrupted, it is good practice to save at frequent intervals the contents of RAM on a magnetic disk or other nonvolatile storage medium.

Figure 1-8 shows a RAM and a ROM connected to buses. Note that the data

Memory Size	Bytes of Storage	Bits of Storage
64 KB	65,536	524,288
1 MB	1,048,576	8,388,608

FIGURE 1-7. Memory storage.

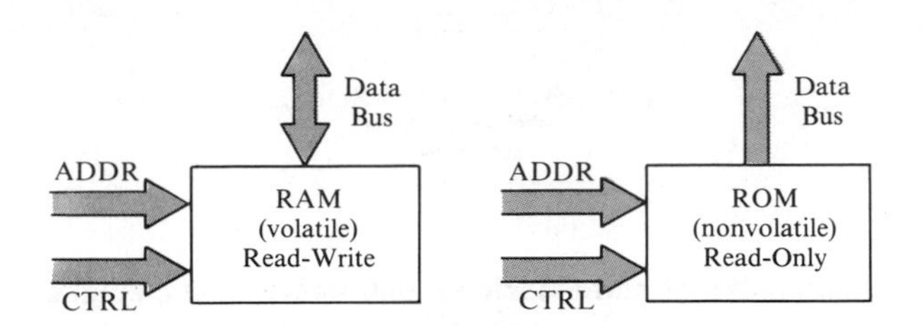

FIGURE 1-8. A RAM and a ROM and their buses.

bus of the RAM is bidirectional, whereas that of the ROM is unidirectional; data can only leave the ROM. Acronyms such as ROM and RAM are common in the literature, and those of this book are listed in the index.

1-2 MACHINES AND CLOCKS

A *digital machine* is a system consisting of both hardware and software designed to perform a specified function. *Hardware* refers to the physical circuits and related equipment, whereas *software* refers to programs that direct the operation of the machine. A program is an example of an *algorithm*, which is a step-by-step procedure for carrying out a task or solving a problem. The steps must be ordered so that when one step is done, the next one is unambiguously defined; they must also be well defined. When executed, the steps must produce a reasonable result in a finite amount of time. All algorithms must satisfy these requirements, including those of programs.

The sequence of instructions of a program is usually prepared as statements written in terms of alphanumeric characters, but such statements must always be encoded into strings of 0s and 1s before they can be accepted by the processing system and stored. This encoding is usually done by software. A general purpose digital processing system functions as different machines at different times, depending on the specific program being executed.

Most software processes can be transformed into equivalent hardware processes, and hardware can often be replaced with software. For example, multiplication of two binary numbers can be accomplished with a suitable program that utilizes the adder, but the need for such a program is eliminated if a hardware multiplier is available. Hardware is fast, but inflexible, and it adds to the cost of the equipment.

Software development is also expensive. In fact, most of the cost of designing a new digital processing system is usually for software. The system designer must weigh the advantages and disadvantages of any trade-offs that can be made between hardware and software. Of course, there must always be software to direct the hardware processes, and there must always be hardware to process software algorithms.

Computer Block Diagram

Figure 1-9 shows a block diagram of a digital computer. Program instructions and data are stored in the memory unit, and data are processed and manipulated in the *processor*. Included in the processor is an *arithmetic-logic unit* (*ALU*). Within the

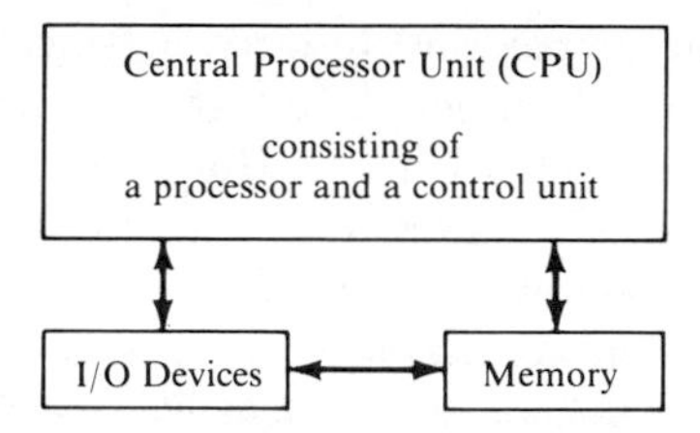

FIGURE 1-9. Block diagram of a digital computer.

control unit is a *master clock generator* that generates a pulse train of alternate 0 and 1 states used to synchronize the timing of the subsystems. At each clock pulse, the control unit provides a set of signals that initiate and control the transfer and processing of data as specified by the program stored in memory. Thus, the controller supervises the timing and execution of the program instructions, with the actual implementation occurring in the processor.

The combination of processor and control unit is called the *central processor unit* (*CPU*), which is the ''heart'' of a computer. The computer's ''brain'' is the program that supervises its operation. A CPU fabricated as a single IC on a small silicon chip is called a *microprocessor*, and a computer using a microprocessor as its CPU is known as a *microcomputer*. Many microprocessors are designed for general-purpose use, whereas others are dedicated to a specific purpose, such as appliance control or perhaps for use as an interface unit that connects two components of a system.

The I/O devices of the block diagram of Fig. 1-9 are *peripherals* that transfer information into and out of the system. Examples are keyboards, paper tape readers and punches, card readers and punches, analog-to-digital (A/D) and digital-to-analog (D/A) converters, magnetic disk units, cathode ray tube (CRT) displays, and teletypewriters. Communication between the computer and its peripherals is aided by interface chips, which are often referred to as *device controllers*.

Figure 1-10 shows a simplified block diagram of a typical personal computer system. The various I/O devices of the figure will be examined later.

The Clock

The sequential circuits of a computer are usually clock controlled. A *clock*, or pulse generator, is a circuit that generates at its output a periodic train of pulses at

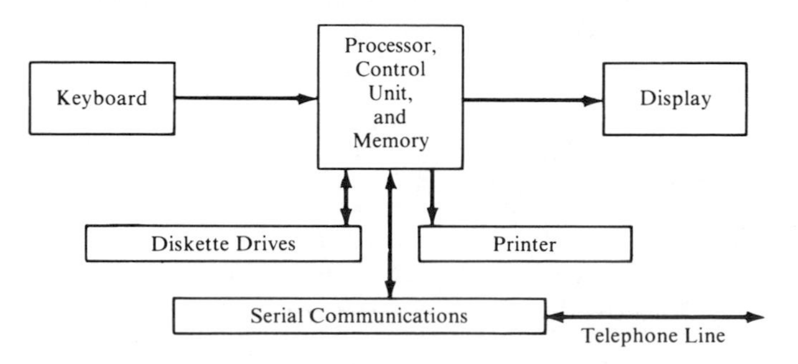

FIGURE 1-10. Main elements of a personal computer.

a fixed frequency. Figure 1-11 shows the output waveform of a typical clock. Sometimes the output of the clock is controlled, or *gated*, so that only a specified number of pulses are supplied, or perhaps only a single pulse. However, the output is normally a pulse train that continues as long as the power is on. The *frequency* in pulses per second, or *hertz*, is the reciprocal of the *period* T in seconds. Clock periods are commonly specified in microseconds or nanoseconds (1 nanosecond = 10^{-9} second). Frequency is often given in kilohertz (kHz) or megahertz (MHz).

During the brief instant when a pulse rises from L to H, the slope of the pulse waveform is positive, defining the *positive* (rising) edge of the pulse. The state change is described as a *positive transition* or a *low-to-high transition*. In function tables of data sheets this transition is indicated by an arrow ↑ or by the symbol ╱. When a pulse falls with a negative slope, which defines the *negative* (falling) edge, the state change is a *negative*, or *high-to-low transition*, which is often represented by ↓ or ╲.

A *positive pulse* is one with state changes from L to H and then to L, whereas a *negative pulse* goes from H to L to H. In both types, the edges are referred to as *leading* or *trailing*, with the leading edge of the pulse occurring first. The pulse train of Fig. 1-11 has alternate positive and negative pulses, and the leading edge of a positive pulse is the trailing edge of a negative pulse.

Pulses do not change state instantaneously. The time required for the voltage of a pulse to rise from 10% of the final voltage to 90% of the final voltage is the *rise time*, and the time taken to drop from 90% to 10% is the *fall time*. With the pulse width measured between the midpoints of the positive and negative edges, the *duty cycle* of a periodic pulse train is the ratio of the pulse width to the period. A square wave has a duty cycle of 50%.

EXAMPLE 1

Suppose that each period of the periodic pulse train of a clock has a 50-nanosecond low state followed by a 75-nanosecond high state. Determine the period, the frequency, and the duty cycle.

SOLUTION

The period is the sum of 50 and 75 nanoseconds, or 125 nanoseconds, which corresponds to a frequency of 8 MHz. Dividing 75 nanoseconds by the period gives 0.6, or 60%, for the duty cycle. ■

Synchronous and Asynchronous Systems. Systems in which continuous time is divided into discrete intervals by means of a clock are said to be *synchronous*. The outputs of the digital circuits are allowed to change only on the *active* clock transition, which is the positive or negative transition as specified. All circuits switch in unison at a rate determined by the clock frequency. Unclocked networks

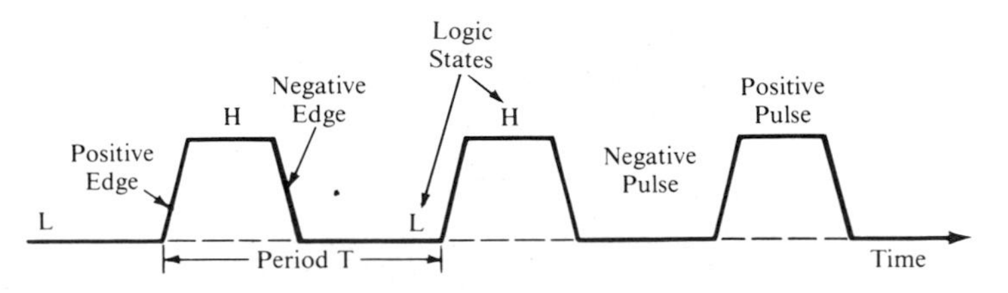

FIGURE 1-11. Output voltage waveform of a clock.

and those that operate with two or more independent clocks are classified as *asynchronous*. Most digital sequential systems are basically synchronous, but *transfer of data between various components of a system is often asynchronous*.

Control signals fed into an IC are synchronous if they become effective only on an active clock transition. In contrast, asynchronous controls are effective without delay, independent of the clock. Both types of controls are usually present.

EXAMPLE 2

The output Q of a certain 1-bit register always takes the value of the bit at input D when the clock input goes low to high, provided inputs CLR and PRESET are both low. At other times, input D has absolutely no effect on the stored bit. Whenever input CLR is high with PRESET low, the cell is clear (Q = 0), regardless of the state of the clock, and whenever input PRESET is high, Q is set to 1.

These conditions are shown in Fig. 1-12. Symbol X denotes a *don't-care state*, defined as one with a value that has no effect whatsoever on the output. Identify inputs D, CLR, and PRESET as either synchronous or asynchronous controls.

SOLUTION

Input D is a synchronous control, effective only on the active clock transition. At all other times, input D is disabled. However, both CLR and PRESET are independent of the clock; therefore, these are asynchronous controls. Note that the design gives PRESET precedence over CLR. Thus, the cell is set whenever PRESET is high. ■

Shift Registers

Data transmission within the CPU is normally along buses that have a separate line for each bit of the transmitted word. The bits are moved simultaneously from the source to the destination. This movement is called *parallel transfer*. However, communication lines often provide only two conductors for data transfer. In such cases, the bits of a word are transmitted *serially*, one bit at a time, with the LSB sent first and the MSB last. In asynchronous communications systems, start and stop bits are used to indicate the beginning and end of a word, and error detection bits may be included.

It is not our purpose here to examine communications systems. This is the subject of Chapter 10. However, because the keyboard and various other I/O devices transmit data serially, an understanding of the hardware that provides parallel-to-serial and serial-to-parallel data conversion is desirable. The conversion is usually done with a *shift register*. Another application of shift operations is encountered

CLR	Preset	D	CLK	Q (after CLK transition)
0	0	0	↑	0
0	0	1	↑	1
0	0	X	not ↑	no change
1	0	X	X	0
X	1	X	X	1

FIGURE 1-12. Register conditions for Example 2.

in multiplication and division of binary numbers, a topic examined in the next chapter.

Figure 1-13 is a block diagram of a shift register assumed to have eight storage cells and appropriate control circuitry. The IC has eight I/O pins connected to the bidirectional data bus, and one for serial input and another for serial output. Bit number 0 is always present on the serial output line, and it is also the LSB of the parallel data bus. The signal ground is connected to pin GND. At the CLK input is the periodic pulse train of the system clock. The shift register is a sequential circuit.

An *active control input* is defined as an input at the voltage level selected by the designer to implement the function of the control, whereas an *inactive control input* has the other level. Unless a control is indicated as active low, it is understood to be active high. All inputs to the control block of the figure are active high.

At the CLK input is a small triangle known as a *dynamic indicator*. It indicates that the active clock transition is low to high. Synchronous inputs can change the value of stored bits only during this positive transition. Omission of the symbol implies that controls are able to implement operations at any time during the high state of the clock. In this case, the CLK input is simply an enable pin of the chip, often labeled EN.

Inputs HOLD, CLEAR, and BUS EN are specified as asynchronous signals for this particular shift register. If HOLD is active, the stored bits are fixed, regardless of the states of other inputs. Input CLEAR is used to clear all 8 bits of the register. Whenever CLEAR goes high, all bits become 0s provided HOLD is inactive low. Normally, BUS EN is active high, in which case the stored bits are present on the eight lines of the data bus.

Three-State Outputs. Each parallel output of the shift register has three states. These are designated 0, 1, and *high impedance*. The high-impedance state effectively disconnects a cell from its data bus line, and the output is said to be *floating*. Many ICs are designed with such three-state, or *tristate*, outputs. The advantage is that several chips can have outputs wired to the same bus and still be electrically isolated, but no more than one chip can be in the nonfloating state at any instant.

Three-state pins are not restricted to those connected to a data bus, but often include address and selected control pins. When the BUS EN input of Fig. 1-13 is active, the stored bits are directly coupled to the lines of the bus, but when inactive, the outputs are floating.

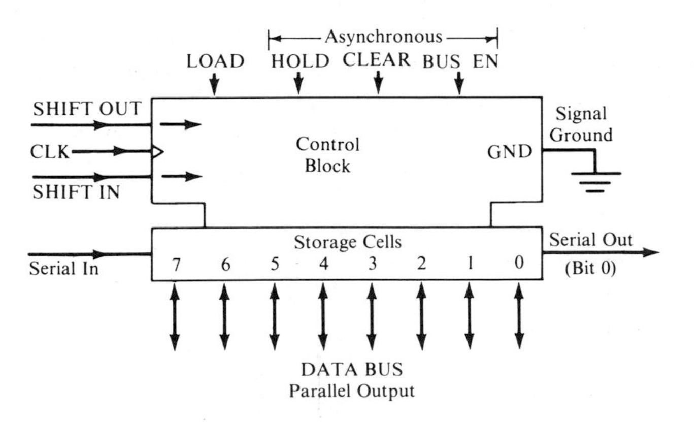

FIGURE 1-13. An 8-bit shift register.

Register Control. Controls SHIFT OUT, SHIFT IN, and LOAD are synchronous, being effective only on the active clock transition. An active SHIFT OUT with both HOLD and CLEAR inactive provides a 1-bit right shift on each positive clock transition. Each stored bit is shifted to the adjacent cell at its right, the original value of bit 0 disappears, and the new value of bit 0 appears on the serial output line. With SHIFT IN inactive, a 0 replaces bit 7. SHIFT IN also provides right shifts, but the bit on the serial input line replaces bit 7, and again bit 0 is lost. With SHIFT IN active, the status of SHIFT OUT is unimportant. Loading of the register from the data bus is done on the positive clock transition when LOAD is active, assuming HOLD and CLEAR are inactive. LOAD has priority over the shift controls.

The small arrow inside the control block adjacent to a shift input is standard notation denoting a right shift, and arrows on control lines are optional provided they do not contact the symbol of the circuit.

To transmit a byte from the CPU to a device connected to the serial data lines, the byte is placed on the data bus and the processor supplies the LOAD bit. This procedure loads the byte into the register. SHIFT OUT is then activated and held for a time interval of eight clocks. Bit 0 is transmitted first, followed by the others in succession. The process is repeated for each byte to be sent. It is assumed that start and stop bits are not used here.

For input from the peripheral, the processor activates SHIFT IN for eight clocks, and the byte enters the register from the serial input line. The LSB is received first and becomes data bit 0 after eight entries. Bits of the stored byte are present on the data bus and are read by the processor. Reading does not affect the stored bits.

EXAMPLE 3

Suppose the shift register of Fig. 1-13 is cleared by means of a positive pulse applied to pin CLEAR. With LOAD, HOLD, and CLEAR inactive and with line SERIAL IN held high, control SHIFT IN is activated for precisely four clock pulses. This operation is followed by an active SHIFT OUT for four clocks. Determine the byte of the data bus and the bit of line SERIAL OUT if BUS EN is active and also if it is inactive.

SOLUTION

After four 1s are shifted in, the storage of the register is that of Fig. 1-14a, and line SERIAL OUT is low, as shown. Following the SHIFT OUT operation, with SHIFT IN inactive, the storage is that of Fig. 1-14b, with line SERIAL OUT high. In both cases, the bits of the cells are present on the data bus if BUS EN is active. Figure 1-14c shows the effective status of the data bus if the storage is that of (b) with BUS EN inactive. ■

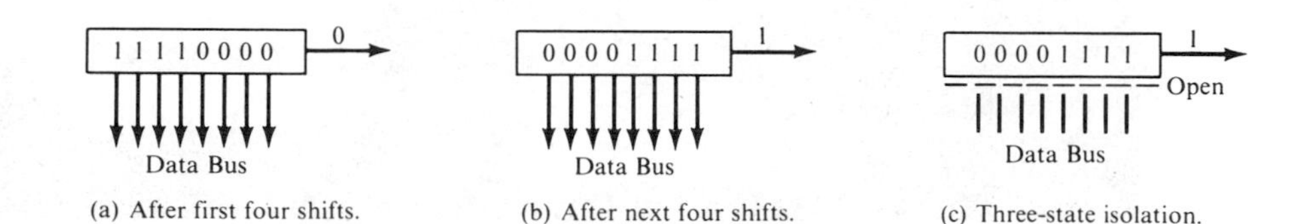

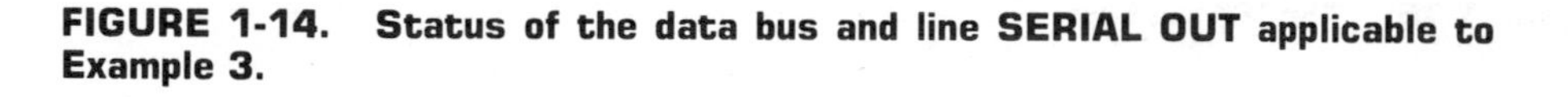
(a) After first four shifts. (b) After next four shifts. (c) Three-state isolation.

FIGURE 1-14. Status of the data bus and line SERIAL OUT applicable to Example 3.

There are different kinds of shift registers. Some have a set of parallel input lines, along with a separate set of parallel outputs. Capability for both right and left shifts may be present, and control inputs often provide for preset, increment, decrement, and other functions. A register that is incremented by 1 on each active clock transition is an *up-counter*, and one that is decremented is a *down-counter*.

The IBM Personal Computer

Many computer concepts, digital functions, and applications are best illustrated by reference to a real system, thereby allowing investigations and designs to be precise. For this purpose, a personal computer has been chosen. Many such computers, although relatively simple, have many of the basic properties of the faster, more powerful, and more expensive *minicomputers* and even *mainframe* machines, which are designed to serve a large number of users concurrently. Furthermore, the personal computer, or PC, is widely used in business establishments and research laboratories, as well as in the home.

There are a number of high quality PCs available. The IBM PC is especially appropriate for our study because it is built around the high performance Intel 8088 microprocessor. In addition, it is widely used. There are numerous personal computers that are compatible with the IBM PC, containing essentially the same hardware and software control. The PC is a specific example, *selected to illustrate basic computer concepts*. Another machine could have been chosen, even a hypothetical one, as is sometimes done.

The Computer System. There are different configurations to meet the needs of various users. The one emphasized in this book consists of the basic computer with 256 KB of user RAM and dual double-sided *diskette drives*, along with a *monochrome display monitor*, a *color monitor*, a *dot matrix printer*, and a *modem* for communications via the telephone. A single diskette provides nonvolatile storage up to 360 KB of data. Except for an internal multifunction board and the Hayes Smartmodem 1200, everything is IBM equipment. The control software, or *operating system*, is MS DOS 2.1. Figure 1-15 is a photograph of the hardware.

At the left is the *system unit*, which contains the computer board, or *motherboard*, and also the two drives for the *floppy disks*. On top of the system unit is the modem, and to the right is the monochrome monitor, which normally displays

FIGURE 1-15. The PC system.

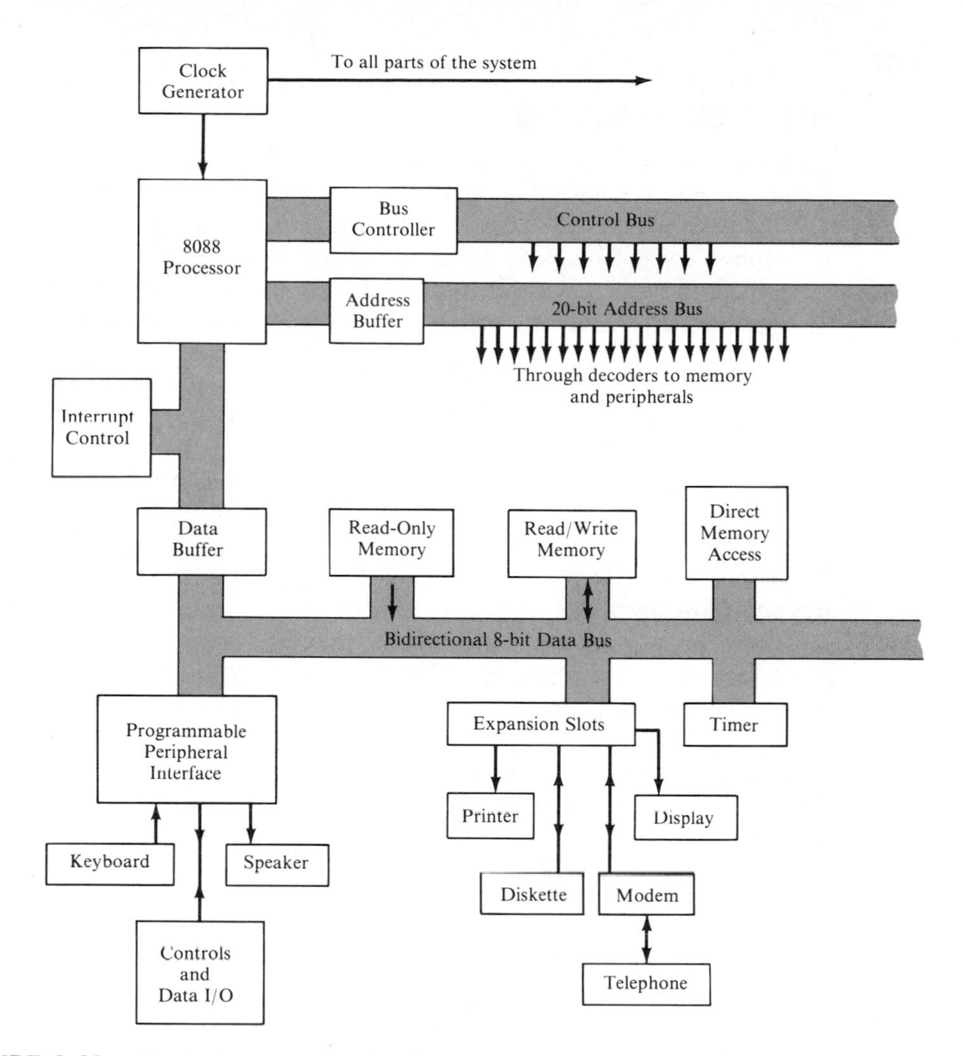

FIGURE 1-16. Block diagram showing the components, buses, and interconnections of a typical PC.

green characters on a black background. It is ideal for word processing and program development, because its high resolution and long-persistence green phosphor screen provide for easy reading with minimum eye strain. In front of the monochrome monitor is the keyboard, and at its right is the RGB (red-green-brown) color monitor. Although not as satisfactory for word processing, it is excellent for color graphics. At the far right is the printer.

Figure 1-16 shows a simplified block diagram of a personal computer system. The Intel 8088 microprocessor is connected to memory and peripherals through data, address, and control buses. Included are support chips, such as the clock generator, the bus controller, and the interrupt controller. The complete system will henceforth be referred to simply as the *PC*. Most of the components are examined in this book, *with the primary purpose being to present basic microcomputer organization.*

NUMBER SYSTEMS

Knowledge of the most common number systems used by computer systems is essential. Let us begin with a review of the familiar decimal system. The base, or *radix*, is 10, and there are 10 digits—0 through 9. In a count starting at 0, the first 10 numbers are 0 through 9, exhausting the available digits. Hence the count proceeds from 9 by placing a 1 in the second column and repeating the digits 0 through 9 in the first column, and so on. The number 402.6 represents

$$402.6 = 4*10^2 + 0*10^1 + 2*10^0 + 6*10^{-1}$$

Because the asterisk is the BASIC symbol for multiplication, it is so employed in this text. From left to right, the digits of 402.6 have *weights* of 100, 10, 1, and 0.1. If each digit is multiplied by its weight, the sum of the results is the value of the number.

Binary Numbers

The *binary system* has a base of 2. Thus there are only two digits—0 and 1. The first two numbers of a count are 0 and 1. When 1 is added to 1, the sum is 0 with a carry of 1 into the second column, giving 10. Next in the count is 11. The addition of 1 to 11 gives 0 in the first column with a carry of 1 to the second column; this combines with the 1 of column 2 to give 0 with a carry into the third column. The result is 100, and the count continues from here, yielding 101, 110, 111, 1000, and so on. These operations are illustrated in Fig. 1-17.

0	1	10	11	100	101	110	111
+ 1	+ 1	+ 1	+ 1	+ 1	+ 1	+ 1	+ 1
1	10	11	100	101	110	111	1000
(1)	(2)	(3)	(4)	(5)	(6)	(7)	(8)

(Decimal values of the results)

FIGURE 1-17. Additions of 1 in binary.

The binary number 101.1 is converted to decimal as follows:

$$101.1B = 1*2^2 + 0*2^1 + 1*2^0 + 1*2^{-1} = 5.5D$$

The letter *B* following a number denotes binary, and *D* denotes decimal. The use of *D* is optional, with no specification interpreted as decimal. Thus, the number 10 is understood to be decimal 10, whereas 10B equals decimal 2. Note that the powers of 2 begin at 0 for the binary digit at the left side of the radix point. A zero digit contributes nothing. Accordingly, 10001001B represents the decimal number $2^7 + 2^3 + 1$, or 137.

An *integer* is defined as a number without a decimal point. The LSB of an integral binary number has weight 1, and the weights double with each leftward bit position. For a nibble the weights are 8421. The decimal equivalent of a binary number is found by multiplying each bit by its weight and adding the results. For

1111B the decimal equivalent equals the sum of the weights, or 15, and for 1010B the sum is 8 + 2, or 10.

Decimal-to-Binary Conversion

An easy way to convert a decimal number to binary is illustrated in Fig. 1-18a. The decimal number, which is 137, is placed at the right and divided by 2, giving 68 with a remainder of 1. These numbers are entered in the adjacent column as shown. Then 68 is divided by 2, giving 34 with a remainder of 0, and these numbers are entered in the next column. The procedure continues until the entry in the top row is 0. The bits of the bottom row form the binary number 10001001, which is decimal 137.

When a decimal number that is to be converted to binary is fractional, the fraction is treated separately. For example, suppose the number is 137.8125. The fraction 0.8125 is placed at the left of the table of Fig. 1-18b and multiplied by 2, giving the integer 1 and the fraction 0.625, which are placed in the adjacent column. The new fraction, 0.625, is multiplied by 2, and the fraction and the integer are entered in the next column. The procedure continues until the entry in the top row becomes 0 or until the number of digits obtained gives the desired accuracy. The binary fraction 0.1101 of the bottom row of Fig. 1-18b equals the decimal fraction 0.8125, and binary 10001001.1101 equals decimal 137.8125.

EXAMPLE 1

Convert 16384.6 to binary.

SOLUTION

Conversion of the integral part of the number, which is the number to the left of the decimal, is shown first. The result is 100000000000000, which is 2 raised to the power 14.

0	1	2	4	8	16	32	64	128	256	512	1024	2048	4096	8192	16384
1	0	0	0	0	0	0	0	0	0	0	0	0	0	0	

Conversion of the fractional part to 12 significant binary digits follows. The complete binary number is found to be 100000000000000.100110011001, which is

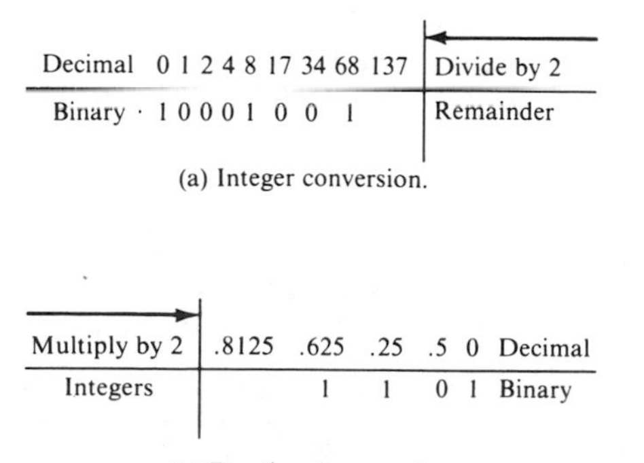

Decimal	0	1	2	4	8	17	34	68	137	Divide by 2
Binary ·	1	0	0	0	1	0	0	1		Remainder

(a) Integer conversion.

Multiply by 2	.8125	.625	.25	.5	0	Decimal
Integers		1	1	0	1	Binary

(b) Fractional conversion.

FIGURE 1-18. Decimal-to-binary conversion.

precisely equal to 16384.599853515625D. The result is not the same as the specified decimal number because there is no exact binary equivalent of decimal 0.6.

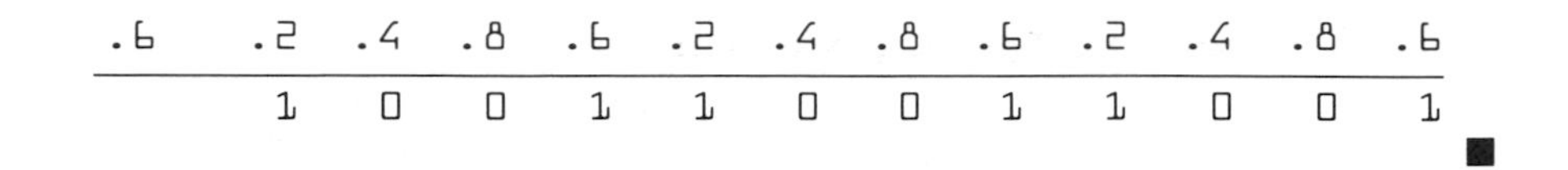

Octal Numbers

The *octal* number system, with base 8, has eight digits, which are 0 through 7. The first eight numbers are 0 through 7. The next eight are 10 through 17, and so on. The octal number 452.6 represents the following:

$$452.6Q = 4*8^2 + 5*8^1 + 2*8^0 + 6*8^{-1} = 298.75D$$

The letter *Q* or *O* following a number denotes octal; *Q* is preferred because *O* is easily mistaken for zero. Three binary digits can represent the octal digits 0 through 7. Binary 000 is octal 0 and binary 111 is octal 7.

To convert a binary number to octal, the binary number is partitioned into groups of 3 bits each, starting at the radix point. *Each group of 3 bits is replaced with the corresponding octal digit*. For example, with spaces ignored, binary 10 111 001. 110 1 is octal 271.64. Digits not shown at the extreme left and right sides of a number are understood to be 0s. The reverse procedure converts octal to binary; that is, replace each octal digit with the corresponding group of 3 bits, starting at the radix point. Although octal numbers are sometimes used in digital systems, the hexadecimal system has many advantages and is emphasized in this book, in addition to the very important binary and decimal systems.

Hexadecimal Numbers

The *hexadecimal* system, with base 16, has 16 digits. These are 0 through 9 followed by A, B, C, D, E, and F. Hexadecimal A equals decimal 10 and digit F is 15. The hexadecimal number FB8.C is given in Fig. 1-19, along with a table relating decimal and hexadecimal numbers.

Four binary digits can represent the hexadecimal digits 0 through F. Binary 0000 denotes hexadecimal 0 and binary 1111 equals hexadecimal F. Henceforth, numbers followed by the letter H are understood to be hexadecimal. Because a hexadecimal digit is equivalent to 4 binary bits, the hexadecimal system is especially convenient. We shall use it extensively.

Decimal	0	1	2	3	4	5	6	7	8	9	10	11	12	13	14	15
Hexadecimal	0	1	2	3	4	5	6	7	8	9	A	B	C	D	E	F

$$FB8.C = F*16^2 + B*16^1 + 8*16^0 + C*16^{-1} = 4024.75$$

FIGURE 1-19. Decimal and corresponding hexadecimal numbers, along with a hexadecimal-to-decimal conversion.

Binary-to-Hexadecimal Conversion

To convert a binary number to hexadecimal, we partition the binary number into groups of 4 bits each, starting at the radix point. *Each group of 4 bits is replaced with the corresponding hexadecimal digit.* For example, with spaces ignored, 1011 1001. 1101B, or 271.64Q, is B9.DH. If the group at either the extreme left or right has fewer than 4 bits, the missing bits are understood to be 0s.

To convert a hexadecimal number to binary, *simply replace each digit with the corresponding group of 4 binary bits*, starting at the radix point. The number 0000H denotes the 16-bit binary number 0000 0000 0000 0000. Spaces such as those used here are optional. The binary contents of a 16-bit register can be specified with 4 hexadecimal digits, whereas 6 octal digits or 5 decimal digits or 16 binary digits are required. This arrangement is a major advantage of the hexadecimal system. Furthermore, the 16 digits are not excessive in number, requiring remembrance of only 6 digits in addition to the 10 digits of the familiar decimal system.

EXAMPLE 2

(a) Convert 101110110011010.111B to octal and also to hexadecimal. (b) Also, convert 100.1 to hexadecimal.

SOLUTION

(a) The conversion of the binary number to octal and hexadecimal is as follows:

```
101 110 110 011 010 .111B        101 1101 1001 1010 .111B
 5   6   6   3   2  . 7 Q         5    D    9    A  .  E H
```

Thus the binary number equals 56632.7Q and 5D9A.EH.

(b) Following the procedures for conversion of decimal to binary, we find that decimal 100.1 is approximately equal to 1100100.000110011B, or 0110 0100 .0001 1001 1000B, or 64.198H. The decimal equivalent of this result is 100.0996. ■

BCD Numbers

Because digital systems operate with only 0s and 1s, the binary number system is the natural one to use for arithmetic operations. Normally, decimal numbers are converted to binary, arithmetic operations are performed, and the results are converted back to decimal. However, when applications involve relatively few calculations, with frequent input and output of decimal numbers, it is more efficient to perform operations directly with decimal numbers in binary code. Electronic calculators often use this procedure, and most computers have this capability.

The first step of decimal arithmetic is to convert each decimal digit into a binary code. Although various *binary coded decimal (BCD)* codes are used, the most common one is the 8421 code, in which each decimal digit 0 through 9 is replaced with its corresponding 4-bit binary number. If the MSB is 1, it has a value of 8, which is 1 times its weight. The weights of the remaining bits from left to right are 4, 2, and 1. Decimal 0 becomes 0000 and decimal 9 is 1001. Word 0111 0101.1000 in the 8421 BCD code, usually referred to simply as *BCD*, is 75.8D, but the same word in binary is 117.5D. In a computer designed to perform both binary and BCD arithmetic, each input word representing a number must be properly interpreted as either binary or BCD.

BCD Addition

A procedure for adding two decimal digits in BCD form is to add the 4-bit words using binary arithmetic. If the sum does not exceed decimal 9, the result is correct. However, if the sum exceeds 9, the binary addition does not give the proper BCD form. For example, the sum of 9 and 9 in binary gives 10010, or 12 in BCD, which is too small by 6.

The correct result is obtained by the addition of 6, which is binary 0110, to any sum that exceeds 9. The number 6 equals the number of 4-bit binary combinations that are skipped when the decimal count goes from 9 to 10. With BCD addition, a computer examines the result of the addition and adds binary 0110 to the result if it exceeds 1001B. A carry is simply added to the next higher 4-bit BCD code. Other arithmetic operations can be performed with decimal numbers encoded into BCD form.

To illustrate, the binary sum of 0101 (5D) and 1000 (8D) is binary 1101 (13D). In BCD the word 1101 has no meaning. Because this binary number exceeds 1001, we take the sum of 1101 and 0110 (6D) and obtain 0001 0011, which is decimal 13 in BCD form. The procedure can be extended to decimal numbers of more than one digit. The least significant BCD digits are added first. If the binary sum exceeds 1001, we add 0110 to the sum, and the carry of 1 is added to the sum of the next BCD digits. The process continues. An obvious disadvantage of BCD arithmetic is the inefficient use of the bits of a word. A 16-bit register can store binary numbers from 0 through 65,535, but the maximum BCD count is 9,999.

In Fig. 1-20 is shown the addition of 29 and 76 in four different number systems. In each case, the result is equal to decimal 105. In the first addition of the BCD numbers, the least significant digit is found by adding 0110 to 1111, giving 0101 and a carry to the high nibble. Because the carry changes the high nibble from 1001 to 1010, it is necessary to add 0110 to this nibble, which gives 0000 and a carry. In the hexadecimal addition, note the carry generated by the sum of D and C.

Decimal	BCD	Binary	Hexadecimal
29	0010 1001	0001 1101	1D
+ 76	+ 0111 0110	+ 0100 1100	+ 4C
105	1001 1111	0110 1001	69
	+ 0110		
	1010 0101		
	+ 0110		
	1 0000 0101		

FIGURE 1-20. Addition of 29 and 76 in four systems.

EXAMPLE 3

Add the binary numbers 01000110 and 00111000. Repeat this operation with the numbers representing binary-coded decimals.

SOLUTION

In binary, the words equal 70D and 56D. The sum is 01111110B, or 126D. In BCD the words represent 46D and 38D. The sum of the least significant nibbles is 1110, which exceeds 1001. Adding 0110 gives 0100 and a carry of 1. The sum of the most significant nibbles is 0111 and the addition of the carry gives 1000. Hence the sum of 46D and 38D is 1000 0100 in BCD, or 84D. ■

Comparison of Number Systems. Figure 1-21 lists various decimal numbers in the left column, including 0 through 16. The other columns show the equivalent numbers in octal, hexadecimal, binary, and BCD systems. In particular, note that the binary and BCD representations of the same decimal number consist of different strings of bits.

Decimal	Octal	Hexadecimal	Binary	BCD
0	0	0	0	0000
1	1	1	1	0001
2	2	2	10	0010
3	3	3	11	0011
4	4	4	100	0100
5	5	5	101	0101
6	6	6	110	0110
7	7	7	111	0111
8	10	8	1000	1000
9	11	9	1001	1001
10	12	A	1010	1 0000
11	13	B	1011	1 0001
12	14	C	1100	1 0010
13	15	D	1101	1 0011
14	16	E	1110	1 0100
15	17	F	1111	1 0101
16	20	10	10000	1 0110
32	40	20	100000	11 0010
63	77	3F	111111	110 0011
77	115	4D	1001101	111 0111
99	143	63	1100011	1001 1001

FIGURE 1-21. Representations of equivalent numbers.

EXAMPLE 4

Convert 60.25D into octal and hexadecimal forms.

SOLUTION

Decimal 60.25 is 111100.01B, found by the method of Fig. 1-18. By partitioning the bits into groups of three starting at the radix point, and replacing each group with its octal digit, we find that the octal number is 74.2Q. A similar procedure with the bits arranged into groups of four gives 3C.4H. ■

EXAMPLE 5

Convert 412Q into hexadecimal and decimal forms.

SOLUTION

Octal 412 is 100001010B, which is 10AH. Direct conversion of the octal number to decimal gives (4 * 64) + (1 * 8) + 2, or 266D. ■

EXAMPLE 6

Convert AB.DH into octal and decimal.

SOLUTION

Hexadecimal AB.D equals 1010 1011.1101B, or 253.64Q, obtained from groups of 3 bits. The decimal equivalent is (10 * 16) + 11 + 13/16, or 171.8125. The slash / is the BASIC symbol for division. ■

1-4 CODES

Before information can be processed by a computer, *it must be encoded into strings of 0s and 1s*. A particular code fed into the system may represent an alphanumeric character such as the letter *k* or the numeric 6. It may be the control word for a line feed instruction to go to the display, or perhaps it is a scan code indicating that a particular key of the keyboard is depressed. Machine instructions are encoded, of course. A certain code may indicate the status of a set of sensors of a system controlled by the computer. Additional examples of codes are BCD and binary numbers, both positive and negative, and there are numerous others.

Clearly, each encoded word must be stored in the system in the correct place at the right time so that the computer will interpret its significance properly. Much of this book is devoted to the various codes and their storage, along with the timing of their processing. In this section, we examine a popular alphanumeric code, followed by a brief study of the PC keyboard and its scan codes.

The ASCII Code

For representing alphabetic characters, decimal digits, and other symbols, a widely used system is the *American Standard Code for Information Interchange*, commonly referred to as *ASCII* (askee). It is the standard code used for exchanging information among data processing systems and associated equipment. Seven-bit words are employed, and they encode a character or a control function for each of the available 128 bit combinations. Of these, 52 represent the uppercase and lowercase letters of the alphabet, 10 identify the decimal digits, and 33 provide special characters that include punctuation marks, arithmetic symbols, and a space. The other 33, with codes ranging from 00H through 20H, are *control codes* for use with printers and other data communication devices.

Figure 1-22 presents 102 of the 128 codes. Included are the 52 uppercase and lowercase letters, the 10 decimal digits, the 33 punctuation marks and special symbols, and 7 of the 33 control codes. These control codes are 00H for null, 07H for bell, 08H for backspace, 0AH for line feed LF, 0CH for form feed, 0DH for carriage return CR, and 20H for space, or blank.

Although code 00H for null normally does nothing when sent to a printer or display, sometimes it is placed at the end of a message as an indicator, and there are many other applications. Unlike null, code 20H prints a blank space. When an LF character is sent to a display, the cursor is advanced one line, and a CR character returns the cursor to the left end of the current row. Usually CR and LF appear in sequence in messages. Form feed 0CH is used with printers to advance the print head to the start of a new form, or page. All numbers are in hexadecimal, with the high bit of the 8-bit word given as 0. This bit is not part of the standard ASCII code.

blank	20	0	30	@	40	P	50	`	60	p	70
!	21	1	31	A	41	Q	51	a	61	q	71
"	22	2	32	B	42	R	52	b	62	r	72
#	23	3	33	C	43	S	53	c	63	s	73
$	24	4	34	D	44	T	54	d	64	t	74
%	25	5	35	E	45	U	55	e	65	u	75
&	26	6	36	F	46	V	56	f	66	v	76
'	27	7	37	G	47	W	57	g	67	w	77
(	28	8	38	H	48	X	58	h	68	x	78
)	29	9	39	I	49	Y	59	i	69	y	79
*	2A	:	3A	J	4A	Z	5A	j	6A	z	7A
+	2B	;	3B	K	4B	[	5B	k	6B	{	7B
,	2C	<	3C	L	4C	\	5C	l	6C	¦	7C
—	2D	=	3D	M	4D	]	5D	m	6D	}	7D
.	2E	>	3E	N	4E	^	5E	n	6E	~	7E
/	2F	?	3F	O	4F	—	5F	o	6F	DEL	7F
null	00	bel	07	BS	08	LF	0A	FF	0C	CR	0D

FIGURE 1-22. Partial ASCII code in hexadecimal.

Among the control codes not included are codes 1CH through 1FH, which move the cursor one space to the right, to the left, up, and down, respectively. Code 0BH moves it to the top leftmost position of the display screen, a location referred to as *home*. All 33 ASCII control codes are given in Fig. 10-2.

From Fig. 1-22, we note that the code for a decimal digit can be obtained by adding 30H to the digit. For example, the code for 6 is the sum of 30H and 6, or 36H. Also, of course, a decimal digit in ASCII is converted to a binary number by subtracting 30H. These methods are often used by programs for conversion from one form to another. The code for a lowercase letter is that of the corresponding uppercase letter plus 20H.

ASCII can be transmitted with 7-bit words. Often the characters are represented by bytes, with the MSB chosen as 0 or 1. Another common choice is to make the MSB a *parity* bit for error detection. This situation is examined later. Programs are frequently stored in memory in ASCII form.

Although ASCII is standard for most data communication networks, there are other alphanumeric codes of importance. A byte can encode 256 characters, and an *extended ASCII code* that includes an eighth bit is sometimes used for this purpose. The PC has such an extended code. With the MSB equal to 0, the characters are those of the standard ASCII set, but an additional 128 characters are provided with the MSB equal to 1. These extended codes represent special symbols and characters used primarily in graphics.

Important is an 8-bit code known as *Extended Binary Coded Decimal Interchange Code* (EBCDIC), which is normally used in IBM mainframes. EBCDIC is a binary representation of the code produced by a punched card. For communications between the PC and an IBM mainframe, software must provide the necessary code translation between the extended ASCII and EBCDIC.

EXAMPLE

The following hexadecimal bytes are stored in order in RAM:

```
0B 0A 0A 20 20 20 20 46 6F 72 20 24 31 30 2C 0D 0A 0A
20 20 20 69 74 27 73 20 79 6F 75 72 73 21 0D 0A 00
```

When the top row from left to right is sent to an ASCII terminal, followed by the bottom row, determine the displayed message.

SOLUTION

The first byte locates the cursor at the home position, the next two are LFs, which move the cursor two lines down, and the four blanks shift the cursor four spaces to the right. Thus, the message starts on line 3 and is indented four spaces. In the middle of the message are a CR and two LFs. These return the cursor to the left side and move it two lines down. The three blanks give a three-space indentation. The CR and LF at the end of the message move the cursor to the left side of the next line, and the null byte signals the end.

The displayed message is

```
    For $10,

   it's yours!
```

Keyboard Codes

Attached to the system unit by means of a coiled cable is the keyboard, which contains the keys, the switch matrix, a one-chip microcomputer, and other circuits. The 6-foot shielded cable has a bidirectional data line for serial transmission of code, a bidirectional signal line used for control, a reset line for initializing the microcomputer, a 5-V power line, and GND (or 0 V). Figure 1-23 is a photograph of the keyboard.

Files. The keys of the keyboard are used for program and data entry, for creating files, and for word processing. A *file* is simply a collection of related information kept in nonvolatile storage, such as a program on a magnetic disk or data on a *hard copy*, which is a readable printed copy. Although text can be created and stored temporarily in RAM, it is not normally considered a file until it is removed from the memory and stored.

The manuscript of this book was originally typed from the keyboard using the word processing program EasyWriter, purchased on a diskette with an accompanying manual. Typed and edited material was stored in memory but moved to a

FIGURE 1-23. The keyboard.

diskette file before the power was turned off. Usually each section was put in one file, and the files for a complete chapter were stored on one side of a diskette, including related material and programs. At any time, a file could be transferred from the diskette to memory for modification. Also, the files could be linked and either displayed or printed in sequence.

Keyboard Scan Codes. There are 83 keys, each of which is assigned a number from 1 through 83. When a key is pressed, it transmits an 8-bit code to a shift register located in the system unit. This code, called the *make scan code*, is the binary number equal to the assigned key number. Hence, scan codes go from 1 through 83D, or 53H. When the key is released, it sends a *break scan code* to the same shift register, with the break scan code equal to the make code plus 80H. Thus, the make and break codes are the same except for bit number 7 of the byte. For example, key A is number 30 with make and break scan codes 1EH and 9EH. The use of scan codes for individual keys, rather than ASCII, gives flexibility by allowing a key to have alternate functions that depend on the closure of another key or perhaps some combination or sequence of key closures.

If a key is pressed and held down, after 0.5 second another make scan code is transmitted, and successive scan codes follow at 0.1-second intervals. Each is processed by the software and treated as a separate key closure. Such keys are said to be *typematic*. In most cases, the software is designed to repeat the function of a key when identical scan codes are processed in succession. Thus, when a letter key is held down, a rapid succession of the letters appears on the display. When more than one key is depressed, the last one pressed is the one that sends the make scan code. All keys behave exactly the same way, except that each has its own scan code. *The function assigned to a particular key is determined entirely by the software*, and software is flexible.

Normally, the scan codes are read and interpreted by a program stored in a ROM chip of the system unit. This 8-KB ROM is called the *Basic I/O System* (*ROM BIOS*), and it contains self-test and initialization routines, as well as numerous routines for control of I/O operations. It is possible, however, for a programmer to substitute a different routine, thereby changing the functions of the keys as desired.

Function Keys. At the left side of the keyboard are 10 function keys, designated F1 through F10. These are called *soft keys*, because they are not dedicated to any specific purpose. A program uses these keys as desired. Frequently, their functions are chosen for display of special symbols or operations selected by the software designer to make the particular program friendly to the user. For example, key F1 is often programmed to display a "help" menu. Keys of the broad center section are dedicated by the routine of the ROM BIOS to have functions similar to those of a typewriter.

Numeric Keypad. On the right side of the keyboard are keys arranged in a pad resembling that of a calculator. These keys can serve as an alternate source of numeric entries, which can be given special significance. The keys are also used for scrolling the display screen and positioning the *cursor*, which is the blinking underline that marks the position at which the next character is to be typed, deleted, or inserted. A HOME key is available for moving the cursor to the upper-left-hand corner of the display, or home.

Shift Keys. For each of eight keys, the software routine in the ROM BIOS sets an associated bit of memory when a make scan code is processed and clears the bit when a break scan code is received. Thus the bit, called a *flag*, serves as an indicator *that signals whether or not the corresponding key is depressed*. The eight keys are called *shift keys* and are given the names shown in Fig. 1-24.

In addition, whenever one of the four keys of the left column (INS, CAPS, NUM, SCROLL) is pressed, *a state-flag bit is changed to the opposite state*; that is, a 0 is changed to a 1 or a 1 becomes a 0. The bit is said to be complemented. The normal condition is state 0. The key must be released and pressed in order to switch the state again. These four shift keys are called *toggle keys*. They can be assigned alternate functions, depending on the state flag, by any particular program.

When a scan code is processed, the routine of the ROM BIOS *checks the shift and state flags of the shift keys before determining the action that should be taken*. For example, if either the LEFT or RIGHT flag is set when a letter key is pressed, an uppercase letter is displayed provided the CAPS state is normal, but a lowercase letter is displayed if the flag of CAPS state is set.

The ALT key allows access to special characters. Any of 256 ASCII and special characters can be printed on the display screen by typing the character code in decimal on the numeric keypad with ALT depressed. This is also accomplished with the BASIC function CHR$(*n*), with *n* denoting the character code. The 8-bit extended ASCII code is used.

Keyboard Processing. Within the keyboard is an Intel 8048 8-bit microcomputer. In addition to the internal CPU, this NMOS chip contains a crystal-controlled clock, an 8-bit timer/counter, a ROM with 1 KB of program storage, a 64-byte RAM for data, and 27 I/O lines. It provides keyboard scanning and transmits the scan codes to the system unit, along with control signals required with each scan code transfer. As many as 20 key scan codes can be stored in a buffer of the RAM of the 8048 microcomputer. A *memory buffer* is an area of memory used by I/O operations for temporary storage of data. Codes are stored in the buffer when the shift register of the system unit is not ready to accept a new code. When power is applied and also upon request, the microcomputer performs a self-test of the keyboard.

Adapter circuits on the computer board of the system unit give the required serial interface, with the shift register that receives the serial scan code converting it into parallel data. When a complete scan code has been received, an output of the shift register sends a request to the CPU for service. An interrupt of the running program is initiated, and the program shifts to a special routine that reads in the byte from the shift register through an 8-bit port of an IC called the *programmable peripheral interface* (*PPI*). The PPI is examined in Section 5-5. It provides several I/O ports, with a *port* defined as a pin or a group of pins used for transfer of data between the CPU and any peripheral unit other than memory.

Toggle Keys		Nontoggle Keys	
INS	(insert)	ALT	(alternate)
CAPS	(caps lock)	CTRL	(control)
NUM	(numeric lock)	LEFT	(left shift)
SCROLL	(scroll lock)	RIGHT	(right shift)

FIGURE 1-24. The eight shift keys.

The KB_BUFFER. Normally, the program moves the scan code from the shift register to the low half AL of a 16-bit processor register called the *accumulator* AX. The high byte of register AX is AH. The code is examined by the program, along with the shift and state flags of the eight shift keys, to determine the character or function represented by the code.

When a character is found, including a space 20H, a CR, an LF, or perhaps a special control function, the scan code and its extended ASCII code are moved to the next available 16-bit location of a memory buffer called KB_BUFFER, *with the scan code put in the high byte location and the extended ASCII put in the low byte position*. If the function has no extended ASCII, then null byte 00H is substituted, which indicates to the program that the character identified by the scan code of AH is special. A program is presented in the next section for examination of codes sent to KB_BUFFER.

Words of KB_BUFFER are transferred to the accumulator in the order received. Each word remains there until processed by the program, after which the next word of KB_BUFFER is sent. The word of a printable character is also transferred via the data bus to the I/O slot that connects to the CRT controller. The controller places the word in the memory buffer of the display monitor and prints the character. Certain scan codes or sequences of code denote operations that are immediately executed. An example is Ctrl-Alt-Del, representing a combination of three keys simultaneously depressed, which initiates a *system reset*. In such cases, the scan codes are not moved to the keyboard buffer of RAM.

KB_BUFFER in main memory can hold up to 15 words. It is the responsibility of the running program to remove and process the scan codes before KB_BUFFER becomes full. If the buffer is full when a character is sent to it, the speaker beeps and the character is rejected. Often a program will move codes from KB_BUFFER to another buffer, and later such keyboard buffers will be encountered. However, the 15-word buffer of memory that first receives the code from the buffer of the keyboard unit will henceforth be identified by the name KB_BUFFER.

The Wait Loop. Whenever power is on, the PC is executing instructions of a program at high speed. The computer never simply halts and stops processing. If the program needs direction from the operator, *it simply waits in a loop for the operator to strike a key*. Such a WAIT loop consists of a short sequence of instructions that do very little except consume time, with the last one of the sequence giving a jump back to the first one.

The program of the WAIT loop is interrupted 18.2 times per second, branching off to provide a tick to the real-time clock that is maintained in memory, but processing returns quickly to the WAIT loop. The sequence of the loop is executed repeatedly, with periodic timer and perhaps other interrupts, until a keyboard entry interrupts the routine. Then the program branches to the routine in the ROM BIOS that services the keyboard. Usually this is done almost instantly, and the program returns to the WAIT loop, *where the computer spends most of its time in normal operation*.

1-5 BASIC

The PC provides for the execution of programs written in *BASIC*, which is an acronym for *Beginner's All-purpose Symbolic Instruction Code*. This is an example of a *high-level* language, which is defined as one that is independent of any particular machine. Other high-level languages often used with the PC, such as PASCAL, FORTRAN, and COBOL, are not investigated in this book. Highly recommended is PASCAL, which is rather simple and easy to learn, and the programs are more compact and powerful than those of BASIC. However, BASIC is provided with nearly all PCs and is quite extensively used.

FORTRAN is popular in scientific work for manipulating numbers, and COBOL excels in the area of business applications. Here we begin the study of BASIC, which continues in later chapters. For readers already familiar with this language, the presentation may serve as a useful review. Because high-level language is closely related to low-level language, it is advantageous to look first at machine and assembly languages briefly.

Assembly Language

Instructions executed by the PC are stored in RAM and/or ROM in binary code referred to as *machine language*. Each instruction consists of an encoded string of 0s and 1s, which have a length of 1 to 5 bytes. Machine language instructions are fetched from memory and executed in sequence, with frequent jumps occurring from one point of a program to another. A jump may be conditional, based on the result of a calculation.

The instruction set is that of the 8086/8088 microprocessor family. A program can be written directly in machine language instead of a high-level language. However, when this is done, the machine language instructions are usually prepared initially in a mnemonic (nee-mon-ik) language known as *assembly language*. Each mnemonic name is chosen to assist the programmer's memory in recalling its function. Unlike the high-level languages, machine and assembly languages are machine dependent. A program in 8086/8088 assembly language will not run on microprocessors other than the 8086 and the 8088, which have identical instruction sets.

To illustrate, the 4-byte instruction that follows, given in both machine language (hexadecimal form) and assembly language, will move the 16-bit word 9AC5H into the 16-bit register AX of the 8088 microprocessor:

```
C7 C0 C5 9A    (machine language hexadecimal form)
MOV AX,9AC5H   (assembly language form)
```

The 4 bytes are stored in order in memory, with byte C7H in the location with the lowest address. After execution, register AX stores 9AC5H. Although there is a one-to-one correspondence between each machine language and assembly language instruction, assembly language programs are obviously easier to write than are programs in machine language.

A program called an *assembler* translates an assembly language program into the machine language that must be placed in memory. This translation is referred to as *assembly*. Although programming is much easier in a high-level language, assembled programs normally use less memory and run considerably faster. Also, many of the more powerful features of a computer are accessible only through assembly language. It is common practice to incorporate assembly language subroutines in high-level languages in order to speed up certain processes or to utilize special features of the computer.

Later chapters contain numerous programs in both assembly language and BASIC. Many of these programs parallel one another, with the assembly language and BASIC accomplishing essentially the same functions. This procedure helps one to understand the relationship between the two languages, as well as their relative advantages and disadvantages.

Compilers and Interpreters

The 8088 microprocessor of the PC is designed to operate only on the machine language instructions of its unique instruction set. With a high-level language such as BASIC, each instruction represents a command or statement that must first be translated into a sequence of 8088 machine-language instructions prior to execution. There are two methods commonly employed. One consists of using a *compiler* program that converts the BASIC program into a machine language program, which is stored on a disk or other nonvolatile storage device. Then the machine language can be loaded into memory and executed. The compiled program can be run as often as desired. Each computer type must have its own special compiler program for each high-level language that is to be compiled.

The second way is to use an *interpreter*. This is a program that translates into a sequence of machine instructions a single BASIC instruction and immediately executes it, provided it is executable. Then the next instruction is translated and executed, and so on. If the program is run a second time, the entire process of translation and execution is repeated, one instruction at a time. Each computer type must have its own special interpreter for the high-level language that is interpreted. Execution of programs using an interpreter is not nearly as fast as execution of the same programs that have been compiled, but *interpreters are easier to use, are convenient for debugging, and are less expensive*. Most PC operators run their BASIC programs on the interpreter.

The BASIC Interpreter

The PC comes with a BASIC interpreter, which has three versions. These are referred to as follows:

1. Cassette BASIC
2. Disk BASIC
3. Advanced BASIC

The *Cassette BASIC* interpreter is stored in four 8-KB ROM chips having absolute addresses from F6000H through FDFFFH, which consist of 32 KB of machine-language instructions. The ROMs are mounted on the 8.5 × 11-inch computer board of the system unit. If the computer is turned on with the diskette latches open, the mode is automatically that of Cassette BASIC. The BASIC prompt OK appears on the display, and program commands can be entered and executed. Cassette BASIC supports most of the BASIC language, but it does not support diskette operation. Thus, when using Cassette BASIC, programs cannot be saved on diskettes and programs on diskettes cannot be loaded and run.

When the computer is started with a properly formatted diskette in drive A with the latch door closed, the mode is that of the *disk operating system* (*DOS*). The prompt is A>, with the letter A denoting that the default is drive A. In Fig. 1-15, this is the one on the left. To get into the *Disk BASIC* mode, one must type BASIC and press the ENTER key. About 16 KB of BASIC are loaded into RAM, supplementing Cassette BASIC, and the BASIC prompt OK appears. Disk Basic provides all features of Cassette BASIC, plus support for diskette I/O operations and serial communications. In addition, it maintains the date and time in the internal time-of-day clock.

If BASICA is entered instead of BASIC, about 26 KB of BASIC are loaded into RAM, and the mode is that of *Advanced BASIC*. This mode has all the facilities of Disk BASIC plus various advanced features primarily related to music, graphics, color, and communications. Most BASIC programs presented in this book require only Cassette BASIC, but the BASICA mode is assumed if needed. Return to DOS from BASIC is accomplished simply by executing the BASIC command SYSTEM, which gives the DOS prompt A>.

Direct and Indirect Modes

BASIC instructions are known as commands or statements. *Commands* generally operate on programs, and *statements* direct the flow of a program. Instructions are given in either the direct mode or the indirect mode. In the *direct mode* the instruction is not preceded with a line number, and when it is entered by pressing key ENTER, the command or statement is executed immediately. In the *indirect mode*, each instruction is preceded with a line number, and the instructions are not executed until the program is run.

Programs are entered in the indirect mode and stored in memory. After the entire program is stored, it is executed by the interpreter by entering the RUN command in the direct mode. This is accomplished by typing RUN and pressing the ENTER key, also referred to as key CR for carriage return. Instructions are executed in order of ascending line numbers. The order in which the lines are typed and entered is unimportant, and line numbers do not need to be consecutive. In fact, one should leave unused numbers between those of the lines so that there is room for adding instructions later should the need arise.

Line numbers must be in the range from 0 to 65,529. Common practice is to number the statements in increments of 10, starting with 10 (or perhaps 100). With this numbering method, a line can be inserted between lines 10 and 20 for example, by giving it number 15. Programs can be typed in either lowercase or uppercase letters, or a mixture. However, all letters are automatically converted to uppercase, except those of string expressions in quotation marks and those of remarks.

A program line can have up to 255 characters, including blanks and the CR used to enter the line. This line is called a *logical line*. Because the screen is only 80 characters wide, a *physical line* is limited to 80 characters. It follows that a logical line of a program may occupy up to four physical lines containing as many as 255 characters. Shown is a logical line of 189 characters of a remark statement, which uses three physical lines:

```
100 REM    This program solves a set of
               3 simultaneous equations
               with 3 unknowns.
```

Each of the first two lines has 80 characters, including blanks, and the third line has 12 leading blanks, 16 characters, and CR. A physical line is allowed to wrap around to the next physical line only by typing characters, not by means of cursor movement keys, but blanks entered with the space bar are treated as characters. Spaces produced within a line by cursor movement are counted as characters. If a logical line has more than 255 characters, those in excess of 255 are simply ignored, and there is no error message.

After the preceding logical line has been typed, it is entered by pressing CR *with the cursor located at any point of the logical line.* If a correction is made after the line has been entered, it is not effective until the line is again entered. The cursor movement keys, along with the backspace, delete, and insert keys, provide for easy editing. Although the three letters of REM are always listed with uppercase letters, the letters within the remark are exactly as typed.

Constants

Constants are the actual values that are used by BASIC during the execution of a program. There are two types of constants, referred to as *string* and *numeric*.

String Constants. A *string constant*, or simply *string*, is a sequence of characters enclosed in double quotation marks. Some examples are given in Fig. 1-25.

A string must not have more than 255 characters, and the double quotation marks are not considered as part of the string. There are numerous applications of the *null string*, just as there are for 0 in arithmetic. The string constant in the command SAVE "B:REPORT" is a file specification, or *filespec*, that signifies the file named REPORT on the diskette of drive B.

Numeric Constants. Positive and negative numbers are *numeric constants*. Because commas in BASIC are used to separate numbers or other items in a series, they are not allowed between digits of a number. Only decimal, hexadecimal, and octal numbers are recognized. Hexadecimal numbers must be preceded with &H and are restricted to four digits without a decimal point. An octal number has prefix &, with six digits allowed. All other numbers are decimal.

To convert a hexadecimal number to decimal, the PRINT statement can be used. For example, execution of the following statement gives 3914 decimal on the display:

```
PRINT &HF4A
```

"Word 'bit' is a 0 or 1."	(avoid " within a string)
"The cost is $500."	
"10"	(this is a string, not a numeric)
" "	(two blank characters)
""	(null string constant)

FIGURE 1-25. Examples of string constants.

For conversion of decimal to hexadecimal, the BASIC function HEX$(*n*) is available, with *n* denoting the decimal number. For example, the display reads F4A when the following statement is executed:

```
PRINT HEX$(3914)
```

OCT$(*n*) is the corresponding function for octal conversion.

Keyboard Program

A *listing* is a readable display or printout of a program. To illustrate a BASIC program and also to provide additional information on the operation of the keyboard, we examine the listing of Fig. 1-26.

```
10 REM Display KB_BUFFER contents in hexadecimal.
20 DEF SEG = &H40
30 PRINT "Offset   Make Scan Code   'ASCII'   Character"
40 FOR I = 30 TO 0 STEP -2
50 LET X = &H1E + I: Y = &H1F + I
60 PRINT HEX$(X), HEX$(PEEK(Y)), HEX$(PEEK(X)), CHR$(PEEK(X))
70 NEXT I
80 PRINT "BUFFER_HEAD at 1AH points to offset ";HEX$(PEEK(&H1A))
90 PRINT "BUFFER_TAIL at 1CH points to offset ";HEX$(PEEK(&H1C))
```

FIGURE 1-26. BASIC program that reads KB_BUFFER.

In the BASIC mode, suppose the program is entered by typing each line, with the Enter key (CR) pressed while the cursor is on the line. To run the program, type RUN or simply press key F2, which is programmed by BASIC to accomplish exactly the same function. The result is a display of the contents of the complete keyboard buffer, consisting of 16 word locations from address 0041EH through 0043CH. Each word location stores 2 bytes.

As shown in Fig. 1-27, there are four columns in the display. The first column, with the heading Offset, gives the addresses relative to the base address, with the actual address equal to the sum of the offset and the base address of 00400H, or 1024 decimal. Next is the high-address byte of the location, with the heading Make Scan Code, and the third column is the low byte with the heading 'ASCII'. The single quotation marks around ASCII are used to indicate that not all keys or key combinations have ASCII codes, and for these the entry in the ASCII column is 0. Double quotation marks should not be included within the double quotation marks of a PRINT statement. The last column shows the character, provided the code is one that can be displayed.

Keyboard entries into KB_BUFFER are removed and processed by the BASIC interpreter about as fast as they are entered. Thus, the buffer display is that of an "empty" buffer. However, when a word is read from a location, it remains there until overwritten. Accordingly, even though the buffer is empty when the program is run, the display *shows the last 16 entries*. The program provides an easy way *to study the behavior of all possible combinations of keys that send characters to the buffer*.

To illustrate, suppose the following 16 keys are pressed and released in sequence: qwerty, Alt-a, F2, 1234, Esc, p, CR, and F2, with Alt-a, F2, Esc, and CR denoting

```
Offset    Make Scan Code    'ASCII'    Character
3C             3             32            2
3A             2             31            1
38             3C            0
36             1E            0
34             15            79            y
32             14            74            t
30             13            72            r
2E             12            65            e
2C             11            77            w
2A             10            71            q
28             3C            0
26             1C            D

24             19            70            p
22             1             1B            ←
20             5             34            4
1E             4             33            3
BUFFER_HEAD at 1AH points to offset 2A
BUFFER_TAIL at 1CH points to offset 2A
```

FIGURE 1-27. Display of KB_BUFFER. The blank row is generated by the carriage return (ODH).

single key codes. The last key closure runs the program, and the resulting display resembles that of Fig. 1-27. The entered sequence is observed to start at offset 2AH, but this initial offset value depends on past history. The Escape key is number 1 with scan code 1 and ASCII 1BH, and function key F2 is number 60, or 3CH, with no standard ASCII. The last entry into the buffer, at offset 28H, is the word of key F2. The next keyboard entry will be into the memory location at offset 2AH.

The buffer *wraps around*; that is, when the entry goes to the highest location at offset 003CH, the next one goes to the lowest location at offset 001EH. BUFFER_TAIL is a memory location at offset 001CH; *it stores the 16-bit offset of the buffer location that is to receive the next keyboard entry*. Whenever a word is entered, the contents of BUFFER_TAIL are incremented by the program by 2 and possibly adjusted for wrap around, so that BUFFER_TAIL *always points to an empty location*. On the other hand, if there is at least one valid entry, location BUFFER_HEAD at offset 1AH *points to the next word of the buffer that is to be removed by the program*. When a word is removed, the contents of BUFFER_HEAD are incremented by 2 and possibly adjusted for wrap around. Anytime an increment of either pointer gives a value outside the buffer, the value of the pointer is adjusted for wrap around.

An empty buffer is indicated whenever BUFFER_HEAD equals BUFFER_TAIL. In this case, there are no words to be removed. Although KB_BUFFER has 16 word locations, only 15 valid entries are possible, because BUFFER_TAIL must always point to an empty location.

Any combination of keys can be activated prior to running the program. Some combinations do nothing, and some provide a function without placing code into the buffer. In any event, *the program always shows the last 16 entries*. The system reset initiated by combination Ctrl-Alt-Del clears all buffer locations and initializes the two buffer pointers to the lowest offset. Ctrl-Break empties the buffer by equating both pointers to the lowest offset 1EH; after this, 0000H is moved into location 1EH, followed by the double increment of BUFFER_TAIL to offset 0020H.

When the 0000H entry is read by DOS, BUFFER_HEAD is incremented by 2, leaving the buffer empty.

Break codes are never sent to the buffer. For all except the eight shift keys, the interrupt routines activated by the break scan codes do nothing useful. However, the break scan codes of the shift keys are important, and *they clear appropriate shift flags without putting data into the buffer*. On occasion the buffer may become full, with 15 active entries. This happens when repeated entries are made at a time when the program is doing something more important than reading the keyboard, or when words are entered faster than they are being processed. As stated earlier, the result is a warning beep on the speaker, and entries in excess of 15 are rejected.

Two-Byte Extended Codes. For some keys or key combinations, the word put into KB_BUFFER has a low byte of 00H, and the high byte is either the scan code or an assigned *pseudo scan code* that identifies the key combination. An example is code 3C00H for function key F2, shown in Fig. 1-27. These codes are referred to as *2-byte extended codes*. Normally, when KB_BUFFER is read, the low byte of a location is obtained first. If it is 00H, a 2-byte extended code is indicated, and a read of the high byte is essential to determine the key combination. The 2-byte extended codes are listed in Appendix G of the *IBM BASIC Manual*. Let us now consider the program statements.

REM. Line 10 of Fig. 1-26 consists of a remark, which is ignored by the computer but displayed in the listing. Remarks can be placed anywhere in a program as often as desired.

DEF SEG. The DEF SEG statement of line 20 of Fig. 1-26 specifies the *segment address* to be 0040H, and this 16-bit number determines the base address for use by BASIC statements that access memory, such as PEEK and POKE. *The actual base address is found by multiplying the segment address by 16*, which converts 0040H to 00400H. Multiplication of a hexadecimal number by 16 simply *shifts the hexadecimal digits one position to the left*, with the least significant digit replaced with 0.

The 20-bit *absolute address*, or *physical address*, of a memory location is the segment address times 16 plus the 16-bit offset. These are usually given in hexadecimal, separated by a colon, with the segment address first, and the combination is called the *logical address*. For example, suppose the segment address is F000H and the offset is E345H, corresponding to physical address 1041221D. Then,

logical address F000:E345 = absolute address FE345

If DEF SEG is used alone, without an address, the segment address becomes that of the data-segment register DS as established by BASIC. *Note that a 20-bit absolute address is specified by means of two 16-bit numbers*, which is a method for using effectively the 16-bit registers of the 8088 processor.

FOR and NEXT. The FOR statement of line 40 of Fig. 1-26 and the NEXT statement of line 70 function together to provide a *program loop*. The sequence of instructions between FOR and NEXT are executed a given number of times, after which the program exits the loop at the line following NEXT.

On the first pass of the loop, the variable I is specified as 30. With STEP given as -2, variable I has a value of 28 on the second pass, and so on. Because the final value of I is given in line 40 as 0, the value of I on the last pass is 0. STEP can be specified as positive or negative, and if omitted, its default value is 1. FOR loops can be nested, with one FOR loop placed within another. Inclusion of variable I after NEXT in line 70 is optional.

LET. The BASIC keyword LET is used to *assign* the value of an expression to a variable. In line 50 is the statement

```
LET X = &H1E + I
```

This statement adds the value of the variable I to the hexadecimal number 1E and assigns the result to the variable X. The word LET is optional, and to save typing effort it is often omitted.

Any number of BASIC statements can be placed on a logical line *provided they are separated by colons*; of course, the total number of characters must not exceed 255. In line 50 is a second assignment statement in which the word LET is omitted. Note that an assignment statement such as X = 2X is acceptable. It gives the variable X a value equal to twice its original value. The initial values of all variables are 0 prior to assignment. Thus, if X has not been previously assigned, the statement X = X + 2 assigns value 2 to X.

PEEK(*n*). Operation PEEK (*n*) looks into a byte location of RAM or ROM at offset *n* from the current segment, as defined by the DEF SEG statement. The offset must be from 0 to 65,535. The byte can be assigned to variable X with the statement X = PEEK(*n*) or displayed with statement PRINT (PEEK(*n*)). In line 60, PRINT applies to each of the four values separated by commas, and the inclusion of HEX$ before PEEK returns the byte in hexadecimal.

PRINT and LPRINT. The statement PRINT used alone simply *creates a blank line on the display*. When followed by a list of numeric and/or string expressions separated by commas, as in line 60, the items are displayed on the same line in print zones of 14 characters each. Each value is printed starting at the beginning of its 14-character zone, as shown in the display of the program results.

If semicolons replace the commas, as in line 80, *values immediately follow one another*. Note the space after "offset" in line 80, which provides a space before the displayed value of HEX$ (PEEK(&H1A)). If the list of expressions is terminated with a comma or a semicolon, the next PRINT statement begins printing *on the same line*, with spacing according to the previous rules. LPRINT is similar to PRINT except that output is to the printer instead of to the CRT display.

The program described here provides for inspection of the keyboard buffer. However, it is easily modified for examination of any region of RAM or ROM (see Prob. 1-21). With a STEP of 1, a column can give consecutive offsets in ascending order, and another column can give the stored byte in either octal, decimal, or hexadecimal.

REFERENCES

BASIC, IBM Corp., Boca Raton, Fla, 1981.

N. GRAHAM, *Programming the IBM Personal Computer: BASIC*, Holt, Rinehart and Winston, New York, 1982.

Personal Computer Technical Manual, IBM Corp., Boca Raton, Fla., 1981.

PROBLEMS

SECTION 1-1

1-1. Define logical 0, logical 1, bit, nibble, byte, word, IC, I/O, LSB, MSB, K, KB, MB, RAM, ROM, preset, clear, set, reset, bus, memory address.

1-2. State the difference between positive and negative logic, states H and L, storage cell and register, volatile and nonvolatile memory, combinational and sequential circuits.

SECTION 1-2

1-3. Define ALU, CPU, CRT, nanosecond, hertz, kHz, MHz, algorithm, clock, positive transition, negative transition, positive pulse, negative pulse, microprocessor, microcomputer, peripheral, synchronous network, tristate output, high-impedance state, floating output.

1-4. The pulse train of a synchronous system with logic level voltages of 0 and 5 has a pulse repetition rate of 10 MHz. The rise time is 8 nanoseconds, the fall time is 4 nanoseconds, and the pulse width is 40 nanoseconds. Determine the period and the percent duty cycle, and sketch and dimension the pulse train for an interval of two periods. Label the positive and negative edges and the positive and negative pulses.

1-5. (a) Explain the basic difference between parallel and serial transfer of data. Which is normally used in communications systems and why? (b) How do synchronous controls differ from asynchronous ones? (c) What is the significance of the dynamic indicator at an input terminal?

SECTION 1-3

1-6. Express binary 11010.101 in octal, decimal, hexadecimal, and BCD form.

1-7. Express 237.1D in binary, octal, hexadecimal, and BCD form.

1-8. Express 1101D in binary, octal, hexadecimal, and BCD form.

1-9. Express FFFFH in binary, octal, decimal, and BCD form.

1-10. Express 2748.8125 in binary, octal, hexadecimal, and BCD form.

1-11. Express 111.3Q in binary, decimal, hexadecimal, and BCD form.

1-12. Express 110010111.1B in octal, decimal, hexadecimal, and BCD form.

1-13. Determine the decimal and hexadecimal equivalents of the word 10010111.0101 if it is a binary number and also if it is a BCD number.

1-14. Without converting to another number system, add the BCD numbers 0010 0110 1000 and 0011 0100 1001. Verify your answer by converting the numbers and their sum to decimal form.

SECTION 1-4

1-15. Twenty-two 16-bit registers, with locations encoded by addresses 100 through 121, store a message in ASCII. In hexadecimal code the bits of the respective registers, beginning at address 100, are as follows: 4120, 4E49, 4242, 4C45, 2049, 5320, 4120, 342D, 4249, 5420, 574F, 5244, 2E20, 4120, 4259, 5445, 2048, 4153, 2038, 2042, 4954, 532E. Determine the message.

1-16. A memory consisting of a number of 16-bit registers, with locations encoded by addresses beginning at 400 and numbered consecutively, stores in ASCII the message LOGICAL 0s AND 1s CAN REPRESENT NUMBERS, VOLTAGE LEVELS, AND SIGNAL ACTIVITY. Number the addresses and give, alongside each address, the stored word in hexadecimal. Also, give the bit string of the first five addresses.

1-17. Repeat the preceding problem using lowercase letters.

1-18. A 16-bit shift register stores 0ED6H. Suppose three left shifts are implemented with 0s shifted in at the right. Give the stored number in both hexadecimal and decimal after each shift. Repeat for three right shifts of 0ED6 with 0s shifted in at the left.

1-19. At a certain instant, at location 0041AH of the IBM keyboard buffer is stored 0426H, and at location 0041CH is stored 043AH. Suppose that five words are moved in sequence to the accumulator while two new words are entered into the buffer by the typist. At this instant, determine the storage of the two locations.

1-20. With all shift states of the PC initially normal, suppose that shift keys INS, CAPS, SCROLL, ALT, and LEFT are pressed, and that only CAPS and ALT are released. After these operations, identify the shift flags and the state flags that are set.

SECTION 1-5

1-21. Write a BASIC program for the PC that will display the bytes stored in RAM in absolute locations 0 through 1023. Use three columns that give the decimal address, the hexadecimal address, and the hexadecimal byte. Although the data will not fit on the screen, the display can be stopped at any time by pressing keys Ctrl-NumLock, and scrolling is resumed when another key is pressed. What modification sends the output to the printer?

1-22. Revise the keyboard program of Section 1-5 with DEF SEG = 0 replacing that of line 20. Add statements that give in the displayed output a descriptive title followed by a blank line, and that also give a blank line between the heading of the display and the columnar data. Give the listing with several lines of appropriate remarks.

1-23. State the difference between (a) machine and assembly language, (b) 8088 assembly language and BASIC, (c) an assembler and a compiler, (d) a compiler and an interpreter, (e) direct and indirect modes of BASIC, (f) logical and physical lines of BASIC, (g) string and numeric constants, (h) "20" and 20 in BASIC, (i) &HF0 and 240, (j) HEX$(100) and &H64.

1-24. Write a BASIC program that will give a hard copy of the entire contents of the ROM BIOS, which occupies the highest 8 KB of the 1-MB memory space. Hexadecimal addresses in ascending order and the hexadecimal bytes should be printed in two columns. Use a segment address of F000H and include headings and remarks.

Arithmetic Operations

Included in the ALU of a data processor of a computer are circuits for implementing arithmetic and logical operations on binary words. Many ALUs have hardware adders, subtractors, multipliers, dividers, and other arithmetic circuits. However, smaller systems utilize the adder, along with suitable programs, to perform the various operations. This procedure substitutes software for hardware, thereby reducing equipment complexity. Some of the basic arithmetic techniques are investigated here, and logical operations are examined in Chapter 4.

2-1 POSITIVE AND NEGATIVE NUMBERS

Most important to arithmetic operations is the *adder*; a block diagram of one with 4 bits is shown in Fig. 2-1. Four-bit adders are available as ICs in 16-pin packages; in addition to the 14 terminals of Fig. 2-1, two pins are required for the voltage supply. Inputs are the 4-bit words A and B plus a carry K, and outputs are the 4-bit sum S and carry C. With C regarded as the MSB of the 5-bit output, this word is the binary sum of A, B, and input carry K. For example, if word A is the binary number 1110 and B is 0111 with K zero, the output is 10101. Because terminals are provided for both input and output carries, four such adders can be cascaded to form a 16-bit adder. The carry out of a stage simply connects to the

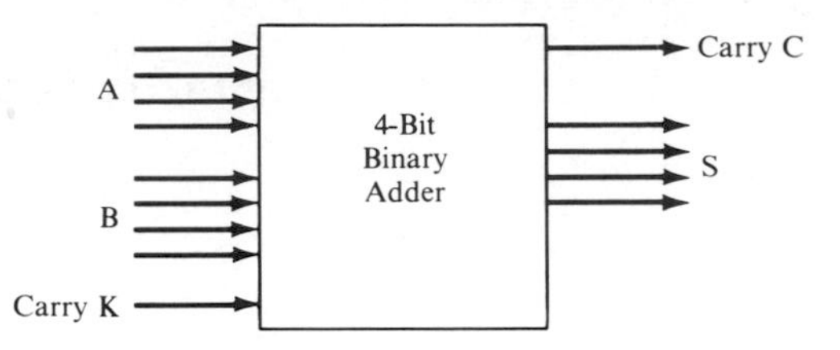

FIGURE 2-1. A 4-bit binary adder.

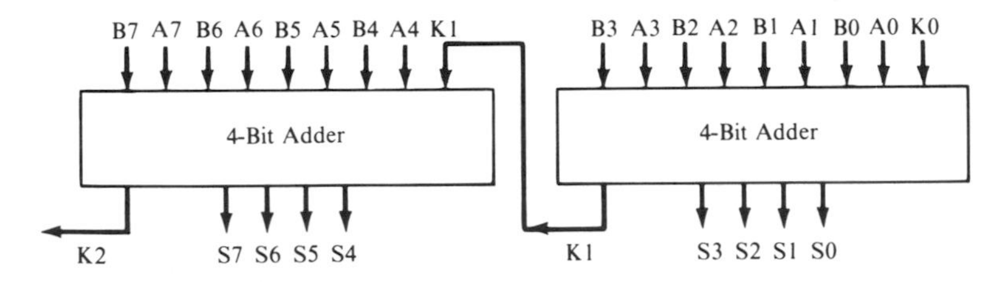

FIGURE 2-2. A cascade of two 4-bit adders.

carry into the next stage of higher order. In Fig. 2-2 is shown a cascade of two 4-bit adders.

Complement

If variable A represents a bit, which at any instant of time has either state 0 or state 1, then the *complement* of A, or NOT A, is defined as the variable that always has a state opposite that of A. The complement of A is written as A′, or $\overline{A}$, or NOT A. If A is 1, then A′ is 0, and vice versa. Letters are also used to represent binary words. The complement of binary word B is NOT B, or B′, or $\overline{B}$, with each bit of B′ being the complement of the corresponding bit of B. For B = 1011, its complement B′ is 0100. The sum of B and B′ is 1111.

Complement is a logical operation normally available in ALUs and included in computer instruction sets. Execution of the 8088 assembly language instruction NOT AX replaces each of the 16 bits of the accumulator with its complement. Examples are given in Fig. 2-3.

Suppose *bit A = 0* and *byte B = 11001010*. Then
complement of A = $\overline{A}$ = A′ = NOT A = 1
complement of B = $\overline{B}$ = B′ = NOT B = 00110101

FIGURE 2-3. Examples of the complement operation.

Signed Numbers

A number system of radix *r* has two types of complements, known as the *(r − 1)'s complement* and the *r's complement*. The (r − 1)'s complement of a number is found by subtracting each digit of the number from r − 1, and the r's complement is found by adding 1 to the (r − 1)'s complement, with any *resulting carry discarded*. The most familiar numbers are decimals, which are considered next.

10s Complement. In the decimal system, radix r is 10 and r − 1 is 9. To obtain the *9s complement*, each digit is subtracted from 9. Thus the 9s complement of 00809 is 99190, and the *10s complement* of the number is 99191. The sum of 00809 and its 10s complement 99191 is 00000, with the carry of 1 discarded. In fact, with the carry of 1 rejected, the sum of any decimal number and its 10s complement is 0.

Because the sum of 00809 and −00809 is also 0, it is often convenient to represent the negative number −00809 by 99191, which is the 10s complement of +00809. When this is done, we must understand that a negative number is indicated by a most significant digit of 9, which is called the *sign* digit. Also, the number of digits must be sufficient so that the sign digit of a negative number is indeed 9. To illustrate, if the result 99191 of the preceding example gave only the three low-order digits 191, it would erroneously be interpreted as a positive number.

In Fig. 2-4 are several six-digit positive and negative numbers in 10s complement form. With the carry of 2 in the 10s complement sum discarded, the resulting two sums are equivalent. Note that the addition of each of the 10s complement negative numbers with the magnitude of its corresponding decimal gives 0 after the carry of 1 is discarded.

10s Complement	Corresponding Decimal
012345	012345
099999	099999
900001	−099999
987654	−012346
999999	−000001
Sum = 999998	Sum = −000002

FIGURE 2-4. Positive and negative numbers in 10s complement form.

The 10s complement representation of a negative decimal number is used in operations with decimal numbers in BCD form. Many hand calculators employ decimal arithmetic, and many computers have the capability. However, we shall confine our attention here to the more important case of binary arithmetic.

2s Complement. A binary number has a radix of 2, and there is a 1s complement and a 2s complement. The *1s complement* is obtained by subtracting each bit of the number from r − 1, or 1. Subtraction of 0 from 1 gives 1 and subtraction of 1 from 1 gives 0. Therefore, the 1s complement is found simply by replacing each bit with its complement. Thus, *the 1s complement is exactly the same as the complement.*

The *2s complement* of a binary number is found by adding 1 to the complement and *throwing away any output carry*. As an example, consider the 5-bit word 01011. Its complement is 10100, and its 2s complement is 10101. The sum of the number and its complement is 11111, and the sum of the number and its 2s complement is 00000, with the carry of 1 discarded. Additional examples are shown

Binary Number	1s Complement	2s Complement
0000	1111	0000
0001	1110	1111
0100	1011	1100
1111	0000	0001

FIGURE 2-5. Binary numbers and complements.

in Fig. 2-5. In the third column of the figure are the 2s complements of the numbers of the first column. Note that the numbers of the first column are also the 2s complements of those of the third column. A binary number is the 2s complement of its 2s complement, and the sum of the two is 0.

Negative binary numbers are usually represented in digital systems by the 2s complement of the corresponding positive number. In this system there must be enough bits so that the leftmost bit, called the *sign bit*, is always 0 for a positive number and 1 for a negative number. To illustrate, decimal 27 is binary 011011 and −27 is 100101, with the logical 1 sign bit signifying that the binary number is negative. The 2s complement of a negative number gives the corresponding positive number, with a sign bit of 0. Negative numbers are supplied to the processing system in 2s complement form.

The attractive feature of 2s complement arithmetic is that the computer does not usually need to differentiate between positive and negative numbers. Algorithms for addition and subtraction are the same for both signed and unsigned numbers. Care must be taken to avoid an operation that erroneously changes the sign bit, such as adding two numbers that produce a sum too large for its storage register.

For example, with 8-bit registers the sum of signed numbers 80H and 80H, each of which represents −128, gives the erroneous result 00H because the true value, −256, has too many bits for 8-bit storage. Thus there is overflow. With 16-bit registers the signed numbers would each be FF80H and their sum is FF00H, representing −256.

When an arithmetic operation such as A − B is encountered, it makes no difference whether A or B or both are positive or negative, and the result automatically acquires the correct sign bit provided there is no overflow. *The 2s complement of a positive number gives a negative number of the same magnitude, and the 2s complement of the negative number gives the positive number.* Negative numbers in the user program are converted by the assembler into 2s complement form when the program is assembled into machine language prior to execution, and negative numbers supplied by a computer to an output device, such as a display monitor, can be decoded into the familiar signed decimal form.

A quick way to form the 2s complement of a binary number is to start with the LSB and progress to the left, leaving all bits unchanged up to and including the first 1 encountered. Other bits are complemented For example, the 2s complement of 11000 is 01000, and the two numbers add to 0 with the carry dropped. For other examples, apply the rule to the first column of Fig. 2-5, obtaining the third column.

Often binary numbers are expressed in hexadecimal form. The 2s complement is easily found by first converting the hexadecimal number into its binary equivalent. However, there is an easier procedure. Start with the least significant digit of the hexadecimal number and move left to the first non-0 digit, which is subtracted from decimal 16. Other digits to the left are subtracted from 15. For example, the

Binary Number in Hexadecimal Form	2s Complement in Hexadecimal Form
0000	0000
0001	FFFF
0ABC	F544
ABCD	5433
FFFF	0001

FIGURE 2-6. Binary numbers and their 2s complements in hexadecimal form.

2s complement of A510H is 5AF0H. Their sum is 0000H, which provides a way to check the conversion quickly. Each represents a 16-bit binary. Examples are shown in Fig. 2-6.

Both signed and unsigned numbers are used in this book, *with all signed numbers understood to be in 2s complement form.* A 16-bit register can store signed numbers through the range from 8000H to 7FFFH, or from −32,768 to +32,767 inclusive. Unsigned numbers are positive regardless of the state of the MSB. For these, the sign bit does not indicate a sign. In particular, unsigned numbers are used for addresses, which are never negative. A 16-bit register stores unsigned numbers from 0000H through FFFFH, or from 0 through 65,535.

The process of addition is the same for both signed and unsigned numbers. For example, if FABCH is added to 0540H, the result is FFFCH. With signed numbers, this is the decimal sum of −1,348 and 1,344, giving −4, but with unsigned numbers the operation is the addition of 64,188 and 1,344 to get 65,532. Processing by a computer does not depend on the interpretation. Examples of signed and unsigned numbers are presented in Fig. 2-7.

In Fig. 2-7 the signed values are easily found by first converting the binary numbers into their 2s complement form using hexadecimals. The hexadecimal result is then converted to decimal with a minus sign added. An alternate method for the negative values consists of subtracting the unsigned decimal value from 65,536, corresponding to 10000H.

Signed Magnitude Numbers. There are other methods for representing negative numbers. Sometimes binary numbers are specified by their magnitudes and a sign bit, with 0 for positive and 1 for negative. The sign bit is located in the MSB position. In this *signed magnitude system*, the 8-bit binary number 00000110 is +6 and 10000110 is −6. With 2s complement arithmetic, the latter binary number denotes −122. We will not use binary numbers in signed magnitude form, and henceforth all binary numbers must be interpreted as either unsigned or signed numbers in 2s complement form.

Binary Number in Hexadecimal Form	Unsigned Decimal Value	Signed Decimal Value
1111	4,369	4,369
ABCD	43,981	−21,555
FFFF	65,535	−1

FIGURE 2-7. Signed and unsigned numbers.

ADDITION AND SUBTRACTION

Computers can add and subtract both signed and unsigned numbers. First, we consider addition. All signed binary numbers are understood to be in 2s complement form.

Addition

The *accumulator A* is the processor register designated to receive the result of an arithmetic or logical operation. Some processors, including the 8088, allow any of their general registers to serve as an accumulator. Assuming 16-bit registers, suppose the operation A ← A + B is to be implemented, with hexadecimal D0F7 stored in A and 7C65 stored in B. This statement is an example of *register transfer language*, and it specifies that the contents of register B are to be added to those of A, with the results stored in A. The word of B is unchanged by the operation.

Associated with the register serving as the accumulator is a flag register that stores status flags and perhaps some control flags. *Status flags* are bits of a memory or register that indicate the results of an operation, and *control flags* exercise control over certain computer operations. One of the status bits of the flag register is the *carry flag CF*, which stores the output carry of an add operation. For the preceding addition, Fig. 2-8 shows the augend A, the addend B, the sum, and the carry flag CF.

Each lowercase *c* of the top row denotes an internal carry of 1. The absence of an indicated carry is considered to be a carry of 0.

For unsigned numbers the augend is 53,495, the addend is 31,845, and the result stored in the register is 19,804, which is erroneous. The error is indicated by CF. The flag is set, indicating an *unsigned overflow*. The true sum is 85,340, but this exceeds the capacity of the register.

On the other hand, for signed numbers the augend is −12,041, and the addend and sum are the same as before. In this case the sum is correct, even though CF is set. The bit of CF is now a signed carry, but *there is no overflow as long as the carry into the sign bit position is the same as that out of the sign bit position*. Both carries are 1s in the example. However, if the carry into the sign bit differs from the carry out, then the sign of the result is erroneously changed by the carry that enters the bit position. This is a *signed overflow*. An example with signed overflow is given in Fig. 2-9.

The augend is decimal 26,035, the addend is 14,713, and the indicated sum is −24,788. There is a carry of 1 into the sign bit, but the carry out is 0. Accordingly, the sign bit is erroneously changed. Within the flag register is *overflow flag OF*, also a status flag. The computer hardware sets this flag whenever a signed overflow occurs, as indicated by unlike carries into and out of the sign bit position of the

CF					
	ccc	c	cc	ccc	
	1101	0000	1111	0111	Augend
	0111	1100	0110	0101	Addend
1	0100	1101	0101	1100	Sum

FIGURE 2-8. Addition example.

	cc	cc	ccc	cc	
	0110	0101	1011	0011	Augend
CF	0011	1001	0111	1001	Addend
0	1001	1111	0010	1100	Sum

FIGURE 2-9. Addition example with signed overflow.

accumulator. In the example, the true sum of 40,748 is out of the signed-number range of the 16-bit register.

In conclusion, *for binary addition with unsigned numbers, an overflow is indicated by a set carry flag CF. With signed numbers in 2s complement form, an overflow is indicated by a set overflow flag OF*; if OF is clear, the result is correct regardless of CF.

Flag Register. There are other bits of the flag register that are affected by arithmetic operations. The *sign flag SF* stores the MSB of the result. For signed numbers this is the sign bit, and for unsigned numbers it is nothing more than the 0 or 1 digit of the MSB. *Zero flag ZF* is set if the result of an operation is 0; that is, ZF is 1 for a word having all bits equal to 0.

An *auxiliary carry flag AF* is required for BCD arithmetic. It is the carry or borrow flag of the low nibble of the result, and its use by program instructions is normally confined to those that adjust the results of decimal operations. The required adjustment is a consequence of the fact that 4-bit combinations are used to represent decimal digits in BCD arithmetic. Then there is the *parity flag PF*, which is set only if the *low byte* of an arithmetic operation contains *an even number of 1 bits*. Flag PF is useful for error detection.

The flags are affected not only by arithmetic results but also by rotate, shift, and logical operations. With an *operand* defined as a sequence of bits to be operated on in some manner, for byte operands the flags that have been described apply to bytes. For example, if 2 bytes are added, flag CF is the carry out of bit 7, flag SF stores a bit equal to that of bit 7, flag OF is set in case of an overflow with signed 8-bit numbers, and flag ZF is set if bits 7 through 0 of the result are 0s.

Subtraction

Next, suppose the operation A ← A − B is to be performed by computer hardware, with A = 1E24H and B = 7A41H. In decimal, minuend A is 7,716 and subtrahend B is 31,297. The binary operation is shown in Fig. 2-10, with each lowercase b of the top row representing internal borrows of 1. A *borrow* is a deficit from subtraction that must be obtained from the next digit, and a borrow of 0 is defined as no borrow.

	bb	bbb	b	bb	
	0001	1110	0010	0100	Minuend A
CF	0111	1010	0100	0001	Subtrahend B
1	1010	0011	1110	0011	Difference

FIGURE 2-10. Subtraction example.

CF					
	bb	bbb	b	bb	
	1001	0101	0010	1100	Minuend
	0010	0011	1001	0011	Subtrahend
0	0111	0001	1001	1001	Difference

FIGURE 2-11. Subtraction example with signed overflow.

For the subtract operation, CF, *which now serves as a borrow flag*, is set when a borrow of 1 is supplied from CF to the MSB of the difference. Also, OF is set when the borrow from CF is not equal to the borrow from the MSB. In the example, CF is set and OF is clear. With signed numbers the difference A3E3H is decimal -23,581, which is correct. There is no overflow. The borrow from CF into the sign bit position is offset by the borrow from the sign bit. The result is within the signed-number range of the accumulator, and the set CF is a signed borrow.

For unsigned numbers the indicated difference is 41,955, which is erroneous. The true result is negative, which is out of the unsigned-number range, and the error is indicated by the set CF. Another example is presented in Fig. 2-11, which illustrates signed overflow.

For unsigned numbers in decimal, the minuend is 38,188, the subtrahend is 9,107, and the difference is 29,081. There is no error. However, for signed numbers the minuend is −27,348 and the correct difference is −36,455, not the indicated difference of 29,081. Flag OF is set because *there is a borrow from the sign bit position but none from CF*. This is signed overflow, often called *underflow*. The correct result is outside the signed-number range of the accumulator.

Subtraction Algorithm

Not all computers have hardware subtractors. For these, subtraction is accomplished with the adder, the complement operation, and program instructions. The operation A − B can be treated as the addition of A and −B. Thus, in 2s complement arithmetic, an algorithm for the subtract operation A − B is as follows:

1. Complement B, obtaining B′.
2. Add B′ and 1, obtaining B′ + 1.
3. Add A and B′ + 1; the result is A − B.

EXAMPLE 1

Signed numbers A and B are 00101101B (45D) and 00110100B (52D), respectively. With the complement procedure, find A − B.

SOLUTION

The difference A − B is A + (B′ + 1). The complement B′ of B is 11001011 and the 2s complement B′ + 1 is 11001100. Adding this to A gives 11111001, which is A − B. Because the MSB is 1, the result is negative. The 2s complement of the result is 00000111, or 7D, and A − B is −7. ■

EXAMPLE 2

Suppose the signed number A is 11000111 and B is 11101110. Determine A + B and A − B. Also, find A, B, A + B, and A − B in decimal.

SOLUTION

Direct addition of A and B gives 10110101 with the carry discarded. The difference A − B is the sum of A and B′ + 1, which is 11011001. Both results are negative. The 2s complements of A, B, A + B, and A − B are 00111001 (57), 00010010 (18), 01001011 (75), and 00100111 (39), respectively. Therefore, A is −57, word B is −18, the sum A + B is −75, and A − B is −39. ■

2-3 MULTIPLICATION AND DIVISION

Basically, multiplication is repeated addition. The product 5 * 3 is 5 + 5 + 5. With unsigned numbers, an elementary multiplication routine consists of adding the multiplicand to the contents of an initially cleared register a number of times equal to the count of the multiplier. Upon completion of the process, the register stores the product.

A similar procedure can be used for unsigned division. The quotient is the number of times that the divisor can be subtracted from the dividend without giving a negative result. What remains of the dividend after the subtractions is the remainder.

Both of the preceding methods are wasteful of computer time. More efficient routines involve shift operations. We shall examine these routines, as well as the ones using repeated addition or subtraction. Initially, only unsigned numbers are considered. For convenience, both flowcharts and pseudo code are used to illustrate the algorithms.

Repeated Addition

Multiplication by repeated addition is illustrated here first by means of a *flowchart*. The chart provides a graphic representation of a procedure, and such diagrams are often useful for designing computer programs and systems.

Flowcharts. A flowchart consists of blocks connected by directed lines that designate the path from one step to the next. Figure 2-12 shows the flowchart for the multiplication routine.

The rectangular blocks are *function* blocks that list one or more functional operations. They may be general or specific, with each showing an action or actions to be taken. The arrow ← means "replace the value on the left with the value on the right." The block below START directs the computer to load 0 into register A, the multiplicand into register B, and the multiplier into register C. Operation A ← A + B specifies that multiplicand B is to be added to the partial product of A, and the remaining functional block tells the computer to decrement the multiplier C. An exit from the loop occurs when C becomes 0, with register A storing the product.

The diamond-shaped *decision* block has two exits. The selected exit depends on the status condition stated within the block, and there may be more than two exits. *Ovals indicate the beginning and the end* of the routine. Blocks other than those illustrated are sometimes used. For example, two circles containing the same number represent a connection, and a *parallelogram denotes input from or output to a peripheral*.

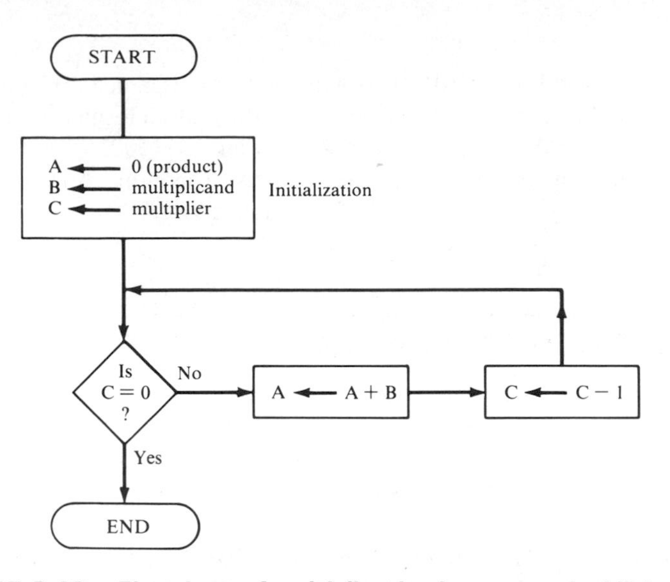

FIGURE 2-12. Flowchart of multiplication by repeated addition.

Pseudo Code. Most programmers use pseudo code in preference to the flowchart to represent the algorithm of a program to be developed. *Pseudo code* is a programming language *with no formal rules of syntax*. It explains how a program will work before coding into an actual computer language, and the designer uses language as desired. There are guidelines, however.

There should be only one entry point and one exit point, and the code should lead to a program that executes in a finite amount of time, with no infinite loops. It should lead naturally into a reasonably well-structured program. The development should be *top-down* so that program code can be verified as correct as it is read from the start, before going to lower levels. The logic must be evident as one reads it from the start, and code should normally be executed in the order in which instructions appear in the program, starting from the top. Briefly, *program flow is vertical*. GOTO statements that give jumps from one section of code to another should usually be avoided.

Branches to program modules are permissible. A *module* is a separate set of instructions given a name and invoked from either the main program or another module. It normally has a single entry and a single exit, and return is to the point of call. Assigned to a module is a single function, and the module is small in size compared with the total size of the program.

Pseudo code is designed to create structured programs. Three basic structures are used: *sequence*, *choice*, and *repetition*. Any logic problem can be solved with combinations of these structures. A sequence structure is one that processes two functions in the order of their appearance in the program. A choice structure, also called IF-THEN-ELSE, provides for selection between two alternatives, with the selection depending on the result of a logical expression. Then the repetition, or DO-WHILE, structure gives looping for as long as some specified condition is true. If the condition is false on entry to the loop, the statements of the loop are never executed.

Figure 2-13 shows pseudo code that gives the algorithm for multiplication by repeated addition. The actual wording of the statements is selected by the designer, without formal rules.

```
PROGRAM Multiplication by Repeated Addition
        initialize product A to 0
        select values for multiplicand B and multiplier C
begin
    do while multiplier (count) C is greater than 0
        replace product A with A plus multiplicand B
        decrement multiplier C
    end do-while
    display product A
end program
```

FIGURE 2-13. Pseudo code for multiplication by repeated addition.

Multiplication by Repeated Addition in BASIC. Step-by-step implementation of the multiplication routine that has been presented is accomplished by execution of the BASIC program of Fig. 2-14.

Suppose key F2 for RUN is depressed, executing the program. After 30 seconds the product, 2185218, is printed on the monitor. If the multiplicand and the multiplier are interchanged, the number of repeated additions is greatly reduced, and execution time is 5 seconds. However, if the instruction PRINT 567 * 3854 is executed in the direct mode, the product appears without any apparent delay. Clearly, the computer uses a more efficient multiplication process.

When WHILE is first encountered, the expression is evaluated, and because C is greater than 0, the sequence of the loop is entered. Upon reaching WEND, the condition is again checked, and if true, the program repeats the numbered statements of the loop. On the last pass, C is decremented to 0, and the test reveals that the expression of WHILE is false. Hence the program progresses to the statement that follows WEND. The total number of passes through the loop equals the initial value of the multiplier C, used as a count. WHILE-WEND loops can be nested, in which case each WEND matches the most recent WHILE. The number of statements that can be placed in the loop is unlimited.

In the assignment statements of lines 30, 50, and 60, the optional keyword LET is omitted. Note the indentations in lines 50 and 60, which are within the loop. Extra spaces in a line are ignored by the interpreter, and program listings are easier to follow if statements of loops are indented. The optional comments at the right are preceded by a single quotation mark, and although printed in the listing, they are ignored during execution. In the statement of line 10, REM can be replaced with a single quotation mark, if desired, and the single quotation mark of line 20 can be replaced with REM.

```
10 REM Multiplication by repeated addition.
20 '
30 A = 0: B = 567: C = 3854 'multiplicand B, multiplier C
40 WHILE C > 0              'begin loop
50   A = A + B              'form the partial product
60   C = C - 1              'decrement multiplier C
70 WEND                     'test WHILE condition
80 PRINT A                  'print product on display
```

FIGURE 2-14. Multiplication by repeated addition.

For signed numbers the signs must first be examined, and if a negative number is found, it is converted into its 2s complement. The sign of the product should be changed only if those of the multiplier and multiplicand are unlike. A few additional steps are added to the program.

Repeated Subtraction

Figure 2-15 presents pseudo code for division of unsigned numbers by repeated subtraction. After execution, the quotient is contained in register A, and the remainder is that of B. The corresponding flowchart is easily developed from the pseudo code (see Prob. 2-11).

PROGRAM Division by Repeated Subtraction

initialize the quotient A to 0
select values for dividend B and divisor C

begin

do while divisor C is less than or equal to dividend B
replace dividend B with B minus divisor C
increment the quotient A
end do-while
display quotient A and remainder B

end program

FIGURE 2-15. Pseudo code for division by repeated subtraction.

Division by Repeated Subtraction in BASIC. The step-by-step implementation in BASIC is given by the program of Fig. 2-16. The program subtracts the divisor of 20 from the dividend of 65,535 until the reduced dividend is less than 20. With each subtraction, the quotient is incremented, starting from 0. Execution begins when key F2 is pressed, and about 30 seconds later the quotient of 3,276 and the remainder of 15 are printed on the monitor. The computer uses a much more efficient division procedure. Execution of the statement PRINT 65535/20 gives 3276.75 with no apparent delay.

For signed numbers the signs must first be examined, and if a negative number is found, it is converted into its 2s complement. The sign of the product should be changed only if those of the dividend and divisor are unlike. A few additional steps are added to the program.

```
10 REM Division by repeated subtraction
20 '
30 A = 0: B = 65535: C = 20
40 WHILE C <= B
50   B = B - C                    'subtract divisor from dividend
60   A = A + 1                    'increment quotient A
70 WEND
80 PRINT "Quotient = "; A;" Remainder = "; B
```

FIGURE 2-16. BASIC division program.

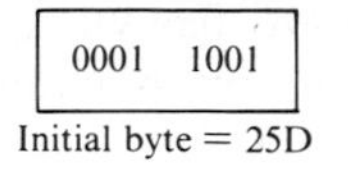

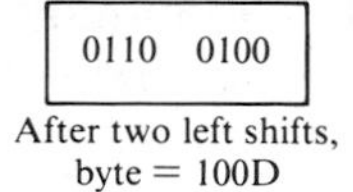

FIGURE 2-17. Two left shifts (SHL or SAL) of a register.

Register Shifts

Efficient multiplication and division routines use shift operations. The *logical left shift SHL* moves the bits of a register to the left by a specified number of bits, *with 0s shifted in on the right.* The number stored in the register is doubled with each bit shift, provided there is no overflow. Thus, a 1-bit logical left shift of the number 7 gives 14, and a shift of −7 gives −14.

For unsigned 8-bit integers, the number in the register before the 1-bit shift must be in the range (0, 127); for signed 8-bit integers, it must be in the range (−64, 63). For unsigned 16-bit integers, the number in the register before the 1-bit shift must be in the range (0, 32767); for signed 16-bit integers, it must be in the range (−16384, 16383). Otherwise there is overflow. Shown in Fig. 2-17 are the contents of an 8-bit register initially, and also after two SHL operations.

A *logical right shift SHR* moves the bits to the right, *with 0s shifted in on the left.* For both unsigned and positive signed numbers, a 1-bit logical right shift *divides the stored number by 2.* If division yields a fraction, it is lost by the shift process; that is, the quotient is truncated to an integer. For example, a 1-bit logical right shift of the number 7 gives 3, and a shift of number 1 gives 0. Normally, logical right shifts are not made on negative numbers because of the sign change that would occur. Figure 2-18 shows the logical right shift of the unsigned decimal number 253 stored in an 8-bit register, giving 126.

There are also arithmetic shifts. These are defined so as to keep the sign bit unchanged unless there is overflow. An *arithmetic left shift SAL* on signed numbers is exactly the same as the logical left shift. Operations SHL and SAL are identical.

The *arithmetic right shift SAR* shifts the bits to the right the specified number of times, *with all bits shifted in at the left being equal to the original MSB.* Bits shifted out at the right are lost. However, for signed numbers an arithmetic right shift *keeps the sign bit unchanged.* SAR and SHR are identical operations when the number is positive. SAR divides the stored number by 2, with any fraction lost. For both positive and negative numbers, the result of the division is truncated to the integer in the direction of minus infinity. For example, a 1-bit SAR of number 3 gives 1, whereas a 1-bit shift of −3 gives −2, as shown in Fig. 2-18.

Some flags of the flag register are affected by shifts. The sign flag SF, the zero flag ZF, and the parity flag PF are normally updated after any shift operation. Auxiliary flag AF is undefined. *The carry flag CF always stores the last bit shifted*

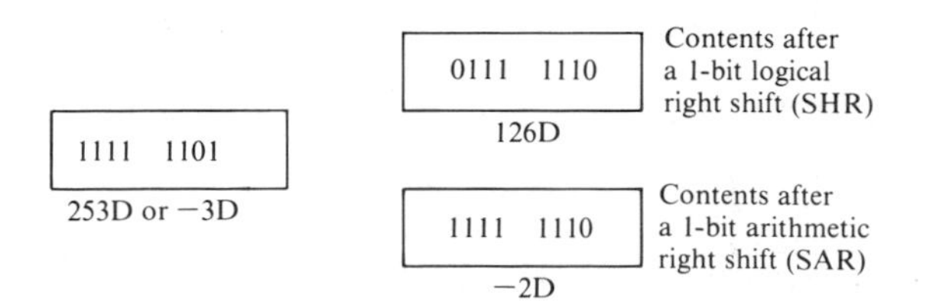

FIGURE 2-18. One-bit logical and arithmetic right shifts of a register.

out for both right and left shifts. In a 1-bit shift, the overflow flag OF is set if the shift changes the state of the MSB. Otherwise, OF is cleared. With multibit shifts, the bit of OF is undefined.

Efficient Multiplication

To illustrate certain principles of multiplication that will help us to understand the algorithms to be presented, let us consider the elementary example of Fig. 2-19. For simplicity, we assume signed numbers in 8-bit registers, with the numbers suitably restricted so that there is no overflow of the product in the 8-bit accumulator A. Binary multiplication is simpler than decimal, because digits are restricted to 0s and 1s, which are the only digits in the multiplication table.

Bits of register A are numbered 7 through 0 from left to right, starting at the vertical line of the figure. Bit 5 of A is the sum of the three 1s of the column plus a carry from the preceding column, giving 100. Thus, there is a carry from column 5 to column 7. There is also a carry from column 6 to column 7, and the sum of the five 1s of column 7 is 101. This carry goes to column 9, which is out of the range of register A. *If the sign bit of the multiplier were extended indefinitely to the left, all product bits to the left of the vertical line would be 1s.*

In the example, the multiplicand B is 19, the multiplier C is −6, and the product stored in the 8-bit accumulator A is −114. There is no overflow. Rather than add columns of numbers as done here, the computer adds two numbers at a time. First, product register A is cleared.

Certain concepts can be deduced from the example. Clearly, multiplication can be done *with add and shift operations*. Each bit of the multiplier is examined in sequence. If it is 0, there is only a shift, because each product of a bit of the multiplicand by 0 is 0. On the other hand, for a bit of 1, the multiplicand is placed in the row where it is added to contribute to the product, along with the proper shift. The addition can be done each time a multiplicand is to be added, with the partial product building up to the total product following the last addition.

The next basic concept to be noted is that *the product of two n-bit numbers requires a register of width 2 * n to hold the product*. For example, an unsigned number of an 8-bit register has a maximum value of 255, and the product of two such numbers is 65025, or FE01H, which requires a 16-bit register. Also, it is evident that right shifts can replace the left shifts of the example, because the relative positions of the numbers to be added are the same. Multiplication of unsigned numbers is considered first.

Multiplicand B		00010011	(19)
Multiplier C		11111010	(−6)
		00000000	Row 1
	0	0010011	Row 2
	00	000000	Row 3
	000	10011	Row 4
	0001	0011	Row 5
	00010	011	Row 6
	000100	11	Row 7
	0001001	1	Row 8
	0010010	10001110	Product A (−114)

FIGURE 2-19. Binary multiplication.

15 14 13 12 11 10 9 8 7 6 5 4 3 2 1 0

FIGURE 2-20. Accumulator register AX with initial values.

Unsigned Multiplication. The first multiplication algorithm presented applies to unsigned 8-bit numbers having a product that can be stored in a 16-bit register *AX*. If the product is sufficiently small to be stored in a byte register, the high byte of the product of AX contains only 0s. The high byte is designated *AH* and the low byte *AL*. These names apply to the accumulator of the 8088 microprocessor of the PC. The register is illustrated in Fig. 2-20.

Initially, byte AH is 00H, and the multiplier is placed in register AL, as shown in Fig. 2-20. The 16-bit product will reside in AX after program execution. *Eight logical right shifts of register AX remove the multiplier*. Prior to each shift, the LSB is examined. *If it is 1, the 8-bit multiplicand is added to the contents of the high-byte register AH*. *If it is 0, there is only a shift*, with a 0 shifted in at the left. As the multiplier is shifted out at the right, the partial product is shifted to the right. Although it started in AH, *it occupies the full register AX after eight shifts*. A little reflection reveals that the addition-and-shift pattern corresponds to the relative positions and values of the product rows of Fig. 2-19. The pseudo code based on these observations is given in Fig. 2-21 (see Prob. 2-12).

A BASIC program that implements the algorithm step by step is given in Fig. 2-22. Although multiplication in BASIC would certainly not be done this way, the program is presented to illustrate further the algorithm.

```
PROGRAM Unsigned Integer Multiplication
        initialize AH to 0 and initialize count I to 8
        select an unsigned 8-bit value for the multiplier
        select an unsigned 8-bit value for the multiplicand
begin
        move the multiplier to register AL
        move the multiplicand to any available register
        do while count I is greater than 0
                decrement count I
                examine the LSB of AL
                if the LSB is 1 then
                        add the multiplicand to AH
                end if
                shift AX logically 1 bit to the right
        end do-while
end program          {the 16-bit product is in AX}
```

FIGURE 2-21. Pseudo code for unsigned multiplication.

```
10 REM  AX is initially the multiplier (AX <= 255).
20 REM  B is the multiplicand.
30 REM  The final value of AX is the product.
30 I = 8: B = 194: AX = 251
40 WHILE I > 0
50    I = I - 1
60    IF AX/2 > INT(AX/2) THEN AX = AX - 1 + B*256
70    AX = AX/2
80 WEND
90 PRINT AX
```

FIGURE 2-22. Unsigned multiplication algorithm program.

In line 60 of the program of Fig. 2-22, the INT function returns a whole number that is less than or equal to the integral value of its argument. It follows that the expression is true if the LSB of AX is 1. In this case, 1 is subtracted from AX, giving an even number. Also, multiplicand B is multiplied by 256, and the product is added to AX: these operations accomplish the required addition of B to the high byte of AX. The logical right shift is implemented in line 70 by dividing AX by 2.

IF-THEN-ELSE. The IF statement is used *to make a decision regarding program flow, based on the result returned by an expression.* With the ELSE clause optional, the format is as follows:

IF *expression* THEN *clause* ELSE *clause*

The word *clause* denotes *a BASIC statement or sequence of statements separated by colons. Clause* can be a line number indicating the location of the next instruction to be executed. If the expression of the IF statement is true, the clause after THEN is executed. Otherwise, the clause after ELSE, if present, is executed. With ELSE omitted, as in line 60 in Fig. 2-22, a false expression causes program flow to proceed to the next line. Nesting of IF statements is allowed.

In the following example, if A is greater than B, the value of A is displayed and B is assigned value 5. However, if the expression is false, the value of B is displayed, and the program branches to line 500. Note that *multiple statements can follow both THEN and ELSE.*

```
IF A > B THEN PRINT A: B = 5 ELSE PRINT B: GOTO 500
```

Signed Multiplication. Without proof, the algorithm for signed multiplication of 8-bit numbers is the same as for unsigned numbers, *except that the final add-and-shift operation is replaced with a subtract-and-shift operation and logical shifts are replaced with arithmetic shifts.* In the pseudo code of Fig. 2-21, the count I should be initialized to 7. Also, following the DO-WHILE loop, the statements of the loop should be repeated, but with omission of the decrement-count statement and with "subtract..from" replacing "add..to". In addition, all shifting must be specified as arithmetic. The corresponding BASIC program is given in Fig. 2-23.

In line 120, if multiplier C is negative, AX is assigned a value equal to the sum of C and 256. This value assigns to AX a high byte equal to 0 and a low byte equal to C. For example, suppose C is −128. Then AX is +128, or 0080H, giving a high byte of 0 and a low byte of 80H, or −128, as required (see Probs. 2-13 and 2-14).

```
100 REM B is the multiplicand and C is the
    multiplier.
110 I = 7: B = 124: C = -117
120 IF C < 0  THEN  AX = C + 256  ELSE AX = C
130 WHILE I > 0
140    I = I - 1
150    IF AX/2 > INT(AX/2) THEN AX = AX - 1 + B*256
160    AX = AX/2
170 WEND
180 IF AX/2 > INT(AX/2)  THEN  AX = AX - 1 - B*256
190 AX = AX/2
200 PRINT AX
```

FIGURE 2-23. Signed multiplication algorithm program.

The signs of the multiplicand and the multiplier are unrestricted, but the numbers must properly fit within byte registers. If the product of the signed numbers is negative and sufficiently small to fit in a byte register, the high byte of AX contains only 1s as expected for a 16-bit negative number.

The unsigned and signed multiplication algorithms that have been described are implemented in the instruction set of the 8088 microprocessor. *Instruction MUL provides multiplication of unsigned 16-bit words, and IMUL gives signed integer multiplication.* The multiplicand and the multiplier can be those of either byte or word (16-bit) registers, with the product placed in word or doubleword registers, as required. For both instructions, the carry and overflow flags CF and OF are reset if the high-order half of the result is simply a sign extension of the low-order half. Otherwise the flags are set.

Division

Decimal 69 divided by 5 yields 13 with a remainder of 4. The division process in binary is illustrated in Fig. 2-24. Basic operations are *comparison*, *conditional subtraction*, and *logical shifts*. Unsigned numbers are assumed.

Initially, the divisor is compared with the first bit of the dividend and found to be larger. Accordingly, the first bit of the quotient is 0. Next, the divisor is compared with the initial 2-bit word 10 of the dividend. Again it is larger, and so another 0 is assigned to the quotient. Comparison with the 3-bit word 100 gives the same result. However, the divisor is not greater than the first 4-bit word 1000. A bit of value 1 is placed in the quotient, and 101 is subtracted from 1000, yielding 11. The bit of value 1 is then brought down. Because the divisor is smaller than the new word 111, another 1 is contributed to the quotient. The process continues

```
      0001101   (quotient = 13)
101 )1000101    (69/5)
      101
       111
       101
        1001
         101
         100    (remainder = 4)
```

FIGURE 2-24 Binary division.

```
PROGRAM Unsigned Integer Division
        initialize count I to 8 and the high byte AH of AX to 0
        select the 8-bit dividend X and the 8-bit divisor Y
begin
        move dividend X to 8-bit register AL
        move divisor Y to any available 8-bit register
        do while count I is greater than 0
                decrement count I
                shift left the contents of AX
                if divisor Y is <= high byte AH of AX then
                        replace AH with AH minus divisor Y
                        add 1 to the low byte AL of AX
                end if
        end do-while
        display the quotient AL and the remainder AH
end program
```

FIGURE 2-25. Algorithm for unsigned integer division.

until the last bit of the dividend has been brought down, with the final subtraction yielding the remainder of 100, or 4.

An algorithm for unsigned division of an 8-bit dividend by an 8-bit divisor is given in Fig. 2-25 in the form of pseudo code. Let us note that *division of a 2n-bit dividend by an n-bit divisor generates an n-bit quotient and an n-bit remainder* (see Prob. 2-16).

Initially, the 8-bit dividend is placed in the low half AL of register AX, and the high half AH is cleared. The algorithm gives left shifts, and after eight 1-bit shifts the dividend has been removed from AL. Following each shift, the divisor is compared with the byte of AH. *If the divisor is less than or equal to this byte, it is subtracted from the byte of AH, which leaves the remainder in AH; also, the 0 bit shifted into AL by the left shift is replaced with a 1.* On the other hand, if the divisor is greater than the byte of AH, there is no subtraction, and the LSB remains 0. *After eight left shifts, the byte of AL is the quotient and that of AH is the remainder.* This process is the same as that of the binary division previously described.

The method used here is similar to the one implemented by the 8088 microprocessor for both unsigned and signed division of integers. However, the dividend can have up to 32 bits and the divisor is allowed 16 bits. *Instruction DIV gives unsigned division, and IDIV gives signed integer division.* Flags are undefined after either operation.

Figure 2-26 shows implementation of the algorithm in BASIC for unsigned integers. The dividend and divisor are restricted to 8 bits (see Prob. 2-15).

The left shift of AX is done by the statement of line 140 of Fig. 2-26. The value of AL after a left shift is fixed in line 150, and that of AH is fixed in line 160. Then the comparison of the divisor Y with AH is made in line 170. If Y is equal to or less than AH, the divisor is subtracted from AH; also, AL is incremented, which sets its LSB. Execution displays the quotient AL and the remainder AH.

```
100 REM  I = count, AL = dividend, Y = divisor.
110 I = 8:  AL = 250:  Y = 69:  AH = 0
120 WHILE I > 0
130   I = I - 1
140   AX = 2 * (256*AH + AL)
150   IF 2 * AL > 255  THEN  AL = 2*AL - 256 ELSE AL = 2*AL
160   AH = (AX - AL)/256
170   IF Y <= AH THEN AH = AH - Y:  AL = AL + 1
180 WEND
190 PRINT AL, AH
```

FIGURE 2-26. Unsigned division.

2-4

OPERATIONS IN BASIC

Various commands are presented here that allow us to save, list, and load programs, and also to perform other essential operations related to programs. In addition, we shall consider arithmetic in BASIC, along with the different types of numbers used. Relational operators are applied to numeric and string variables, and a program is examined.

Commands

The IBM BASIC MANUAL lists 22 commands. Command RUN for program execution and command SYSTEM for return to DOS have already been mentioned. Others are briefly examined. Although commands may be given line numbers and included in programs, they are most often entered in the direct mode.

SAVE filespec. Suppose a program has been entered into memory and one wishes to save it on a diskette in drive B. The command

```
SAVE "B:REPORT.BAS"
```

will store the program on the diskette, assigning it the name REPORT. The string expression in parentheses is the *filespec*, with B denoting the *device* (drive B), REPORT being the assigned *filename*, and BAS representing the *extension*. If "B:" is omitted from the filespec, the default drive is understood. Normally this is drive A, although the default drive can be changed. For example, drive B becomes the default if "B:" is entered in the DOS mode.

The name REPORT is arbitrarily selected, but it must not exceed eight characters, which may consist of letters, the decimal digits, and characters $ and @, in any order. The extension may be no more than three characters long. If none is included in the filespec, the extension BAS is automatically added, becoming part of the filespec. This indicates a BASIC file. The second double quotation mark is optional.

LOAD filespec. As before, filespec must be in quotation marks, which identify it as a string expression, with the second quotation mark optional. The command

```
LOAD "REPORT
```

will load into memory from the default drive the program REPORT.BAS, assuming such a program is present on the diskette. Otherwise an error message appears on the CRT of the display. If the filespec is followed by a comma and then R, the program is loaded and immediately run. An example is LOAD "REPORT",R.

LIST and LLIST. The LIST command causes the entire program stored in memory to be listed on the display, with the lines in numerical order. If the program is too large to fit on the screen, pressing simultaneously the keys Ctrl-NumLock stops the listing, which is resumed by pressing any other key. The LIST 20 command displays only line 20, and LIST 20–100 gives all lines from 20 through 100. There are various other options. The LLIST command is identical to LIST except that *the listing is printed on hard copy* (paper) by the printer rather than on the screen.

KILL filespec. The KILL command is used to erase a file from the diskette. The proper extension must be included in the filespec. An example is KILL "B:A1.XYZ".

FILES. Execution of FILES displays the names of all files on the diskette of the default drive. The command FILES "B:*.*" gives the names of all files of drive B, and FILES "B:*.XYZ" provides names of those files with extension XYZ. In both the DOS and BASIC modes, the symbol "*" can be used in a filespec as a *global* character signifying that *any character can occupy its position and all remaining positions of the filename or extension.* For example, command FILES "R*.*" will display names of only those files of drive A with names beginning with R.

NEW and DELETE. The NEW command deletes the entire program currently in memory, thereby preparing memory for a new program. DELETE is used to delete program lines. For example, execution of DELETE 5–36 eliminates all lines with numbers 5 through 36.

AUTO and RENUM. For generating line numbers automatically, use the AUTO command. Line 10 appears, and when its statement is typed and entered, line 20 appears, and so on in increments of 10. After entry of the last line, press key combination Ctrl-Break to terminate the automatic numbering. The command AUTO 100 starts numbers at 100, and AUTO 50,5 starts at 50 and gives increments of 5. Other choices can also be made.

At any time the RENUM command can be executed. It renumbers lines, starting at 10 with increments of 10, and the starting number can be specified. For example, RENUM 100 renumbers all lines, starting with 100. Many times when programs are created, lines must be inserted between others, and the numbers become irregular, such as 10, 12, 17, and so on. Although not essential, RENUM arranges lines in a nice sequence and automatically adjusts all numbers referenced by program statements.

Other Commands. Twelve commands have been examined. These and others are described in detail in the IBM BASIC manual. The additional commands are given in Fig. 2-27, and some of them are discussed later. When commands and statements are presented in this book, only the options considered to be the most important are given.

BLOAD	Load machine language program into memory.
BSAVE	Save a portion of the computer's memory.
CLEAR	Clear program variables and set the memory area.
CONT	Continue program execution after a break.
EDIT	Display a program line for editing.
MERGE	Merge the diskette program with one in memory.
NAME	Rename a diskette file.
RESET	Close all diskette files.
TRON	Turn the program trace on for debugging.
TROFF	Turn the trace off.

FIGURE 2-27. Selected BASIC commands.

Arithmetic Operators

The arithmetic operators of BASIC, listed in order of precedence, are given in Fig. 2-28.

In *integer division* each operand *is rounded to the nearest whole number before the division is done, and the result of the division is truncated* by dropping the fractional part. For the example shown in the figure, the rounding process gives 11\4. With ordinary division the result is 2.75, but truncation yields the answer 2.

The *MOD* operation gives *the integer value of the remainder of an integer division*. Thus 11.4 MOD 3.6 equals 11 MOD 4, which has a remainder with an integer value of 3.

A *numeric expression* with various operators evaluates to a single value. Parentheses can be used freely in expressions, and all parts within parentheses are evaluated first. Thus PRINT 6 * (3 + 2) gives 30, but without the parentheses the result would be 20.

Numbers

Numbers in BASIC are divided into the three following types:

1. Integers
2. Single precision numbers
3. Double precision numbers

Operator	Operation	Example
^	Exponentiation	1.5^2.0 = 2.25
−	Negation	−5
* and /	Multiply, divide	2.3*5; 5/6
\	Integer division	11.4\3.6 = 2
MOD	Modulo arithmetic	11.4 MOD 3.6 = 3
+ and −	Add, subtract	5+3; 4−6

FIGURE 2-28 Arithmetic operators.

Integers are signed numbers written without decimal points that can be stored in 2s complement form in 16-bit registers. Accordingly, a number from −32,768 through 32,767 is automatically considered an integer. Numbers in octal or hexadecimal form are restricted to integers. Execution of the statement PRINT &HFFFF gives −1 on the display, and PRINT &H8000 gives −32768.

A *single precision* constant is a decimal number that is not an integer and is written with 7 or fewer digits, or is written in exponential form using E, or is followed with an exclamation point !. Examples are as follows:

10.0   3456789   −1.789E05   .89E-8   −.987654321!

In the exponential form, the letter E denotes "times 10 to the power." Thus, 5E−2 equals .05, and −2E3 equals −2,000. Up to 7 digits are printed. A single precision constant requires 4 bytes of storage space, whereas an integer uses only 2 bytes.

A *double precision* constant is a decimal number written with 8 or more digits, or written in exponential form using D, or followed with the number sign #. The interpretation of D is the same as that of E, but up to 16 digits are printed, and storage requires 8 bytes. Examples are as follows:

3.333333333333   5.4D−2   234#

When an expression is evaluated, all operands are converted to the precision of the most precise operand, and the result is found to this degree of precision. However, if assigned to a variable of lesser precision, the result is rounded. Thus, for A = 20/3 followed by PRINT A, the displayed value of the single-precision variable is 6.666667, whereas for "A# = 10/3# the display is 6.666666666666667. Numbers such as these, containing a decimal point, are called *fixed point*, or *real*, numbers. Constants in exponential form are known as *floating point* numbers.

In most cases, the programmer does not need to be concerned with number types. The higher precision numbers require more storage space and processing time is increased, but these factors are not usually of major importance.

Numeric Variables

A *variable* is a *name used to represent a value*, and the value assigned to the variable can change as the program progresses. There are both numeric and string variables. The value of a numeric variable is a number. Up to 40 characters are allowed in the variable name. These may be letters, numbers, and the decimal point, but the first character must be a letter. The decimal point is often used for separating two words. For example, TOTAL.AMT is more meaningful than TOTALAMT. BASIC has a list of reserved words that cannot be used as variable names. The list includes all BASIC commands, statements, function names, and operator names. In order, examples of each are RUN, IF, SIN, and MOD.

Unless otherwise specified, all numeric variables are assigned single precision numbers requiring 4 bytes of storage. It is possible to declare a numeric variable as either integer, single precision, or double precision by following the name with %, !, or #, respectively. For example,

SUM% is an integer variable (2 bytes)
ABC! is a single precision variable (4 bytes)
AMT# is a double precision variable (8 bytes)

When a higher precision value is assigned to a lower precision variable, BASIC always rounds the value rather than truncating it. For example, the statement C% = 10.6 assigns the number 11 to variable C%.

BASIC statements DEFINT, DEFSNG, and DEFDBL can also be used to declare variable types as integer, single precision, and double precision, respectively. Examples are as follows:

```
DEFINT A-Z    'all variables are integers
DEFSNG AB     'names starting with AB are single precision
DEFDBL PI     'names starting with PI are double precision
```

When these statements are used in a program, they must precede the declared variables. Type declaration characters %, !, and # take precedence over the DEF statement. Sometimes the statement DEFINT A-Z is placed at the beginning of a program, with type declaration characters used for the few variables that must be of higher precision. This procedure uses a minimum of storage, and programs run faster.

String Variables

The rules for names of numeric variables also apply to string variables, but the last character of a string variable name must be the dollar sign $. *This sign declares the variable as a string*. The character string value within double quotation marks can have up to 255 characters. An example is as follows:

```
NAME$ = "Smith"
```

The value of NAME$ can be changed at any time by the program. Prior to assignment, string variables are assumed to be null, with a length of 0.

An alternate way to declare a string variable is provided by the DEFSTR statement. An example is as follows:

```
DEFSTR PA   'Names starting with PA are string
             variables
```

After execution, names such as PAY and PAYMENT are string variables unless overridden by a type declaration character.

Relational Operators

A numeric or string expression is either a constant or a variable, or it is a combination of constants, variables, and operators that give a single numeric or string

=	is equal
>< or <>	is not equal to
<	is less than
>	is greater than
<= or =<	is less than or equal to
>= or =>	is greater than or equal to

FIGURE 2-29. Relational operators.

value. Values of constants, variables, and expressions are compared in BASIC by means of six *relational operators*. These operators and their meanings are shown in Fig. 2-29.

When values are compared, the result is either true or false. *If true, the BASIC program returns an integer value of FFFFH, or −1, and if false, the returned 16-bit value is 0000H*. To illustrate, execution of the PRINT statement that follows gives on the display the results that are shown immediately below the statement:

```
PRINT  3 < 6;   4 >= 6;   5 = 10
-1  0  0
```

As indicated, *a printed number is always followed by a space, and a positive number is preceded by a space*. Each negative number is preceded by a minus sign in place of the leading space.

String comparison is based on ASCII codes. Characters are examined one by one in order from each string *until their ASCII codes differ*. When one code is found to be less than the other, the smaller string is identified. If the end of one string is reached before the other, the shorter one is the smaller. Strings are equal only if all codes are equal. Each of the following relational expressions is true (−1):

"ABB" < "ABC"	"Name" < "Name "
"abbb" < "abc"	A$ < "12" if A$ = "110"

Strings can be joined together, or *concatenated*, by means of the plus symbol (+). Suppose A$ = "Cost", B$ = " is", and C$ = " $5". Execution of the statement that follows gives the result on the right:

```
PRINT A$ + B$ + C$                Cost is $5
```

There are various string functions in BASIC, each of which returns a string result. Some examples follow, with the displayed results shown at the right:

```
PRINT CHR$(65)                        A
PRINT HEX$(242)                       F2
DATE$ = "5/29/84": PRINT DATE$        05-29-1984
TIME$ = "7:45": PRINT TIME$           07:45:00
```

Function CHR$(*n*) returns the character having the ASCII code *n*, and as explained earlier, HEX$(*n*) gives the hexadecimal string representing the decimal argument. The last two examples set the current date and time, and then immediately retrieve them.

```
100 PI = 3.1416                         'assign variable PI
110 WHILE R <> 0
120   READ R                            'get value of radius
130   A = PI * R^2                      'calculate area
140   PRINT "R ="; R, "Area ="; A       'display results
150 WEND                                'check condition
160 GOTO 190
170 DATA 1, 2, 3.456                    'table of values
180 DATA 4, 5, 0                          of R
190 END
```

FIGURE 2-30. Program with READ and DATA statements.

Program 1

The program shown in Fig. 2-30 calculates and prints the area of a circle for several different radii. It illustrates the use of READ and DATA statements for inserting data into a program.

READ and DATA. The READ statement reads constants from a DATA statement and assigns the value read to the single precision variable R. The nonexecutable DATA statements may be placed anywhere in the program, and each may contain as many constants as will fit on a line. The constants can be read only once unless a RESTORE statement precedes the READ statement that is to fetch the data a second time. For example, the END statement of line 190 could be replaced with RESTORE, with another READ statement on line 200 that would start reading the same data again.

GOTO. In general, GOTO statements should be used sparingly in structured programs, because they make programs more difficult to trace and debug. Except for a line number that might be in a distant part of the program, the statement does not indicate where the program is going. However, there are some cases in which its use is justified. One of these is for a jump to the end of a program, as in the routine of Fig. 2-30. However, if the GOTO statement is simply deleted, the program executes just as well.

A running program stops whenever the statement STOP is encountered. This characteristic is useful for debugging, which is examined later. GOTO can be used in the direct mode to reenter the program at any desired point.

END. The END statement of line 190 terminates program execution, closes all files, if any, and returns to the command level. An END statement can be placed anywhere in a program, and whenever one is encountered, processing of the BASIC instructions ends. The use of END at the end of a program is optional, because a program always stops when the last numbered line has been executed. In the previous routine, the combination of GOTO and END provides a way to jump the nonexecutable data.

```
10 WHILE R <> 0
20   INPUT "What is the radius"; R
30   A = 3.1416 * R^2
40   PRINT "R ="; R, "AREA =";A
50 WEND
60 END
```

FIGURE 2-31. Program with keyboard entry.

Program 2

Use of the INPUT statement is another way to supply data to a program. The program of Fig. 2-31 is almost the same as the previous one except that data are entered into the program by means of the keyboard rather than by READ and DATA statements.

INPUT. When line 20 is executed, the optional prompt string is displayed, followed by a question mark, and the program waits for the numerical entry from the keyboard. Thus, the display reads

```
What is the radius?_
```

After a number is typed and entered, the program assigns the entry to the variable of the INPUT statement and displays both the radius R and the area. The program loops, again pausing with the displayed prompt. A keyboard entry of 0 ends the program, or it can be terminated with Ctrl-Break. BASIC programs can always be stopped with this key combination.

If the prompt string is omitted, the prompt is only a question mark. *By replacing the semicolon before the variable with a comma, the question mark is eliminated.* For a string variable name in the input statement, the keyboard entry is understood to be a string, and double quotation marks are optional.

Any number of variables can be included in an input statement. To illustrate, suppose one desires to calculate the volume of a box with length L, width W, and height H. These values can be entered as follows:

```
10 INPUT "Length, Width, Height"; L, W, H
```

The string prompt followed by a question mark is displayed during the pause, and the operator must type from the keyboard three numbers, separated by commas and then followed by Enter. The values entered are assigned to variables L, W, and H, in order. Entries must correspond in number with the number of variables. Spaces of the entry line have no effect.

Program 3

The program presented in Fig. 2-32 illustrates some of the concepts that have been discussed and also shows a useful way to convert units from one system to another.

All variables except GRAMS! are declared to be integers. In particular, note that the value of INTEGER.OUNCES is rounded to the nearest whole integer. The

```
10 'Conversion of grams to integer pounds and ounces
20 DEFINT A - Z
30 INPUT "Enter the number of grams: ", GRAMS!
40 INTEGER.OUNCES = GRAMS! / 28.35
50 INTEGER.POUNDS = INTEGER.OUNCES \ 16
60 INTEGER.OUNCES.REMAINING = INTEGER.OUNCES MOD 16
70 PRINT
80 PRINT GRAMS!; "grams equal";
90 PRINT INTEGER.POUNDS; "pounds and";
95 PRINT INTEGER.OUNCES.REMAINING; "ounces"
```

FIGURE 2-32. Unit-conversion program.

remark of line 10 begins with an apostrophe instead of REM, the choice being optional. In line 30 the comma between the string prompt and the variable suppresses the question mark, and without the blank at the end of the string, there would be no space between the number entered and the prompt. Both integer division and the MOD operation are used to convert whole ounces into integer values of pounds and ounces. The three PRINT statements at the end can be combined into a one-line statement. However, by terminating lines 80 and 90 with semicolons as shown, *the printout of the display will be on a single line.*

For an entry from the keyboard of 5678, the display after program execution is as follows:

```
Enter the number of grams: 5678
 5678 grams equal 12 pounds and 8 ounces
```

The PRINT of line 70 gives the blank line between the two lines of the display. The result is accurate to the nearest whole ounce.

REFERENCES

J. K. HUGHES and J. I. MICHTOM, *A Structured Approach to Programming,* Prentice-Hall, Inc., Englewood Cliffs, N.J., 1977.

C. KAPPS and R. L. STAFFORD, *Assembly Language for the PDP-11,* Prindle, Weber & Schmidt and CBI Publishing Co., Boston, 1981.

G. M. SCHNEIDER, S. W. WEINGART, and D. M. PERLMAN, *An Introduction to Programming and Problem Solving with Pascal,* John Wiley & Sons, Inc., New York, 1982.

M. E. SLOAN, *Introduction to Minicomputers and Microcomputers,* Addison-Wesley Publishing Co., Reading, Mass., 1980.

PROBLEMS

SECTION 2-1

2-1. Find the 2s complement of the following 8-bit binary numbers: 00101000, 00000000, 11111111, 11011101, 10000001, and 11110000.

2-2. Without conversion to binary, find the hexadecimal form of the 2s complement of each of the following hexadecimal representations of 16-bit signed numbers: 10D0, 85EC, ABCD, FFF0, 8000, and 4321.

2-3. Determine both the signed and unsigned binary number range of a 32-bit register that stores integer numbers.

2-4. Sketch the circuit of four 4-bit adders cascaded to form a 16-bit adder. If input words are hexadecimal 4ABC and ED49, with input carry K = 0, find the sum and the output carry.

SECTION 2-2

2-5. Using 8-bit 2s complement arithmetic, find A + B, A − B, B − A, and −A − B, with A = 10010110 and B = 00010101.

2-6. In 10s complement decimal arithmetic, five-digit numbers A and B are 99277 and 99414, respectively. Using 10s complement arithmetic, evaluate A + B, A − B, −A − B, and B − A. In addition, give A and −B in BCD form.

2-7. Hexadecimal bytes E9, 7A, and 86 are stored in 8-bit registers A, B, and C, respectively. Determine the flags CF, OF, SF, ZF, AF, and PF after each of the following independent operations: A + B, A − B, B + C, and C − B.

SECTION 2-3

2-8. Show the word of an 8-bit shift register that stores the binary number representing decimal 26. (a) Two 1-bit SALs are made. After each shift show the register words and their decimal equivalents. (b) Repeat (a) but assume SARs. (c) Repeat both (a) and (b) for the decimal number −26.

2-9. Suppose the 8-bit register AL stores EDH. The operation SAR AL, 1 is implemented. For signed numbers, determine the decimal content of AL before and after the shift, and give the final states of CF, OF, SF, ZF, and PF. Repeat if AL stores 13H.

2-10. Suppose the 16-bit register AX stores the unsigned number F2B5H. The operation SHR AX,1 is implemented. Determine the decimal content of AX before and after the shift and give the final states of CF, OF, SF, ZF, and PF.

2-11. Design the flowchart for division by repeated subtraction from the pseudo code of Fig. 2-15.

2-12. For the algorithm of the pseudo code of Fig. 2-21, trace the 16-bit content of AX for a multiplier of 250 and a multiplicand of 19. Show successive binary values in a column. Repeat for a multiplier of 19 and a multiplicand of 25.

2-13. In a form similar to that of Fig. 2-21, give the pseudo code for signed multiplication of 8-bit numbers. Then trace the contents of AX for multiplier/multiplicand values of −6 / 19, 19 / −6, and −19 / −6. For each case, show successive binary values in a column.

2-14. Trace the binary content of AX for the signed multiplication BASIC program of Fig. 2-23. Show successive binary values in a column.

2-15. Repeat Prob. 2-14 for the unsigned division BASIC program of Fig. 2-26.

2-16. Trace the binary content of AX for the unsigned division pseudo code of Fig. 2-25, for a dividend of 250 and a divisor of 17.

SECTION 2-4

2-17. Give the value of each of the following expressions, and compare the results of (a) and (b), and (c) and (d).

(a) 4 * 2^ 8 / 2 − 2 * 4 + 3

(b) ((4 * 2) ^ (8 / 2) −2) * (4 + 3)

(c) 10 \ 3 + 10 MOD 3 (d) 9.5 \ 3.4 + 9.5 MOD 3.4

2-18. Write a single program that prints each variable and its value, and give the results shown on the display.

A = 1.278 − 2.34^3 and A% = A
B = 289 − 12^3 + 279.39 / 1.679 and B# = B
C = (9.37 + $5.6^{2.3}$)/16.9 and C% = C
D = ($\sqrt{73.1234}$) * 4.5^4 / 2.33 and D# = D
CUBE = 2^3 + 3^3 + 6^3 + 8^3
INTEGER = 5.3\2.6; REMAINDER = 5.3 MOD 2.6

2-19. Using a READ statement for data input, and using DEFINT for declaration of all variables as integers, write a program that will print Y and its value for (ABC) values of (123) and also for values (456) and (789), with Y equal to A + B^2 + C^3. Repeat, using an INPUT statement for data entry.

2-20. Write a program that converts a total number of seconds to hours, minutes, and seconds, following the format of program 3 of Section 2-4.

2-21. 15 A# = 11/3: A! = A#: A% = A#: A = A#
20 PRINT A#, A!, A%, A
Give the display when the program is run.

2-22. 100 PRINT 2 = 3, 2 >= 1, 5 < 2; 2 >< 2 : PRINT
200 PRINT "ABCD" > "ABCC", "Jodi" > "Jodie",
300 PRINT "4" + "bits", "4" + " bits"
Give the display when the program is run.

2-23. What is accomplished when each of the following BASIC commands is entered in the direct mode?

(a) `SYSTEM`
(b) `SAVE "REPORT"`
(c) `LOAD "A:MENU",R`
(d) `LLIST 100 - 200`
(e) `KILL "B:*.BAK"`
(f) `DELETE 50`
(g) `DELETE 0 - 200`
(h) `DELETE 75-`
(i) `DELETE -75`
(j) `FILES "B:X.*"`
(k) `FILES "*.*"`

The 8088 Microprocessor and Assembly Language

There are numerous vendors of many different types of microprocessors. Associated with each type are specially designed interface units, peripherals, an assembler and perhaps numerous compilers, program development aids, and possibly a single-board microcomputer for training. In addition, there are supporting manuals and in-house workshops. Some of the programmable interface chips may be as sophisticated as the microprocessor itself. The computer designer must choose a processor, and the most appropriate choice depends on various factors, such as cost, capabilities and functions provided, the detailed requirements of the application, and the availability of supporting hardware and software packages.

A major objective of this book is to introduce features and basic concepts that are typical of microprocessors in general. These include hardware architecture, control signals, operating characteristics, addressing modes, use of the instruction set, and the relationship between the machine instructions and the statements of a high-level language. Clearly, the features vary considerably from one device to another. Rather than attempt to generalize, we shall confine our attention primarily to a single microprocessor—the Intel 8088. By doing so, our study can be specific and meaningful. Furthermore, it is the CPU of many different kinds of PCs, and familiarity with the 8088 processor is essential for later examination of the complete PC system, including interface procedures for I/O.

In this chapter the 8088 is introduced, with emphasis on the organization of internal registers and addressing of memory. A number of assembly language instructions are examined, along with a few short programs, and the study continues in the chapters that follow. Also, more BASIC is presented, with discussions of stacks and arrays.

3-1 MACHINE ORGANIZATION

The general-purpose 8088 microprocessor, with 16-bit registers, is basically a chip that contains both the control and data path units of a computer in a 40-pin dual in-line package, often called a *DIP*. The circuit supplies addresses to a 20-bit address bus and data to an 8-bit data bus, and either 8-bit or 16-bit words can be processed internally.

The required power supply is 5 V, with a current less than 340 mA. Maximum *source* and *sink* currents are 0.4 and 2 mA, respectively. Accordingly, the processor can supply as much as 0.4 mA to loads connected to output pins, and it can sink up to 2 mA from sources connected to input pins. The 20-bit address can access up to 1 MB of memory, and output status lines are available for control of large systems.

The clock frequency must be between 2 and 5 MHz, with voltage levels of 0 and 5 V and a 33% duty cycle. In the PC, the clock is provided by the 8284 clock generator chip. Cells of internal registers of the microprocessor store an electric charge, and the magnitude of the charge differentiates between the two types of bits. The circuitry is such that the stored charge leaks to ground, and within a few milliseconds the stored bit is lost unless the charge is restored. Such circuits are said to be *dynamic*. *Refreshing* is a process that restores the charges of dynamic cells to their full values, and internal refresh operations are implemented periodically. A clock frequency greater than 2 MHz is required for providing an adequate refresh rate. Thus, single stepping of instructions, which is useful in program debugging, cannot be accomplished simply by disabling the clock.

Dynamic circuits are also found in the read/write memory and in the registers of various peripherals of the PC. These too are refreshed periodically, with the charge of each cell restored at least once every 2 milliseconds. The operating frequency of the PC is 4.77 MHz.

Functional Block Diagrams

As shown in Fig. 3-1, the microcomputer has two separate processing units, called the *execution unit* (*EU*) and the *bus interface unit* (*BIU*). The EU is a combined data processor and computer control unit, and the BIU controls all operations of the external data and address buses. Data transfer between the two units is along the 16-bit internal data bus, and instructions are supplied to the EU from the BIU in byte sequences over the internal 8-bit Q bus.

Execution Unit

Within the EU is the data processor and its timing and control circuits. There are four 16-bit data registers with separately addressable high and low bytes. Because registers and internal buses are 16 bits wide, a word for this machine is understood

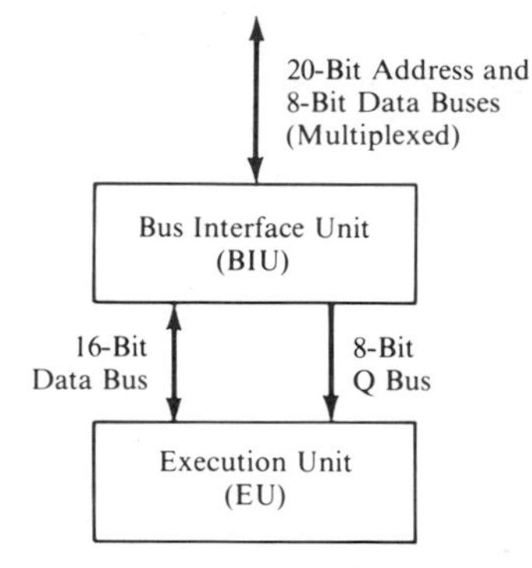

FIGURE 3-1. Basic subdivisions of the processor.

to consist of 16 bits. Henceforth, *word* is used to denote 2 bytes, unless otherwise specified, and a *doubleword* is 4 bytes. Each of the registers AX, BX, CX, and DX stores a word, and the bits are numbered such that the MSB of a register is bit 15 and the LSB is bit 0.

The Data Registers. For the *accumulator AX*, bits 15 through 8 constitute the high byte AH, and bits 7 through 0 are the low byte AL. The *base register BX* has bytes stored in BH and BL, and the *count register CX* and the *data register DX* are similarly divided. The four data registers and their designations are as follows:

Accumulator	AX	(AH, AL)
Base	BX	(BH, BL)
Count	CX	(CH, CL)
Data	DX	(DH, DL)

Most computers have a special register called the *accumulator*. Some are designed so that one of the operands to be processed by the ALU always comes from the accumulator, with this same register automatically receiving the result of the operation. To illustrate, suppose a string of numbers stored in memory are to be added. The accumulator is cleared, and the first number is transferred from memory into a processor register referred to as the *memory data register* (*MDR*). It is then added to the zero contents of the accumulator, with the sum stored in the accumulator. Then the next number is brought in and added. The process continues until completion. The register "accumulates" the sum as the addition progresses, which accounts for the name.

Other examples of accumulation are the multiplication and division routines that were investigated in Chapter 2. An *operational register* is one that is capable of modifying its stored data, with operations such as shift, increment, and clear. Often an accumulator is designed as an operational register that can perform many of the functions of an ALU.

In the 8088 the accumulator AX is simply one register of a four-register array used primarily for temporary storage of an operand. The base (BX), count (CX), and data (DX) registers are similar. However, AX commonly serves as an accumulator, and BX often contains the base address of a table of data or words of an array stored in memory. CX frequently holds a count, and DX may have general data. The registers are not restricted in any way to special functions. For example,

CX can operate as an accumulator, and AX may maintain a count. The programmer uses the registers as desired, except that some instructions relate implicitly to certain registers.

CPU Block Diagram. The data registers are often referred to as *scratchpad* registers, because they are available for temporary storage as needed and are especially useful for storing intermediate results of ALU operations. These and others are shown in the simplified functional block diagram of the CPU in Fig. 3-2. The BIU is above the dashed horizontal line, and the EU is below it. Numbers alongside the flow paths indicate the number of conducting lines.

Pointer and Index Registers. Also within the EU are two *pointer* registers and two *index* registers. These are designated as follows:

Stack pointer	SP
Base pointer	BP
Source index	SI
Destination index	DI

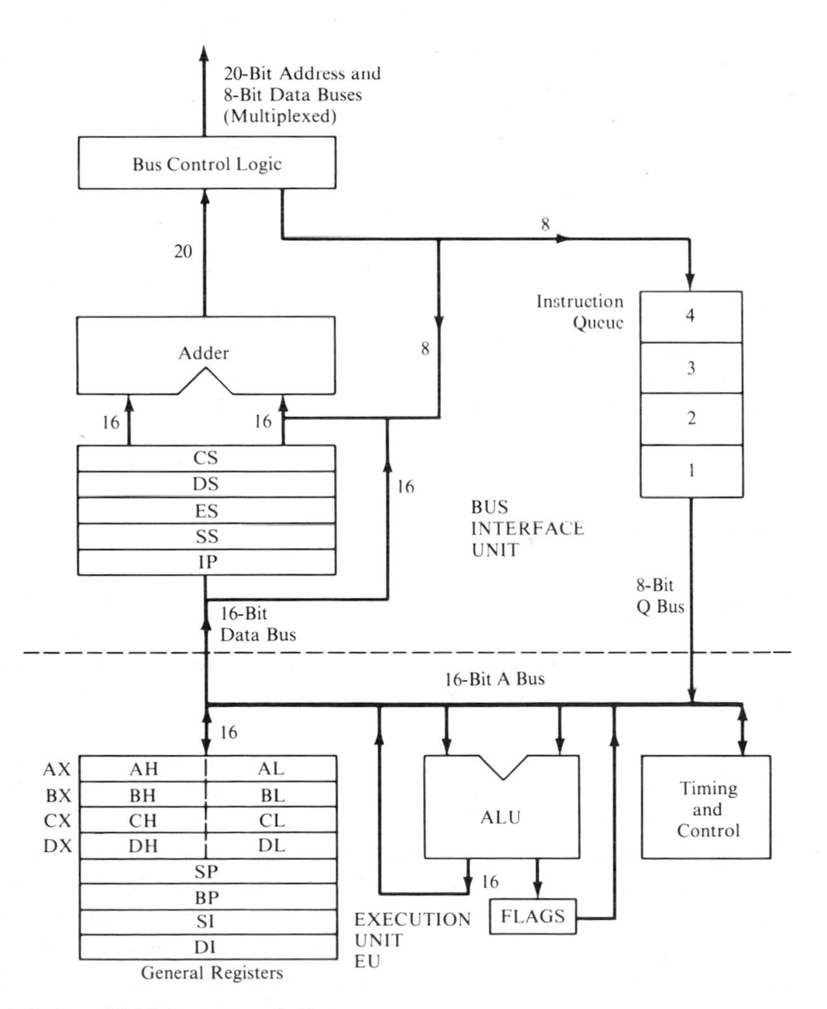

FIGURE 3-2. 8088 functional diagram.

15	14	13	12	11	10	9	8	7	6	5	4	3	2	1	0
				OF	DF	IF	TF	SF	ZF		AF		PF		CF

FIGURE 3-3. Storage of flag bits.

They are accessed as 16-bit units only and are normally used to determine the effective offset address of a memory location. The four data registers and the four pointer and index registers are collectively called the *general registers*. Any of the set of eight general registers can be used to store data, to hold a count for a program loop, or to serve as an accumulator. The assigned names simply reflect their implicit usage in some instructions and their most common usage.

Flag Register. There are 9 bits in the flag register. Six of these bits are status flags that reflect certain properties of the result of an arithmetic or logical operation. These flags are CF, PF, AF, ZF, SF, and OF, denoting the carry, parity, auxiliary carry, zero, sign, and overflow flags, respectively. Flag PF is set if and only if the *low byte* of the result of an arithmetic or logical operation has an even number of 1s. Other status flags are based on the result of either a byte or word operation, and the numbers may be signed or unsigned. Flag AF is set when there is a carry out of the *lowest nibble* or a borrow into this nibble. The three control flags are the *interrupt enable flag IF*, the *direction flag DF*, and the *trap flag TF*. Their functions are examined later.

Flags can be moved to and from other processor registers and memory. When placed in a 16-bit register, the 9 flag bits are positioned as shown in Fig. 3-3, and the remaining seven locations store undefined bits. Flag bits are numbered as indicated. Undefined bits can be cleared by ANDing the flag word with 0FD5H.

A timing and control unit is present in Fig. 3-2. This unit supplies control bits to the various parts of both the EU and the BIU at the proper time and, in addition, feeds a number of control signals to package pins of the 8088 for connection to external circuits. Signals that control external bus operations, such as READ and WRITE, are supplied by the bus control logic of the BIU.

Bus Interface Unit

All operations involving the external address and data buses are supervised and controlled by the BIU. The discussion of this unit here includes the registers that contain data for logical addresses. One of these registers is used for addressing special areas of memory referred to as *stacks*. The organization of such stacks is examined.

As shown in Fig. 3-2, the major components of the BIU are the bus control logic, an adder, an instruction queue with 4 byte-wide registers, the 16-bit *instruction pointer IP*, and four 16-bit *segment registers*. Transfer of data between the CPU and memory or peripherals is done by the BIU upon demand from the EU.

In addition, when the EU is busy executing instructions, the BIU fetches bytes from memory and stores them in the register array called the *instruction stream queue*. This queue holds up to 4 bytes, as shown in the block diagram. Usually, when the EU is ready for another instruction word or immediate operand, it is present in the queue, having been prefetched. Thus, the EU does not normally have to wait for a FETCH cycle, which is the bus cycle that fetches from RAM the first byte of an instruction.

A byte is fetched from memory whenever at least one location of the queue is empty and the EU is not using the external buses. If a jump occurs in the program, the queue is cleared, the instruction at the new address is fetched, and the BIU again begins to refill the queue. Although 16-bit instructions and operands are fetched in bytes via the 8-bit data bus, the processing speed is only a little less than that of the 8086 microprocessor, which is identical to the 8088 except for its 16-bit data bus and its two additional byte-wide registers in the instruction stream queue. Both processors would be considerably slower, and the difference between their processing speeds would be much greater, were it not for prefetched instructions stored in the queue.

Many computers fetch an instruction, with the data path unit idle, and then execute it with the buses idle. This procedure is followed by the fetch and execution of the next instruction, and so on. Such a procedure is inefficient, because parts of the machine not directly involved in the particular cycle of fetch or execution are inactive. A better design allows several instructions to be in the machine pipeline simultaneously. For example, while one instruction is being processed by the ALU, the following one could be moving operands into appropriate registers, and the one after this might be in the FETCH cycle. This process is referred to as *pipelining*. *Throughput* is defined as the amount of processing that can be done in a given time interval. The pipelining that is provided by the separate program control unit constituting the BIU considerably enhances throughput.

The adder of the BIU provides a 20-bit absolute address at its output, which is an unsigned number. If leading bits or digits are omitted in an address, they are understood to be 0s. For example, address 0H is 00000H, and 0A5H is 000A5H. The total address space of the PC is 1 MB, with the lower 640 KB available for user programs. Space above this area is reserved for the video monitors, the ROMs, and other special purposes.

Without doubt, the major advantage of the 8088 processor in comparison with the earlier 8-bit devices is the 1 MB address space. The 8-bit machines with 16 address lines are able to address only up to 64 KB of memory. This restriction seriously limits the development of microcomputer programs. With the advent of the 16-bit microprocessor and its ability to handle 20-bit addresses efficiently, more powerful and sophisticated programs have become available. No longer is the software designer hemmed in by the 64-KB restraint.

Throughput of the 8088 is considerably greater than that of earlier 8-bit chips. This results from the higher clock frequency and from pipelining provided by the BIU. Other improvements are the inclusion of multiply and divide instructions for both signed and unsigned numbers. The powerful instruction set and associated addressing modes are especially well suited for accommodating high-level language processing. Also, the 8088 is designed for use in a multiprocessor environment.

Segment Registers

Because the 20-bit physical address must be obtained from operations on 16-bit words, the megabyte of memory space is divided into *logical segments* of 64K, and the 20-bit base address, or starting address, of a segment is identified by the content of a 16-bit segment register. This was discussed briefly in Section 1-5. With X denoting a hexadecimal digit, segment address XXXXH defines the ab-

Logical Address	Absolute Address
0000:F897 400:567 F214:9ABC	0F897 04567 FBBFC

FIGURE 3-4. Comparison of logical and absolute addresses.

solute base address as XXXX0H. *The base address of a logical segment always starts at an absolute address that is a multiple of 16.*

It is important to understand that *segment address* denotes *the 16-bit word of a segment register*, whereas the absolute base address of a segment is the segment address multiplied by 16. A 20-bit physical address can be expressed as an offset from a segment address. The absolute address is computed by the adder of the BIU, which first shifts the segment address 4 bits to the left and then adds the result to the 16-bit offset. A 4-bit left shift, with 0s inserted at the right, multiplies the unsigned number by 16.

Unless otherwise stated, all logical addresses are given in hexadecimal. The segment address is first, followed by a colon and then the offset. The absolute address is the sum of the shifted segment address and the offset. Three examples are given in Fig. 3-4. Note the omission of leading 0s in the second logical address.

Segment addresses are stored in the four 16-bit segment registers of the BIU. These registers are as follows:

Code segment	CS
Data segment	DS
Extra segment	ES
Stack segment	SS

The first of the listed segment addresses is the *code segment CS*. Instructions are always fetched from CS, and the instruction pointer IP is the offset. For example, suppose the segment address of CS is 0600H, and IP stores 0400H. Then the next instruction is fetched from CS:IP equal to 0600:0400, or absolute location 06400H. Now suppose it is necessary to jump from this location to a routine at absolute location 20000H. Because this location is outside the range of the present code segment, the jump instruction must load CS with a new segment address, in addition to specifying IP. A possible choice is the jump address 2000:0000. An alternate choice is 1A00:6000, and there are many others.

Whenever a program branches to a location outside the present code segment, register CS must be loaded with a new segment address. This is an *intersegment jump*, or FAR jump, whereas an *intrasegment jump* is a NEAR jump. Both FAR and NEAR jumps are common. For a program with many FAR jumps, the contents of CS are changed frequently during program execution.

Let us consider an example. In the ROM BIOS at logical address F000:E724 is the instruction

```
JMP BOOT_LOCN
```

with BOOT_LOCN denoting the far location 0000:7C00. Values of registers CS and IP corresponding to the location of the JMP instruction and also to the target

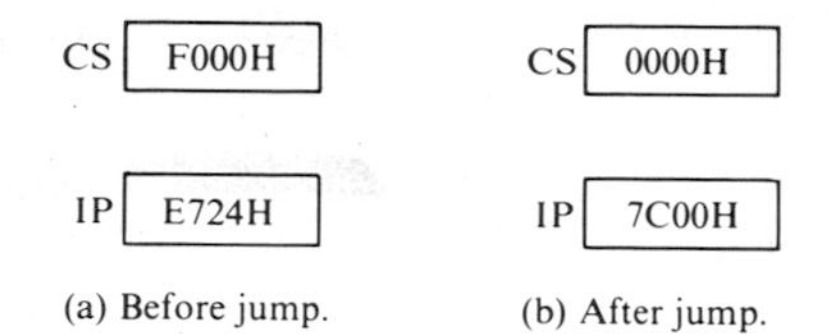

FIGURE 3-5. Register values before and after a FAR jump.

of the jump are shown in Fig. 3-5. The unconditional jump is from absolute address FE724H to 07C00H. Note that the new CS segment address is 0.

The *data segment* register *DS* and the *extra segment* register *ES* refer to the storage of variables and operands. For each of these registers, the 16-bit offset is called the *effective address EA*. EA is either included in the instruction or computed by the EU. In Fig. 3-2 the EA is fed into the right side of the adder from the A bus of the EU or directly from the instruction, and the segment address of DS or ES enters the left side. The adder output is the sum of EA and the shifted segment address.

There can be as many data and extra segments as desired. For example, it may be desirable to store data in memory locations that are more than 64K apart. However, the current data segment is the one identified by DS, and the current extra segment is defined by ES. Although these addresses can be changed at any time by the program, most programs use a single 64-KB data area, leaving the contents of DS unchanged from its initialized value.

Stack Organization

Before discussing the *stack segment*, let us examine the organization of typical stacks. A *stack* is a storage region, consisting of either registers on a chip or a region of external memory. Its purpose is to hold data and addresses temporarily.

A memory stack is a special group of storage registers in external memory having consecutive addresses, with addressing by a processor register called the *stack pointer* (*SP*). Initially, with only meaningless bits present (sometimes called *garbage*), the stack is empty, and the SP contains an address that is one greater than the highest address assigned to the stack. The assembly language instruction "PUSH word" moves the specified word to the stack. First, the SP is decremented so that it points to the highest address of the valid stack, and then a write-memory machine cycle transfers the word from its register or memory location via the data bus to the location in memory addressed by the SP. If another PUSH is encountered, the SP is again decremented, followed by storage of the new word. Henceforth, instructions are understood to be in assembly language unless clearly indicated otherwise.

Removal of a word from the stack is normally accomplished with the instruction POP. *The SP always points to the last word that was pushed on the stack*, except when the stack is empty. A POP operation transfers this word via the data bus to the appropriate register or memory location, and then SP is incremented. A second POP transfers the word from the new location addressed by SP, and again SP is incremented. Because the word that was last written into the stack is the first one to be read out, the memory region is called a *LIFO stack* (last in, first out). Although words popped out still remain stored in the stack space, PUSH operations write over them, and so they are of no importance.

The *top of the stack* (*TOS*) is defined as the location addressed by SP. Meaningful words of a nonempty stack are stored only from the location with the highest assigned stack address down to and including the TOS. *As words are pushed on the stack, the TOS moves to lower addresses, and as words are popped, it moves to higher addresses.*

A stack is full when register SP contains 0000H. If a PUSH occurs when the stack is full, SP is decremented by 2, becoming FFFEH (65534D), and the word is written into this offset from the segment address. The location may be outside the designated stack space, but if it is not, the operation overwrites a valid word of the stack. If a POP now occurs, the word is read and SP becomes 0. Thus, *the stack wraps around the 64K segment*, which should be avoided. A POP instruction when the stack is empty, with SP pointing to the word location immediately above the highest assigned address, should also be avoided, because the word addressed by SP is outside the stack.

There are other stack architectures. The one described here is a PUSH-down POP-up stack with the SP of an empty stack pointing to the location just above the highest assigned address. Some are organized as PUSH-up POP-down stacks, with the SP of an empty stack pointing to the location just below the lowest assigned address. Often a stack is included on the chip of a processor or control unit, in which case it is called a *register stack* or simply a *file*. Many stacks will not wrap around, as in the 8088, with SP restricted to proper values of the designated stack.

At times a program may call a *subroutine*, or *procedure*. This is a routine that can be called and executed repeatedly by the same program. When it is called, a branch to the procedure occurs, its instructions are executed, and another branch returns processing to the main program. It is necessary to save the return address on the stack. Also, any registers used in the procedure must have their contents pushed on the stack if these contents are important to the main program. At the end of the subroutine, the words are popped from the stack and returned to the proper registers. Then the return address is popped into the instruction pointer, and processing of the main program is resumed.

When processing a subroutine, calls to other subroutines may occur, and again the return address and the contents of processor registers must be saved on the stack. Nested subroutines are allowed. Furthermore, at times the main program, or a subroutine or nested subroutine, may be interrupted so that a request of high priority can be serviced. As before, the return address and words of processor registers are pushed on the stack. Interrupts of interrupt routines are common. Clearly, a LIFO stack containing many memory locations may be essential.

Instructions that automatically refer to stack operations, such as PUSH and POP, *always use the segment address of the stack segment register SS, with the offset*

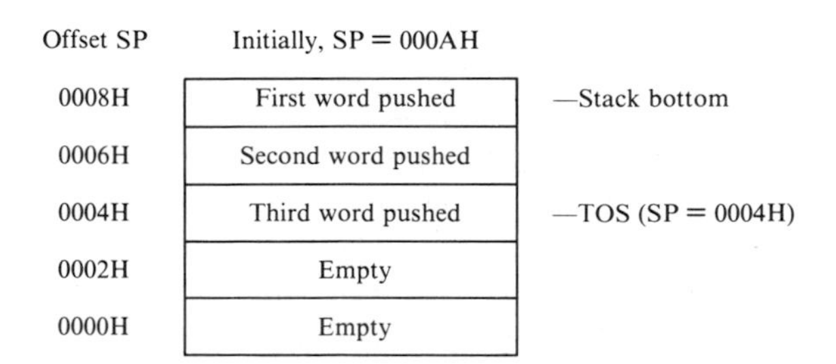

FIGURE 3-6. Stack after three PUSH operations.

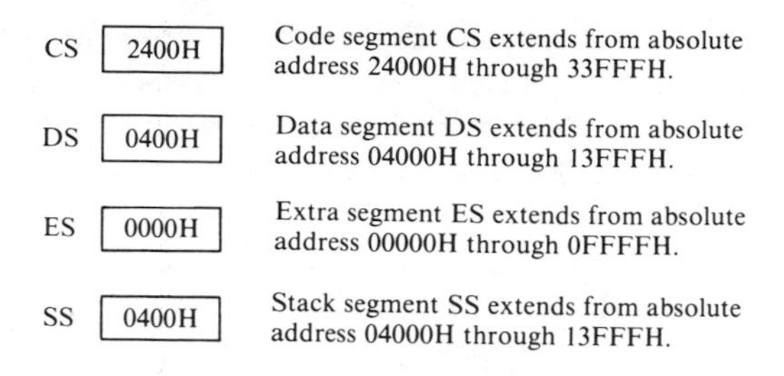

FIGURE 3-7. One choice for current segment addresses.

being the stack pointer SP of the EU. There is no limit to the number of stacks that can be used, and each may be up to 64K bytes long. However, the current stack that is automatically addressed by special stack instructions is the one having the segment address of SS. This address can be changed at any time by the program. Stacks other than the current one addressed by SS can be used by the program as desired. The address of SS multiplied by 16 is the absolute base address of the current stack, and the top of the stack TOS is located at logical address SS:SP.

Only words (16 bits) are stored, each requiring two memory locations. A word is pushed onto the stack by decrementing SP by 2 and then writing the 2-byte word into two memory locations. *The high-order byte is stored first, at the higher address.* A POP reads the word and then increments SP by 2. Words of the stack are neither moved nor erased, but a write operation replaces a word.

To illustrate, suppose a small stack is organized with a segment address of 0500H and an initial SP of 000AH. In this case, 10 bytes of memory are allocated to this stack, and 5 words can be stored. After three PUSH operations, the TOS is at absolute location 05004H, and the word stored at the bottom of the stack is at absolute location 05008H. This is shown in Fig. 3-6. After five push operations, the stack is full, with the TOS at 05000H. Note that the bottom of the stack is not the base address. In fact, it has an address greater than the address of the TOS. This is characteristic of a PUSH-down stack, *which grows downward.*

Segment registers can be loaded with any desired addresses, and the program can change them at any time. Segments can overlap. In fact, if all four segments are loaded with 0s, each segment extends from address 0 to absolute address 64K. Sometimes this arrangement is convenient when all code and data, including the stack, are located in the 64-KB low end of memory. In Fig. 3-7 is shown a possible choice for the four current segment addresses.

Note that each segment extends upward from its base address for 64 KB of storage, provided RAM or ROM is present at all indicated locations. The range of a segment can include regions of no memory. Segments DS, ES, and SS overlap.

3-2 ASSEMBLY LANGUAGE

A major objective of this book is to introduce fundamentals of assembly language programming. Although most programs are written in a high-level language, some are prepared in assembly language in order to achieve the maximum possible speed of operation with the machine language occupying the minimum amount of memory. Furthermore, some important features of a computer are accessible only through the use of assembly language. Many programs written largely in a high-level language use one or more assembly language subroutines.

Each type of processor has its own instructions. However, different instruction sets are usually quite similar, and in any event, programming principles are applicable to all. We will concentrate entirely on the instruction set of the Intel 8086/88 microprocessor. There are about 90 different mnemonic instructions, all of which are listed in Appendixes A and B and included in the Index.

Rules for conversion of assembly language into machine language are presented in Appendix A, and a brief summary of the instruction set is given in Appendix B. *Frequent reference to this material in conjunction with the study of assembly language should prove helpful.* Of course, assembly is usually done with an assembler, and both the IBM MACRO Assembler and the mini-assembler of the DEBUG program are discussed in detail in Chapter 7. The mini-assembler is especially convenient for assembly of short routines such as the ones in this chapter. We begin with a program that shifts a block of data in memory.

Example: String Move

Suppose that one wishes to move a 32K byte string from a continuous region of ROM, starting at the absolute base address F6000H, to a region in RAM starting at base address 05000H. Because the source and destination locations are separated by more than 64 KB, different segment addresses are required. Let us choose logical address F000:6000 for the base address of the source and logical address 0000:5000 for the base address of the destination. These are arbitrary choices; other selections would be just as good. The program flowchart is shown in Fig. 3-8.

The assembly language routine describes a few instructions and their format, introduces several addressing modes, uses segment registers, establishes a counter, and shows a way to create a program loop. Hexadecimal machine code is given at the right (found from Appendix A). An alternate and easier way to obtain the code is through the assemble command of program DEBUG, described in Chapter 7. The sequence is given in Fig. 3-9.

When the code is loaded into memory and run, using the DEBUG program of the PC, the 32 KB of ROM are moved to RAM in less than 1 second. Let us examine the instructions.

Initialization. The first instruction moves decimal number 32,768 into the 16-bit register CX, which is to be used as a counter. Any of the data registers can be used for this purpose. The number can also be written as 32768D, or 8000H, or 100000Q. The instruction format is *label*: MOV *destination, source* ;*comment*

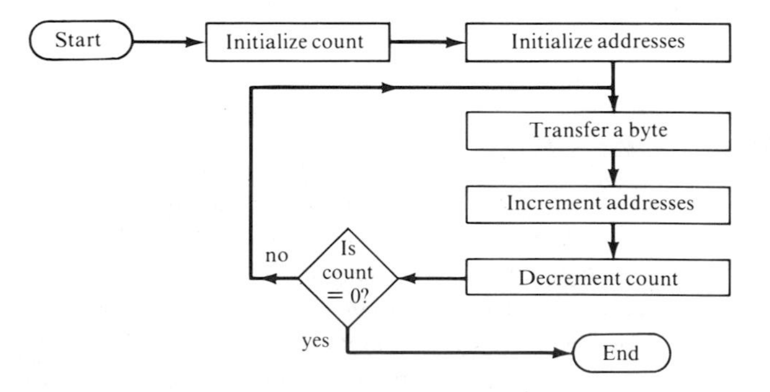

FIGURE 3-8. Program flowchart.

```
   MOV CX, 32768       ;initialize CX with 32K           (B90080)
   MOV AX, 0F000H      ;F000H is DS address              (B800F0)
   MOV DS, AX          ;establish DS address             (8ED8)
   SUB AX, AX          ;clear AX                         (2BC0)
   MOV ES, AX          ;ES segment address = 0           (8EC0)
   MOV SI, 6000H       ;source string offset             (BE0060)
   MOV DI, 5000H       ;destination string offset        (BF0050)
A: MOV AL, [SI]        ;move source byte to AL           (8A04)
     MOV ES:[DI], AL   ;move byte to destination         (268805)
     INC SI            ;increment source pointer         (46)
     INC DI            ;increment destination pointer    (47)
     DEC CX            ;decrement the counter            (49)
   JNZ A               ;repeat if count is not 0         (75F6)
```

FIGURE 3-9. Program that transfers a block of data.

Spaces are ignored by the assembler, but there must be at least one space after the mnemonic MOV. The label is optional; it is normally used for identifying targets of jumps and calls. Instructions do not have to be aligned. *The destination always precedes the source* in instructions of this set. Comments are preceded by a semicolon. Although ignored by the assembler, they appear in the listing of the program, and they should be used freely. The immediate number 32,768 is the source operand, and register CX is the destination. Source addressing is said to be *immediate*, and that of the destination is *register addressing*.

In the machine code B90080H, byte B9 is the operation code, or opcode, which identifies an immediate move of a word to CX, and 0080 is the immediate hexadecimal data, with the low byte 00 first, followed by the high byte 80. The immediate datum 0080 represents the operand 8000H, which is the count. Because code is placed in memory and also fetched from memory at locations having addresses that progressively increase, opcode B9 has the lowest address, followed by the operand 8000H. Bytes of word operands are always located in memory as follows:

1. low byte at the lower address
2. high byte at the higher address

This convention is standard in Intel processors. Move instructions do not affect processor flags.

The next two instructions move the segment address F000H into DS by means of an immediate move to AX, followed by a register-to-register transfer from AX to DS. The instruction set does not allow an immediate move into a segment register. Note that the 16-bit operand F000H is preceded in the operand field of the instruction by a 0. This is absolutely essential. The assembler distinguishes numbers from names by the first character of a string, which *must always be a digit for a number and a letter for a name*. Although the leading 0 is ignored, it must be there.

Instruction SUB AX,AX clears register AX by subtracting the contents of AX from those of AX. The instruction MOV AX,0 can also be used to clear AX, and there are other ways. After AX is cleared, its contents 0000H are transferred to ES. Although 0000H and 0 are the same, the form of the hexadecimal number clearly indicates a 16-bit word, just as 00H denotes a byte. A transfer of data from a source register to a destination register does not change the word of the source.

Instructions MOV SI, 6000H and MOV DI, 5000H move the immediate data to the respective index registers. After the transfer, the source index register SI holds the initial offset into segment DS of the first source operand, and the index register DI contains the initial offset into segment ES of the first destination. Registers BX and BP could have been used, and in fact, the roles of SI and DI can be reversed.

After execution of the seventh instruction, the initial addresses are stored as follows:

DS:SI = initial logical address of the source

ES:DI = initial logical address of the destination

This procedure completes the initialization of the counter and the addresses.

Byte Transfers. The last six instructions constitute a program *loop*. Label A is the target of JNZ (jump-if-not-zero). With each pass through the loop, the offsets of the source and destination are incremented by 1 and the count is decremented by 1. When the count reaches 0, flag ZF is set, the conditional jump is not made, and the program exits the loop, going to the instruction that follows JNZ. This instruction was not included. It is not needed when the program is run, provided a breakpoint is set at the location just after the JNZ instruction. A label consists of up to 31 letters, numerals, and some special characters, but spaces are not allowed. The underline is used for separating words, as in KB_BUFFER.

The loop is similar to a WHILE-WEND structure, but there is a difference. Here the condition "Is count <> 0?" is tested at the end of the loop, whereas the expression of a WHILE-WEND loop is tested at the start. To illustrate, if the count is initialized to 0 ahead of the loop, the sequence of a WHILE-WEND structure is bypassed. However, if CX has 0 on entry to the loop starting at label A, the first decrement changes the count to FFFFH. Thus, *the sequence of the loop would be executed 65,536 times*, which equals the sum of FFFFH and 1 for the initial pass.

Although unconditional jumps, which are implemented by "JMP target", are unrestricted in range, conditional jumps are restricted to SHORT jumps, which means that the target must be within the byte range (-128,127) of an 8-bit signed displacement from the value of IP following the jump instruction. After execution of JNZ, register IP holds the address of the first byte of the next instruction. Examination of the machine code within the loop reveals that IP must be decremented by 10. This means that -10D, or F6H, is to be added to IP. In code 75F6H for the JNZ instruction, the second byte F6 is the 8-bit signed displacement. Jump instructions do not affect the status flags.

The instruction JMP is not restricted to SHORT jumps. The following sequence can be used for longer jumps, replacing JNZ A:

```
    JZ  B
    JMP A
B: (next instruction)
```

Here the *jump-on-zero* instruction *JZ* branches around the unconditional JMP instruction if the zero flag ZF is set, thereby indirectly giving a jump to target A when ZF is not set. There are many other conditional jumps in the set. All are listed in Appendixes A and B.

The instruction MOV AL, [SI] with label A has register addressing for the destination, but that of the source is called *register-indirect* addressing. Brackets around a register indicate that it is a pointer. Thus, the effective address EA of the source operand is equated with the word contained in SI, and the operand is fetched from segment DS. The segment was not specified, but DS is the default segment for register-indirect addressing, when either BX, SI, or DI is used. Register BP is also available for register-indirect addressing, but its default segment is SS. However, *default segments can always be overridden* with either CS, DS, ES, or SS placed ahead of the source and/or destination operand, as in the instruction MOV ES:[DI], AL. The destination address is now ES:[DI].

Thus, the two MOV instructions of the loop transfer a byte from the source to AL and then from AL to the destination. *A direct transfer from memory to memory is not possible*. Register AL in each instruction identifies the transfer as that of a byte. If AL were replaced by AX, a word would be transferred.

The structure of a WHILE-WEND loop, with the condition "do while count of CX is not 0," can be implemented with the sequence of Fig. 3-10. The PUSH and POP instructions save and restore the contents of CX, and the ADD instruction adds the contents of CX to itself. The destination of PUSH and the source of POP are understood implicitly to be the stack, which is *implied* addressing. The ADD instruction sets ZF if (CX) is 0, with parentheses around a register name denoting the contents of the register. The first three instructions of Fig. 3-10 can be replaced with the single instruction AND CX, CX, which is described in the next chapter.

PUSH and POP. PUSH always transfers a word to the current stack at SS:SP, and POP retrieves a word. No flags are affected. A single byte cannot be put on the stack. Instructions PUSHF and POPF transfer the flag word. Sometimes it is desirable to move the flag word to a register, but there is no form of MOV that accomplishes this. However, it is readily done with PUSHF followed by POP *r*, with *r* denoting any 16-bit processor register except CS. All allowed instructions are indicated in Appendix A, and frequent reference to them when writing programs helps one to avoid using program statements that are not allowed.

ADD and SUB. The instruction ADD replaces the destination operand with the sum of the 2 bytes or words. It makes no difference whether the operands are signed or unsigned binary numbers, ASCII, BCD, or other codes. All are treated exactly alike. *Add-with-carry ADC* is the same as ADD except that the 0 or 1 bit of carry flag CF is added to the sum. Addition of the carry from a previous operation allows the programmer to write a routine that adds numbers with more than 16 bits. All status flags are updated.

Subtract SUB is similar to ADD. *Subtract-with-borrow SBB* corresponds to ADC.

```
   PUSH  CX                       ;save count
   ADD   CX, CX                   ;set ZF if (CX) = 0
   POP   CX                       ;restore count
A: JZ     B
     (instructions of loop)
     DEC CX
   JMP A
B: (next instruction)
```

FIGURE 3-10. Assembly language WHILE-WEND structure.

Both the source and CF are subtracted from the operand of the destination field, and SBB allows the preparation of programs that can subtract numbers having more than 16 bits. The instruction *NEG dest* subtracts the byte or word operand of the destination from 0, thereby *converting a signed number into its 2s complement.*

String Instructions

The preceding program can be simplified with a string instruction. With reference to assembly language programming, a *string* is simply a sequence of related data bytes or words. The most common type of string encountered is a sequence of bytes representing ASCII characters.

There are five basic string instructions, which replace normal program loops. Their advantages are greater speed and simplified programs. Operations are implemented on individual elements, one at a time, that belong to memory strings consisting of byte or word sequences up to 64K in length. String instructions are called *primitives.*

If a source operand is present in a string instruction, excluding CMPS, it is assumed to reside in the current data segment at the offset given by the word of the source index register SI; however, a segment override prefix is allowed. Again excluding CMPS, a destination string must be in the extra segment, and its offset is given by the word of the destination index register DI. Each individual string operation *automatically updates SI and/or DI* in preparation for the next one, with the change DELTA depending on the state of the direction flag DF. If DF is 0 (direction UP), the index registers involved are automatically incremented by 1 for byte strings and by 2 for word strings. They are automatically decremented when DF is set (direction DOWN).

For the special case of *CMPS dest, source*, the source operand is addressed by ES:DI and the address of the destination string is DS:SI. As before, segment ES cannot be overridden. CMPS is the only string instruction that uses DI to address the rightmost source operand. Source and destination addresses and the direction flag control are summarized in the following enumerated statements:

1. Excluding CMPS, the default source string address is DS:SI. A segment override prefix is allowed.
2. Excluding CMPS, the destination string address is ES:DI. No alternate segment is allowed.
3. If DF = 0, index registers are incremented. If DF = 1, index registers are decremented. The increment or decrement DELTA is 1 for byte strings and 2 for word strings. Instructions CLD and STD, respectively, clear and set DF.
4. For *CMPS dest, source,* the source address is ES:DI and the default destination address is DS:SI.

Each basic string instruction implements an operation on one string element and then updates the index registers SI and DI by DELTA. Repetition is accomplished automatically by hardware only if the instruction is preceded by a special prefix. In Fig. 3-11 are listed the string instructions and their optional prefixes, with comments at the right. Two of the instructions affect the indicated flags OSZAPC.

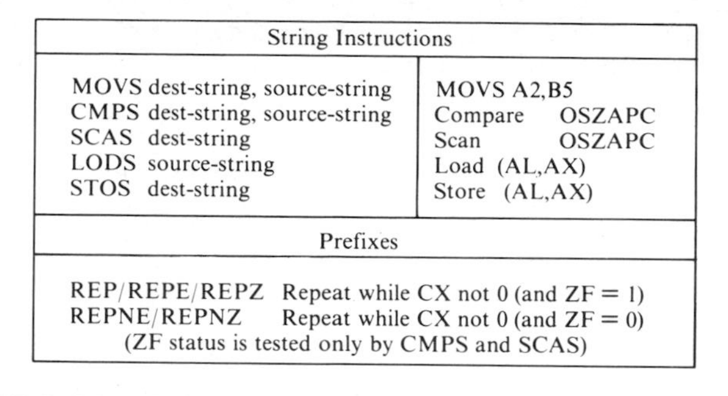

String Instructions	
MOVS dest-string, source-string	MOVS A2,B5
CMPS dest-string, source-string	Compare OSZAPC
SCAS dest-string	Scan OSZAPC
LODS source-string	Load (AL,AX)
STOS dest-string	Store (AL,AX)

Prefixes	
REP/REPE/REPZ	Repeat while CX not 0 (and ZF = 1)
REPNE/REPNZ	Repeat while CX not 0 (and ZF = 0)
(ZF status is tested only by CMPS and SCAS)	

FIGURE 3-11. String instructions and prefixes.

MOVS. The instruction "MOVS destination-string, source-string" *moves 1 byte or word from the source string to the destination string and updates both SI and DI.* The state of the direction flag determines whether the index registers are automatically incremented or decremented by DELTA. Instructions CLD and STD with no operand are used to clear and set the flag. From the specified destination string and source string operands, the assembler determines whether the strings consist of bytes or words. However, these operands are not used for addresses, and therefore they are not really essential. For this reason, there are alternate forms of the MOVS instruction that employ no operands. These forms are MOVSB and MOVSW, with B for byte and W for word.

For example, if variable names STRING_A and STRING_B denote byte strings, the following two instructions are identical:

```
MOVS STRING_A, STRING_B      MOVSB
```

Both have the same 1-byte opcode A4H. *The longer form allows the use of a segment override prefix,* if desired, and its format is more informative. If the variables are words, the equivalent instruction is MOVSW.

REP/REPE/REPZ. The optional prefixes *repeat REP*, *repeat-while-equal REPE*, and *repeat-while-zero REPZ* have identical code (F3H) and functions. The code is placed ahead of the opcode of the string primitive. When a primitive is preceded by REP or an alternate form, *the operation is repeated over and over until the count of CX, which is automatically decremented with each string operation, becomes 0. For CMPS and SCAS the zero flag is also tested, and unless it is set, the operation terminates, regardless of the count.* Prior to execution of a repeat string instruction, register CX should be loaded with the desired count. Interrupts between repetitions are allowed.

REPNE/REPNZ. The prefixes *repeat-while-not-equal REPNE* and *repeat-while-not-zero REPNZ* are identical, both having the same code (F2H). They function as do REP/REPE/REPZ, except that *repetition of CMPS and SCAS occurs only when flag ZF is 0.* Of course, a count of 0 in CX also terminates repetition.

Only the MOVS primitive is considered here. The other string instructions are examined later. Let us now reconsider the previous program used to transfer 32 KB from ROM to RAM, given in Fig. 3-9. The six instructions of the loop can be replaced with the two that follow:

```
CLD          ;clear DF for automatic increment   (FC)
REP MOVSB    ;move the 32K bytes                 (F3A4)
```

These 2 lines with 3 bytes of code replace six lines with 10 bytes. Execution of the new program requires considerably less time and eliminates the program loop.

3-3

MORE PROGRAMS

The instructions MOV, ADD, ADI, SUB, SBB, CLD, STD, JZ, JNZ, INC, DEC, PUSH, POP, PUSHF, POPF, and MOVS were described in the routines and discussions of the previous section, along with four addressing modes, referred to as register, register-indirect, implied, and immediate. Two programs are presented here, which introduce numerous additional instructions. Program loops with conditional exits are again used. No new addressing modes are encountered.

Example: String Compare

When displayed on the color monitor in the graphics mode, a character with an 8 × 8 dot pattern can be represented by a sequence of 8 bytes. The first byte gives the dot pattern of the top row of eight dots, the second one gives the next row, and so on. A zero is a blank, and a one generates a dot. To illustrate, the sequence here (30, 78, CC, CC, FC, CC, CC, 00) of hexadecimal bytes produces the letter A, which is easily verified by dots placed on an 8 × 8 grid, as shown in Fig. 3-12.

Within the BIOS program of the ROM is a 1-KB table that contains the 8 bytes for each of 128 characters. The start address is F000:FA6E, labeled CRT_CHAR_GEN, and the 8-byte dot codes are arranged in order of ascending ASCII codes 0 through 127D. Thus, each of the first 8 bytes are 00H for the null character with ASCII code 0, the next 8 bytes give the special character with ASCII code 1, and so on.

The CRT *character generator table* is used primarily in the graphics mode for converting from one code to another. With this table, a program can transform an 8-byte dot character stored in the video RAM of the color graphics adapter into its ASCII code and print the character on any device that accepts ASCII. Also, a program can utilize the table to transform an ASCII character entered from the keyboard into its 8-byte dot code, which can then be sent to the video RAM for display in the graphics mode.

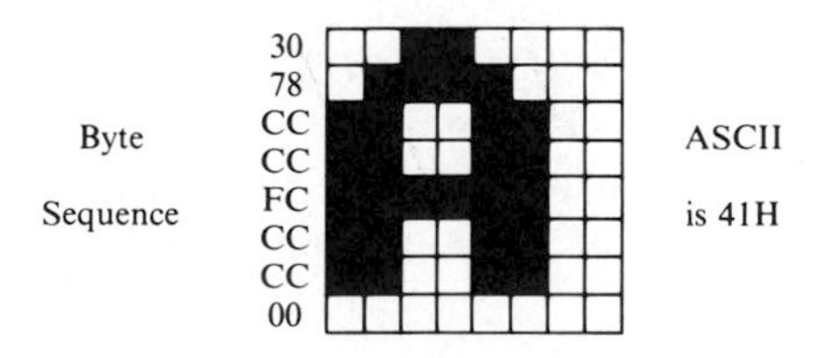

FIGURE 3-12. Character 8 × 8 dot pattern.

The Requirement. Suppose a sequence of 8 bytes representing the dot pattern of an unknown character is stored in the data segment of RAM at start location DS:8000. The problem here is to determine if this character is one of the 128 byte strings of the table at F000:FA6E, and if so, to determine its ASCII code and place it in AL. If no match is found, byte 00H should be put in AL. Because the compare string CMPS instruction will be used, we shall now examine it.

CMPS. Operation "CMPS dest, source" *subtracts the source byte or word from that of the destination, but changes neither operand.* The status flags and index registers SI and DI are updated. The address of the (rightmost) source operand is ES:DI and that of the (leftmost) destination operand is DS:SI, and DS can be overridden. The use of pointers DI and SI *is contrary to normal procedure.* CMPSB and CMPSW may be used without operands in place of CMPS and its required source and destination operands. The operation is often followed by a conditional jump that depends on the result of the comparison. CMPS is often preceded with REPE to compare two strings, with the comparison continuing while the bytes or words are equal and CX is not 0. The program to be described uses this combination.

A related instruction is CMP *dest, source*, with any of the available addressing modes used. CMP subtracts the source operand from that of the destination without changing either. Only the flags are affected. To illustrate, CMP AL, 41H followed by JZ A gives a jump to target A if and only if register AL contains 41H.

Pseudo Code. The algorithm for implementing the string comparison is shown in Fig. 3-13. In the initialization section, the pointer ES:DI to the CRT character generator is assigned, and its value is F000:FA6E. The maximum number of

```
PROGRAM String Compare

    initialize pointer to base of character table
    initialize count to number of characters (128)
    initialize AL to ASCII code 0

begin

    do while count > 0
       compare 8-byte source with 8 bytes of table
           if byte strings are equal   then
               set flag ZF
           else
               increment ASCII code of AL
               increment table pointer to next 8 bytes
               decrement count
           end-if
    end do-while
    if AL is 128 then
         clear AL
    end-if

end program
```

FIGURE 3-13. Pseudo code for the string compare program.

```
        MOV AX, 0F000H      ;initialize ES:DI (table address)
        MOV ES, AX
        MOV DI, 0FA6EH
        MOV BX, 128         ;count of 128 byte strings
        MOV AL, 0           ;initial ASCII code
        CLD                 ;direction up
A1:     MOV CX, 8           ;number of bytes of a string
          MOV SI, 8000H     ;RAM string offset
          PUSH DI           ;save DI
          REPE CMPSB        ;compare 8-byte strings
          JZ A2             ;exit loop if strings match
          INC AL            ;increment ASCII code
          POP DI            ;restore DI
          ADD DI, 8         ;update DI to next string
          DEC BX            ;decrement count
        JNZ A1              ;loop for next string comparison
        MOV AL, 0
A2:     (Continue)          ;AL has ASCII code or 00H
```

FIGURE 3-14. Program illustrating string comparison.

8-byte character strings is initialized to 128, and the initial ASCII code is fixed at 0. The 8-byte strings are compared until a match is found or until the count reaches 0.

If a match is found, the loop ends with the ASCII code of the 8-byte character of the table contained in AL. Otherwise AL has 128, indicating no match. Because AL is incremented after each string comparison, at the start of a comparison it always contains the ASCII code of the 8-byte string of the table. On exit from the loop, AL is cleared if the count is 0.

The Program. If compared strings are found to be unequal after 1 or more bytes have been examined, the repeat operation is automatically aborted as a result of the cleared flag ZF, and DI and SI then have incorrect values for the next comparison. For this reason, initialization of SI must be done within the loop, and DI will be saved on a stack and recovered after the comparison. Incrementing DI by 8 gives the pointer to the next 8-byte string of the table. The program is shown in Fig. 3-14.

During the CMPSB operations of each loop in which the strings do not match, DI is incremented with a value from 1 through 7. The PUSH saves the value of DI just prior to the comparisons, and the POP followed by the addition gives the offset to the next character of the table. If a match is found, the program jumps to label A2, with AL storing the ASCII code. Suppose the 8 bytes of RAM are FE6268786860F000. After the routine is executed, AL stores 46H. This is ASCII for the letter *F*.

Example: Table Scan

At one end of a keyboard is a numeric pad with the 10 decimal digits, a plus and a minus sign, and a period. When a key is depressed, a scan code byte is placed in AL. Assume that the 13 scan codes for the pad are stored in a table with label C5, and the corresponding ASCII codes are tabulated in the same order, with label C6. The tables are in the current code segment. A routine is needed to determine if the byte of AL is from the pad, and if so, to return its ASCII code to AL. Let us first examine SCAS, which is used in the program to scan the table.

SCAS. The *scan-string* "SCAS dest-string" *subtracts its operand with address ES:DI from the contents of the accumulator* without altering either the destination string or the accumulator. Both DI and the flags are updated. For a byte string the register is AL, and for a word string it is AX. The direction is controlled by DF. *Scanning with a repeat prefix is a convenient way to search for the first occurrence of a particular character in a string or to search a table for a specified byte or word.*

Program. The table scan program is shown in Fig. 3-15. *There are two tables in the program*, with labels C5 and C6. Table C5 contains 13 bytes given in hexadecimal form, and the one with label C6 has the corresponding ASCII bytes in the same order. The first five instructions, starting below the tables, give the required initialization. The OFFSET directive of the second instruction *tells the assembler to use only the 16-bit offset of the logical address of label C5*, which denotes a logical address of the form CS:offset. Note the PUSH and POP instructions that transfer the address of CS to ES.

Suppose AL holds scan code 4BH when the routine is processed. After execution of REPNE SCASB, flag ZF is set, index register DI stores the sum of OFFSET C5 and 5, and the count of CX is 8. Because a match is found, the conditional jump JNZ is not taken. Subtraction of 47H from the scan code 4BH of AL gives 04, which is the offset from the base of scan code table C5. The first byte of the table has offset 0. Thus, the ASCII code is located at offset 4 from the base of table C6, and this offset is stored in AL. Instructions LEA and XLAT transfer the ASCII value 34H to AL. Then the program branches to A3 for processing of the code.

LEA, LDS, and LES. The instruction *load-effective-address* "LEA dest, source" *transfers the offset of the source operand in memory, but not the operand itself, to any of the 16-bit general, pointer, or index registers specified in the destination field.* The source operand must be one of memory, and the destination must be a

```
START: JMP A1
  ;
  C5    LABEL   BYTE
         DB      47H, 48H, 49H, 4AH, 4BH, 4CH, 4DH
         DB      4EH, 4FH, 50H, 51H, 52H, 53H
  ;
  C6    LABEL   BYTE
         DB      '789-456+1230.'
  ;
  A1:   MOV  CX, 13             ;decimal byte count of table
        MOV  DI, OFFSET C5      ;table C5 offset to index reg
        CLD                     ;direction up
        PUSH CS
        POP  ES                 ;transfer CS to ES
        REPNE SCASB             ;scan string C5 for match
        JNZ  A2                 ;jump to A2 for no match
        SUB AL, 47H             ;get offset to string C6 in AL
        LEA BX, C6              ;load BX with C6 base offset
        XLAT CS:C6              ;move ASCII to AL
        JMP  A3                 ;go to service routine
   A2:  (continue)              ;not pad entry, look elsewhere
```

FIGURE 3-15. Table scan program.

register. In most cases, LEA can be replaced with MOV; this is true for the LEA instruction in the program of Fig. 3-15. LEA allows the source operand to be subscripted, whereas this is not allowed using MOV with the OFFSET operator.

The LDS and LES instructions *load a doubleword pointer variable of memory, with the offset loaded in the 16-bit specified register*. For LDS, the segment word of the 32-bit pointer goes to DS, and for LES it goes to ES. These instructions are especially helpful when used in conjunction with string instructions. For example, LES DI, [BP] fetches from the stack segment, at an offset equal to the content of BP, a 32-bit word. The 16-bit word from the lower address goes to register DI, and the 16-bit word from the higher address goes to register ES. Recall that BP refers to the stack segment SS unless it is overriden with a segment prefix.

XLAT. The *translate* instruction "XLAT source" *replaces the byte of AL with a byte from a table*. The segment address of the table is that of the source operand, the base of the table has offset equal to the word of register BX, and the index into the table is the byte of AL. Thus, *the sum of the contents of BX and AL gives the effective address EA*. The source operand must be included in the instruction. Its only purpose is to allow the assembler to determine the segment address, which is that of the default segment of the source operand unless a segment override prefix is included. The offset part of the source address is ignored.

The instruction is normally used to translate values in one code into those of another, using a table of values stored in memory. Register BX is first loaded with the offset to the base of the table, and the byte of AL becomes the index, having values from 0 through 255. In the table scan program, the keyboard scan code placed in AL, assumed to be 4BH, was translated into its corresponding ASCII code 34H by using XLAT and the tables of C5 and C6. If the scan code of AL is not within the range of the data of table C5, the conditional jump gives an exit to target A2, with the scan code remaining in AL. On the other hand, if a match is found, the program branches to a service routine starting at label A3.

If there is any doubt about the identity of the default segment of a source or destination, a segment override prefix should be inserted immediately ahead of each appropriate operand. In the program of Fig. 3-15, prefix CS precedes the source operand C6, which identifies the table as one of the current code segment. Execution requires 15 clock periods. A common application of XLAT is to change the ASCII code of a character into EBCDIC, or the reverse.

The two tables of the program use the *directives* LABEL and DB (*define byte*), *which are instructions to the assembler*. These and various other directives are examined next. Alphanumeric characters enclosed within single quotation marks, as in the second table, are converted into ASCII code by the assembler. Spaces are important, *with each blank space of such an enclosed string replaced with its code 20H*.

3-4 DIRECTIVES AND ADDRESSING MODES

In addition to a description of several directives to the assembler, four new addressing modes are introduced. These are known as *direct addressing, based addressing, indexed addressing,* and *based indexed addressing*. Also, more instructions are encountered, mainly in conjunction with two short assembly language routines.

The study of assembly language, addressing modes, and assembler directives is continued in the chapters that follow. Before long, we will learn how to prepare a program in a form suitable for assembly and how to assemble and run a program. We begin with several directive statements that help the assembler translate source code into the machine language 0s and 1s of the object code.

Data Definition Statements

The *equate* statement *EQU* is used to define symbolic names that represent values or other symbolic names. A value can be that of an expression. Examples are as follows:

```
PORT_A     EQU    60H
COUNT      EQU    CX
DEL_KEY    EQU    4*20 + 1
```

In the PC, the address of port A of the PPI chip is 00060H, and the first of the equate statements is included in the ROM BIOS program. An instruction such as IN AL,60H can be written as IN AL, PORT_A, which is equivalent and more informative. The EQU statement simply equates the name on the left with the value or name on the right. Names and label identifiers can have up to 31 characters. EQU statements can be placed anywhere within a program, but it is usually convenient to put them at the beginning.

Several data definition directives allocate memory space for one or more data items and may associate a symbolic name with the memory address. Initial values can be assigned if desired. Some examples, including EQU and ORG statements, are given in Fig. 3-16.

The first statement specifies a *program segment* named XYZ. When the machine language of the assembled program is loaded into memory, the loader program assigns a value to XYZ; this value is the segment address. The EQU statement, which can be placed anywhere in the program, is not associated with the segment.

DB, DW, and *DD* denote *define byte, define word,* and *define doubleword,* respectively. The assembler reserves three byte locations starting at offset 0 in segment XYZ, with name THREE_BYTE denoting the first of the stored variables. *The question marks indicate that no initial values are specified.* These values are to be supplied by the program during its execution. Although no identifying name is specified in the next statement, 128 word locations are reserved for use by the program without initialization. These locations can be used for a small memory stack or for some other purpose.

```
XYZ  SEGMENT
   PI  EQU  3.14
   THREE_BYTE     DB  ?,?,?          ;3 bytes
                  DW  128 DUP(?)     ;128 words
   POINTER        DD  12340000H      ;address 0000:1234
  ORG   512
   D1             DB  'PARITY CHECK 2', 10, 13, 0
XYZ   ENDS
```

FIGURE 3-16. Examples of data definition statements.

The doubleword for POINTER might be used to store both the 16-bit offset and the 16-bit segment address of a pointer to a special routine. This statement is followed by an ORG statement, *which specifies that the next storage location of the segment will have offset 512*. The last data statement identifies name D1 with the first of 17 byte locations that have ASCII codes for the 14 characters of the message within the two single quotation marks, followed by LF, CR, and a byte location initialized to 0. Such a character sequence is an assembly language string constant. If desired, an initial value can be given in the form of an expression for evaluation by the assembler, and comments can be placed after semicolons, as shown. *Characters in front of a string in quotation marks should be avoided,* but characters following a string, as shown, are acceptable. The ENDS directive *notifies the assembler of the end of segment XYZ.* It is essential.

Allocation of memory by the assembler places the data items in sequence within the defined segment with no gaps, unless addresses are specified by ORG. When ORG is encountered, the assembler assigns the next location to the specified offset value, and the sequence continues from this point. Because the data of a location can be changed by the program, the symbolic names associated with the data items are called *variables*. If a variable has more than 1 byte, such as a word or a doubleword, its address is defined as that *of the lowest of the several byte locations*. For example, if XYZ is assigned address 500H by the loader, the address of the first byte of the data at D1 is 500:200, which has offset 512D.

Direct Addressing

Operands may be included in the instruction itself, or they may be stored in registers, memory, or peripherals connected to I/O ports. The flexibility of the instruction set is considerably enhanced by the various addressing modes. Although addresses and offsets are always unsigned numbers, signed increments and displacements help determine their values.

Of the various memory addressing modes, *direct addressing* is perhaps the one most often used, and it is straightforward. The *effective address EA* of an operand stored in memory is given directly in the instruction. *EA is defined as the 16-bit offset that gives the physical address when added to the shifted word of the base segment register*. The default segment is the data segment DS, but others can be specified with a prefix added to the instruction. Offsets are specified by giving the name of a data variable, and EA is the offset address of the named variable. Refer to Table A-2 of Appendix A with mod = 00 and r/m = 110. Inclusion of more than one memory address in an instruction is never allowed. Some examples are as follows:

```
MOV DS, SEGMENT_BASE        INC BAUD_RATE
MOV BETA, 6                 DEC SS:COUNT
MOV AL, BYTE PTR BETA
MOV WORD PTR ALPHA,AX
```

The first MOV transfers to segment register DS the word stored in two consecutive memory locations identified by the name SEGMENT_BASE, which is a defined DW variable. To illustrate, suppose DS originally stores 0040H, and SEGMENT_BASE is defined as the variable associated with the location at offset 006CH in the data segment. The offset is the EA of the source operand. Execution of the instruction fetches the word at address 0400:006C and places it in DS. In this case, the code is 8E1E6C00. The object code is obtained from Appendix A, noting that in the mem-to-SR move *mod* is 00, *SR* is 11, and *r/m* is 110, with *disp* 6C00 corresponding to offset 006C.

The second MOV transfers either byte 06H or word 0006H to the location associated with variable BETA, depending on the designation of BETA as DB or DW. Assuming a word location at offset 0123H in segment DS, the 6-byte object code is C70623010600. The immediate-to-mem MOV has word $w = 1$, $mod = 00$, $r/m = 110$, $disp = 2301$, and $data = 0600$. For a byte location, C7 of the code becomes C6.

In the next two MOV instructions, ALPHA is assumed to be either a variable or a label of a byte location, and BETA is a variable or label of a word location. *Operator PTR is used to override the variable type.*

BAUD_RATE in the INC instruction must be defined in the program as either a DB or a DW variable, and either a byte or a word is incremented, depending on the definition. The 16-bit offset of BAUD_RATE in the default data segment DS is the EA. In the decrement instruction is prefix "SS:" preceding COUNT. This prefix indicates that the offset associated with COUNT is relative to the stack segment SS, not the default data segment DS. Thus, the logical address of the operand is SS:OFFSET COUNT. Variable COUNT must be defined in the program as a byte or word operand. If the offset of COUNT is 0123H, the code is 36FE0E2301 for a byte location, and byte FE becomes FF for a word location. The segment override prefix is 36H.

To illustrate direct addressing, suppose variable ALT_INPUT has been defined by a DB statement. Assume that we want to move the stored byte to AL and the offset address to BX, and then clear ALT_INPUT. The following sequence can be used:

```
MOV AL, ALT_INPUT            ;variable contents to AL
MOV BX, OFFSET ALT_INPUT     ;EA to BX
MOV ALT_INPUT, 0             ;clear storage of ALT_INPUT
```

Example: Clear Display

The requirement here is for a routine that clears the monochrome display of the PC. There are 2000 characters, including blanks, in the 25-row × 80-column display. With the characters numbered from 0 through 1999, left to right from the top row to the bottom, the numbers correspond to the offsets of words in a read/write display buffer having an absolute base address of B0000H. The first 2000 words of the 4-KB video memory are applicable. Each word stores the ASCII code of the character at the byte even address and an *attribute* code at the byte odd address. An attribute of 07H specifies the normal display of green characters on a black background.

The 4-KB chip is on the display adapter card. *The video circuits of the adapter continually scan the video memory, and the 2000 stored ASCII characters are printed on the high persistence screen 50 times per second.* A character generator in a ROM of the adapter card provides the conversion from ASCII into the 126 dots of the character box, which has a height of 14 dots and a width of 9 dots. This gives 350 vertical lines (14 × 25) of resolution and 720 lines of horizontal resolution (9 × 80). Thus, when a word is written into one of the buffer locations,

```
CLD                     ;direction up
MOV   CX, 2048          ;word count of entire buffer
MOV   AX, 0B000H
MOV   ES, AX            ;base address to ES
SUB   DI, DI            ;establish initial 0 offset
MOV   AX, 0720H         ;word for storage in buffer
REP   STOSW             ;store the word in all locations
```

FIGURE 3-17. Program for clearing the display.

the character represented by the ASCII code appears on the display at the corresponding position and remains there until removed. The string instruction STOS, which is used in the program, is examined next.

STOS. *Store-string* "STOS destination" *transfers a byte or word from AL or AX to the destination string at address ES:DI, and updates DI by DELTA to point to the next location.* Alternate forms are STOSB and STOSW without an operand. With prefix REP, the same element is transferred from the accumulator to each string element. This form has numerous applications, such as deleting the storage of a buffer in memory by replacing the elements with 0s, blanking out a print line, and feeding an identical byte to all memory locations for testing for parity errors.

Program. To clear the display, each word location of the buffer should be written with 0720H, which gives the ASCII code 20H for blank in the low byte position at the even address and the normal attribute code at the odd address. The program is shown in Fig. 3-17.

When the routine is executed on the PC, the monochrome display is cleared. If byte 20H in the MOV instruction is changed to 41H, the entire screen becomes filled with A's.

Example: Divide String

Let us suppose that locations starting at 0400:0000H store hexadecimal words A0B0, 7061, 4F02, 7AB3, and D123. The objective is simply to examine the program that is given in Fig. 3-18 and to determine what it accomplishes. Assume that DS stores 0400H and ES stores 0800H, with registers SI and DI and the direction flag DF initially cleared. Another string instruction is now described, for use in the program.

LODS. Although instruction *load-string LODS* has a source string operand, LODSB and LODSW without operands are allowed. Unless a segment override prefix is used, the source string has address DS:SI. *The addressed byte or word is loaded into the accumulator, and SI is updated by DELTA* to point to the next element of the string. Repeat prefixes are not normally used, because a repetition would overwrite the prior element that was loaded into AL or AX. However, the instruction is found in software loops, providing the convenience of automatically updating SI with each move from memory to the accumulator.

Divide String Program. The divide string program is that of Fig. 3-18. On each of the five passes through the sequence, index register SI is incremented by

```
    MOV DH, 5                ;loop count = 5
A1: LODSB                    ;load AL from DS:SI, inc SI
    STOSB                    ;store low byte at ES:DI, inc DI
    LODSB                    ;load AL from DS:SI, inc SI
    MOV ES:[DI+1FFFH], AL    ;high byte to memory
    DEC DH                   ;decrement the count
    JNZ A1                   ;loop if count is not 0
```

FIGURE 3-18. Program that divides a memory string.

2 and index register DI is incremented by 1. The low byte of a word is loaded and stored at ES:SI, as indicated in the comments. Then the high byte is loaded and stored at an address increased by 2000H. At the end of the routine, the low bytes B0, 61, 02, B3, and 23 are in the extra segment at offsets 0 through 4, and the high bytes A0, 70, 4F, 7A, and D1 are at offsets 2000H through 2004H. Note that the low byte of a word, which is stored at the lower address, is transferred first.

The program illustrates a way to divide a string, with even-address bytes put in one string and odd-address bytes in the other. Operations such as this are not unusual. The MOV instruction uses indexed addressing, which is examined after a brief discussion of arrays and records.

Arrays and Records

Based addressing, indexed addressing, and a combination of the two are especially useful for accessing elements of strings, arrays, and records. A sequence of related data items of the same type and identified by a single variable name is an *array*. Individual items are *elements*, or *fields*, and all elements contain the same number of bits. Each item has an *index*, which is a number that identifies its relative position within the array. Thus, the location of an element is specified by its index and the array name corresponding to the base address. Strings of character data can be defined as byte arrays. Data items of an array are normally stored in a continuous block of memory. An array can be designed as a look-up table in RAM, perhaps storing the characters of a message.

A *record*, or *structure*, is also a collection of related data fields, but unlike the array, the fields are not necessarily of the same type. Each field is associated in some way with the other fields. The record and the individual fields are assigned names. An example is the record EMPLOYEE, having fields SSN, YR_HIRED, DEPT, and SALARY. Each element is identified by the program as a data byte DB, a data word DW, or a double word DD. A field of a record may be an array; likewise, an element of an array may be a record.

Memory storage is not allocated to a record. The record is associated with a particular memory space only when a base value is given in an instruction, along with a field name. The base value, which locates the record, may be either the word of BX or BP or a variable name assigned a value that can be modified as desired. Changing the base value associates the record with another area of memory.

Based Addressing

Only base register BX and base pointer BP are used in *based addressing*. In this addressing mode the effective address EA is *the sum of the word of either BX or*

BP and a byte or word displacement included with the instruction. Refer to Table A-2 of Appendix A (mod = 01 and 10 and r/m = 110 and 111). If BP is specified, the BIU is directed to obtain the operand from the current stack segment SS. Clearly, the two pointer registers BP and SP are intended for convenient access to the stack. When addressing based records, *the register holds the offset to the start of the record, and the displacement gives the distance into the record*. Shown are four examples:

```
MOV AL, [BP + 8]        MOV DEPT [BX - 8], 3F8H
MOV DX, [BP - 12]       INC NUMBER [BX]
```

In the two instructions on the left, the displacements are included in the brackets, and the total offsets are from SS. The codes are 8A4608 and 8B56F4. As indicated in Table A-2 of Appendix A, with register BP and a zero displacement, it is necessary to use mod = 01 with byte 00 included in the object code for *disp8*. Instructions such as these are convenient for accessing stack data; with BP cleared, the displacement is the offset address from the stack segment address, and the word at this location is moved.

The two instructions on the right have displacements specified by the names DEPT and NUMBER. Because the operand 3F8H is too large for a byte location, the assembler requires that variable DEPT be defined in the program as a word. This variable is used by the assembler to allocate memory space for the operand. If DEPT has offset 3 in DS, the immediate operand 03F8H is moved to the location having an offset found by adding the net displacement of −5, or FFFBH, to the word of BX. Because an 8-bit displacement is sign extended by the processor, the immediate-to-memory MOV can use disp8 = FB, and the code is C747FBF803, with mod = 01 and r/m = 111.

Fields of a record are conveniently accessed with based addressing. *The base register is first loaded with a word that points to the base of the record, and elements are then addressed with their displacements from this base*. The word DEPT of the previous MOV instruction might be the field of the record EMPLOYEE, with DEPT identifying the displacement from the record base. Identical records at different memory locations are accessed with the same instruction simply by changing the base register. A program loop with the base value properly incremented on each pass is able to access many records rapidly.

Indexed Addressing

In *indexed addressing*, the sum of a displacement and the word of an index register is the effective address. The two index registers are the source index SI and the destination index DI. Refer to Table A-2 (mod 01 and 10 with r/m = 100 and 101).

This addressing mode is commonly used to access elements of an array, *with the displacement selected as the offset of the first element*. Then an index of 0 selects this element; for word storage, an index of 2 selects the next one, and so on. *By incrementing the index register within a program loop, elements can be accessed in sequence*. The method is the reverse of that of based addressing, in which the displacement is incremented relative to a fixed address in a base register. The MOV instruction of the program of the preceding string division example utilizes indexed addressing, as do each of the instructions that follow:

```
MOV CX, COUNT [DI + 200H]          MOV ARRAY [SI], AL
MOV AX, M3 [SI - 3]                INC NUM_FIVE [SI]
```

The displacement variables are COUNT, ARRAY, and NUM_FIVE. In the program, COUNT must be defined as a word (DW), and ARRAY is assumed to be a byte (DB). With hexadecimal offsets of 00E4, 00AB, and 0075 for COUNT, ARRAY, and NUM_FIVE, respectively, and with OFFSET M3 denoting 1234H, the codes in order are 8B8DE402, 8B843112, 8884AB00, and FE4475.

Consider the first of the codes, with reference to Appendix A. The instruction is mem-to-reg MOV with a 16-bit displacement 02E4; direction d is 1, word w is 1, reg CX is 001, and from Table A-2 the mod is found to be 10 with 101 for r/m. Thus, mod-reg-r/m gives 8D for the second byte. With the PC the hexadecimal code can be entered into the DOS debug program and unassembled. This procedure converts the machine language back into assembly language and prints it on the monitor, providing a check on the original translation. The mini-assembler of the DEBUG program provides an easier way to get code, and an assembler for the computer is available on a diskette. Thus, hand assembly is unnecessary. The assemblers are examined later.

Based Indexed Addressing

An especially flexible and convenient addressing mode is *based indexed addressing*. The effective address EA is the sum of a base register BX or BP, an index register SI or DI, and a displacement. Three instructions utilizing this mode are as follows:

```
MOV ALPHA [BX] [SI], 0ABCDH
MOV AL, ES:ARRAY [BX][SI]
INC BYTE PTR [BP - 4] [DI]
```

Note the segment override prefix. In addition to based indexed addressing, the first one uses immediate addressing, in which the operand is included in the instruction, and the second one has register addressing. The variable ALPHA must be the name of a word location, and ARRAY must be a byte variable; otherwise the assembler gives an error message. The code for the INC instruction is FE43FC, with the offset in SS.

Suppose it is desired to address an array on a stack. *Base pointer BP can hold the offset of a reference point on the stack, and the displacement can provide the offset of the beginning of the array from the reference point. Then the index register can be incremented to access the array elements in sequence.* Based indexed addressing is also often used to access arrays within records and two-dimensional arrays of matrices.

It should be noted that the memory addressing methods specify offsets relative to the values contained in the segment registers. This method is known as *relative addressing*. It allows programs to be placed in convenient regions of memory after they are written by adjusting only the words of the segment registers. An inactive program in memory can be saved on a diskette, and the memory space that was occupied can be used for other purposes. If the program is needed later, it can be read back into memory at any available region. Furthermore, a program in memory can be relocated dynamically to different physical addresses by means of a block move of the code, with new segment addresses specified.

BASIC STACK AND ARRAYS

The BASIC interpreter must have a stack in memory. It is used primarily for storage of return addresses and for saving the contents of registers when subroutines are called or when interrupts occur. By default the reserved stack space for BASIC is 512 bytes. This is usually adequate unless a program has many FOR-NEXT loops or subroutines or PAINT statements. The CLEAR command can be used to override the default value and for other purposes.

The work area for BASIC starts above the region of memory used to store interrupt jump addresses (1KB), the DOS programs, and the Disk or Advanced BASIC routines. By default, it extends for 64 KB or to the end of available memory, whichever is smaller, but it never exceeds 64 KB, regardless of the size of memory.

At the low end of this work area is space reserved for use by the interpreter, followed in order by regions for storage of the user program, BASIC variables, arrays, strings, scratchpad data, and a stack. The stack is placed at the high end of the work area, extending downward toward the programs and data. Note that storage of a scalar variable precedes that of arrays and strings. Thus, when a numeric variable is defined in a program, *previously defined arrays and strings are pushed up in memory* to make room for the variable. Blocks of data are transferred quickly with the machine language string instructions.

CLEAR. The CLEAR command is usually entered in the direct mode. Its purpose is to set all numeric variables to 0 and all string variables to null values, along with two options. One option allows a limit to be set on the size of the BASIC work area. Another option allows the BASIC stack space of the work area to be specified. Consider the command:

```
CLEAR ,32768,1000
```

This command clears the numeric and string variables; specifies that a maximum of 32 KB is allowed for storage of data, programs, and the stack in the BASIC work area; and reserves 1000 bytes for the stack. The optional numbers can be replaced by expressions. Formats with one or both of the options omitted are as follows:

```
CLEAR ,32768     CLEAR ,,1000     CLEAR
```

One reason for sometimes limiting space for BASIC is to set aside memory for machine language subroutines. If such subroutines are placed at addresses sufficiently high to be outside the limit set for BASIC by the CLEAR command, there will be no problem with BASIC programs or data replacing code of the subroutines. There are other ways to protect such machine language programs. One way is to put the machine code into defined arrays in the BASIC data area, a method that is examined in Chapter 6. Another way is to use a DEF SEG statement to put the code in memory above the BASIC work area, provided there is sufficient memory.

Arrays

In BASIC, an *array* is a group or table of values referred to by a single name. Each value is called an *element*. A one-dimensional array has the form NAME (*n*), with the number *n* being the *subscript*. Arrays are normally created by the dimension statement DIM, such as

```
10 DIM A(3)
```

This statement defines a one-dimensional numeric array A with 4 single precision elements, which are A(0), A(1), A(2), and A(3). Each element is an *array variable*. The subscripts are 0, 1, 2, and 3, and the initial value of each element is 0. Variables not associated with arrays are called *scalar* variables.

The following BASIC statement creates a two-dimensional array of single precision values:

```
10  DIM  B(1,2)
```

The elements are identified as follows:

```
B(0,0)   B(0,1)   B(0,2)
B(1,0)   B(1,1)   B(1,2)
```

The first subscript has two values and the second one has three, for a total of 2 X 3, or 6 elements. Each element has its own unique pair of subscripts. An array can have up to 255 dimensions, or subscripts, with 32K elements per dimension. READ and DATA statements can assign values to elements of an array, and type declaration characters can be used to change the precision to either integer or double precision values.

A one-dimensional *string array* with nine elements is created by the statement

```
10  DIM  C$(8)
```

Each element is a variable-length string with an initial value of null. Strings can be assigned to the elements with READ and DATA statements, with quotation marks in the DATA statements optional. If a variable name of a numeric or string array is used without a DIM statement, the maximum value of a subscript is 10 by default. The example shown in Fig. 3-19 illustrates one way to assign values to the variables of arrays. In this example, elements with subscript 0 are not used, this being an optional choice.

```
10  DIM A(5), B$(5)
20  FOR I = 1 TO 5
30    READ A(I), B$(I)
40  NEXT
50  PRINT A(1), A(2), A(3), A(4), A(5)
60  PRINT
70  PRINT B$(1); B$(2); B$(3); B$(4); B$(5)
80  DATA 23.7, "The ", &HFE, time, 5.2E-3
90  DATA " is ", -8, 12, &HA000, " noon."
```

FIGURE 3-19. Assignment of values to arrays.

When the program is run, the display shows

```
23.7      254      .0052      -8      -24576

The time is 12 noon.
```

Note that a string in a DATA statement can have no blank spaces at the beginning or end *unless the blanks are enclosed in double quotation marks*. Also note that the 16-bit number A000H is interpreted as negative. INPUT statements provide another method often used to assign values to variables of arrays, replacing READ and DATA statements.

REFERENCES

MCS-86 Assembly Language Reference Manual, Intel Corp., Santa Clara, Calif., 1978.

S. P. Morse, *The 8086 Primer,* Hayden Book Co. Rochelle Park, N.J., 1978.

The 8086 Family User's Manual, Intel Corp., Santa Clara, Calif., 1979.

The IBM Personal Computer MACRO Assembler, IBM Corp., Boca Raton, Fla., 1981.

PROBLEMS

Section 3-1

3-1. (a) Identify all general registers of the 8088, and within the group give the names of all byte, word, pointer, and index registers. (b) If the stored flag word is 04C4H, give the state of each status and control flag.

3-2. Suppose the CS, SS, DS, and ES registers of the BIU are initialized with hexadecimal words 0200, 0500, 1500, and 2AC4, respectively. Sketch a diagram showing the segment locations in the total memory space. On the diagram, show the lowest and highest hexadecimal addresses within each segment. How many distinct memory byte locations are contained within both CS and SS?

3-3. Suppose SP is initialized with 5ABCH and segment register SS stores BF3AH. (a) After five words are pushed onto the stack, determine the physical address of TOS and also that of the word stored at the bottom of the stack. (b) If words are pushed until SP stores 0000H, how many words are in the stack?

3-4. Give the absolute addresses in both hexadecimal and decimal corresponding to each of the following logical addresses: F600:7ABC, 0040:FFFF, A95B:F87D, and FFFF:0000.

3-5. The instruction ADD CX,DX adds the word of DX to that of CX and stores the result in CX. All status flags are affected. Suppose CX and DX store DA97H and BCB8H, respectively. Using the form of Fig. 3-3, sketch the flag bit storage and give the hexadecimal word, assuming control and undefined bits are 0s. Repeat if the instruction is changed to ADD CL,DH and also give the final word of CX.

SECTION 3-2

3-6. (a) Prepare an assembly language program that moves 1000 words from the stack, beginning at offset 100H, to memory at absolute address 06543H. Any changed segment addresses must be restored. Use REP. (b) For your program, determine the hexadecimal machine code.

3-7. (a) The 16-bit displacement ABCDH is added to the offset address B859H of a segment. Determine the new offset, assuming the displacement is signed and also assuming it is unsigned. (b) An 8-bit signed displacement 9EH is sign extended to 16 bits and added to the offset address B859H of a segment. Find the new offset.

3-8. Unassemble the following MOV and INC instructions given in hexadecimal object code: (a) 8904; (b) FEC2; (c) 2688A50020; (d) B93000; (e) 26884501; (f) 2E8B5600; (g) FE43FF

SECTION 3-3

3-9. Decimal scan codes 16 through 25 represent the respective characters QWER TYUIOP. A byte placed in AL is known to be the binary scan code for one of these characters. Write an assembly language routine that replaces the scan code of AL with the ASCII code. Use a table and XLAT.

3-10. If AX = B8D5H and BX = AC76H with CF = 1 and all other flags cleared, determine AX in hexadecimal and the status flags after execution of (a) ADD AX,BX; (b) ADC AX,BX; (c) INC AX; (d) SUB AX,BX; (e) SBB AX,BX; (f) DEC AX. Tabulate the results.

3-11. Determine the hexadecimal machine code for the example-string-compare routine and give the IP offset for each instruction, starting at 100H.

3-12. (a) For the example-table-scan routine, suppose CS = DS = 0800H, with table C5 at offset 100H in CS. The locations are continuous. Assume that AL contains scan code 49H, and trace the program step by step, giving the new hexadecimal contents of registers and memory following each instruction execution. Include each step of the repeat sequence. (b) Sketch a flowchart of the program. (c) Determine the hexadecimal machine code for the routine and give the IP offset for each instruction.

SECTION 3-4

3-13. Write a routine that clears all bits of even addresses and sets all bits of odd addresses from 0800:1000H through 0800:1FFFH. Restore the original contents of any segment registers used. Include the machine code.

3-14. (a) Suppose 1000 words are stored in memory, with the initial address at 0800:1000H. Write a routine that transfers the low bytes of each word to consecutive locations starting at 0700:0100H and the high bytes to consecutive locations starting at 0800:0200H. Restore the contents of any segment registers used. (b) Give the machine code.

3-15. Suppose DS, ES, SS, BX, and BP store 2400H, 1500H, B2A5H, 063AH, and 674BH, respectively. After execution of each of the following instructions, give in hexadecimal the physical addresses to which the source operand is moved, and alongside each address state the stored data:

```
MOV ES:[BP], BL     MOV WORD PTR [BX], 5
MOV [BP], BX        MOV BYTE PTR [BX], -8
```

3-16. Assume that registers DS, BX, and DI each store A96CH, registers ES, SS, BP, and SI each store 8A2EH, the value of OFFSET ALPHA is 2EH, and that of OFFSET ARRAY is 3F28H. For the three instructions of the subsection Based Indexed Addressing, determine the physical address of each source and destination operand of memory.

3-17. Using the minimum number of bytes, deduce the hexadecimal object code for the following assembly language instructions, and for each, identify both the source and destination addressing mode:

```
(a) INC AH                              (c) MOV [BP], AL
(b) MOV AX, 0                           (d) MOV AH,[SI+BX+200]
(e) MOV AX, 78*256 + '+'
(f) DEC BYTE PTR A1                     with OFFSET A1=004A
(g) MOV SP,OFFSET TOS                   with TOS=0030:0100
(h) MOV AL,CS:M7 [SI]                   with M7=F000:F0F4
(i) MOV ES:ALPHA+2,OFFSET BETA
    with OFFSET ALPHA = 0020 and BETA = 1234:ABCD
```

3-18. In the program segment at 0040H, following the statement ORG 0B3H is the statement BETA DD 12345678H. Registers DS, ES, SS, BX, and BP store 40H, A432H, EDCBH, 0, and 100H, respectively. State precisely the result after assembly, followed by execution if no error is reported, of each of the following instructions:

```
LEA BX, BYTE PTR BETA
LDS BX, BYTE PTR BETA
LES BX, BYTE PTR BETA
MOV BL, BYTE PTR BETA
JMP BETA
```

3-19. Sketch flowcharts of the Example-Clear-Display and Example-Divide-String programs.

SECTION 3-5

3-20. Suppose the BASIC work area starts at address 07400H. The statement POKE *n,m* writes byte *m* into offset *n* of the segment defined by DEF SEG. (a) Write a program that writes bytes +1, &HFF, +2, &HFE, and &H80 into an area of memory outside the BASIC work area, assuming a total RAM of 64 KB. Also, the program should read the bytes from memory and print them in decimal on the display. Include statements DATA, FOR, and NEXT. (b) Repeat for a memory of 128 KB, without using CLEAR.

3-21. Without using the statement CLS, write a BASIC program that uses POKE to clear the screen of the monochrome display.

3-22. Write a program that defines a two-dimensional numeric array with subscripts *i,j* from 1,1 to 2,4. Assign values to the array elements from the keyboard with the statement INPUT, including a prompt. The program should display the assigned values in a 2 × 4 table.

Logic Functions

Logic is the science of the formal principles of reasoning, dealing with validity in thought and action. Binary logic is restricted to the classification of all information as either true (logical 1) or false (logical 0). For signals that activate circuits, more appropriate names for the two categories are *active* and *inactive,* or *asserted* and *not asserted*.

The deductive mathematics that treats the processes of logic variables is known as *Boolean algebra,* and the variables are often called *Boolean variables*. A Boolean variable representing a digital signal is usually assigned a mnemonic name. This is a short identifier that is spelled so as to help one to recall its function. For example, the mnemonic EN might represent ENABLE. Somewhat different are acronyms, such as NMOS and ASCII, which are short names commonly used in the language as normal replacements for longer phrases.

The major objectives here are to investigate Boolean algebra and to develop basic logic concepts that provide a foundation for digital circuit design. Actual design is examined in Appendix C, which requires the material in this chapter as a prerequisite.

THE INVERTER

The most widely used logic circuit, and perhaps the easiest to understand, is the *inverter*. Closely related to it is the *buffer* circuit. Both are examined here. However, we will first look at the mixed logic system, comparing it with positive and negative logic.

Prior to defining the inverter, we need to become familiar with the *static indicator symbols* used on logic circuit diagrams. We have encountered the dynamic indicator. Let us recall that one was shown at the clock input of the shift register of Fig. 1-13. As explained in Section 1-2, it shows symbolically that the dynamic signal affects the circuit only on a state change from low to high, or perhaps from high to low. For a dynamic signal, the active state is defined as the one to which it goes on the active transition.

The presence or absence of a static indicator symbol at an input or output terminal, including those with dynamic signals, reveals the active state of the signal. The symbol is defined here for the positive, negative, and mixed logic systems, and it is always present at one terminal of the logic symbol of an inverter.

Mixed Logic

In the positive logic system, a logical 1 represents a high voltage level; an active signal is either 1 or 0, depending on whether the assertion is active high or low. Practical systems have some signals asserted high and others asserted low. System design often requires less imagination if a variable is specified simply by its physical state, either high H or low L. Furthermore, this practice reduces the problems encountered when certain subsystems are designed on the basis of positive logic along with others specified with the negative logic convention. In the PC, a memory bit at the low voltage level is read and printed as a logical 0, and one at the high level is read and printed as a logical 1. This is positive logic. On the other hand, transmission via the communications interface uses negative logic, with the low state designated as 1.

In view of these considerations, the use of a mixed logic system has considerable merit. In this system, a logical 1 is chosen to imply an active signal and a 0 to imply one that is inactive. *These designations are independent of the actual voltage level*. When signals are expressed in terms of voltage, we shall use *state H* and *state L* to denote the high and low levels, respectively. There is no uniformity in the industry and no prescribed standard favoring one logic system over another. All are used extensively, and the computer engineer should understand each system.

Static Indicator Symbols

A *logic symbol* is a graphic representation of a logic function. It may have the form of a rectangle containing a special symbol, or *qualifier*, identifying the function. Alternately, it may have a distinctive shape that uniquely specifies its function. In general, there are both inputs and outputs, and indicator symbols are placed at terminals as needed to identify the state or voltage level of the active signal.

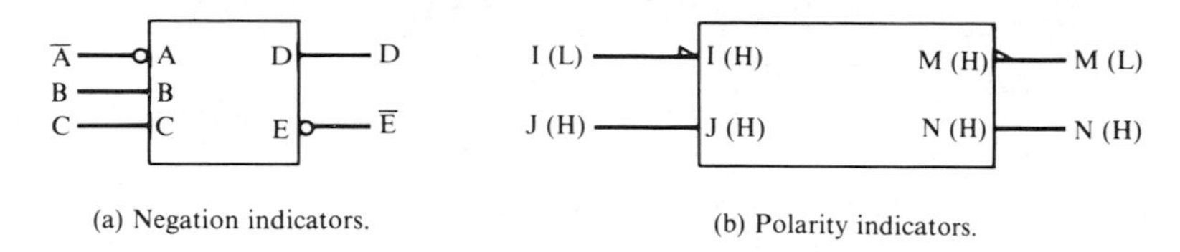

(a) Negation indicators. (b) Polarity indicators.

FIGURE 4-1. Logic circuits with indicator symbols.

There are two static indicator symbols. One of them is the *negation indicator,* used only in positive and negative logic systems. It consists of a small circle, often referred to as an *inversion circle,* placed at an input or output terminal, as shown in Fig. 4-1a. The other is the *polarity indicator,* used only in mixed logic systems. It resembles a half arrowhead, as shown in Fig. 4-1b.

With the logic function unspecified, the circuit of Fig. 4-1a has two negation indicators. The presence or absence of a negation indicator provides for signal representation *independent of its physical value.* If present, the 0 state of the signal is the active 1 state of the circuit represented by the symbol, and the 1 state of the signal is the inactive 0 state of the circuit. On the other hand, if the indicator is absent at an input or output, the 1 state of the signal is the active 1 state of the circuit. Thus, *the negation indicator inverts the state,* and this is true regardless of the actual voltage level.

As indicated in the figure, in order for the internal signal A to be in active state 1, the signal applied to the input terminal must be in state 0, which implies the complement $\overline{A}$ of A. Inputs of 1 at B and C give internal states of 1, because no inversion circles are shown at these inputs. The meaning at an output is the same. The 0-state of output $\overline{E}$ corresponds to the 1-state of the internal signal E. Negation indicators are used only in positive logic and negative logic systems, *not in the mixed logic system.*

Now let us examine the definition of the polarity indicators of Fig. 4-1b. *The presence of this symbol implies that the indicated 1 state of the signal, when referred to the logic function, is the low voltage.* Its absence indicates that the 1 state of the signal is the high level. Thus, the polarity indicator provides for representation of logic function inputs and outputs *in terms of their physical values,* and its use is *restricted to mixed logic systems.* The half arrowhead always points in the direction of signal flow. Variables of the figure are followed by either (H) or (L), used here to denote the active signal level. For both input I and output M, the indicated 1 state is low L, and for signals J and N the indicated 1 state is high H.

Whenever a logic symbol with one or more negation indicators is encountered in this book, positive logic is implied unless negative logic is clearly stated. One or more half-arrowhead polarity indicators in a diagram automatically signifies mixed logic. *Polarity and negation indicators must not be used in the same system,* because the three logic systems are mutually contradictory.

Signals in the positive and negative logic systems are conceived or thought of in terms of their abstract logical state, referred to as state 1 or state 0. The translation into real physical quantities depends on the choice of positive or negative logic. On the other hand, *signals in the mixed logic system are conceived or thought of in terms of their physical state,* referred to as state H or state L, and the active level is defined as the 1 state. A variable that is active low is normally identified as such by placing a bar over its name, or a prime immediately following its name,

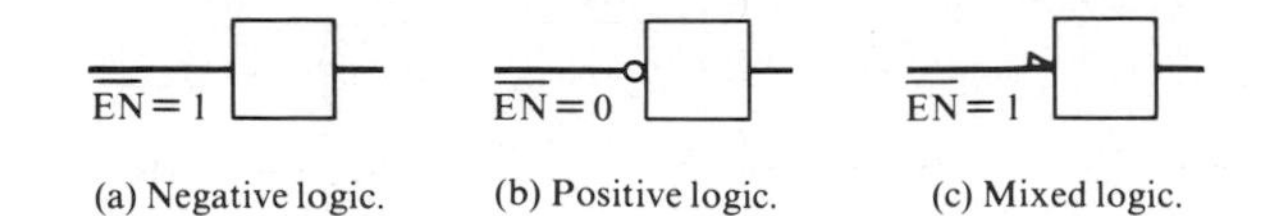

(a) Negative logic. (b) Positive logic. (c) Mixed logic.

FIGURE 4-2. Identical logic circuits with low-level $\overline{EN}$.

or the letter L at its end. The following mnemonics for the active low I/O READ signal are equivalent:

$$\overline{IOR} = IOR' = IOR(L)$$

When this special notation is observed on the names of one or more variables of a logic diagram, then those without the notation are understood to be active high.

Suppose a logic circuit has a terminal with label $\overline{EN}$ that is active when the applied voltage is low. A high voltage disables the circuit, perhaps floating its terminals. In a negative logic system there would be no negation indicator at this input because a 1-state input represents a low voltage. However, a negation indicator must be present in a positive logic system. In a mixed logic system, a polarity indicator is essential. It may appear that the negation indicator of positive logic and the polarity indicator of mixed logic are the same, but this is not so. The active enable signal of the positive logic system has state 0, whereas that of the mixed logic system has state L, which corresponds to state 1. These three cases are illustrated in Fig. 4-2.

Now suppose that the terminal of the preceding circuit is defined by the mnemonic DIS for disable, rather than by $\overline{EN}$. There is no change in the hardware, only in the name. Being the complement of EN, disable DIS is active high. Accordingly, a negation indicator must be placed at the input of the symbol in a negative logic system, but there is no negation indicator with positive logic, nor is a polarity indicator present in the mixed system.

Each of the static indicators can be used in conjunction with the dynamic indicator, as shown in Fig. 4-3. A dynamic clock input with an inversion circle is active on the high-to-low, or negative, transition in a positive logic system. This is shown in Fig. 4-3b. With negative logic, the opposite is true. The use of the bar over CLK conforms to the definition of the active voltage level of a dynamic signal, which is *the level to which the signal goes on the active transition*. The combination of a polarity indicator and a dynamic indicator, as in Fig. 4-3c, denotes that the active clock transition is high to low. This corresponds to the indicators of a positive logic system. However, in the mixed logic system, the active low state of the dynamic clock is state 1.

Sometimes it is desirable to convert a logic diagram from one system to another. Conversion from a positive to a negative logic convention, and vice versa, is

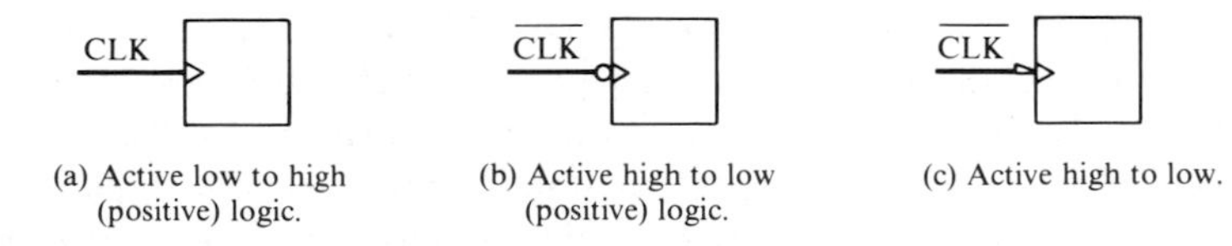

(a) Active low to high (positive) logic. (b) Active high to low (positive) logic. (c) Active high to low.

FIGURE 4-3. Indicators for a dynamic clock input.

accomplished by the following algorithm:

1. Add one additional inversion circle to each terminal.
2. Delete any pairs of inversion circles.

For conversion from positive to mixed logic, simply replace each inversion circle with a polarity indicator. The reverse procedure is applicable to a transfer from mixed logic to positive logic. Data sheets published by manufacturers usually show symbols in the positive logic convention, but these are readily converted into the other logic systems.

The Inverter Function

A most important and very common combinational circuit is the inverter. Many ICs contain hundreds or even thousands of inverters. Normally, the circuit has one input and one output terminal, although some have an additional three-state control.

The function of the inverter is simply to change the voltage level. A high input voltage is transformed into a low voltage at the output. Conversely, a low input becomes a high output. With 0 and 1 denoting the voltage levels of a positive or negative logic system, an inverter with an input of 1 gives an output of 0, and an input of 0 provides an output of 1. This is the *complement* operation described in Section 2-1. It is also called *negation,* or *NOT*. If the input variable is A, the output is NOT A, or A′, or $\overline{A}$. An inverter circuit is also called a *NOT gate*.

An inverter changes the active level of a signal. An active high input becomes an active low output. If a chip is to be enabled with an active high variable EN and its enable pin is active low, an inverter must be inserted.

Truth Table

Many combinational networks have numerous inputs and outputs, and each output variable is a function of the inputs. A logic function that relates the output variables to the input variables can be represented graphically by a *truth table,* which gives in tabular form the state of each output for every possible combination of input states. For *n* inputs the number of different combinations is 2 raised to the power *n*. Thus, the truth table for a circuit with three inputs would have 8 rows, one with four inputs would have 16 rows, and so on. Because the inverter has only one input and one output, the truth table has only two rows, one input column, and one output column. The table and standard symbols are shown in Fig. 4-4 for positive and negative logic.

The rectangular figure and the distinctive triangle are alternate symbols with identical functions. The 1 of the rectangle is the standard qualifier denoting inver-

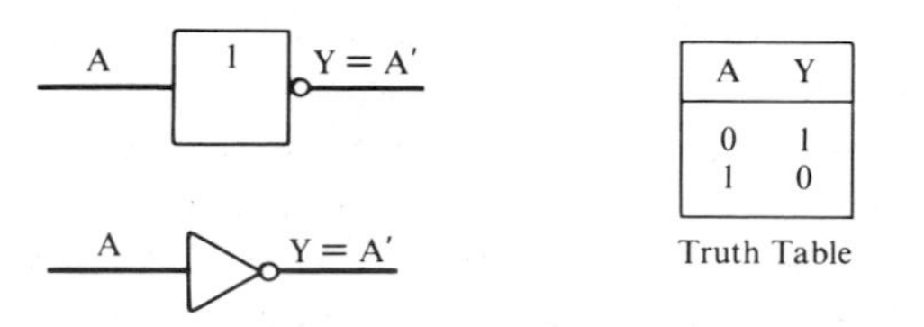

A	Y
0	1
1	0

FIGURE 4-4. Inverter symbols and a truth table for positive and negative logic systems.

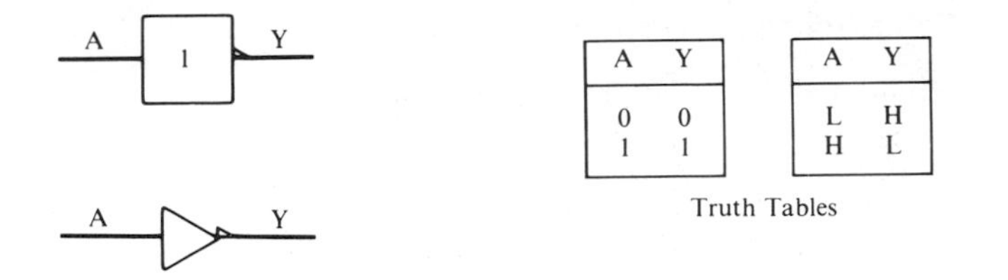

FIGURE 4-5. Inverter symbols and truth tables for the mixed logic system.

sion. Its placement within the triangle is optional. Because there is international acceptance of rectangular symbols for representation of logic functions, these symbols are often used when diagrams are expected to have international applications. If an inverter converts an active low input signal into an active high output, the negation indicator is often placed at the input terminal rather than at the output. This position is optional, but recommended. It is symbolic only, having no effect on the function.

In a mixed logic system the effect of an inverter is to change the voltage level, but not the logic state. Figure 4-5 shows equivalent symbols and two truth tables, one giving the 1 and 0 states and the other giving the voltage levels.

For the symbols of Fig. 4-5, the indicated 1 state at the input is the high voltage, and the indicated 1 state at the output is low, with these indicated 1 states determined by the absence or presence of the polarity indicator. Note that a logical 1 input gives an output of 1. There is no inversion of the logic states, and the NOT operation is not performed on the 1 and 0 bits.

However, with respect to states H and L of the voltage levels, the inverter does implement the NOT function, and *design in the mixed logic system is based on the physical voltage levels,* not the abstract logical 1 and 0 states. Thus, the output Y of each of the symbols of Fig. 4-5 can be expressed as NOT A or $\overline{A}$, *provided the voltage levels are understood.* The polarity indicator may be moved from the right side to the left side of the rectangle or triangle. Although the function is not affected, at the input the indicated 1 state is now the low voltage, and at the output the indicated 1 state is high.

The Buffer

Another digital circuit having only one input and one output is the amplifier, which is normally referred to as a *buffer*. Although logic states and voltage levels are not affected by the noninverting buffer, the circuit provides *increased capability for current sourcing to a load and current sinking from a load.* Also, it restores logic levels to optimum voltage values, improves noise immunity, and minimizes changes in switching threshold voltages caused by load currents.

A buffer is sometimes connected between a logic circuit and its load in order to isolate them, thereby preventing the load current from adversely affecting the logic circuit. Many chips have internal buffers, and a complex system such as the PC has buffers at various points of the address and data buses, as well as numerous other buffers throughout the system. As stated earlier, the term *buffer* also denotes a region of memory reserved for temporary storage of I/O data.

Figure 4-6 shows three symbols that represent buffers. Standard symbols with identical functions are the triangle and the rectangle containing the triangular qual-

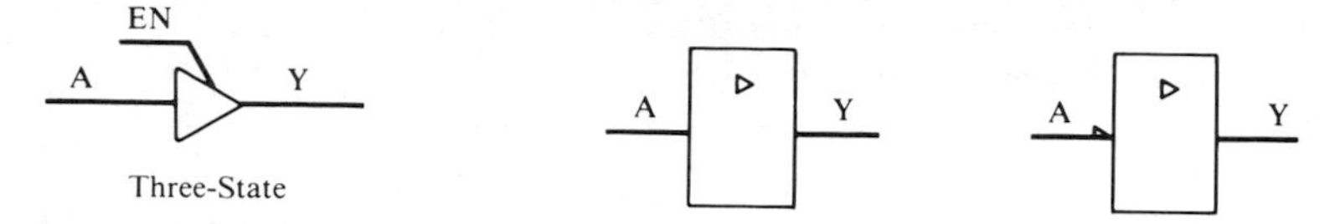

FIGURE 4-6. Noninverting and inverting buffers.

ifier. A negation or polarity indicator can be placed at either the input or output, thereby providing inversion in addition to buffering. The output of the three-state buffer floats when EN is low.

With EN high, for each of the two noninverting buffers of Fig. 4-6 the output Y equals the input A. There is no change in either the state or voltage level. The inverting buffer at the right side changes the voltage level. For this mixed logic function, the output assumes its indicated 1 state at the high level if and only if the input assumes its indicated 1 state at the low level.

EXAMPLE

In the network of Fig. 4-7, determine the states and levels of variables B, C, D, and E if input A has its indicated 1 state. Repeat with A = 0.

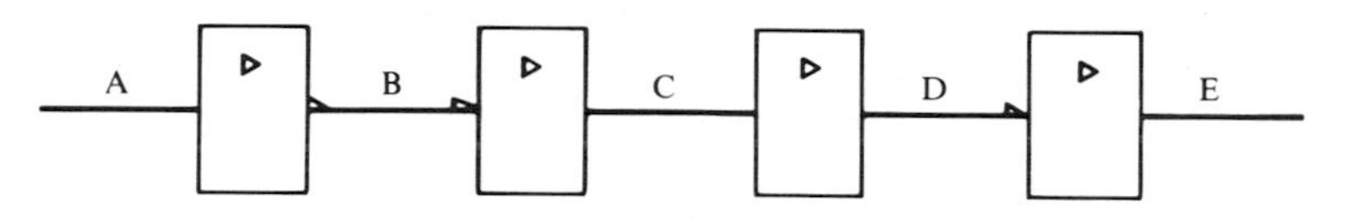

FIGURE 4-7. Buffer cascade.

SOLUTION

The first, second, and fourth stages are inverting. Therefore, with A high, variable B is low and C and D are high. Output E is low. Signals A, B, and C are logical 1s. Because D is high, it is 1 relative to the buffer at its left, but it is 0 relative to the inverter at its right. The indicated 1 state of output D is high, and the indicated 1 state of input D is low. Although not common, such mismatches do occur. Because the indicated 1 state of output E is high, the low-level E has state 0.

For A = 0 (low level), variable B is high, C and D are low, and output E is high. Signals B and C are logical 0s. Variable D is 0 with respect to the buffer at its left, and it is 1 with respect to the output inverter. Output E has its indicated 1 state. ■

Logic Families

It is not necessary to concern ourselves with the internal circuitry or the detailed terminal characteristics of an inverter or a buffer. These features depend on the basic logic system selected. The six *logic families* in greatest use today are as follows:

NMOS	N channel MOSFET logic
PMOS	P channel MOSFET logic
CMOS	Complementary MOSFET logic

TTL Transistor transistor logic
IIL Integrated injection logic
ECL Emitter-coupled logic

The three MOSFET systems use *field-effect transistors,* known as *MOSFETs,* and the TTL, IIL, and ECL systems use *bipolar junction transistors*, or *BJTs*.

A popular chip is the TTL hex inverter IC, with six independent inverters in a 14-pin package. Figure 4-8 shows two of these ICs, identified as type 7404. The actual circuits are contained in a small silicon wafer within the illustrated *dual in-line package (DIP)*, which is about 2 cm long. Note the conventional pin numbering system, with pin 1 located adjacent to the dot on the top side. On some DIPs, a small notch is present in place of the dot.

For the 7404 IC, five volts should be connected between pins 14 and 7. The positive side of the voltage supply goes to pin 14, which is designated *VCC*, and the negative side goes to pin 7, which is referred to as *ground (GND)*. The other 12 pins provide one input and one output for each of the six inverters. Pins to unused inverters can be left disconnected. The high and low voltages of the logic signals are typically about 4.8 V and 0.2 V, depending on the load. The chip belongs to the widely used TTL "7400" series of digital integrated circuits.

CMOS 4049A and 4050A ICs are inverting and noninverting hex buffers, respectively. Voltage VCC can be selected between 3 and 15 V with respect to ground, and the high and low voltages are VCC and 0. ICs of the CMOS 4000 series have many applications.

The ICs of the PC are either NMOS or TTL. The NMOS 8088 processor and other NMOS circuits have thousands of transistors, and their DIPs have 24 to 40 pins. The TTL chips provide buffering, inverting, and various logic functions. Most belong to the 74LS00 series, with the letter L denoting *low-power* and S denoting *Schottky-type* transistors, which are designed for extra high speed.

In comparison with discrete circuits, ICs have smaller size, lower cost, less power dissipation, higher reliability, and higher speed. For these reasons, digital circuits are integrated, and systems such as that of the PC consist of numerous interconnected chips.

FIGURE 4-8. TTL 7404 hex inverter in a DIP.

AND, OR, and XOR

A *logic gate* is a hardware circuit that manipulates binary information supplied to its inputs, generating combinational logic functions at its outputs. Each output depends only on the binary bits present at the input terminals. There are different types of basic functions, and each has a name, one or more graphic symbols, and a Boolean expression describing the operation. The I/O relationships can be presented in a truth table. We investigate here the AND, OR, and Exclusive-OR (XOR) logical operations.

The AND Function

The *AND function,* with two or more inputs and one output, is defined as follows:

The output has its indicated 1 state if and only if all inputs have their indicated 1 states.

In the positive logic convention, the 1 state is always the high level. However, *with a negation indicator present at an input or output terminal, the indicated 1 state of the AND function is present when the terminal has a low-voltage, logical 0 state.* With a negation indicator at a terminal of a negative logic AND symbol, the indicated 1 state of the function corresponds to a terminal high-voltage, logical-0 state.

With mixed logic, the indicated 1 states may be either high or low, depending on the polarity indicators of the logic symbol. Figure 4-9 shows the two standard logic symbols for the three-input AND gate. The qualifier of the rectangle is the ampersand &, and its placement within the distinctive shape symbol is optional.

FIGURE 4-9. AND logic symbols.

The symbols of Fig. 4-9 apply to each of the three logic conventions. However, in the mixed system, there may be polarity indicators, such as those of Fig. 4-10.

For inputs A and B of the rectangular symbol of Fig. 4-10, the indicated 1 state is low and the indicated 1 states of input C and output Y are high. Accordingly, Y is high if and only if A is low, B is low, and C is high.

For the distinctive shape symbol, there are two outputs. Logic elements may have more than one output, provided each output implements the function. With

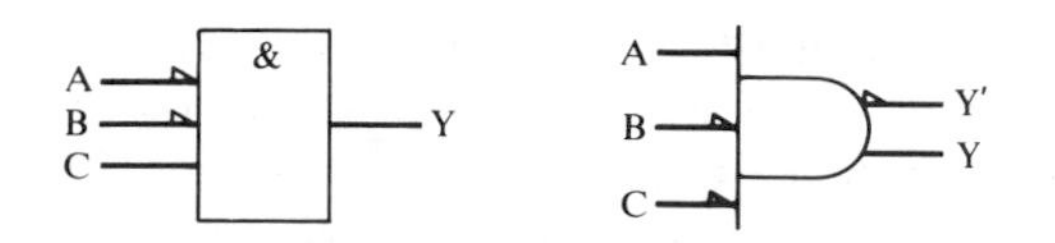

FIGURE 4-10. Additional AND symbols for mixed logic.

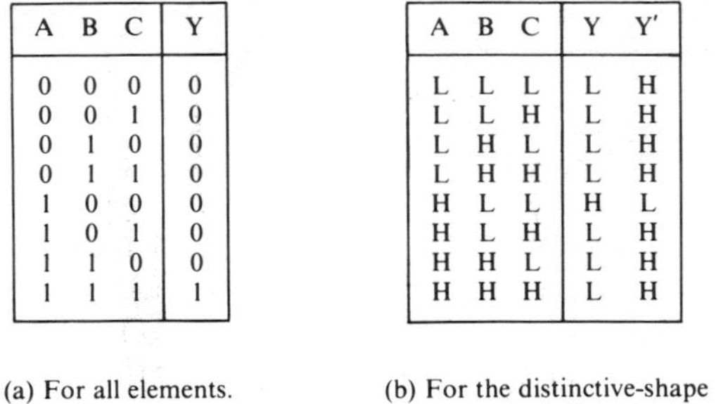

A	B	C	Y
0	0	0	0
0	0	1	0
0	1	0	0
0	1	1	0
1	0	0	0
1	0	1	0
1	1	0	0
1	1	1	1

(a) For all elements.

A	B	C	Y	Y′
L	L	L	L	H
L	L	H	L	H
L	H	L	L	H
L	H	H	L	H
H	L	L	H	L
H	L	H	L	H
H	H	L	L	H
H	H	H	L	H

(b) For the distinctive-shape element of Fig. 4-10.

FIGURE 4-11. AND function truth tables.

respect to voltage levels, one output is the complement of the other. Output Y has its indicated high-level 1 state and Y′ has its indicated low-level 1 state if and only if A is high and both B and C are low. Note the symbol extension at the input, which is allowed only for distinctive shape symbols.

For each of the logic elements of Figs. 4-9 and 4-10, the AND function truth table with 1s and 0s is given in Fig. 4-11. Also shown is the table in terms of the H and L states for the distinctive shape element of Fig. 4-10.

For any number of inputs, the output Y is 0 if at least one of the inputs is 0. As shown by the logical truth table, the AND function conforms to ordinary multiplication, referred to as *Boolean multiplication*. Thus, in terms of the logical 1s and 0s of the truth table, Y is ABC, read as "Y is equivalent to A and B and C." If the three inputs of the AND element of Fig. 4-9 are connected together, forming a single input, it becomes a buffer. The TTL 74LS21 chip contains two positive-logic, four-input AND gates in a 14-pin DIP.

It is useful at times to give logic expressions in the mixed logic system in terms of the H and L states. This is done *by representing the H state with 1 and the L state with 0,* corresponding to positive logic. Thus, in terms of voltage levels, the logic of the two functions of Fig. 4-10 is given by

$$Y = A'B'C \text{ (rectangle)} \qquad Y = AB'C' \text{ (distinctive shape)}$$

Boolean algebraic equations are those relating logical expressions. Several Boolean identities involving the AND and complement operations are as follows:

$$\begin{aligned} ABC &= (AB)C \qquad & AA &= A \qquad & A1 &= A \\ AB &= BA \qquad & A\overline{A} &= 0 \qquad & A0 &= 0 \end{aligned} \tag{4-1}$$

The identities are obvious, recognizing that ordinary multiplication applies and that the variables are restricted to values of 1 and 0. The AND operation AB is often written A · B.

When a logic function is applied to words, it operates on their corresponding bits. For example, if word A is 0110 and word B is 1011, then A AND B is 0010. With words, A AND B is usually written as $A \wedge B$.

Suppose it is desired to clear all bits of the 8088 byte register AL except bits 7 and 6. This *mask operation* is accomplished by ANDing the byte of AL with binary 11000000. Somewhat similar to the mask operation is *selective clear*. This clears the bits of a word that correspond to the 1s of a second word. The operation $A \leftarrow A \wedge B'$ implements selective clear. For example, if A is 1101 and B is 0100, A becomes 1001 after the selective clear operation.

FIGURE 4-12. OR logic symbols

The OR Function

The logical *OR function* is defined as follows:

The output has its indicated 1 state if and only if there is at least one input with its indicated 1 state.

Figure 4-12 shows the standard symbols for the OR function. The qualifier indicates that one or more inputs must be 1 to give an output of 1. It is sometimes placed within the distinctive shape symbol. There may be more than two inputs. The TTL 7432 chip has a quadruple of two-input positive logic OR gates in a 14-pin DIP.

The symbols of Fig. 4-12 apply to positive, negative, and mixed logic systems. In the positive and negative logic systems, negation indicators can be present at any of the terminals, in which case the indicated 1 state of the function is the 0 state of the terminal. For the mixed logic system, polarity indicators may be added, as shown in Fig. 4-13.

FIGURE 4-13. Additional OR symbols for mixed logic.

The output Y of the rectangle in Fig. 4-13 is low if and only if input A is high or input B is low. For the distinctively shaped OR symbol, output Y is high if and only if A is low or B is high. The truth table for the OR gates in Figs. 4-12 and 4-13 is shown in Fig. 4-14. Note that the output is 1 if at least one input is 1.

Boolean addition is accomplished by using the OR function. The output is 0 only when all inputs are 0s. With three inputs, the output Y is expressed as A + B + C, often read as "Y is equivalent to A or B or C." Boolean addition is similar to ordinary addition except that 1 + 1 equals 1. The OR function is also known as *Inclusive-OR*, because an output of 1 for a two-input gate includes the case in which both inputs are 1. In contrast, the XOR function excludes this case.

A	B	Y
0	0	0
0	1	1
1	0	1
1	1	1

FIGURE 4-14. OR function truth table.

A B	A′	A′B	A + A′B	A + B
0 0	1	0	0	0
0 1	1	1	1	1
1 0	0	0	1	1
1 1	0	0	1	1

FIGURE 4-15. Truth table verification of a logic relation.

The following Boolean identities involving the OR, AND, and NOT functions are useful in design:

$$\begin{array}{ll} A + A = A & A + B + C = (A + B) + C \\ A + 0 = A & A + \overline{A}B = A + B \\ A + 1 = 1 & A + BC = (A + B)(A + C) \\ A + AB = A & A(B + C) = AB + AC \end{array} \qquad (4\text{-}2)$$

The expression A + AB equals A(1 + B) and 1 + B = 1; one of the identities is verified. By similar reasoning, the sum of AB and ABCD is AB. All relations that have been presented are easily verified by application of the rules of Boolean addition and multiplication.

An alternate procedure for verifying Boolean identities, which is rigorous and direct, is to form a truth table with columns developed for expressions that are supposedly equal. For example, consider Fig. 4-15. The two columns on the left contain all possible combinations of A and B. The column for A′ is found from A, and A′B is determined from the AND operation. Bits of the remaining columns are found by Boolean addition. Because columns A + A′B and A + B are identical, this logic expression of (4-2) is verified.

The OR operation on binary words A and B is written as $A \vee B$. If A is 1001 and B is 0100, then $A \vee B$ is 1101. Note the difference between the notation for AND and OR word operations:

$$A \text{ AND } B = A \wedge B \qquad A \text{ OR } B = A \vee B$$

The *selective set* operation sets the bits of register A that correspond to 1s of a second word, B. This is performed by the operation $A \leftarrow A \vee B$. Suppose it is desired to set bits 2 and 1 of register AL of the 8088 processor. Setting is accomplished by ORing the contents of AL with 6, which is binary 00000110.

It is sometimes necessary to insert 0s and 1s into selected cells of a register without affecting the other stored bits. This can be done by first masking, or clearing, the selected cells and then utilizing selective set. To illustrate, suppose we want to change the high nibble of AL to 0110 without affecting the low nibble. ANDing the byte of AL with 0FH masks the bits of the high nibble, and then ORing the result with 60H completes the *insert operation*.

Exclusive-OR

Another logical operation of considerable importance is the two-input *Exclusive-OR function*, or *XOR*. It is defined as follows:

The output has its indicated 1 state if and only if one and only one of the two inputs has its indicated 1 state.

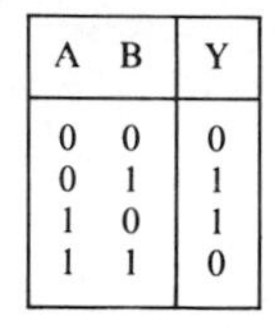

A	B	Y
0	0	0
0	1	1
1	0	1
1	1	0

$$Y = A \oplus B = AB' + A'B$$
$$= (A + B)(A' + B')$$

FIGURE 4-16. XOR truth table and logic expressions.

In other words, the output is 1 if and only if the inputs are 1 and 0. The truth table and several equivalent logic expressions for the two-input function are shown in Fig. 4-16. The TTL 74L86 chip contains four positive logic XOR gates.

The circled plus sign of the expression in Fig. 4-16 is conventional for the XOR operation. The two-input function is similar to that of OR, but when both inputs are 1, the XOR output is 0. The standard XOR logic symbols for two-input functions are shown in Fig. 4-17.

FIGURE 4-17. XOR logic symbols.

The qualifier of the rectangle, which may be placed in the distinctive shape symbol, indicates an output of 1 when one and only one input is 1. In positive and negative logic systems, negation indicators may be present, and with the mixed logic convention, any of the terminals of either gate of Fig. 4-17 may have polarity indicators.

XOR gates with more than two inputs are not defined by IEEE standards. However, the International Electrotechnical Commission has defined a three-input XOR function as one having an output of 1 if and only if one and only one of the inputs is 1. This is often called a *one-and-only-one function with three inputs*. Its symbol is that of Fig. 4-18, and polarity indicators may be added.

A common computer operation is *character comparison*. Characters stored in registers or memory can be compared using the XOR function. The operation $A \leftarrow A \oplus B$, with A and B denoting words, produces all 0s in A if and only if the words are equal. Suppose that two ASCII codes have been loaded into registers AL and AH of the 8088 processor. Execution of the assembly language instruction XOR AL,AH places the result of the XOR operation in AL and sets appropriate flags of the flag register. If the zero flag ZF is set, the two characters are the same. When sequences of characters that represent instructions, such as XOR, are entered from a keyboard, a program can compare them with the characters of mnemonic instructions stored in a table in memory, with a match determining the instruction. Execution of XOR AL,AL clears register AL.

When addition and subtraction of signed numbers give a result that exceeds the capacity of the accumulator, the overflow flag OF is set. This can be done by an XOR gate. Inputs are the carry into or borrow from the MSB and the carry from or the borrow into the MSB. If these inputs are unequal, the output of the gate is 1, which indicates an overflow. This bit feeds into flag OF.

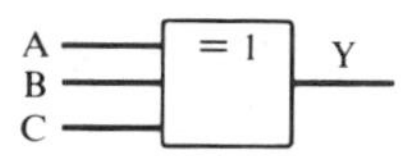

Y is 1 only if inputs
are 0, 0, and 1
in any order.

FIGURE 4-18. One-and-only-one function.

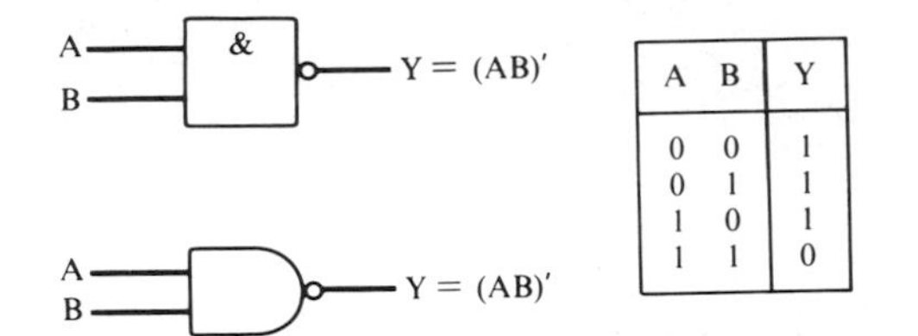

A	B	Y
0	0	1
0	1	1
1	0	1
1	1	0

FIGURE 4-19. NAND function symbols and truth table.

4-3

NAND, NOR, and XNOR

Here we examine several functions that are closely related to the AND, OR, and XOR functions of the preceding section. We begin with NAND.

The NAND Function

By connecting an inverter to the output of an AND element in either the positive or negative logic system, the result is a NOT AND, or *NAND, function*. In the mixed logic system this is still the AND function, with a polarity indicator at the output, and the logical truth table of the AND function remains valid. Accordingly, *the NAND function has no representation in the mixed logic system*. It is included within the general definition of AND.

In positive and negative logic systems, the symbols and truth table for a two-input NAND element are shown in Fig. 4-19. The negation indicator at the output negates the output of the AND symbol, changing the 1 state to 0 and the 0 state to 1. Directly from the symbol we deduce that the output is 0 if and only if inputs A and B are 1s. Output (AB)′ is the complement of the Boolean product AB. A NAND function with all input terminals connected to form one input is an inverter.

The NOR Function

As in the case of NAND, in the mixed logic system the *NOR function* has no representation, because an OR function with a polarity indicator at the output is still OR. However, in the positive and negative logic systems the NOR function is defined as NOT OR, which is obtained by inverting the output of an OR function. For two inputs A and B, the output is the complement of A + B, or (A + B)′. The standard NOR symbols are those of the OR element with a negation indicator

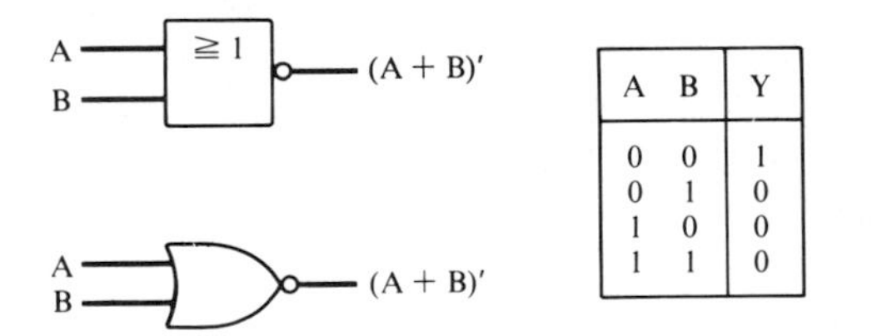

A	B	Y
0	0	1
0	1	0
1	0	0
1	1	0

FIGURE 4-20. NOR function symbols and truth table.

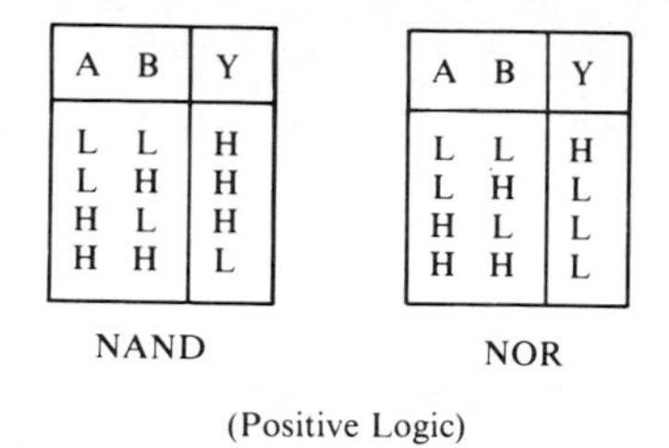

A	B	Y
L	L	H
L	H	H
H	L	H
H	H	L

NAND

A	B	Y
L	L	H
L	H	L
H	L	L
H	H	L

NOR

(Positive Logic)

FIGURE 4-21. Truth tables for positive logic NAND and NOR gates.

at the output, as shown in Fig. 4-20. The negator in conjunction with the OR symbol clearly reveals that the output is 0 if and only if A or B is 1, and this is shown in the truth table of the figure. A NOR element with all inputs connected together to form a single input is an inverter.

The TTL chip 7400 contains a quadruple of positive-logic, two-input NAND gates. Each of the four gates has its two inputs and one output connected to package pins, and the other two pins of the 14-pin DIP are for the 5 V supply. An eight-input NAND gate is provided by the 7430 chip, and there are others. A positive logic quadruple of two-input NOR gates is present on the 7402 IC. The truth tables in terms of voltage levels are given for two-input, positive logic NAND and NOR gates in Fig. 4-21. Although the levels do not depend on the choice of the logic convention, the function designation does.

Suppose the voltage levels of the NAND truth table of Fig. 4-21 are replaced with logical 1s and 0s of a negative logic system and those of the NOR truth table are replaced with 1s and 0s of a positive logic system. The results are identical. It follows that the positive logic NAND gates of the 7400 chip are NOR gates in a negative logic system. Conversely, a positive logic NOR gate is a negative logic NAND gate. In a similar manner, it can be shown that a positive logic AND gate is a negative logic OR gate, and a positive logic OR gate is a negative logic AND gate. Gate descriptions are usually given on data sheets in the positive logic convention.

The truth tables of Fig. 4-21 are implemented in the mixed logic system by the elements shown in Fig. 4-22. Although the one at the left side of the figure provides the NAND function in terms of voltage levels, its logical truth table in terms of 1s and 0s is that of AND. Even though the mixed logic function is implemented with a positive logic NAND gate, the circuit is regarded and referred to as an AND gate. The output is low (state 1) if and only if A and B are high (state 1).

The OR element of Fig. 4-22 has the voltage-level truth table of the positive logic NOR gate, but the mixed logic function is that of OR. The output is low (state 1) if and only if A or B is high (state 1). This is the OR function, not NOR.

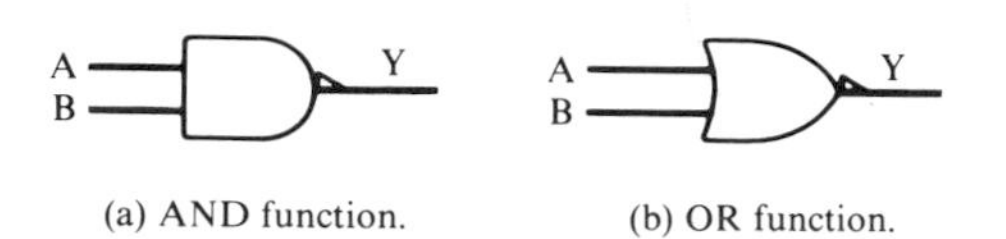

(a) AND function. (b) OR function.

FIGURE 4-22. Mixed logic AND and OR functions.

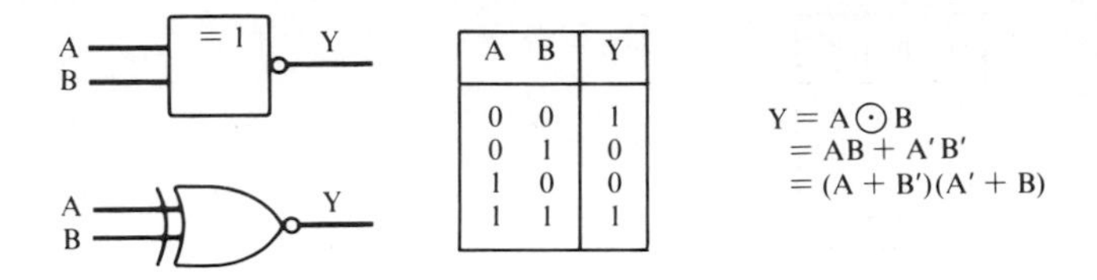

A	B	Y
0	0	1
0	1	0
1	0	0
1	1	1

FIGURE 4-23. XNOR symbols, truth table, and logic expressions.

Exclusive-NOR

The two-input *Exclusive-NOR (XNOR) function* in positive and negative logic systems is a combination of NOT and XOR. There is no representation of XNOR in the mixed logic convention, because the function is included within the defined XOR operation. Figure 4-23 shows the symbols and truth table.

For the two-input XNOR function, the output has the active indicated 0 state if one and only one of the inputs has the active indicated 1 state; otherwise the output has the inactive 1 state. A hardware logic gate designed to give the XNOR function in positive logic gives the XOR function in the negative logic convention, and one designed for positive logic XOR provides XNOR in negative logic. The XNOR function is often called the *equivalence function,* because the output is 1, or true, only when the two inputs are both true or both false, a condition referred to as *equivalence*.

For comparison, symbolism often used to denote the XOR and XNOR operations is given in (4-3):

$$Y = A \oplus B \qquad Y = A \odot B = (A \oplus B)' \tag{4-3}$$

TTL 74S135 provides a quadruple of XOR/XNOR gates. Each of the four gates has the configuration of Fig. 4-24. For positive logic the XOR function is implemented when control C is 0, and XNOR is implemented when C is 1. The state of C is reversed for negative logic.

The XNOR function can, of course, operate on words. To illustrate, suppose word A is binary 1101 and B is 0110. Then A ⊕ B is 1011, and its complement is 0100, which is A ⊙ B.

DeMorgan's Laws

By means of a truth table it is easy to verify that

$$(A + B)' = A'B' \tag{4-4}$$

Also, $(A' + B')' = (A')'(B')'$, which is (4-4) in terms of variables A' and B'. The right side is simply AB, because a variable is not changed by a double com-

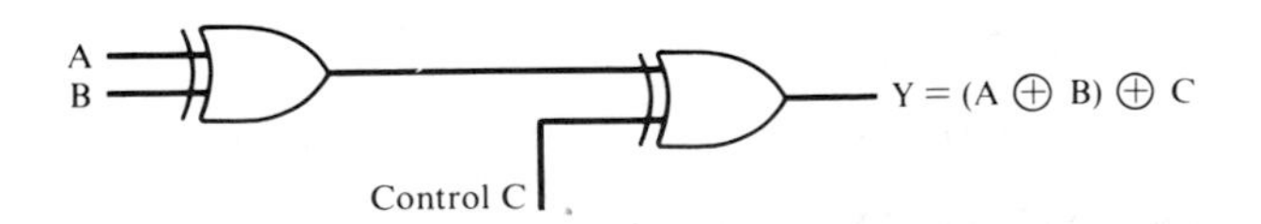

FIGURE 4-24. The XOR/XNOR function of TTL 74S135.

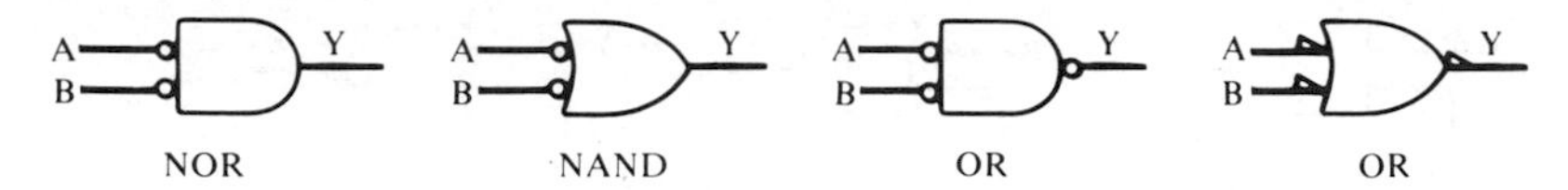

FIGURE 4-25. Logic functions for the example.

plement. With the two sides of the equality interchanged and complemented, the result is

$$(AB)' = A' + B' \tag{4-5}$$

Relations (4-4) and (4-5) are *DeMorgan's laws*. They are quite useful in simplifying logic expressions and logic diagrams.

The relation $(A + B + C)' = A'(B + C)'$ is (4-4), with variable B replaced with the new variable $B + C$, and $(B + C)'$ equals $B'C'$ by application of (4-4). The result is (4-4) extended to three variables. The procedure can be repeated again and again to include any number of variables. Replacement of B in (4-5) with the new variable BC leads to the extension of (4-5) to three variables, and this procedure can also be repeated as often as desired. Accordingly, DeMorgan's laws in terms of an unspecified number of variables are as follows:

$$(A + B + C + \cdots)' = A'B'C' \cdots \tag{4-6}$$
$$(ABC \cdots)' = A' + B' + C' + \cdots \tag{4-7}$$

EXAMPLE

Verify the indicated logic function for each of the symbols of Fig. 4-25.

SOLUTION

For the labeled NOR function, the inputs to the AND symbol are A' and B' because of the negation indicators. Thus, Y is $A'B'$, or $(A + B)'$ by DeMorgan's law, and this is the NOR function. The symbol clearly indicates that the output is 1 if and only if A and B are 0. The next one, labeled NAND, has an output 1 state if and only if A or B is 0; thus Y is $A' + B'$, which equals $(AB)'$.

The AND symbol with the three negators signifies that Y is 0 if and only if A and B are 0. Accordingly, Y is $(A'B')'$, or $A + B$. The output of the mixed logic function is low (state 1) if and only if A is low (state 1) or B is low (state 1). In terms of voltage levels, the function is AND. However, it is classified as OR because it gives the OR function in terms of 1s and 0s. ■

Gating Digital Signals

The word *gate*, when used for a hardware circuit, refers to the gating of digital signals. *To gate* a signal means to control the passage of the signal through a digital circuit. The verb usually refers to the control of the train of voltage pulses from a pulse generator, or clock. Signals can be gated by AND, NAND, OR, NOR, and other types of functions. Figure 4-26 shows three gating elements. For the AND function, output X equals input A when control G is 1, and X is 0 when G is 0; this is true regardless of the state of A. Operation of the other two gates is indicated by the truth tables.

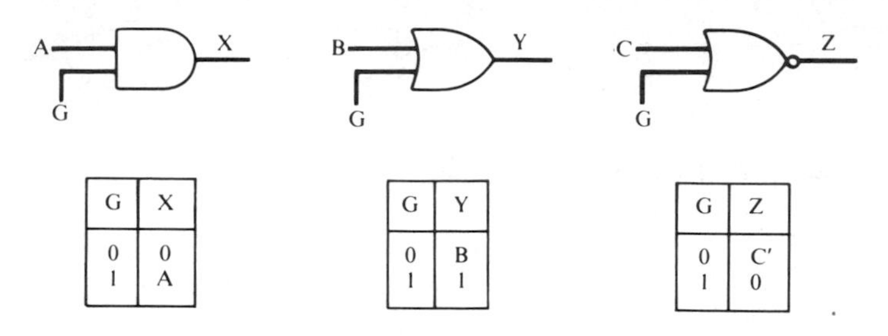

G	X
0	0
1	A

G	Y
0	B
1	1

G	Z
0	C′
1	0

FIGURE 4-26. Gating circuits with truth tables.

Somewhat similar to the verb *gate* are *strobe* and *enable*. *To strobe* is to activate a specific operation, and *to enable* is to activate a circuit by removing a disable, or suppression, signal. *Strobe* and *enable* are often used interchangeably.

The gates that control the passage of digital signals act very much like electrical switches. Accordingly, digital circuits are sometimes called *switching circuits*. Binary signals control the electronic switches. Gates are also called *logic circuits* because they perform logical operations on their input variables. Application of Boolean algebra to the analysis and design of switching circuits is known as *switching theory*.

Integration Levels

Four integration levels based on chip complexity are defined as follows:

SSI — A small-scale-integrated circuit having 1 to 12 gates and memory cells on the chip.

MSI — A medium-scale-integrated device with 13 to 100 gates and cells.

LSI — A large-scale-integrated device with more than 100 gates and cells.

VLSI — A very-large-scale-integrated device that has thousands of gates and cells. There is no standard definition. Usually a VLSI circuit is regarded as an LSI chip with 10,000 or more gates.

For the CMOS, TTL, and ECL logic systems, the gates that have been described in this chapter are available in SSI IC packages. The number of package pins limits the number of elements per IC. For example, a 14-pin package allows 4 two-input NAND gates with two pins available for VCC and GND, or 3 three-input NAND gates, or 2 four-input gates, or 1 eight-input gate, with each of the gates having a single output.

Complex logical functions are implemented by various arrangements of the basic AND and OR functions, along with inverters. These include, of course, the NAND and NOR functions, and they can be combined to give XOR and XNOR. The types of gates used often depend on the logic family employed and the choice of the logic convention. ICs with interconnected gates are common, and a number of them are examined in later chapters. Although not obvious at this point, it is possible to implement all circuits of a large data processing system, including those of ALUs, clock pulse generators, timing and control units, and memories, using only two-input NAND gates, or NOR gates. Accordingly, these are called *universal gates*.

System design usually begins with an algorithm at the highest level, prepared without regard to specific circuitry. Pseudo code describing system behavior is useful. The process then proceeds to lower levels where details are supplied. Actual implementation with hardware or software is considered only at the lowest level. This is the top-down structured design procedure. Thus, complex systems are divided into subsystems, and logic implementation with hardware is made in the final stages of the design with the aid of logic diagrams, circuit diagrams, and various design tools.

Systems are often designed with *random logic,* using commercially available MSI and LSI circuits whenever possible. The basic logic gates are interconnected to implement combinational logic and to build memory cells that form registers for storage of encoded digital words. The early microprocessors contained much random logic.

However, as electronics moves into VLSI systems with very large circuit densities, minimization of complexity becomes essential. This is accomplished by careful selection of logic gates and judicious arrangements of the elements on the silicon surface so that interconnections between components and the layout of control, data, and power lines are simplified as much as possible. Thus, topology becomes a major consideration. Emphasis is given to circuit design leading to a geometric layout that is repetitive and organized to give a regular interconnection topology and minimum surface area. Still, it is essential that the electronics engineer know how to design random logic circuits with the minimum number of chips and logic gates, employing Boolean algebra, mapping techniques, and other aids. Such design is investigated in Appendix C.

4-4 LOGICAL OPERATIONS IN ASSEMBLY LANGUAGE

Now that the logic functions have been studied, it is appropriate to present logical operations in assembly language. For illustration, a sequence of instructions in a keyboard subroutine of the ROM BIOS is selected. Also considered here briefly are closely related shift, rotate, and conditional jump instructions.

Logical Instructions

The logical instructions are often understood to include those that shift and rotate bytes and words, as well as those that implement Boolean functions. In general, the 12 instructions of this group manipulate bits. They are listed in Fig. 4-27, subdivided into *logicals, shifts,* and *rotates*. Included are brief comments and examples, with affected status bits indicated at the right side of the figure.

NOT. NOT simply complements the bits of the destination operand; it has no effect on the flags. For example, suppose CX stores 89D. Execution of NOT CX replaces 89 with −90, which is the complement of 89, and the flags are unchanged.

AND, OR, XOR. The logical functions AND, OR, and XOR operate on corresponding bits of bytes or words, always clearing OF and CF and leaving AF undefined. Flag ZF is set if the result is 0, flag PF is set if the low byte has even parity, and flag SF is set if the high-order bit is 1.

INSTRUCTION	COMMENTS/EXAMPLES	
	Logicals	
NOT destination	Logical NOT; NOT WORD_A	
AND dest, source	AND AX, 101000B	OSZPC
OR dest, source	OR CMD_WORD[BX], AX	OSZPC
XOR dest, source	XOR OLD_CODE, 0D0ABH	OSZPC
TEST dest, source	AND (no operand change)	OSZPC
	Shifts	
SHL/SAL dest, count	Shift left; SHL AL, 1	OC
SHR dest, count	Shift logical right	OC
SAR dest, count	Shift arithmetic right	OSZPC
	Rotates	
ROL dest, count	Rotate left; ROL SI, CL	OC
ROR dest, count	Rotate right; ROR BETA, 1	OC
RCL dest, count	Rotate left through carry	OC
RCR dest, count	Rotate right through carry	OC

FIGURE 4-27. Logical instructions.

The 8088 microprocessor does not have an XNOR instruction. However, the XNOR operation on words of registers is done with the following sequence:

```
XOR AX, BX
NOT AX
```

The first of these implements AX ← AX ⊕ BX, and instruction NOT complements the bits of AX. Both registers store 16 bits. NAND and NOR operations are obtained similarly.

TEST. The TEST instruction is precisely the same as AND insofar as the flags are concerned, but neither operand is changed. With it, bits of an operand can be tested, and status flags give the results. TEST is frequently used to control program jumps based on the bits of a word. For example, consider the sequence

```
TEST AH, 0C0H
JNZ  LABEL_C2
```

JNZ (jump-if-not-zero) specifies a jump only if flag ZF is not set. The TEST instruction sets ZF provided the two high-order bits of AH are 0s, but otherwise ZF is cleared. Accordingly, if either bit 7 or bit 6 of AH is 1, flag ZF is not set, and the conditional jump is taken. For a byte of 00XXXXXX in AH, the jump is ignored, and the program continues to the next instruction in sequence. TEST can be replaced with AND provided the loss of the byte in AH is unimportant.

Example: Keyboard

In order to illustrate the applications of logical instructions in a program, a portion of a keyboard subroutine in the ROM BIOS is examined. In particular, we will consider the processing of the four toggle keys (INS, CAPS, NUM, and SCROLL). Recall that there are eight shift keys, including the toggle keys, which were defined in Section 1-4 and listed in Fig. 1-24.

Address
0040:0017

KB_FLAG DB ?

INS state	CAPS state	NUM state	SCROLL state	ALT shift	CTL shift	LEFT shift	RIGHT shift

Equates:

```
INS_STATE      EQU  80H  ; For toggle of Insert state
CAPS_STATE     EQU  40H  ; For toggle of CapsLock state
NUM_STATE      EQU  20H  ; For toggle of NumLock state
SCROLL_STATE   EQU  10H  ; For toggle of ScrollLock state
ALT_SHIFT      EQU  08H  ; Alt key depressed indicator
CTL_SHIFT      EQU  04H  ; Ctrl key depressed indicator
LEFT_SHIFT     EQU  02H  ; Left shift key depressed
RIGHT_SHIFT    EQU  01H  ; Right shift key depressed
```

FIGURE 4-28. Memory variable KB_FLAG and related equates.

Shift Key Flags. Whenever a system reset takes place, the initialization routine in the ROM BIOS reserves two byte locations in RAM for storage of flag bits associated with the shift keys. The locations are assigned variable names KB_FLAG and KB_FLAG_1. Figure 4-28 shows the address and the designated bits of KB_FLAG.

A bit with *state* included in its name, as shown in Fig. 4-28, is used *to indicate the state of a toggle key* (INS, CAPS, NUM, SCROLL). For example, bit 7 of KB_FLAG is identified as the INS *state*. If this bit is 0, the insert mode is normal (inactive), but if it is 1, the mode is active. Of course, the particular program determines the significance, if any, of this mode. With *shift* in the name, *the bit indicates whether or not the indicated key is depressed*. Here, a logical 1 means that the last scan code received from the key was a make code. When a break code is received from the key, the program clears the bit. Each of the eight shift keys has an associated bit in KB_FLAG, with four of these being state bits of toggle keys and the other four being shift bits of nontoggle keys.

Also shown in Fig. 4-28 are eight EQU directives that relate each bit name to a value that can be used to set, clear, or complement (toggle) the bit. For example, the INS state is set by execution of

```
OR KB_FLAG, INS_STATE
```

The bit is cleared by execution of

```
MOV AL, INS_STATE
NOT AL
AND KB_FLAG, AL
```

Finally, it is toggled by

```
XOR KB_FLAG, INS_STATE
```

Note that the preceding instructions have no effect on any bits other than number 7.

Address
0040:0018

KB_FLAG_1 DB ?

INS shift	CAPS shift	NUM shift	SCROLL shift	HOLD state	X	X	X

Equates:

```
INS_SHIFT      EQU   80H
CAPS_SHIFT     EQU   40H
NUM_SHIFT      EQU   20H
SCROLL_SHIFT   EQU   10H
HOLD_STATE     EQU   08H
```

FIGURE 4-29. Location KB_FLAG_1 and related equates.

Figure 4-29 shows the address, designated bits, and related EQU statements for variable KB_FLAG_1. The high nibble provides the shift bits for the four toggle keys. *A set shift bit indicates that the toggle key is depressed.* Bit 3 is the HOLD state bit, which is activated by the key combination Ctrl-NumLock. It is cleared when any nonshift key is depressed. *Most programs use this key combination to allow the keyboard operator to introduce a pause in program execution.* When a two-key combination such as this one is given, it implies that the first key is depressed initially and *held down until the second one is pushed.* The three low-order bits of KB_FLAG_1 are not used. For the four toggle keys, there are flag bits for both the state mode and the shift mode.

Scan Code and Mask Search. The subroutine sequence for processing a toggle key has four major parts. In the first part, the processor determines the scan code, which is specified here to be that of a toggle shift key. We assume that the scan code has been read from an input port of the PPI chip and put in both AL and AH.

In the ROM BIOS are the scan code table and the *mask value* table shown in Fig. 4-30. Each has eight entries, one per shift key. The offset of any mask value from the base of table T2 equals the similar offset of its corresponding scan code in table T1. At the end of table T1 is an EQU statement equating T1_NUMBER to $-T1. The assembler recognizes the dollar sign followed by a hyphen and the table label as denoting *the number of elements of the table,* and hence, it assigns decimal 8 to name T1_NUMBER. Both tables are in the code segment of the routine of the ROM BIOS.

The first part of the subroutine sequence is shown in Fig. 4-31.

The AND instruction of the routine of Fig. 4-31 turns off the break bit (if there is one) so that the scan code of a possible shift key will correspond to its value in

```
T1  LABEL  BYTE                  ; scan code table
   DB   82, 58, 69, 70   ;INS, CAPS, NUM, SCROLL scan codes
   DB   56, 29, 42, 54   ;ALT, CTL, LEFT, RIGHT scan codes
T1_NUMBER  EQU  $-T1

T2  LABEL  BYTE                  ; table of mask values
   DB  80H, 40H, 20H, 10H  ;INS, CAPS, NUM, SCROLL masks
   DB  08H, 04H, 02H, 01H  ;ALT, CTL, LEFT, RIGHT masks
```

FIGURE 4-30. Tables of scan codes and mask values for the shift keys.

```
    CLD                         ;direction up
    AND AL, 7FH                 ;turn off the break bit
    PUSH CS
    POP ES                      ;code segment address to ES
    MOV DI, OFFSET T1           ;point DI to table T1 base
    MOV CX, T1_NUMBER           ;byte count 8 of table T1 to CX
    REPNE SCASB                 ;search T1 for scan code match
    MOV AL, AH                  ;recover scan code in AL
    JZ  A1                      ;jump if a match is found
    JMP X1                      ;exit to label X1 for no match
A1: SUB DI, OFFSET T1 + 1       ;put index to the match in DI
    MOV AH, CS:T2[DI]           ;move mask value to AH
```

Figure 4-31. Scan code and mask search routine.

table T1. Then the segment address of table T1 is put in ES, and the T1 offset is placed in DI. This procedure establishes ES:DI as the address of T1, as required for the destination of the scan string instruction SCASB. Register CX is assigned the count of 8.

Instruction REPNE SCASB searches the table for a match. Repetition occurs as long as flag ZF is clear and count CX is not 0. If the scan code is not that of a shift key, ZF is 0 after each scan, and there are eight scans before exit to the next instruction. However, if a match is found, ZF is set, and DI points to the offset of the next element of the table. Because the scan code of AL was possibly changed by the initial AND instruction, the true value is recovered so that the register has the proper code when the jump occurs. The program branches to A1 if a match is found, as assumed here. With no match, there is a jump to exit X1 for further scan code searching.

At label A1, instruction SUB subtracts OFFSET T1 + 1 from the contents of DI and stores the result in DI. Before execution, DI stores the table offset address plus the increments to DI during the scan. These increments are greater by 1 than the offset from the table base to the match, because DI was incremented when the match was found. Thus, the subtraction *gives the index value from the base of the table*. For example, if the key is NUM, the value in DI is 3. The last instruction uses indexed addressing to move the mask value from table T2 to AH. A segment override prefix is used because T2 is in segment CS rather than the default segment DS. *At the end of this part of the routine, AL has the scan code and AH has the mask value.*

Toggle Key Determination. In part 2 of the program, the code of AL is found to be one of the eight shift keys. We have assumed that the code is a make code of a toggle key. Figure 4-32 shows the sequence that follows part 1, and it determines if our assumption is correct.

```
    TEST AL, 80H           ;is this a break scan code?
    JNZ X2                 ;if so, go to another routine
    CMP AH, 10H            ;is the key a toggle key?
    JNC A2                 ;if so, go to A2
    OR  KB_FLAG, AH        ;not toggle key, set shift bit
    JMP RETURN             ;go to subroutine return
A2: (continue)
```

FIGURE 4-32. Toggle key determination routine.

The TEST instruction ANDs the contents of AL with 80H without changing AL. The flags are affected, however. If flag ZF is clear after the test, the key code is a break scan code indicating that the key was released. For the assumed make scan code, the jump to X2 is not taken, and the compare instruction CMP is executed next.

CMP is the same as SUB in that the source operand is subtracted from the destination operand. However, whereas SUB replaces the destination operand with the result, CMP has no effect on either. *Only the processor flags are affected.* The four mask codes of the toggle keys are 80H, 40H, 20H, and 10H, whereas those of the nontoggle keys are 8, 4, 2, and 1. After CMP, the carry flag CF is set only if the key scan code is less than 10H. Thus, for an assumed toggle key, CF is clear, and the conditional jump *JNC (jump-on-no-carry)* is taken to A2.

For a nontoggle shift key, no jump occurs and instruction OR is executed. With the mask code in AH, this operation sets the shift flag bit of the key without affecting other bits of the location. Then the unconditional jump to RETURN restores the contents of those registers saved on the stack and gives a return from the subroutine to the main program. We now proceed to part 3 of the routine, starting at A2. Register AL contains the make scan code of one of the toggle keys INS, CAPS, NUM, and SCROLL, and the corresponding mask code is in AH.

A Few Tests. Figure 4-33 shows the routine of part 3, which looks only for the scan code of the INS key. If it is not found, the program proceeds to part 4. Assuming that the scan code in AL is found to be that of the INS key, location KB_FLAG is examined to determine if the ALT key is depressed, giving an active ALT shift. The combination ALT-INS sends a 0 to memory location ALT_INPUT. For an inactive ALT shift, tests are made to determine if the INS key should generate ASCII code 30H for character "0". This determination requires examining location KB_FLAG for the status of the NUM state and the LEFT and RIGHT shifts.

The first test of part 3 of the routine determines if the Ctrl key is depressed. If so, the program jumps out of the portion of the subroutine examined here to a routine at label X3 that processes keyboard codes with CTL_SHIFT active. We assume that flag CTL_SHIFT of location KB_FLAG is inactive.

```
A2: TEST KB_FLAG, CTL_SHIFT   ;AND KB_FLAG with 04H
    JNZ X3                    ;jump if CTL_SHIFT active
    CMP AL, 52H               ;INS key scan code
    JNZ A5                    ;jump if not INS key
    TEST KB_FLAG, ALT_SHIFT   ;AND KB_FLAG with 08H
    JZ A3                     ;jump A3, no ALT_SHIFT
    JMP X4                    ;jump X4 on ALT_SHIFT
A3: TEST KB_FLAG, NUM_STATE   ;AND KB_FLAG with 20H
    JNZ A4
    TEST KB_FLAG, LEFT SHIFT + RIGHT_SHIFT
    JZ  A5
    JMP X5
A4: TEST KB_FLAG, LEFT_SHIFT + RIGHT_SHIFT
    JZ  X5
A5: (continue)
```

FIGURE 4-33. Tests for the INS key.

Next, the scan code of AL is compared with that of the INS key. If the toggle-key scan code is that of INS (52H), it is necessary to determine whether or not the key represents the insert mode or *simply character "0" with ASCII code 30H*. The ASCII code applies if the NUM state is inactive with an active left or right shift, or if the NUM state is active with inactive left and right shifts. The tests are indicated in the flowchart of Fig. 4-34, with targets of jumps labeled. A jump to X4 indicates ALT-INS and a jump to X5 indicates the INS key in the mode that puts ASCII 30H in the current low byte location of KB_BUFFER, with the scan code in the high byte position. An exit at A5 signifies only that none of the exit conditions that provided branches to X3, X4, or X5 were valid.

Set Flag Bits. The final part of the routine considered here starts at label A5. If the make scan code of the toggle key is not a repeat code, the routine sets the appropriate shift bit of KB_FLAG_1 and toggles the corresponding state bit of KB_FLAG. In addition, if the key is INS, word 5200H is sent to KB_BUFFER.

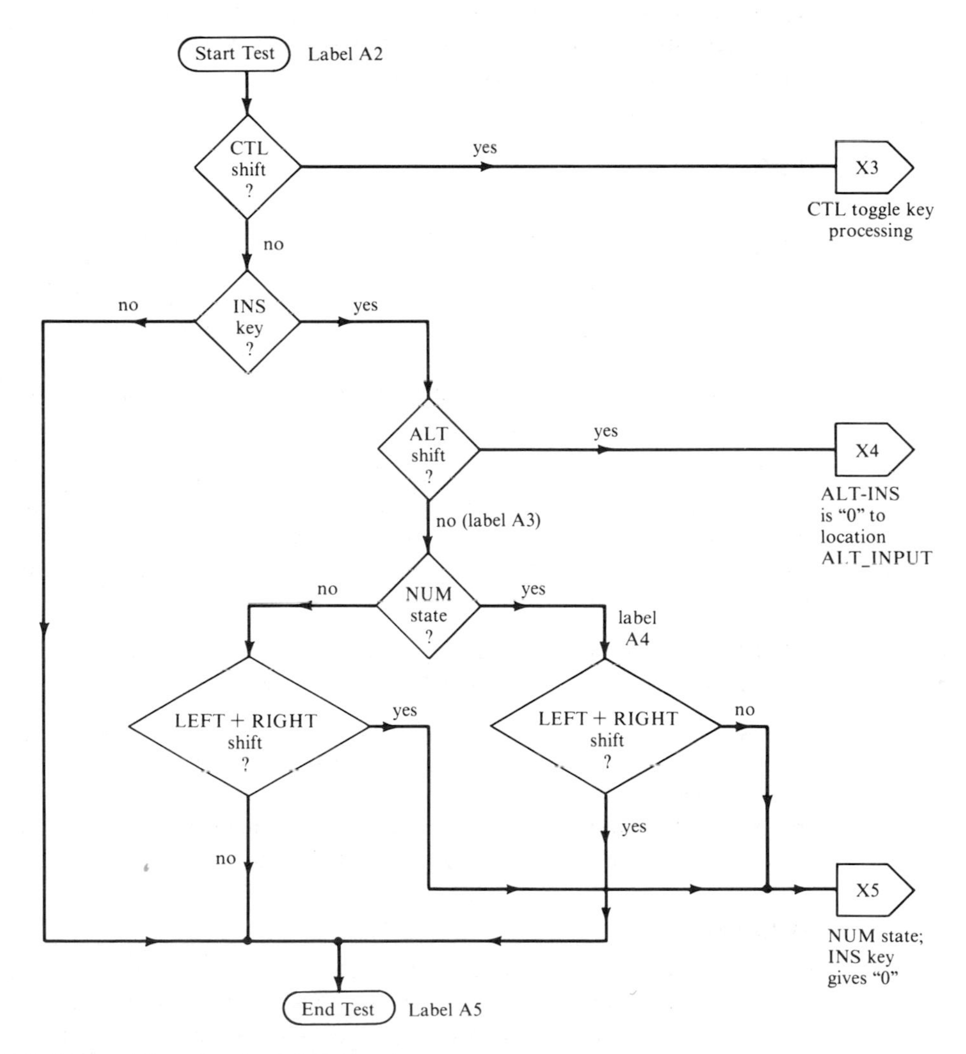

FIGURE 4-34. Flowchart of INS key tests.

```
A5: TEST AH, KB_FLAG_1   ;is this a repeat scan code?
    JNZ  RETURN          ;if so, take no action
    OR   KB_FLAG_1, AH   ;set shift bit, key is depressed
    XOR  KB_FLAG, AH     ;complement the state bit
    CMP  AL, 52H         ;is the toggle key the INS key?
    JNE  RETURN          ;if not, return from subroutine
    MOV  AX, 5200H       ;INS scan code to AH, 0 to AL
    JMP  OUTPUT          ;put 5200H in KB_BUFFER
```

FIGURE 4-35. Flag-bit routine.

Register AH has the mask value and AL has the scan code. The instruction sequence is shown in Fig. 4-35.

The keyboard flags are updated. For toggle keys CAPS, NUM, and SCROLL, the program terminates, and nothing is put into KB_BUFFER. However, for key INS, its scan code (52H) and byte 00H are sent to the buffer, where the main program will examine the word and take appropriate action. The action taken depends entirely on the software of the particular program being run, and in some cases there may be no action. Byte 00H is the low byte of the output buffer whenever a key has no standard ASCII code.

Shifts

For all three shift instructions of Fig. 4-27, the count can be specified either as the constant 1 or as the byte of CL. Using CL allows a shift constant from 0 through 255. Following a shift operation, flag AF is undefined, flags PF, SF, and ZF are updated, and CF always contains the last bit that was shifted out of either the left or the right side. In a 1-bit shift, flag OF is set if the sign bit changes, but in a multibit shift, OF is undefined. The operand can be a byte or a word, and the number of bit shifts equals the count.

SHL/SAL. *SHL* denotes *shift-logical-left,* and the equivalent form *SAL* denotes *shift-arithmetic-left.* In 2s complement arithmetic they are identical. Either mnemonic can be used, and both have the same object code. Added bits at the right side are 0s. Each 1-bit left shift multiplies the operand by 2 unless a significant bit is shifted out at the left side.

SHR. *SHR* is *shift-logical-right* which inserts a 0 into the left side for each bit shift that occurs. The last bit shifted out at the right side is stored as CF.

SAR. *Shift-arithmetic-right SAR* preserves the sign by shifting in bits at the left equal to the original sign bit. Each bit shift divides the operand by 2, with the fractional part lost. For example, a 1-bit right shift of 7 gives 3, and a 1-bit arithmetic left shift of -7 yields -4. For positive numbers, the logical and arithmetic right shifts are identical. Flag AF is undefined following SAR.

EXAMPLE 1

Variable Y is the 16-bit signed number FDA7H, or -601, and CL stores 04H. (a) If Y is moved to AX and the instruction SAR AX,CL is executed, determine the new hexadecimal and decimal contents of AX. Compare the result with Y/16. (b) Repeat with SHL replacing SAR, comparing the result with 16 * Y. (c) Repeat part (a) using SHR, but compare with the unsigned value of Y.

SOLUTION

(a) After the 4-bit arithmetic right shift, AX has FFDAH, or −38. Each 1-bit shift implements division by 2, with the fraction truncated. Variable Y divided by 16 is −37.5625. (b) The 4-bit left shift gives DA70H in AX. This is −9616, or exactly 16 * Y. (c) The logical right shift gives 0FDAH, or +4058. This is the unsigned value of Y/16 with the fraction truncated. ■

Rotates

As for shifts, the source field of the rotate instructions must be either 1 or CL. In multibit shift operations, bits shifted out are lost. This is not so with the rotate instructions, *which circle the bits shifted out of one end back into the other end.* The carry flag can be excluded from the loop, or it can be included. In each case, *the bit of CF is always equal to the value of the last bit shifted out.* For the two instructions that use CF as part of the loop, the flag is treated as an extension of the operand of the destination. The rotate instructions affect only flags CF and OF, but for multibit shifts, OF is undefined.

The two rotate instructions that exclude CF from the loop are *rotate-left ROL* and *rotate-right ROR*. Those including CF are *rotate-through-carry-left RCL* and *rotate-through-carry-right RCR*. To illustrate, with CF clear, byte 10111110 in AL, and byte 00000010 in CL, execution of ROR AL,CL gives a 2-bit right shift. The new byte of AL is 10101111, and CF is set because the last bit shifted out was 1. If ROR in the instruction had been RCR, the new word of AL would have been 00101111, with CF set as before.

EXAMPLE 2

Memory variable BETA is 6AD4H and flag CF is 1. Determine BETA and CF after execution of ROL BETA,1. Repeat using ROR, RCL, and RCR, with BETA having its initial value for each rotate.

SOLUTION

For ROL the result is D5A8H, and for ROR it is 356AH. Although flag CF is not included in the loop, it equals the bit shifted out, which is 0 for both cases. Operation RCL gives D5A9H, and RCR gives B56AH. Flag CF is included in the loop, and CF is again 0 for both cases. ■

Conditional Transfers

All conditional jumps are SHORT, which means that the target must be within the range of an 8-bit signed displacement from *the value of IP after the jump instruction has been fetched.* Each is a 2-byte instruction that is appropriate for position-independent routines. Suppose a conditional jump instruction is located at an IP offset of 1000D. After the fetch, IP is 1002. An 8-bit signed displacement has a decimal range from −128 through +127. Therefore, the target must be within the range from 874D through 1129D. Otherwise the conditional instruction is not allowed. Note that JZ 0 passes control to the next instruction in sequence regardless of the state of the flag.

Mnemonic	Jump If Condition Is True
JC/JB/JNAE	CF = 1
JNC/JAE/JNB	CF = 0
JZ/JE	ZF = 1
JNZ/JNE	ZF = 0
JP/JPE	PF = 1
JNP/JPO	PF = 0
JS	SF = 1
JNS	SF = 0
JO	OF = 1
JNO	OF = 0
JBE/JNA	(CF or ZF) = 1
JNBE/JA	(CF or ZF) = 0
JL/JNGE	(SF xor OF) = 1
JNL/JGE	(SF xor OF) = 0
JLE/JNG	[(SF xor OF) or ZF] = 1
JNLE/JG	[(SF xor OF) or ZF] = 0

FIGURE 4-36. Conditional jump instructions.

There are 16 conditional transfer instructions listed in Fig. 4-36. Many have two or three equivalent forms, and it makes no difference to the assembler which is chosen. For each mnemonic of the table, the jump is taken if and only if the stated condition is true. The letters A and B in the mnemonics, denoting *above* and and *below,* refer to the relationship of two *unsigned* values, whereas G and L, for *greater* and *less* refer to the relationship between *signed* values. Execution of a conditional jump uses 16 clocks if the jump is made and 4 clocks if there is no jump, assuming the instruction has been prefetched. The last six of the listed jumps are conditioned on the result obtained from a logical expression.

4-5

LOGICAL OPERATIONS IN BASIC

Just as there are logical operations in assembly language, there are similar ones in BASIC. We examine here the Boolean operators and some examples of their use in various BASIC statements. Because the operators are often included in the expressions of IF and WHILE statements, along with arithmetic and relational operators, these are investigated. Also, two programs that sort data are presented.

BASIC Logical Operators

There are six logical, or Boolean, operators available in BASIC. From left to right in order of precedence, they are

```
NOT     AND     OR     XOR     IMP     EQV
```

Operands of logical operations, such as variable A of NOT A, are restricted to *signed 16-bit integers,* with a range from −32,768 to +32,767. Numbers are understood to be in 2s complement form. Although the operand may be a single or double precision variable, the operator converts it to an integer, and if the value is not in the required range, an "Overflow" error message is generated. *The result returned by the operation is an integer*. Each of the six operators is now examined in order of precedence.

```
10 A1 = 0: A2 = -1: A3 = 8: A4 = - 8192
20 B1% = NOT A1: B2% = NOT A2: B3% = NOT A3: B4% = NOT A4
30 PRINT B1%, B2%, B3%, B4%
40 V3 = VARPTR (B3%)
50 PRINT V3, HEX$ (PEEK(V3 + 1)); HEX$ (PEEK(V3))
```

FIGURE 4-37. Routine with NOT operations.

NOT. Let A be a BASIC variable. It can be declared as integer, single precision, or double precision, but we assume here that it is within the range of a 16-bit signed number. The statement B% = NOT A assigns to the declared integer B% a value equal to the logical complement of word A. Variable A is converted to an integer, and each bit of A is complemented, just as in the corresponding assembly language NOT instruction.

Integer B% is stored in two byte locations, with the low-order byte at the lower address. If B% is replaced with the single precision variable B, it is stored in BASIC's floating point format in four byte locations, even though NOT A represents a 16-bit integer. In an assignment statement the precision is that of the target.

We know that $-A$ equals the 2s complement of A, or $A' + 1$, and A' is NOT A. Accordingly, we have

```
TWOSCOMP.A = -A = (NOT A) + 1     (4-8)
NOT A = -A - 1                    (4-9)
```

The program of Fig. 4-37 illustrates the concepts discussed.

The BASIC variable pointer function VARPTR of line 40 *returns the address in memory of the stored variable,* which may be a number, a string, or an array element. The address is the 16-bit offset into BASIC's data segment of the first byte of data, and function VARPTR is not affected by the DEF SEG statement if one is used.

When the program was run, the results observed on the display were

```
-1        0        -9    8191
 2277    FFF7
```

Decimal 2277 of the display is the offset of the low-order byte of integer B3% in segment DS, and the stored word is is FFF7H, or -9. Note that the statement of line 50 prints the high byte first. The reason for declaring B3% as an integer was to allow us to examine the hexadecimal value of the 2-byte result rather than the 4-byte floating point format. Of course, memory is also conserved, but this is not important for the short program here. Variables A1, A2, A3, and A4 are single precision by default.

AND, OR, and XOR. The logical operators AND, OR, and XOR are essentially the same as the corresponding assembly language instructions. To illustrate, execution of the instructions of Fig. 4-38 gives the values shown at the right.

Each decimal operand is stored in 2s complement form in a 16-bit register, the operation is performed, and the result is sent to the display. For example, -5 AND 6 is FFFBH AND 0006H, and this evaluates to 0002H. Normally, values are shown in decimal, but the last statement of Fig. 4-38 specifies the hexadecimal character string corresponding to the result, which is decimal -2.

Statement	Display
PRINT 5 AND 6	4
PRINT 5 OR 6	7
PRINT 5 XOR 6	3
PRINT -5 AND 6	2
PRINT -5 OR 6	-1
PRINT HEX$ (-5 XOR 6)	FFFD

FIGURE 4-38. Logical operations and results.

BASIC has no NAND and NOR operators. However, these can be implemented by combining NOT with AND or OR. For example, NOT(− 5 AND 6) equals − 3, which is the same value as that of the disallowed expression − 5 NAND 6.

IMP. The *implication operator IMP* is the following combination of NOT and OR:

$$\text{A IMP B} = (\text{NOT A}) \text{ OR B} \tag{4-10}$$

The parentheses around NOT A are not required, because NOT has precedence over OR. However, it is good practice to use parentheses freely to avoid possible confusion.

EQV. In Section 4-3 it was pointed out that XNOR is often called *equivalence. The BASIC operator for Exclusive-NOR is EQV.* We saw that 5 XOR 6 is 3. With identical numbers, the EQV function is

$$\text{5 EQV 6} = \text{NOT (5 XOR 6)} = \text{NOT 3} = -4 \tag{4-11}$$

This is the same value as that of the disallowed relation 5 XNOR 6.

Operators in Expressions

The arithmetic, relational, and logical operators are commonly used in expressions that must be interpreted as either true or false. *If a relational expression is true, the BASIC interpreter always returns a value of − 1 (FFFFH). If it is false, the returned value is 0.* The examples of Fig. 4-39 illustrate this condition.

It has been shown that logical operation NOT (− 1) is 0, and NOT 0 is − 1. Thus, the NOT operator converts a true value to false and a false value to true. For example, the operation NOT 5 = 5 returns 0, and NOT 5 = 6 is − 1, or true. Logical operations often return values other than − 1 and 0. *BASIC regards any*

Statement	Display	True/False
PRINT 5 * 2 = 10	-1	true
PRINT 5 = 6	0	false
PRINT 5 ^ 3 >= 125	-1	true
PRINT 5 / 2 < 2	0	false

FIGURE 4-39. True/false expressions.

Statement	Display
`IF NOT (-5) THEN PRINT 5 < 5 ELSE PRINT 5 = 6`	0
`IF 4 THEN PRINT 5 = 6 ELSE PRINT 5 = 5`	0
`IF 0 THEN PRINT 5 = 6 ELSE PRINT 5 = 5`	-1
`IF 5 AND 10 THEN PRINT 5 AND 10 ELSE PRINT 8`	8
`IF 5 OR 6 THEN PRINT 5 OR 10 ELSE PRINT 8`	-1
`IF NOT "A C" > "AB" THEN PRINT 2 ELSE PRINT 3`	2

FIGURE 4-40. Some IF-THEN-ELSE statements.

non-0 result as true. Only a value of 0 is false. Consider the IF-THEN-ELSE statements of Fig. 4-40.

The NOT (-5) of the first statement of Fig. 4-40 is 4. Although this is true (-1), value 0 is printed because $5 < 5$ is false. In the second row the "expression" is simply 4. As in the first statement, this is true, but the display is 0 corresponding to the false expression $5 = 6$. The next one has 0 after IF, which is false. Accordingly, the display is -1 because of the true statement following ELSE. Expression 5 AND 10 of the fourth row equals 0, which is false.

In the fifth row, 5 OR 6 is 7, which is true. Therefore, 5 OR 10 is evaluated, giving 15, and -1 is printed. The last IF statement compares two string constants. The inequality is invalid, because the ASCII code 42H for B is greater than code 20H for the space. Thus, the result of the inequality is 0 and NOT 0 is -1. PRINT 2 is executed.

The BASIC statements of Fig. 4-41 further illustrate the use of logical operators. Suppose HE is 65 and SHE is 25. Then "HE > 60" is true, and the program gives -1, or FFFFH. However, "SHE < 20" is false and the program gives 0000H. The AND of these two signed integers is 0000H, or false. The program branches to line 30. The others are left as exercises.

As with AND and OR, operator XOR usually serves to connect two or more relations and return a true or false value. The value can then be used in a decision. In the example that follows, both relational expressions are true, and so the XOR statement is false.

```
He = 65: she = 19: if he > 60 XOR she < 20,
   then 30 else print NOT (he XOR she)
```

The computer prints -83, which is the complement of the XOR of 65 and 19. Typed characters may be either lowercase or uppercase, or a mixture as in the example.

Consider next the following BASIC statements:

```
he = 65: she = 18: if he < 60 EQV she > 20,
   then print "True" else print "False"
```

```
IF HE > 60 AND SHE < 20 THEN 20 ELSE 30
IF HE > 60 OR SHE < 20 THEN 20 ELSE 30
IF HE > 60 AND NOT SHE < 20 THEN 20 ELSE 30
IF NOT (HE > 60 OR SHE < 20) THEN 20 ELSE 30
```

FIGURE 4-41. Logical operators in IF statements.

Execution gives "True" on the monitor, because both relations are false, giving a *true equivalence*. The meaning is unchanged if EQV is replaced with XOR and NOT is put ahead of the expression, which must be enclosed in parentheses.

Example: Sort

The *bubble sort* program that follows illustrates the use of both WHILE-WEND and FOR-NEXT loops. Its purpose is to arrange the elements of an array in order of ascending values. The array can be either numeric or string, but a string is selected. Only four elements are assigned to the array in order to simplify the discussion. Statement SWAP is introduced. *It simply exchanges the values of the two variables that follow the keyword SWAP*. The program is presented in Fig. 4-42, followed by a flowchart in Fig. 4-43 used to explain the operation.

Although the dimension statement reserves space for A$(0), it is not used in the program. This element can be removed with OPTION BASE 1 placed ahead of line 100. After execution of line 110, ASCII codes 63H, 62H, 64H, and 61H are assigned to the elements of the array in the order listed. In line 120, variable W can be any number not equal to 0. We could use W = 5 if desired, or some other number.

Nested within the WHILE loop is a FOR loop. When this loop is entered, the IF statement is processed three times. If the expression after IF is true at least once, variable W is assigned the value −1, and following the exit from the FOR loop, the WHILE loop continues again. Exit from the WHILE loop occurs at the end of the FOR loop *only when the four elements have the proper values at the entry of the FOR loop*. Accordingly, after the loop that completes the sorting of the elements, there is one final pass of the WHILE and FOR loops. When the program is run, the display reads

```
a   b   c   d
```

We now examine the flowchart.

Flowchart. In each complete run of the FOR loop of the flowchart of Fig. 4-43, *the three passes within the loop move the highest element value that is not in its proper place to its proper place in the array*. This is done by comparing the first two elements on the first pass and swapping their values if and only if the first one is greater than the second one. On the next pass, with element 2 now greater

```
100  DIM  A$(4)
110  READ  A$(1),  A$(2),  A$(3),  A$(4)
120  W = NOT 0
130  WHILE W
140      W = 0
150      FOR I = 1 TO 3
160          IF A$(I) > A$(I + 1) THEN
                 SWAP A$(I), A$(I + 1):  W = NOT 0
170      NEXT
180  WEND
190  PRINT  A$(1),  A$(2),  A$(3),  A$(4)
200  DATA  c,  b,  d,  a
```

FIGURE 4-42. Routine for sorting strings.

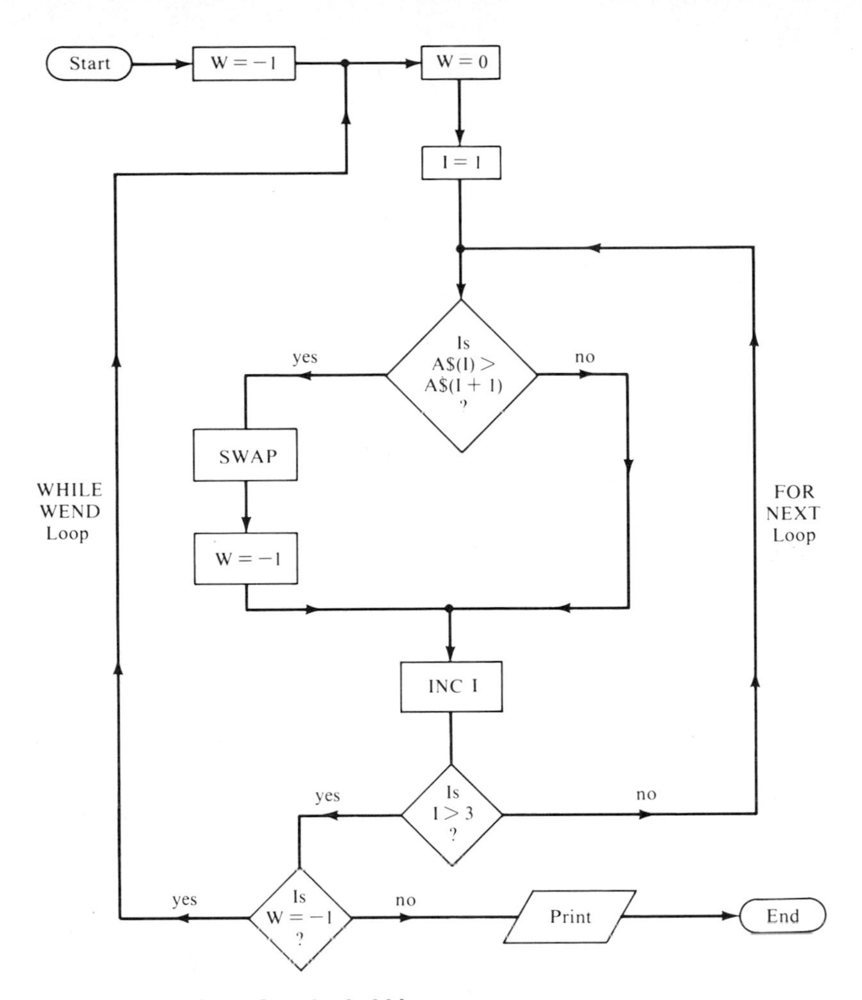

FIGURE 4-43. Flowchart for the bubble sort.

than element 1, the values of elements 2 and 3 are ordered. Then on the third and final pass, with element 3 now greater than both elements 1 and 2, the values of elements 3 and 4 are ordered.

The process is repeated until the sorting is done. During the last pass of the WHILE loop, the values are in order, and no swapping occurs in the three passes of the FOR loop. *Variable W keeps its initial value of 0, or false, and the exit from the WHILE loop occurs.*

To illustrate, there are four passes through the WHILE loop of the preceding example. At the end of pass 1, the element values are changed from *cbda* to *bcad*. They become *bacd* on the second pass and *abcd* on the third pass. There is no change on the final pass.

As shown in the flowchart, at the end of the FOR-NEXT loop the count I is incremented and then tested. This is accomplished by the NEXT statement. The WEND statement that ends the WHILE loop tests the expression, which is simply W in this case. If it is found to be true, the loop is repeated. Note the parallelogram used in the flowchart for PRINT. This is the standard flowchart symbol for I/O operations.

```
100  DEFINT A - Z
110  INPUT "Sort how many values"; NUMBER
120  DIM A(NUMBER)
130  RANDOMIZE
140  PRINT
150  FOR I = 1  TO  NUMBER: A(I) = INT(1000 * RND):  NEXT
160  FOR I = 1 TO NUMBER:  PRINT A(I);  : NEXT
170  TIME$ = "00:00:00"
180  W = 1
190  WHILE W
200    W = 0
210    FOR I = 1 TO NUMBER - 1
220      IF A(I) > A(I + 1) THEN
           SWAP A(I), A(I + 1): W = 1
230    NEXT
240  WEND
250  PRINT: PRINT: PRINT TIME$: PRINT
260  FOR I = 1 TO NUMBER: PRINT A(I); : NEXT
```

FIGURE 4-44. Sorting of random numbers.

In a string sort, the ASCII codes are compared, 1 byte at a time. If the data of line 200 are changed to d, a, 49, and 9, the sort gives 49, 9, a, and d. String 49 is less than string 9, and both are less than string a.

There are more efficient sort routines, which can be found in many books treating BASIC exclusively. The bubble routine is simple but very slow when a large number of elements are present, as demonstrated by the next program.

Another Bubble Sort. The program of Fig. 4-44 is designed to allow the operator to specify a number of numeric elements up to 32,767, which is the maximum allowed size of an array. Initial entries to the array are numbers from 0 to 1000, which are inserted by the program randomly into the array. Displayed are the initial values, the sorted values, and the expired time in hours, minutes, and seconds.

Let us now examine the new BASIC keywords of the routine of Fig. 4-44.

RANDOMIZE and RND. BASIC can generate 65,536 sequences of pseudo-random numbers, with a *pseudo-random number* defined as a calculated number that appears to be randomly selected. As is customary, we refer to these as *random numbers*. A program using random numbers will have the same sequence each time it is run unless the random number generator is *reseeded* by specifying a sequence number. The numbers generated have seven significant digits in the range from 0 to 0.9999999. They can be introduced into a program by means of the function RND.

To illustrate the various formats of RND, suppose the following sequence is executed:

```
A = RND: B = RND: C = RND(34): D = RND(0):
              E = RND(-123)
```

Variable A is assigned the first number of the sequence and B the second number. In spite of decimal 34 after RND, the third number of the sequence goes to C, because a positive number in parentheses after RND has no effect. Variable D equals C. The 0 in parentheses specifies that the number returned will be the same as the previous one. However, E is the first number of a new sequence, identified by -123. *A negative number in parentheses reseeds the random number generator.*

The number n in RND($-n$) is unrestricted. However, not all sequences are independent. In particular, sequences identified by RND($-n$) and RND($-2n$) are exactly the same. For example, sequences with n equal to 1, 2, 4, 8, and so on are identical.

Another way to reseed the random number generator is provided by the statement RANDOMIZE. When RANDOMIZE is encountered without an argument, as in line 130 of Fig. 4-44, program execution halts and on the display appears the message

```
Random Number Seed (-32768 to 32767)?
```

In hexadecimal, thc range is from 0 to FFFF. Each of these positive and negative numbers activates a new sequence. Program execution resumes when a number in the allowed range is entered. In the preceding program, the sequence of random numbers is changed each time the program is run by changing the seed each time.

Also, statement RANDOMIZE (n) is allowed, with n being a seed number within the preceding range. There is no pause in the program when the seed is specified in this manner.

INT Function. The statement Y = INT(x) assigns to Y the largest integer that is less than or equal to x. For example, INT(5.4) returns 5, and INT(-5.4) returns -6.

The purpose of the statements of lines 130 and 150 in Fig. 4-44 is to assign random numbers to the elements of array A(I). The function INT(1000 * RND) converts the 0-to-1 range to 0-to-1000. The sequence of random numbers converted to this new range is displayed by the statements of line 160.

TIME$ and DATE$. In order to obtain the time required by the sort routine, TIME$ is placed in the program. In line 170 is the statement TIME$ = "00:00:00", which sets the internal time-of-day clock at 0 hours, 0 minutes, and 0 seconds. The allowable ranges are up to 23 hours and up to 59 minutes and seconds. A leading 0 can be omitted. Frequently, the clock is set to the correct time when the computer is first turned on, but the clock can be reset to any desired value by a program.

In line 250 is the statement PRINT TIME$. Here, TIME$ is treated as a variable. The updated time of the clock is displayed in the same form as that entered. The time shown is that consumed during execution of the sequence of the sort routine from line 170 to line 250.

Although TIME$ is normally used as a time-of-day clock, it has many other applications. For example, in a game in which speed is important, the PRINT TIME$ statement can be put in a loop so that the operator can observe the seconds passing by. There are numerous engineering applications as well.

Closely associated with TIME$ is DATE$. Two examples showing alternate forms of the statement that sets the date are

```
DATE$ = "11-27-85"     DATE$ = "6/8/1986"
```

Note that the month is first, followed by the day and then the year. PRINT DATE$ returns the updated value. TIME$ and DATE$ cannot be used in cassette BASIC.

Results. In the program of Fig. 4-44, the various PRINT statements with no operands are included to provide spacing. The statements of the last line display the sorted numbers. Because of the semicolon after PRINT A(I), the numbers are displayed side-by-side until a line is full, then going to the next line.

For 20 numbers, the sort time was 2 seconds. For 40 numbers it was 10 seconds, and for 300 numbers it was 9 minutes and 31 seconds. With the DEFINT statement of line 100 omitted, these times were increased 40%. For large arrays, a more efficient routine should be used. A program that is compiled executes much faster than does one that is run on an interpreter.

REFERENCES

W. I. FLETCHER, *An Engineering Approach to Digital Design,* Prentice-Hall, Inc., Englewood Cliffs, N.J., 1980.

IEEE Standard Graphic Symbols for Logic Diagrams (Two-State Devices), approved by the American National Standards Institute, published by IEEE, New York, 1973.

R. S. SANDIGE, *Digital Concepts Using Standard Integrated Circuits,* McGraw-Hill Book Co., New York, 1978.

H. TAUB, *Digital Circuits and Microprocessors,* McGraw-Hill Book Co., New York, 1982.

PROBLEMS

SECTION 4-1

4-1. Convert the logic diagram of Fig. 4-7 into an equivalent diagram for negative logic and give the 1 and 0 states of each variable, assuming A is 0. Repeat for positive logic with A = 1. Use the triangular symbol.

4-2. An odd number of inverters in a closed loop gives a pulse generator. Five inverters are so connected. When the input voltage level of an inverter switches, assume a delay of 10 nanoseconds before the output level switches. With voltage levels of 0 and 5 V and positive logic, suppose that the input of inverter 1 switches from state 0 to state 1 at time zero. Immediately thereafter, the logic states at the outputs of inverters 1 through 5 are 1, 0, 1, 0, and 1, respectively. Sketch and dimension the output voltage of inverter 1 and determine the frequency. Neglect rise and fall times.

SECTION 4-2

4-3. (a) With negation indicators on terminals A, D, and Y, sketch an AND gate and give its truth table, with inputs A, B, C, and D, and output Y. Also, give the logic expression for Y. (b) Repeat for an OR gate.

4-4. (a) With polarity indicators on all terminals, sketch a four-input AND gate and give its truth table in terms of voltage levels. Also, give the logic expression that conforms with the truth table. (b) Repeat for an OR gate.

4-5. Using Boolean identities, simplify as much as possible the following expressions: (a) $A'B'C' + AB'C + A'B'C + AB'C'$; (b) $(A + B)(A + B')$; (c) $A' + AB'$; (d) $A(A + B')$.

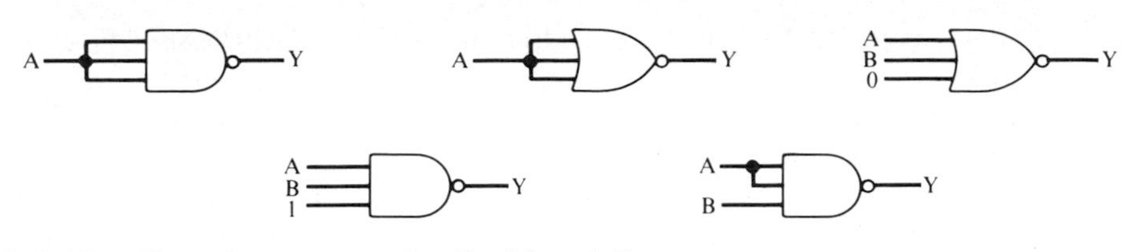

FIGURE 4-45. Three-input gates for Problem 4-7.

4-6. Determine if AB + AB′C + A′BC equals AB + AC + BC and also (A + B) (A + C). Use a truth table.

4-7. Give Y in terms of the inputs for each of the three-input gates of Fig. 4-45.

4-8. For the network in Fig. 4-46, find Y4 in the simplest form. Use a truth table and Boolean identities.

4-9. For the circuit in Fig. 4-46, eliminate the four negation indicators at the outputs of the gates, connect input D to C, and find Y4 in terms of inputs A, B, and C in the simplest form.

4-10. (a) Use the 8088 instruction AND *reg,data* to mask bits 5, 3, and 0 of register AL. (b) Also, use OR *reg,data* to set bits 6, 4, and 1. (c) Give the two instructions that insert 0s into cells 7 and 4 and 1s into cells 3, 2, and 0.

4-11. (a) If 16-bit registers A and B store −199D and +23270D, respectively, what is the signed number content of A after the operation $A \leftarrow A \wedge B$? (b) Repeat (a) for the operation $A \leftarrow A \wedge B'$.

4-12. An XOR gate is sometimes used as a controllable inverter. Let one input be designated as the control C. In terms of input A, determine output Y when the control C is 0 and also when C is 1. Sketch the network and label the terminals.

SECTION 4-3

4-13. With inverters and positive logic AND and OR gates, sketch a circuit that implements a positive logic XNOR function with inputs A and B. Convert the result to an equivalent mixed logic XOR function.

4-14. For the AND and OR functions of Fig. 4-22, give the truth tables in terms of states 1 and 0 and also in terms of states H and L.

4-15. (a) Using the negation indicator and AND and OR functions, sketch a circuit that

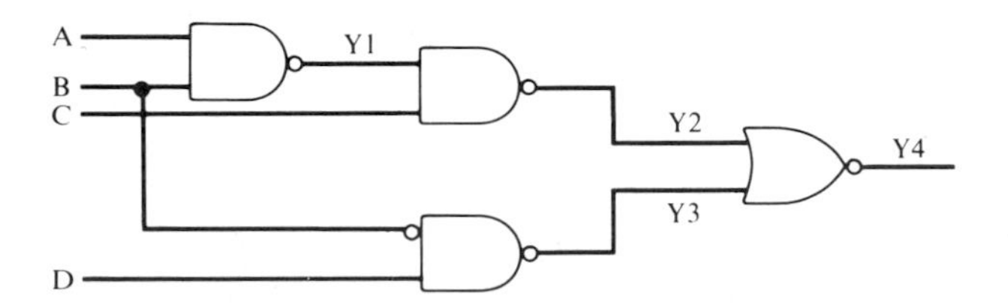

FIGURE 4-46. Network for Problems 4-8 and 4-9.

implements XNOR. (b) Give the truth table for the XOR/XNOR function of Fig. 4-24.

4-16. If AX stores FBD6H and BX stores 5AF7H, determine the decimal contents of AX after execution of assembly language instructions XOR AX,BX followed by NOT AX. Repeat if BX stores 9536H. Numbers are signed.

SECTION 4-4

4-17. In the section Example: Keyboard, assume that the code being processed is the first make scan code of the INS key, with the NUM state active and with the LEFT shift key depressed. (a) List each instruction, giving at its right the hexadecimal contents of each register named in the operand fields of the instruction after execution. (b) Also, identify the addressing mode of each operand (destination and source).

4-18. Suppose AX stores B8D5H with all flags clear. Determine AX in both hexadecimal and decimal, and find the status flags, after execution of (a) NOT AX; (b) SHR AX,1; (c) SAR AX,1; (d) ROL AX,1; (e) RCL AX,1; (f) ROR AX,1; (g) RCR AX,1. Tabulate the results.

4-19. If AX = B8D5H and BX = AC76H, determine AX in both hexadecimal and decimal, and find the status flags, after execution of (a) AND AX,BX; (b) TEST AX,BX; (c) OR AX,BX; (d) XOR AX,BX. Tabulate the results.

4-20. Using Appendix A, hand-assemble the 12 instructions of the sequence of the scan code and mask search program, starting with CLD at assumed address F000:E9AE. In hexadecimal, give both the offset address and machine-language code of each instruction. Assume offsets T1, T2, and X1 are E99EH, E9A6H, and EA4AH.

SECTION 4-5

4-21. In the string sort of Fig. 4-42, give the display if the data of line 200 are changed to z, A, 799, and 8. Also, give the order after each pass through the FOR loop. Repeat for data of "ROM", a, +89, and −2.

4-22. Identify the true expressions (a) NOT 1; (b) NOT −1; (c) NOT −2 < 1; (d) NOT "5 + 4" < "5 − 4"; (e) NOT "5.0" < "5.00"; (f) NOT "5/17/83" < "5/2/83".

4-23. For the routine of Fig. 4-37, with A1, A2, A3, and A4 equal to respective values 25H, −5.25, 1, and −2.6E4, give the displayed results, assuming V3 is unchanged.

4-24. With each occurrence of digit 5 and digit 6 of the routine of Fig. 4-38 changed to 5000 and 6000, respectively, give the display.

4-25. Suppose the following statement is executed:

```
IF NOT A THEN 62 ELSE PRINT A
```

Assume that execution gives printed characters on the monitor. Determine the printout, the 16-bit word A, and the binary word of NOT A. Repeat if NOT A of the statement is changed to NOT (A + 28176).

4-26. (a) Determine the printout on the display when the following BASIC statements are executed:

```
HE = 50: SHE = 22: IF NOT (NOT HE > 60 AND SHE < 20)
   THEN PRINT HE OR NOT SHE ELSE PRINT HE OR SHE
```

(b) Repeat if SHE = 22 is changed to SHE = 18.

4-27. Repeat Prob. 4-26 with the AND and both OR operators changed to XOR.

4-28. (a) With a two-input OR function and negation indicators as needed, sketch an element that implements A IMP B and give the truth table. (b) He = 60: She = 25: If He < 60 IMP She > 30, then print NOT (He IMP She) else print He IMP She. Deduce the printout upon execution.

4-29. Giving results in decimal, evaluate 1709 AND 2931, 1709 OR 2931, NOT 1709 OR 2931, NOT (1709 OR 2931), NOT −1 IMP 0, NOT −1 AND −1, and NOT 0.

4-30. Giving results in decimal, evaluate the following BASIC expressions: −1 EQV −1; −1 EQV NOT −1; NOT 5678 AND −27; NOT (−87 OR −64); NOT (−87 XOR −64); and −87 EQV −64.

CHAPTER 

Bus Control and Input/Output

The 8088 microprocessor was introduced in Chapter 3, where its registers were described. Here we extend our study of the hardware, emphasizing those signals of both the CPU and associated chips that control the flow of data to and from peripherals. Investigated are the various bus cycles of an instruction and the timing of signals within those cycles.

The two methods commonly used for addressing peripherals are known as *standard I/O* and *memory-mapped I/O*. These are defined here, with examples of each. I/O data pass through interface circuits called *ports*. To illustrate the use of latches as ports, a simple chip tester is presented. Also examined are the 8255 programmable peripheral interface (PPI) and the 8253 timer/counter, along with applications.

5-1 BUS CONTROL CIRCUITS

A pin connection determines one of two basic modes of operation. Pin 33 of the 8088 microprocessor, labeled $MN/\overline{MX}$ for *minimum/maximum*, configures the processor in the minimum mode when connected to 5 V. When tied to ground, the mode is maximum, as indicated by the bar over MX. In the minimum mode the CPU has nine pins used for bus control of elementary systems, whereas in the maximum mode the functions of these pins are redefined so as to provide signals necessary for large systems that include coprocessors.

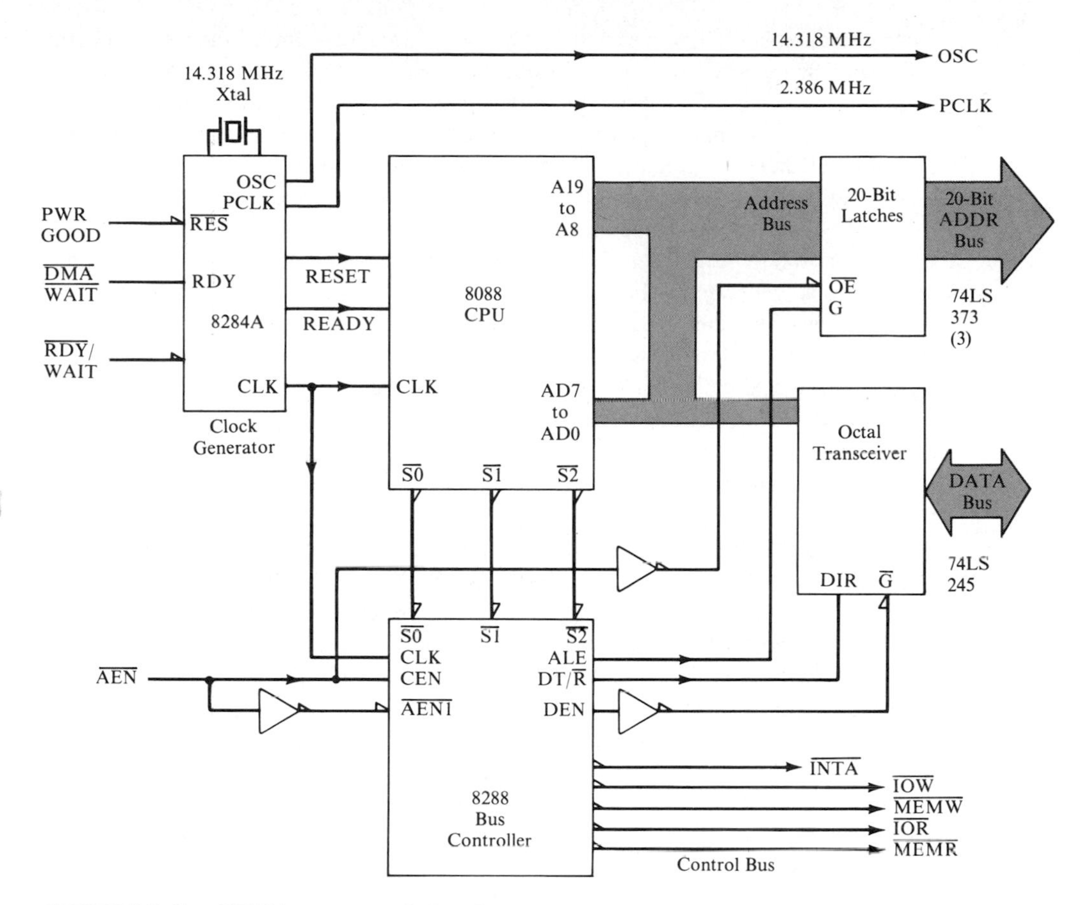

FIGURE 5-1. CPU bus control circuits.

Because the PC has the CPU configured in the maximum mode, this mode is emphasized. With fewer pins available for bus control, the controls are encoded, and they must be decoded into the required signals by another chip. The 8288 *bus controller* is designed for this purpose. In Fig. 5-1 is a diagram of the 8088 microprocessor, the 8284A clock generator, the 8288 bus controller, a data bus octal transceiver, and three chips providing 20-bit address latches. These functions and others are provided by a single chip currently available from several vendors. The illustrated system is that of the original PC. The latches and transceivers, which are described in detail later, provide essential buffering.

Clock Generator

Designed for the 8086/88 microprocessors, the 8284A *clock generator* package supplies periodic pulse trains for system use. It contains bipolar junction transistors in an 18-pin DIP. In addition, the chip has logic for RESET and READY operations. Only those controls important to our study of the PC system are considered here.

The system clock CLK of the computer has a 33% duty cycle and a period of 210 nanoseconds, corresponding to a frequency of 4.77 MHz. This is the crystal frequency of 14.32 MHz divided by 3. The frequency is also divided by 6, giving a peripheral clock PCLK output at 2.386 MHz with a 50% duty cycle for peripheral use. Specifically, PCLK is used by the keyboard control and the timer chip. In addition, there is a 3.58-MHz color burst signal, which is needed when the monitor is a color TV set; this frequency is one-fourth that of the crystal. The 14.3 MHz output labeled OSC is connected to the I/O channel. Memory bus cycles for reading or writing data require four clock periods, or 840 nanoseconds, and I/O cycles use five clocks, or 1.05 microseconds.

When power is turned on, the voltage supply provides the PWR GOOD signal to the clock generator. The signal is held low for more than the required 50 microseconds and then goes high. This input generates a positive pulse at the RESET output. As long as RESET is high, the processor is dormant. When RESET switches to the low state, a *reset* sequence is triggered within the processor. The circuit of Fig. 5-2 is used to give the PWR GOOD signal. Upon closure of switch S, the voltage across capacitor C rises from 0 to 5 V at a rate determined by the R and C parameter values, with the low state below 1 V maintained for the required time.

Before considering the reset sequence, let us examine the READY output of the clock chip, which is fed directly into the READY input of the CPU. A processor *bus cycle* normally consists of four clock cycles, referred to as *T states* and labeled T1, T2, T3, and T4. As stated earlier, I/O devices are allowed five clocks for reading or writing, with *clocks* denoting clock periods. Peripherals are not generally as fast as memory, and a *wait state TW* is inserted between states T3 and T4.

To accomplish this, it is necessary to deactivate the READY line prior to the end of state T2. Then the READY line is reactivated during the wait state TW, and the next state becomes T4. If the READY line is held low, there will be a succession of wait states, each having a duration of one clock period. However, for normal I/O, only one wait state is needed. A bus cycle always ends with state T4, regardless of the number of wait states inserted after T3.

The READY line is controlled by the RDY input of the 8284A, which is itself controlled by the $\overline{\text{RDY}}$/WAIT input. If $\overline{\text{RDY}}$/WAIT is low, the RDY input is enabled, but if $\overline{\text{RDY}}$/WAIT is high, the input is disabled and output READY is inactive. Normally, the input to RDY is high, and $\overline{\text{RDY}}$/WAIT is used to insert the wait state. A wait-state generator on the system board generates a positive-pulse $\overline{\text{RDY}}$/WAIT signal at the proper time during the I/O bus cycle. This procedure disables RDY, pulls READY low, and inserts a wait state into each bus cycle of a peripheral read or write operation. READY then goes high.

At input RDY is a signal labeled $\overline{\text{DMA WAIT}}$. The bar signifies that the signal is active low, but we note from the symbolism of the figure that the RDY input is active high. With input RDY enabled, this notation implies that RDY is activated when $\overline{\text{DMA WAIT}}$ is inactive; that is, there are no wait states introduced into a bus cycle by the inactive signal.

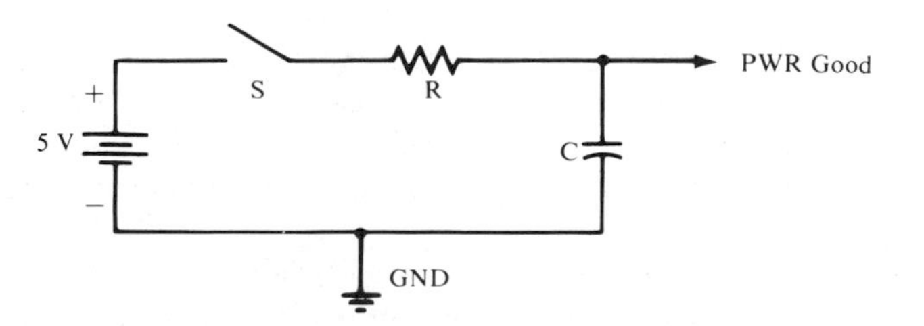

FIGURE 5-2. PWR GOOD generation.

Signal $\overline{\text{DMA WAIT}}$ is driven active low by the *wait-state generator* during direct memory access (DMA). This method transfers words between memory and a peripheral directly, without passing the word through the processor. When DMA is in progress, the CPU continues execution of instructions in the queue, but when a bus cycle is initiated, the inactive READY line causes wait states to be inserted following T3. The wait continues until the DMA controller chip returns use of the buses to the CPU, followed by reactivation of READY.

Reset

The reset sequence is an orderly way to start or restart an executing system. All processing halts when the RESET input to the CPU goes low to high, and the processor remains dormant until the control returns to the low state. For a system reset, the minimum high-state interval is 50 microseconds after a power turn-on, but it is only four clock periods at other times. At the instant RESET goes low, initialization of the system begins. After about seven clocks, the CPU operates normally, and the first instruction of the program is fetched from memory.

As a result of the initialization, the CPU state at the end of the reset is as follows:

- All status and control flags are clear.
- The instruction stream queue is empty.
- Segment registers SS, DS, and ES store 0s.
- The code segment address is FFFFH.
- The instruction pointer IP is 0.

Because the *interrupt-enable control flag IF* is 0, the interrupt system is disabled. With FFFFH in code segment CS and 0 in IP, the logical address of the first instruction is FFFF:0000. This memory location, at absolute address FFFF0H, is reserved for the first instruction after a reset. There is storage space for 16 bytes to the end of memory, and in the PC this region is in the ROM BIOS. Stored is a 5-byte instruction that gives a program jump to absolute location FE05BH, also in ROM.

The PC program sequences through the *firmware operating system* of BIOS. Extensive tests of all parts of the system are performed, the various peripherals are initialized, and important data are placed in RAM. Then the program branches to a *bootstrap loader* in BIOS that loads the *boot record,* or *bootstrap program,* from a diskette with a suitable operating system, and transfers control to its absolute start address of 07C00H in RAM. The boot record reads the main parts of the *disk operating system (DOS)* from the diskette into memory.

DOS is a collection of programs designed to aid the user in creating and managing files, running programs, and transferring data between the CPU and peripheral devices. All resources, both hardware and software, are monitored. There are several different DOS systems designed for the PC, but the most common one is known as *MS DOS,* or *PC DOS*. It is used in many different personal computers, and henceforth will be referred to simply as *DOS*. Its major parts are examined in this and later chapters.

After executing the initialization routine of DOS, the system is ready for use. The process of placing a copy of DOS in memory is called *booting DOS*. If the DOS diskette is not present in the drive, program control is transferred to the absolute start address F6000H of the BASIC interpreter in ROM, providing cassette BASIC facilities.

In order to boot DOS, the operator simply has to turn on the power with the DOS diskette in the proper drive. A *power-on reset* is implemented. Sometimes it is necessary to reset the system when the computer is already on. This can be accomplished by turning the switch off and then on. There is a better way, however, that does not involve the power supply. Whenever the key combination Ctrl-Alt-Del is activated, the software gives a jump to absolute address FE05BH of the BIOS program, and a system reset is implemented. Two hands are required to press these three keys simultaneously, thereby making an accidental reset unlikely. The reset with keys is faster than the power-on procedure, because the memory check is omitted. Both clear all memory locations, eliminating programs of memory.

The greater portion of both the disk and firmware operating systems constitutes the *monitor,* defined as a coordinated collection of programs that *continuously supervise, control, and verify operations of the complete system.* Included in the monitor are numerous interrupt routines and subroutines. Part of the monitor is in ROM and part is loaded into RAM from a diskette. Commands to the monitor can be entered from the keyboard. In general, *an operating system has programs in addition to the monitor.* Typical are loaders, text editors, simulators, emulators, assemblers, compilers, and diagnostic programs.

The CPU

The 8088 microprocessor shown in Fig. 5-1 has 12 three-state address pins labeled A19 through A8 and 8 three-state address/data pins labeled AD7 through AD0. During the first part of a bus cycle, the CPU places the low byte of the address on pins AD, but for the remainder of the cycle the pins serve as a bidirectional data bus. This system is called *time multiplexing.*

The 4 highest bits of the address are time multiplexed with status information. Status bits S3 and S4 on address lines A16 and A17 are encoded to indicate the segment register that computes the address, and bit S5 of address line A18 is the state of the interrupt-enable flag IF. Bit S6 of address line A19 is always low. The other 8 address bits (A15-A8) are valid throughout the bus cycle.

During state T1 of each bus cycle, the 20-bit address is supplied by the processor and latched in the three 74LS373 chips indicated in Fig. 5-1. These chips feed the address bus, holding the bits until a new address is received during the next T1 state. Latch outputs can be floated in the three-state mode when an external unit needs to use the address bus.

The T2 interval is used primarily for fixing the direction of data transfer along the bidirectional data bus. Actual transfer occurs during T3 and T4, and wait states (TW) can be inserted between these states to extend the transfer interval. Inactive clock cycles that may be present between bus cycles are called *idle states.*

In the maximum mode configuration, there are three status bits, $\overline{S2}$, $\overline{S1}$, and $\overline{S0}$, that are fed through output pins of the processor into corresponding input pins of the 8288 bus controller chip. These bits identify the type of bus transaction. In the bus cycle state preceding T4, which is either T3 or a wait state TW, each bit goes inactive high and remains high until the end of T4. If another bus cycle follows, the bits change during T4 and remain active until the next T3 or TW state preceding T4. The status bits are always inactive when the READY input is held low, which gives wait states.

Decimal values 0 through 7, corresponding to the encoded word formed by $\overline{S2}$, $\overline{S1}$, and $\overline{S0}$, represent respective bus transactions as follows: interrupt acknowledge,

$\overline{S2}$	$\overline{S1}$	$\overline{S0}$	Bus Cycle
0	0	0	Interrupt acknowledge
0	0	1	I/O read
0	1	0	I/O write
0	1	1	Halt
1	0	0	Fetch (instruction)
1	0	1	Memory read
1	1	0	Memory write
1	1	1	Inactive

FIGURE 5-3. Status bits and corresponding bus cycles.

I/O read, I/O write, halt, instruction fetch, memory read, memory write, and no bus cycle. Interrupt acknowledge is examined in the next chapter. The status bits and their bus cycles are shown in Fig. 5-3.

The *halt cycle* follows a fetch and execution of the assembly language instruction HLT. In the cycle the three status bits are sent to the bus controller, and processing then stops. The bus controller emits a pulse from pin ALE, with this pulse used for latching the 3 status bits into an external latch, if desired. The computer is now in the halt state, and a permanent exit is by either an interrupt or a system reset.

With the interrupt exit, the interrupt routine is executed first. Then processing continues with the instruction that follows HLT. For many computers *the main use of HLT is to provide a way to idle while waiting for an interrupt,* perhaps waiting for the operator to depress a key. Also, it is often used to stop operations when an event, such as a parity error, prevents the system from functioning properly.

In the PC there are periodic timer interrupts at a rate of 18.2 per second. If a program is run with only HLT to stop it, the first timer interrupt of the halt state causes processing to continue with the code following HLT. Unless a breakpoint is set by a debug routine, HLT should not be used.

5-2 TIMING AND CONTROL

The 8288 bus controller of Fig. 5-1 is a bipolar IC in a 20-pin package, designed to provide command and control signals for 8086/8088 systems. Buffered three-state command outputs can source up to 5 mA and sink up to 32 mA while maintaining satisfactory voltage levels. The figure shows four of the seven command outputs, these being I/O write $\overline{IOW}$, memory write $\overline{MEMW}$, I/O read $\overline{IOR}$, and memory read $\overline{MEMR}$. Because each is active low, the mnemonics are usually written with a bar. The interrupt acknowledge $\overline{INTA}$ command is considered in the next chapter, and the other two (IOWC and MWTC) are not used in the PC.

READ Cycles. Suppose the processor enters bus cycle memory read. The status bits become 101 at the start of state T1, signaling the bus controller that a memory read is commencing. Near the beginning of state T2, command $\overline{MEMR}$ is driven active low by the controller, remaining there until the start of state T4. The signal is fed via the data bus to all memory units, activating the read circuits of addressed chips. The other three commands function in a similar manner. A timing diagram is shown in Fig. 5-4.

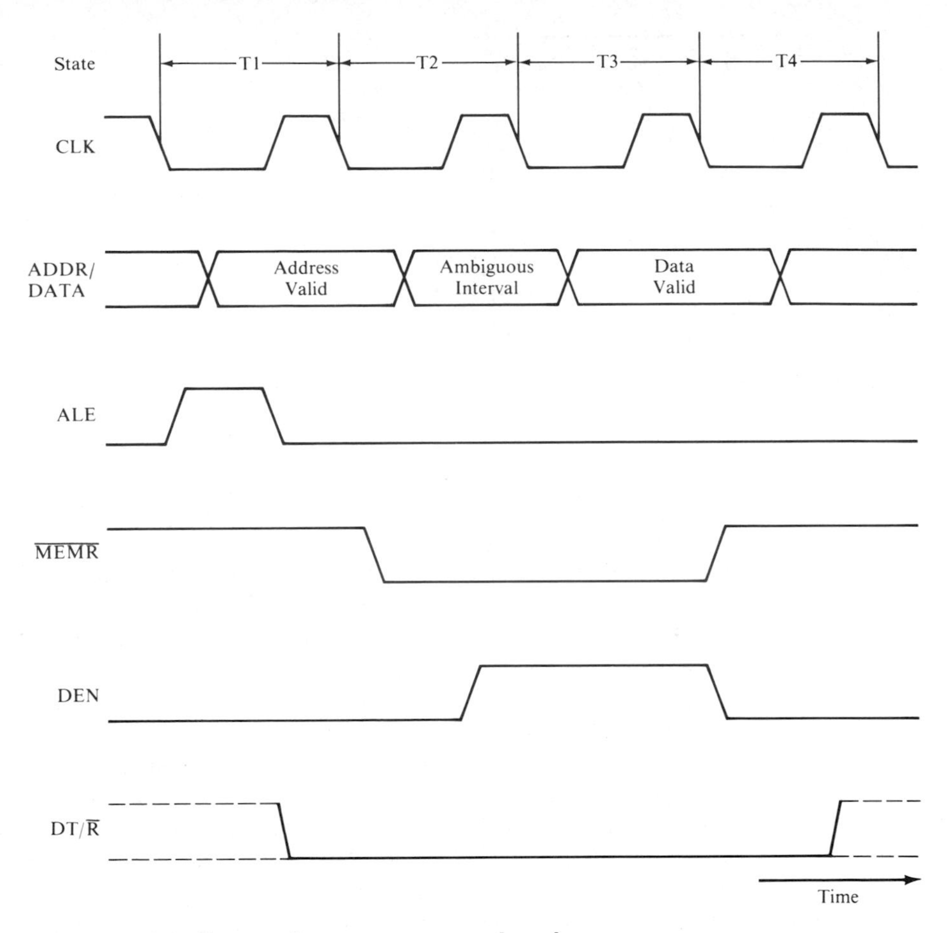

FIGURE 5-4. Timing diagram for a memory read cycle.

At the top of the timing diagram are the four clocks of a memory read cycle, and the other waveforms are relative to the clock. All 20 bits of the address are valid during most of T1 and for a portion of T2. The address is latched while the address-latch-enable ALE control signal is high. Signal ALE is generated by the bus controller and fed into the load-enable input G of the latches. Normally, the outputs are also enabled, so that the latches are *transparent* during pulse ALE. The output of a transparent latch equals its input at all times, just as though the latch was not present. When ALE goes low, the latch inputs are disabled, but the latched address bits are still present on the 20-bit address bus, and the address from the processor is no longer needed.

Just after the address is latched, signal $\overline{\text{MEMR}}$ activates the read circuitry of memory, and a brief delay occurs while the memory is read. The time interval from the instant $\overline{\text{MEMR}}$ becomes active to the instant the valid data byte from memory appears on the data bus, with all bits stable, is the memory *access time*. For the dynamic RAM of the PC this is about 250 nanoseconds, which is not much greater than a clock period.

The data-enable DEN output control of the 8288 enables the data transceivers when high, as indicated in Fig. 5-1. The direction of data flow through the transceivers is established by control line DT/$\overline{\text{R}}$, or data-transmit/receive. A high-level

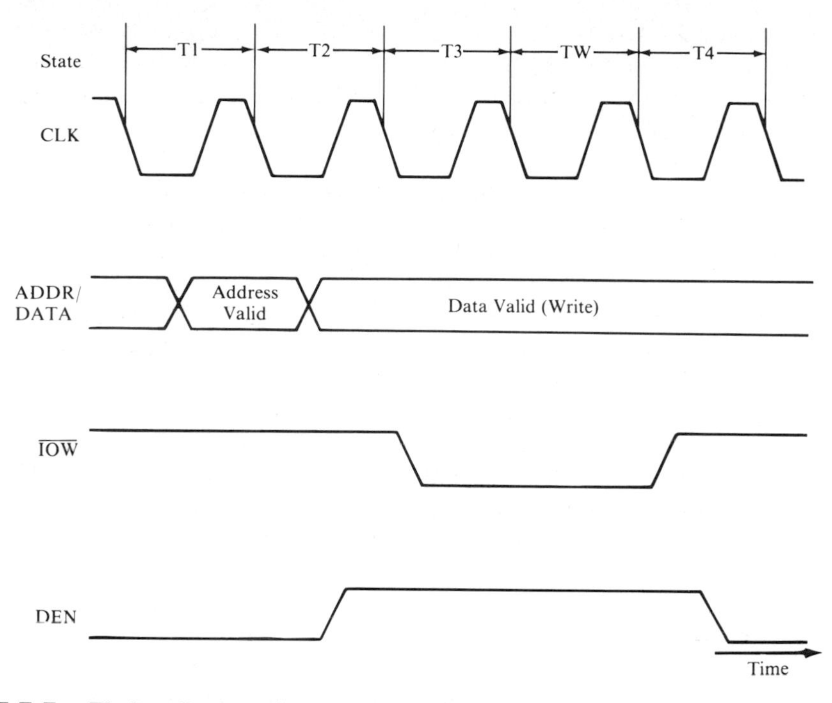

FIGURE 5-5. Timing diagram for an I/O write cycle.

DT/$\overline{\text{R}}$ allows data to flow from the CPU, which is a write operation (transmit), and a low state indicates that the CPU will receive data, which is a read operation.

The timing diagram for I/O read is the same as for memory read except that signal $\overline{\text{IOR}}$ replaces $\overline{\text{MEMR}}$ and a wait state is present between T3 and T4. Controls ALE, $\overline{\text{IOR}}$, DEN, and DT/$\overline{\text{R}}$ from the bus controller are extended one clock period from those of Fig. 5-4, as is the duration of the status signals fed to the controller from the processor.

WRITE Cycles. The memory write cycle differs somewhat from memory read. The negative pulse for $\overline{\text{MEMW}}$ is similar to the $\overline{\text{MEMR}}$ pulse of Fig. 5-4, but it does not become active until the end of state T2. Also, the valid data byte appears on the data bus as soon as the valid address is removed at the end of T1, and the positive data-enable DEN pulse of the figure is extended by the bus controller to include all of T2 and T3 and the low state of T4. Also, DT/$\overline{\text{R}}$ is high, setting the transceiver for transmit. The I/O write cycle is similar, but with a wait state. Figure 5-5 shows the I/O write timing, with one wait state and signals ALE and DT/$\overline{\text{R}}$ omitted.

DMA Bus Control

On the left side of the bus controller in Fig. 5-1 are the command-enable CEN and the address-enable $\overline{\text{AEN1}}$ pins. When input CEN is inactive, all command outputs and the control output DEN are inactive, and when pin $\overline{\text{AEN1}}$ is inactive, the command outputs float in the three-state mode. Both of these are fed by the $\overline{\text{AEN}}$ input signal. Because $\overline{\text{AEN}}$ is normally high, pins CEN and $\overline{\text{AEN1}}$ are normally active.

Control $\overline{\text{DMA WAIT}}$. Signal $\overline{\text{AEN}}$ is low only when the DMA controller has requested control of the buses for direct memory access and after the bus status bits indicate the passive bus condition. The request from the controller activates control signal $\overline{\text{DMA WAIT}}$, which drives the RDY input of the clock generator to the inactive state. These signals are shown in Fig. 5-1. Then the CPU bus cycle has wait states inserted, and the condition continues as long as DMA is in progress. In most cases, this is only for four or five clocks. Because the READY input to the processor is held low, the bus status bits are low during the wait states.

Control $\overline{\text{AEN}}$. The low-level $\overline{\text{AEN}}$ that results from the DMA request disables the three-state outputs of the address latches by driving the output-enable $\overline{\text{OE}}$ pin high. Also, the command outputs of the bus controller are floated by the high input to pin $\overline{\text{AEN1}}$, and the octal transceivers are floated by the inactive DEN. Thus, the address, data, and control buses are placed in their three-state modes, thereby allowing the DMA controller full use of the buses. Both $\overline{\text{DMA WAIT}}$ and $\overline{\text{AEN}}$ are inactive except during DMA transfers. DMA operations are examined in detail in Chapter 8.

Handshake

A computer can supply data to or receive data from an external device without prior communication other than an address. For example, when an address is sent to memory on the address bus, along with a signal specifying that memory is to be read, the word at the addressed location is placed on the data bus and transmitted to a processor register, with timing controlled only by the clock pulses. This procedure is called *synchronous transfer,* with no control signals other than periodic clock pulses. An alternate procedure uses *handshake* lines between the processor and the peripheral. *These lines provide communication in the form of request and acknowledge signals.* Because data transfer is not controlled directly by the clock, it is said to be *asynchronous.*

Suppose a CPU is to supply data asynchronously to a peripheral, using handshaking. When the CPU is ready, it places a word on the data bus, uses the output handshake line to signal the device, and waits for a response. When the peripheral receives the signal on the handshake line, it stores the word that is present on the data bus and then activates the input handshake line. This signal notifies the processor that the peripheral is now ready to receive another word of data. Upon receipt of the input handshake signal, the next data word is transmitted by the processor, following the same procedure.

In the input mode, the peripheral activates the input handshake line after placing a word on the bus. When the word is received, the CPU signals the device that transfer has been accomplished and that it is waiting for another transfer. There are alternate ways. In general, *handshaking refers to the direct two-way communication that supervises the asynchronous transfer of data between two systems.* Communication is accomplished using lines of the control bus.

8088 Pin Diagram

Most of the 8088 pin functions used in the specified PC system have been examined. The pin numbers, with labels indicating their functions, are shown in Fig.

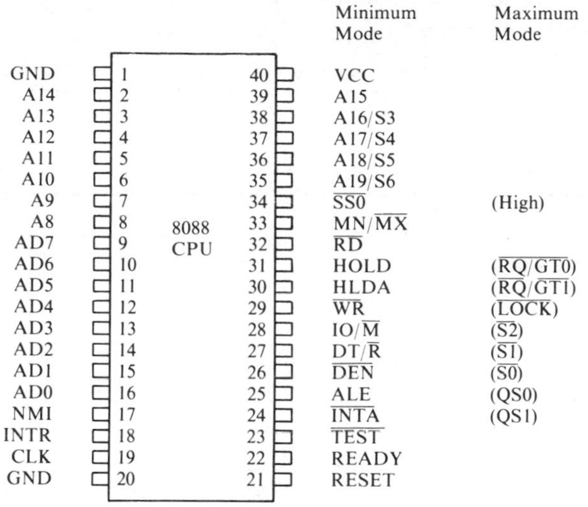

	Pin	Pin	Minimum Mode	Maximum Mode
GND	1	40	VCC	
A14	2	39	A15	
A13	3	38	A16/S3	
A12	4	37	A17/S4	
A11	5	36	A18/S5	
A10	6	35	A19/S6	
A9	7	34	$\overline{SS0}$	(High)
A8	8	33	MN/$\overline{MX}$	
AD7	9	32	$\overline{RD}$	
AD6	10	31	HOLD	($\overline{RQ}$/$\overline{GT0}$)
AD5	11	30	HLDA	($\overline{RQ}$/$\overline{GT1}$)
AD4	12	29	$\overline{WR}$	($\overline{LOCK}$)
AD3	13	28	IO/$\overline{M}$	($\overline{S2}$)
AD2	14	27	DT/$\overline{R}$	($\overline{S1}$)
AD1	15	26	$\overline{DEN}$	($\overline{S0}$)
AD0	16	25	ALE	(QS0)
NMI	17	24	$\overline{INTA}$	(QS1)
INTR	18	23	$\overline{TEST}$	
CLK	19	22	READY	
GND	20	21	RESET	

8088 CPU

FIGURE 5-6. 8088 pin diagram.

5-6. At the right side are nine pins with alternate labels. One set of functions applies to the minimum mode, and the other applies to the maximum mode. As mentioned earlier, pin MN/$\overline{MX}$ is tied to ground in the PC, giving the maximum mode.

In the maximum mode, pins 23, 29, 30, and 31 provide functions for the control of coprocessors that might be connected to the system and used in conjunction with the 8088. There is no instruction in the 8088 set that performs operations on floating point numbers. Software routines are used for this purpose, but their speed is slow compared with that of suitable hardware. On the system board of the PC is an empty socket alongside the 8088, wired for use with an 8087 mathematical coprocessor. This chip is designed to handle floating point numbers at high speed. Additional software support is essential.

Although important, the subject of coprocessors is not investigated in this book. Thus, the assembly language instructions WAIT and ESC and the prefix LOCK, are not discussed, because they are designed to control the use of the buses by more than one processor. Outputs of pins 24 and 25 labeled QS1 and QS0 provide status that allows external tracking of the internal instruction queue. These signals are usually used with systems containing coprocessors and will not be discussed further.

Pins 17, 18, and 24 with respective labels NMI, INTR, and $\overline{INTA}$ provide functions required by the interrupt system, which is described in Chapter 6. Pin 34 is always high level in the maximum mode, having no particular use. When the processor is used in the minimum mode, the additional control signals available allow operation without the 8288 bus controller.

Latches and Transceivers

An even number of inverters connected in a loop forms a memory cell called a *latch*. Figure 5-7 shows such a cell. It has a loop of two inverters, along with control circuitry for entering data.

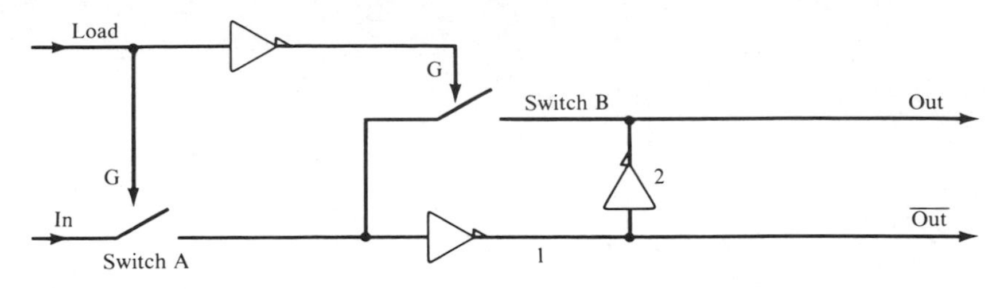

FIGURE 5-7. A 1-bit D-type latch.

Each of the electronic switches is designed to be closed when control G is 1 and open when G is 0. Such switches are easily designed in electronics, frequently consisting of a single transistor. Three-state outputs can be implemented with such switches.

Let us suppose that control LOAD is high. Switch A is closed and switch B of the inverter loop is open. The bit at the input enters inverter 1 and is supplied to the output line from inverter 2. Also available to the output is the complement of the bit. After entry, LOAD is driven low. This procedure opens switch A and simultaneously closes the one in the feedback loop. Relative timing is not critical, because the stored bit remains for a while after the input to the inverter is opened. Now the two inverters are connected in a loop, with the output of one feeding the input of the other. The stored bit is latched in the cell, and it remains there as long as LOAD is low and power is on.

The cell is called a 1-bit *D-type latch,* or *data latch.* When input switch A is closed, there is *direct coupling* between the input and output. In this case, the output equals the input, and the output responds to a change in the state of the input without appreciable delay. *Direct coupling distinguishes latches from registers with dynamic clock controls.* Such registers are never directly coupled. The addition of a third switch at the output of the cell in Fig. 5-7 gives a three-state latch. Sixteen cells can be grouped to give a 16-bit register, and 256 cells can be organized on a chip as an array of sixteen 16-bit registers.

Octal Latches. The TTL 74LS373 IC has eight D-type latches with three-state buffered outputs in a 20-pin package. All outputs float when the output enable $\overline{\text{OE}}$ is inactive high. However, with $\overline{\text{OE}}$ active, the stored bits are supplied to the output pins. Bits on the input pins are loaded, or written, into the cells when enable-input G is active high. Loading is independent of control $\overline{\text{OE}}$. When G goes inactive low, the bits are latched, and a change in the data fed to the input pins now has no effect. The latched bits are stored until new data are loaded or until power is turned off. Whenever both input and output enables are active, the latch is transparent, with the input lines directly coupled to the output lines.

Three chips provide the 20 address latches, with 4 latches of one chip not connected. In small systems the 12 high-order bits of the address do not require latches, but in more complex systems the buffering is essential.

Bus Transceivers. For the bidirectional *bus transceivers* the TTL 74LS245 chip is used, primarily for buffering. In addition, the transceivers provide asynchronous two-way communication between two data buses, and the direction of transmission is determined by the state of the direction DIR control input. Three-

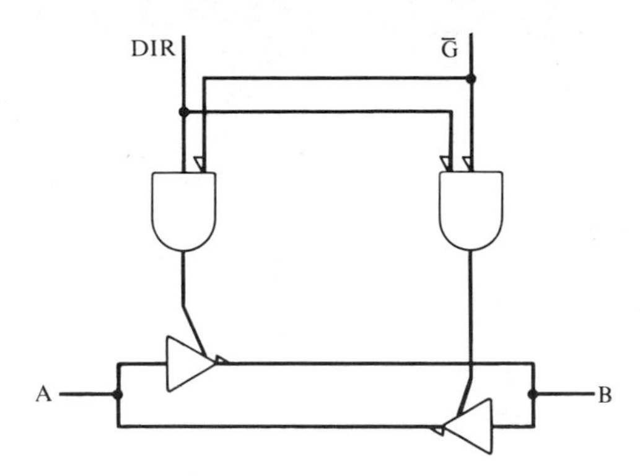

FIGURE 5-8. A bus transceiver.

state outputs of the buffers are controlled by the active-low enable input $\overline{G}$. In Fig. 5-8 is shown the circuit of one transceiver, and the AND gate outputs also control the other seven. Transceivers are not used in small systems.

5-3 INPUT/OUTPUT

As we have seen, the operation of a general-purpose computer is governed by a program, with execution of the program implemented and controlled by the CPU in conjunction with various peripheral elements. The program consists of a sequence of instruction words and associated data words. Each machine language word of 0s and 1s is transferred to RAM from an input device such as a magnetic disk. After the program is stored, the individual instructions are fetched from memory and executed, one at a time. Intermediate data obtained from execution of the program are stored in processor registers or in memory, but when the complete program has been executed, the results are delivered to a suitable output unit.

Input/output (I/O) devices include paper tape readers and punches, card readers and punches, printers, CRT monitors, plotters, magnetic disks, keyboards, optical readers, communications lines, digital instruments, and digital-to-analog (D/A) and analog-to-digital (A/D) converters. Each I/O device is normally connected to the data bus of a computer by means of an *adapter* circuit that interfaces the characteristics of the device with those of the CPU. The adapter consists of one or more ICs that are controlled by the CPU. Several I/O devices and their adapter circuits are examined in some detail in later chapters. Our study here treats assembly language I/O addressing and I/O ports.

I/O Addressing

There are two basic ways to address I/O devices, which are known as *standard I/O* and *memory-mapped I/O*. Although both are used in the PC, most peripherals are addressed with standard I/O.

Standard I/O. Standard I/O is also called *I/O-mapped I/O,* or *isolated I/O*. Control signal $\overline{IOR}$ from the bus controller enables an input device for reading,

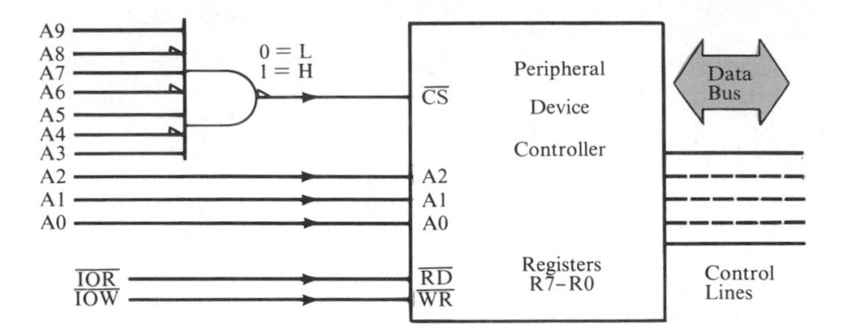

FIGURE 5-9. Peripheral controller and addressing.

and signal $\overline{IOW}$ enables an output device for writing, but neither activates memory. On the other hand, controls $\overline{MEMR}$ and $\overline{MEMW}$ enable memory without affecting the peripherals. In the PC, 16 address lines can be used to address up to 64K of input peripherals and an equal number of output peripherals.

Figure 5-9 shows a peripheral device controller that is addressed through a seven-input AND gate. Chip select CS is active when bits A9 through A3 are 1010101. The AND gate is an *address decoder* that activates the chip if and only if the input bits of the address have the proper values. When CS is high, the data bus is assumed to be floating in the three-state mode, thereby isolating the chip from the system. In the PC, only 10 address bits are used for peripherals.

Within the chip are eight registers, designated R7 through R0 that can be both loaded and read from the data bus. These registers are individually addressed by the 3-bit address word A2-A1-A0, with 111 addressing R7, and so on. The registers may store commands, control words, initialization data, results, or other information. If the precise function of the chip depends on data written to one or more registers, the chip is said to be *programmable*. Although a read/write register has only one address, it is both an input port and an output port.

Assume the chip is addressed, giving an active $\overline{CS}$, and suppose A2-A1-A0 is 100 with $\overline{IOR}$ low. Then the contents of register R4 are placed on the 8-bit bidirectional data bus. On the other hand, if $\overline{IOW}$ is active in place of $\overline{IOR}$, the bits of the data bus are loaded into the register. If both read/write controls are low, the bus floats.

I/O instructions of the processor are restricted to 16-bit addresses. The address space is from 0 to 64K, and segments are not involved. Input and output devices can have identical addresses, because they are enabled by different signals. Parallel input and output operations are accomplished *only with the instructions IN and OUT*, which transfer bytes or words between AX or AL and the selected device. A word transfer on the 8-bit AD bus is, of course, a 2-byte sequence.

Address bits A15 through A10 of the possible 16-bit address are don't-cares, with a don't-care bit defined as one that simply has no effect. Selecting 0s for the don't-care bits, which will always be done with addresses, the eight addresses used to access the registers of the chip are 02A8H through 02AFH. These are the port addresses. An *I/O port* is defined as either a single pin or a group of pins of a peripheral that are used for transfer of data to or from the CPU.

Memory Mapped I/O. The alternate method is *memory-mapped I/O*. It treats I/O devices the same as memory, with the total address space of the processor divided between them. Each peripheral is assigned an addressable memory location, which is enabled with the same read and write control signals used for memory. IN and OUT instructions are usually invalid. In the PC these instructions can use only 16-bit addresses, whereas memory addresses are 20 bits. Input and output are accomplished with the more flexible memory reference instructions applied to input and output ports. For example, if BETA is a DW variable that identifies an input port location, the instruction MOV BX,BETA moves the word from the port into BX. Each device has a logical address and a 20-bit physical address.

With memory-mapped I/O, insofar as addressing is concerned, *each port simply becomes, in effect, another memory location*. Thus, the full power and flexibility of the instruction set are available for data transfer. There are disadvantages, of course. Addresses chosen for peripherals cannot also represent memory locations, and the memory address space is reduced. In large systems, address decoding and memory refresh are more complicated when devices are included in the memory address space. This is not so in smaller systems, which often use this I/O method.

IN and OUT. In the PC all I/O, except that of the display buffers of the monochrome and color adapters, is standard I/O. These buffers are in the CPU 20-bit address space, and they will be referred to as memory rather than as ports. The instruction "IN port" moves from the addressed port *either a byte to register AL or a word to AX*. Ports with addresses from 0 through 255 can be addressed with an immediate byte constant. However, these ports, as well as those with higher addresses, can be addressed by first placing the 16-bit word in register DX. Output port addressing is similar. Some enumerated examples are as follows:

```
1. IN AL, PORT A  3. IN AX, 0A4H       5. MOV DX, 0A42H
2. IN AL, DX      4. OUT 0A4H, AL         OUT DX, AL
```

In (1), PORT_A is assumed to have been equated to the 8-bit address of an input byte port, and in (2), register DX has the port address. In (3) and (4), byte address A4H is that assigned to a word input port and also to a byte output port. There is no conflict, because one is activated by the read control and the other by the write control. In the sequence of (5), first the 16-bit port address is moved into DX and then the byte of AL is transferred to the port. Object codes for the two OUT instructions are E6A4 and EE.

BASIC Input and Output. There are two BASIC instructions that correspond to the assembly language instructions IN and OUT. For input of a byte from a port, the BASIC function is INP(*n*), with *n* representing a port number from 0 to 65,535. The returned value is the byte read from port *n*. Two statements using this function are

```
X = INP (1012)          PRINT HEX$ (INP(&H40))
```

The one on the left reads port 1012, or 3F4H, assigning the value of the byte to variable X. The port is the status register of the diskette controller chip, described in Chapter 8. The statement at the right reads the count register of counter 0 of the 8253 chip described in Section 5-6. A different hexadecimal value is displayed each time the statement is executed, because the count changes at a rapid rate.

For writing to a peripheral in BASIC, the statement OUT *n*,*m* transmits byte *m* to port *n*. An example is

```
OUT &H3F2, 33
```

The port is the *DOR* register, which is described in Chapter 8; writing byte 33 to it turns on the diskette motor of drive B. After running for no more than a few seconds, the motor stops. Although no harm will be done, generally one should not write to ports without knowing precisely what the transmitted byte will do. Note the similarity between the OUT statements of BASIC and the processor.

A number of peripherals are connected to the processor through ports of interface chips mounted directly on the computer board of the system unit. We now examine this board briefly.

The Computer Board

Figure 5-10 is a photograph of the 8.5 × 11-inch computer board mounted horizontally in the base of the system unit. Included on the board are the 8088 microprocessor, ROM chips with 40 KB of code, and dynamic RAM chips that provide the user with 64 KB of read/write memory, along with parity bits for error detection. To the right of the microprocessor is a wired socket for an auxiliary 8087 coprocessor.

Among other chips of the system unit are the clock generator, the bus controller, an interrupt controller, a DMA controller, a shift register, bus transceivers, buffers and latches, registers, decoders, multiplexers, and logic gates. Included are keyboard interface circuits. At the left rear of the system board are five slots. These slots contain four adapter cards, and one slot is empty.

On each of two DIPs are eight switches, designated SW1 and SW2. Only 13 of the available 16 switches are used. These are set either on or off to indicate to the system the number of diskette drives, the amount of memory, and what options have been added. The settings, which can be read by the processor, must usually be changed when an option is added or removed. The two DIP switches are located below the processor and coprocessor sockets.

FIGURE 5-10. PC computer board.

At the rear of the system board behind the processor is a five-pin connector that provides for attachment of a 6-foot coiled cable from the keyboard. At the right edge alongside the auxiliary processor socket is the system board power input from a 63.5-W switching regulator, which supplies +5, −5, +12, and −12 V. Cooling is provided by a small fan. When power is initially turned on, the system unit runs a self-test to verify its readiness.

A 2.25-inch, 8-ohm audio speaker mounted within the system unit has control and drive circuits on the computer board. The connector to the speaker is directly in front of the RAM section. The drive can supply 0.5 W of power, from 37 Hz to 32 kHz, and various methods are employed to drive the speaker. A pulse train is supplied by toggling with software one of the bits of a processor register, and the toggled bit is fed through the PPI chip to the driver circuit.

In addition, a second bit of a program-controlled processor register, also fed through the PPI, can modulate the clock input of one of the counters of the board; the output of the counter is ANDed with that of the first bit, and the result is fed to the speaker. It is also possible to program the counter to generate a specified waveform for driving the speaker. In fact, all three methods may be implemented simultaneously.

The keyboard is, of course, a source of data input. When a mechanical switch opens or closes, the metallic contacts usually bounce a number of times, giving repeated opens and closures, before coming to rest. In an interval of several milliseconds, there may typically be 10 to 50 bounces. Thus, it is essential that keyboards of both computers and calculators be debounced, either by software or hardware. In the PC, *switch debounce* is accomplished with programmed time delays. When a key is depressed, a program loop delays reception of the scan code until sufficient time has elapsed for elimination of switch bounce. Hardware switch debounce is used in many systems (see Prob. 5-13).

In the sections and chapters that follow are numerous examples of standard I/O. In the next section, we consider an example of I/O with memory mapping. The two programs presented there use the display buffer of the monochrome monitor, and the buffer is in a chip on the display adapter. For preparation, we briefly describe here the adapters of the PC, especially that of the monochrome display.

Adapter Cards

A *computer card,* or *adapter card,* consists of a group of ICs wired to implement some useful function. The system of the card is often referred to as a *module*—a name commonly used to refer to a related group of items, either hardware or software. Connections to the card are made through an edge having a number of terminals, and the terminal edge is inserted into one of the slots of the motherboard of the system unit. The cards provide system expansion and interfacing.

Located at the left rear of the system board of the PC is a section consisting of five 62-pin card edge sockets. As many as five adapter cards can be inserted into these slots for the purpose of interfacing with I/O options. In Fig. 5-10, four of the slots are occupied. The card in the second slot from the left is for interface with the display monitor and parallel printer.

There are different versions of memory adapters for expanding system memory. With a suitable adapter inserted into one of the five expansion slots, user memory up to 640 KB is feasible. The remainder of the 1 MB of addressable space is either reserved for future developments or used for the buffers of the monochrome and color displays and for the ROMs.

A communications card provides for RS232 asynchronous, serial communications with another computer, either directly or over telephone lines. Additional adapter cards are available for game control and other purposes. If more than five slots are needed, there is a card that connects to an external connecting board with extra slots. Later versions of the PC have more than five slots.

Monochrome Display Adapter

The card that interfaces with the display monitor contains a Motorola 6845 CRT controller, an 8-KB ROM *character generator,* and 4 KB of RAM. The ROM generator has fonts for 256 alphanumeric and block graphic characters, similar to the ROM BIOS character generator used in the graphics mode and examined in Section 3-3.

Each character box is 9 dots wide and 14 dots high. With a screen display of 25 rows of 80 characters per row, the vertical resolution is 25 × 14, or 350 lines, and the horizontal resolution is 80 × 9, or 720 lines. Normal characters are 7 by 9 dots, excluding the descenders, in a 9 by 14 character box. The resolution, or possible number of dots per unit area, is much higher than that of a standard TV. The lowercase letters have descenders, giving letters a nice appearance.

The 4-KB RAM constitutes the display buffer that stores a code word for each character. The first byte of a word identifies the character, using the IBM extended 8-bit ASCII code. Normal characters are represented with a logical 0 MSB, and additional special characters use a logical 1. The second byte is an *attribute* providing optional underline, blinking, high intensity, nondisplay, and reverse image. For example, attribute 07H is that for a normal display, with the character in green on a black background, and F8 is that for a blinking character in reverse video (black on green).

Within the 2K words of the buffer are the 2000 characters, including blanks, of the 25 rows of 80 characters each. Buffer addresses are from B0000H (704K) to B1000H (708K) in the CPU address space. Location B0000H stores the word of the first character of row 1 of the display, location B0002H stores that of the next character of the row, and so on. The last character of row 25 has address B0F9EH, corresponding to the initial address plus 3998 decimal.

Whenever a word is inserted into a location within this address space, the corresponding character or blank appears in the appropriate row and column of the display. *The circuitry continually scans the buffer at a rapid rate, converting each ASCII code into character dots using the character generator, and then displaying the character*. Blank spaces have code 20H.

The display monitor contains an 11.5-inch diagonal 90-degree deflection CRT, along with necessary analog circuitry. The screen has a green, slow-changing phosphor coating with an etched surface that reduces glare, and it is refreshed at the rate of 50 Hz. Because of its high retentivity, the phosphor provides considerable eye comfort, but a disadvantage is that it negates the use of graphics or a light pen. Of course, other displays can be used, including ordinary black-and-white and color TV sets. These devices are not very satisfactory for most purposes. Two 3-foot cables connect the monitor with the system unit. One of these cables transmits signals, using a nine-pin "D" connector to the adapter card, and the other supplies AC power to the monitor.

MEMORY-MAPPED I/O

To illustrate memory-mapped I/O, we examine an assembly language program that displays on the monochrome monitor all 256 characters of the extended ASCII set of the PC. The display starts at the leftmost position of line 10. This is accompanied with a BASIC program that does the same task, but starting at the top of the screen, with instructions that correspond rather closely to those of the assembly language routine. The two programs are compared.

The 80 characters per line require 80 word locations in the buffer, or 160 bytes. Therefore, line 10 of the CRT starts at offset 1600, or 640H. The assembly language is given at the left side of Fig. 5-11, and the corresponding BASIC program is shown at the right. Initially, we examine the assembly language.

The first two assembly language instructions establish the segment address DS at B000H, which points to the base of the buffer at physical address B0000H. Index register DI is cleared with XOR, and the total word count of 2000, or 7D0H, is moved to CX. Then a loop clears the display by placing the space code 20H in each even location. At the same time, the normal attribute code is put into the odd locations. For the immediate operand 0720H, note that the low byte 20H is placed in the low part of the word location at the even address. Instruction LOOP A is precisely the same as the sequence DEC CX followed by JNZ A.

Directive WORD PTR in the instruction with label A tells the assembler that [DI] points to a word location. The 16-bit immediate operand is unacceptable for this purpose. However, the directive is not required in the instruction with label B because the operand of AL is, without question, a byte.

After the screen is cleared, register AL is initialized with a subtract operation to the value 0 of the first ASCII code. Also, DI is initialized to the offset of the first character of line 10, and CX is loaded with the character count of 256. The loop that starts at label B increments AL on each pass to give the ASCII code for the next pass. On the last pass AL is incremented to 256, but the program then ends with no output. All 256 characters are displayed, with codes from 0 through 255. The first and last ones are blanks, as is the one with code 20H.

```
   MOV  AX, 0B000H                     100 X$ = "00:00:00"
   MOV  DS, AX                         110 TIME$ = X$
   XOR  DI, DI                         120 CLS
   MOV  CX, 2000                       130 DEF SEG = &HB000
A: MOV  WORD PTR [DI], 0720H           140 FOR I = 1 TO 511 STEP 2
   INC  DI                             150   POKE I, 7
   INC  DI                             160 NEXT
   LOOP A                              170 J = 0
   MOV  DI, 1600                       180 FOR I = 0 TO 510 STEP 2
   MOV  CX, 256                        190   POKE I, J
   SUB  AL, AL                         200   J = J + 1
B: MOV  [DI], AL                       210 NEXT
   INC  AL                             220 PRINT:PRINT:PRINT:PRINT
   INC  DI                             230 PRINT TIME$
   INC  DI
   LOOP B
```

FIGURE 5-11. Two routines that display 256 characters.

The BASIC Program. As in the assembly language program, the BASIC program of Fig. 5-11 clears the screen and then displays the 256 characters. The first character, which is a blank, is located at home. At the start of the program, the internal clock is set to 0, and the time upon completion is displayed.

BASIC statement CLS clears the screen and places the cursor at home. When the interpreter receives the code for CLS, it locates the keyword in a table and then branches to a machine language routine such as that of the assembly language loop at label A. Upon completion, the interpreter returns to its command level and seeks the next statement. There is considerable overhead, which reduces processing speed.

The FOR loop from line 140 to line 160 pokes the attribute byte into the first 256 locations with odd addresses, and the second FOR loop pokes the extended ASCII codes into the first 256 even-address locations. The first three PRINT statements preceding PRINT TIME$ give three blank lines below the top of the screen, thereby preventing the display of time from overwriting the characters. Then the fourth PRINT provides a blank line that separates the time from the three lines of characters. The time shown was 2 seconds. The procedure of the BASIC routine corresponds closely with that of the assembly language.

Assembly Language. The assembly language is easily hand-assembled with the aid of Appendix A, or an assembler can be used. The hexadecimal code is found to be B800B0 8ED8 31FF B9D007 C7052007 47 47 E2F8 BF4006 B90001 28C0 8805 FEC0 47 47 E2F8, with spaces separating the instructions.

There are 34 bytes that must be stored in memory. They can be loaded from the keyboard in about 1 minute, using the enter command E of the DEBUG program. An alternate method is to enter the mnemonic code using the mini-assembler. These procedures are described in detail in Chapter 7. A breakpoint must be set to stop execution when the program is run. Execution clears the screen and displays the characters.

Each of the four instructions of the first loop is executed 2,000 times, and each of the five of the second loop is executed 256 times. Along with the seven instructions not in loops, the total is 9,287. From timing information given in the Intel 8086/8088 manual, it is found that they are processed in 90,240 clocks, or 19 milliseconds. The average time per instruction is about 10 clocks, or 2 microseconds.

In most cases, it is a good idea to save some of the registers, especially DS. To save DS, simply add PUSH DS at the start and end the routine with POP DS. Flags are saved with PUSHF and POPF.

Comparison. In the BASIC routine the execution time was given as 2 seconds, but the TIME$ statement does not give fractions. By replacing the PRINT TIME$ of line 230 with the following two statements, a more precise time is obtained:

```
230  DEF SEG = &H40
240  PRINT (PEEK(&H6C)/18.2)
```

There are 18.2 ticks per second. The PEEK (&H6C) function reads the number of ticks stored at byte location 0040:006C, which is the low byte of the timer storage. The number of ticks is divided by 18.2 and printed, giving an accurate execution time. The result was 2.42 seconds. In comparison with the 19 milliseconds of the assembly language run, this is 127 times longer. Thus, we see that the BASIC interpreter is relatively slow.

However, if a DEFINT statement is inserted at the start of the BASIC routine, the time is reduced to 1.65 seconds, which is 87 times longer than that of the assembly language. Note, however, that a compiled BASIC program runs much faster than does one on an interpreter.

The BASIC program is certainly simpler to create and execute on the PC. Furthermore, it can be run on any personal computer having an appropriate interpreter, not just one using the 8088 processor.

For fun, an attempt was made to find the time of execution of the assembly language by reading the timer ticks from location 0040:006C, as was done in BASIC. With repeated executions of the program, the ticks read from the byte location were either 0 or 1. At a rate of 18.2 ticks per second, the interval between ticks is 55 milliseconds, and the execution time of the assembly language is only 19 milliseconds. At most, no more than one tick can occur within a 19-millisecond interval. Whether or not a tick occurs depends on the random start time of the routine.

PEEK

In the BASIC statement x = PEEK(n), with n denoting an integer from 0 through 65,535, the function PEEK(n) reads the decimal byte at the location with offset n and assigns the value to variable x. The segment address is the current segment as defined by the last DEF SEG statement. With PEEK we can examine any byte location in the 1-MB memory space, including both RAM and ROM and the display buffers of the monitors. By putting the PEEK function within a FOR-NEXT loop, a region of memory can quickly be examined, with the stored bytes either displayed or printed on the line printer.

Iteration Instructions

Iteration instructions control the repetition of software loops, treating register CX as a counter. All transfers of the program are SHORT, with an 8-bit displacement that specifies a target relative to IP before the jump. As in the case of conditional jumps, the target range is from −128 through +127 bytes. No flags are affected. The instructions of this group are shown in Fig. 5-12.

Each of the iteration control instructions can be replaced with a sequence of instructions, as shown in Fig. 5-13, except that flags are not affected. Note that LOOPE and LOOPNE are similar to LOOP except for a conditional jump around LOOP. The AND instruction of (d) sets flags without changing CX. The sequences should help to clarify the word explanations that follow.

Iteration Controls	
LOOP short-label	Loop until CX = 0
LOOPZ/LOOPE short-label	Loop if ZF = 1, CX not 0
LOOPNZ/LOOPNE short-label	Loop if ZF = 0, CX not 0
JCXZ short-label	Jump if CX is 0

FIGURE 5-12. Instructions for control of iterations.

```
DEC CX        JNZ  X        JZ   X        AND CX,CX
JNZ A         LOOP A        LOOP A        JZ  A
           X: (next)     X: (next)
```

(a) LOOP A. (b) LOOPE A. (c) LOOPNE A. (d) JCXZ A.

FIGURE 5-13. Iteration instructions and their equivalent sequences, except that flags are unaffected and the final values of CX may differ.

LOOP. LOOP decrements the count register by 1 and transfers the program to the target if CX is not 0. When CX becomes 0, the jump is not taken, and the next instruction in the normal sequence is processed. Execution takes 17 clocks when the jump is made and 5 clocks when not made. *LOOP is the assembly language equivalent of the FOR-NEXT loop of BASIC,* with the initial count put into CX rather into the BASIC count variable. It is often used to introduce a time delay in a program. An example is given in Fig. 5-14, with machine code included.

The conditional jump in the routine of Fig. 5-14 gives two passes through the delay loop. During the first pass, the loop instruction gives a jump to itself 18,432 times, which is 4800H, and CX is reduced to 0. In the first execution of LOOP in the next pass, CX is decremented to FFFFH. Thus, there are 65,536 executions of LOOP in this pass, and LOOP is executed a total of 83,968 times. The total delay is the product of the clock period, the 17 clocks per repetition, and the number of repetitions. For a clock period of 210 nanoseconds, the result is 300 milliseconds.

Clocks of the other instructions are negligible. The first two move instructions require 4 clocks each, DEC uses 2 clocks for each of two passes, and JNZ takes 16 clocks for the jump and 4 for the exit. The clocks for execution of instructions are included in instruction set tables provided by Intel.

LOOPZ/LOOPE. *Loop-while-zero LOOPZ* and *loop-while-equal LOOPE* are the same. Both mnemonic instructions decrement CX by 1 and give a jump to the target if CX is not 0 and if ZF is set. Thus, LOOPZ is similar to LOOP, except that the jump is allowed only while ZF is set. A conditional exit based on ZF is provided. The jump takes 18 clocks, with 6 for no jump.

LOOPNZ/LOOPNE. *Loop-while-not-zero LOOPNZ* and its equivalent *LOOPNE* are the same as the preceding instructions, except that flag ZF must be clear for a repeat. Also, of course, CX must not be 0. Thus, LOOPNZ is similar to LOOP, except that the jump is allowed only while the zero flag is not set. Clocks are 19 and 5.

Suppose the CPU has requested an input device connected to the bit 4 terminal of byte PORT_C to supply serial data. Initially, the device is inactive, and a high voltage is present at the input. The first transmitted bit of the serial data will be a

```
        MOV  BX, 2       ; code is C7C30200
        MOV  CX, 4800H   ; C7C10048
DELAY:  LOOP DELAY       ; E2FE
        DEC  BX          ; FFCB
        JNZ  DELAY       ; 75FA
```

FIGURE 5-14. A time delay sequence.

```
       MOV     CX, 28000
   A1: IN      AL, PORT_C          A1: IN   AL, PORT_C
       TEST    AL, 10H                 TEST AL, 10H
       LOOPNZ  A1                      JNZ  A1
```

FIGURE 5-15. Illustration of a timed exit with LOOPNZ.

low-level start bit, for which the processor must wait. Consider the two routines of Fig. 5-15.

While waiting, each program loops. Instruction IN uses 10 clocks to move the byte from the port to AL, and TEST uses 4 clocks to determine if bit 4 is 0. The test is an AND operation that affects the flags but not the operands. The difference in the routines is in the timed exit provided by LOOPNZ. If there is no response from the device within 194 milliseconds, the routine at the left proceeds to the instruction that follows LOOPNZ, whereas the one at the right waits indefinitely. The possibility of an endless loop such as this should be avoided.

JCXZ. *JCXZ* is the mnemonic for *jump-if-CX-zero*, and the jump is made only if CX stores word 0. The instruction does not affect the content of CX. It is sometimes placed ahead of a loop to bypass the loop if CX is initially 0.

Directive PTR

In Section 3-4 the PTR assembler directives were explained. The instructions that follow illustrate further the use of these directives.

```
MOV BYTE PTR [BX], 5          MOV DWORD PTR [BX], 5
MOV WORD PTR [BX], 5
```

The one with BYTE PTR moves 05H to the byte location pointed to by the contents of BX. The WORD PTR directive causes word 0005H to be transferred to the word location, and the DWORD PTR gives a transfer of 00000005H. Note that instruction MOV [BX],5 is ambiguous. If it is put into a program, the assembler will generate an error. As mentioned earlier, in a memory reference the PTR directive will override a prior declaration of a variable as byte, word, or doubleword.

5-5 PROGRAMMABLE PERIPHERAL INTERFACE

Some I/O devices require sophisticated interface controllers, whereas many others need only a buffered terminal, or port, that can be addressed and used for writing to a device or reading from it. For example, an octal latch such as the TTL 74LS373 shown in Fig. 5-1 might be used as an output port. The decoded address can be ANDed with control $\overline{\text{IOW}}$ to enable input loading from the data bus. The outputs to the load might be enabled at all times. Thus, when an OUT instruction to the port is executed, during $\overline{\text{IOW}}$, the latch is transparent, and the byte that is transferred to the port from register AL passes to the load. When $\overline{\text{IOW}}$ goes high, the byte is latched, and its inputs are isolated from the data bus until it is again addressed with an OUT instruction.

Latches as Ports

In Fig. 5-16 is an application of latches connected as ports. There are two ports, with latch A used for output and latch B for input. *"Input" and "output" are always referenced to the CPU.* The circuit is a chip tester, with a 7404 TTL hex inverter shown, but it can be used for any combinational logic gate with no more than eight inputs and eight outputs. In this example, the hex inverter chip is the peripheral.

In Fig. 5-16 the input signal ADDR feeds from an address decoder. It is active high only when the address has the assumed port number 055H, which applies to each port. The read/write controls come from the bus controller. At the left side of the data bus is the CPU, and the bus is connected to the various units of the system. Note that *latch A is transparent during the negative write pulse of an OUT instruction, and latch B is transparent during the negative read pulse of an IN instruction.* The outputs of latch A feed into the inverters at all times, and the outputs of the inverters feed into latch B at all times.

To illustrate the operation of the chip tester, the sequence of Fig. 5-17 is presented. Following the last instruction shown should be a sequence that prints "Chip OK" on the display, and the ERROR routine should display the message "Chip defective".

The 74LS373 latches and the hex inverter are sufficiently fast to respond to the OUT and IN instructions without error produced by propagation delays within the chips. However, in a system such as that of the PC, data transfer between the processor and a peripheral may pass through several buffers and transceivers, all of which produce delays. In addition, there may be appreciable delay within the peripheral itself. Consequently, it is often necessary to insert delay loops between the OUT and IN instructions in order to allow time for the response to become stable.

In order to implement the chip tester with the PC, there must be a way to access

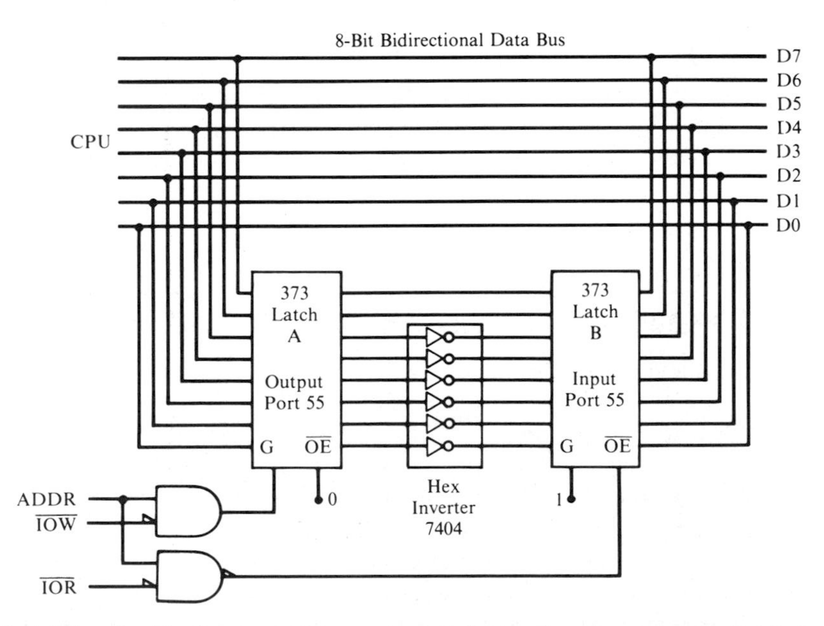

FIGURE 5-16. A chip tester with input port latch A and output port latch B. Each port number is 55H, and this address activates input signal ADDR.

```
XOR AL, AL        ;clear AL
OUT 55H, AL       ;send byte 00H to output port latch A
IN  AL, 55H       ;get byte from input port latch B
XOR AL, 3FH       ;input byte should be 3FH
JNZ ERROR         ;if not 0, go to ERROR routine
MOV AL, 0FFH      ;move FFH to AL for next test
OUT 55H, AL       ;send test byte to output port
IN  AL, 55H       ;get byte from input port
AND AL, 3FH       ;input byte should be C0H
JNZ ERROR         ;if not, go to error routine
continue
```

FIGURE 5-17. Routine for test of the hex inverter chip.

the internal address, data, and control buses. This can be done with a special adapter card that brings the signals from a slot of the system board to an external project board.

The 8255A PPI

Special interface chips that provide a number of general purpose ports are available. We shall examine one of these, the *PPI*, with emphasis on its use in the PC. The hardware and applications presented here are reasonably typical.

The 8255A-5 PPI is an NMOS chip in a 40-pin DIP designed for interfacing peripheral devices with a CPU. Basically, it consists of 24 I/O lines, along with buffers and a timing and control unit. Except for the 5 V supply lines, the pins are identified in Fig. 5-18, which shows the inputs and outputs of the PPI of the PC.

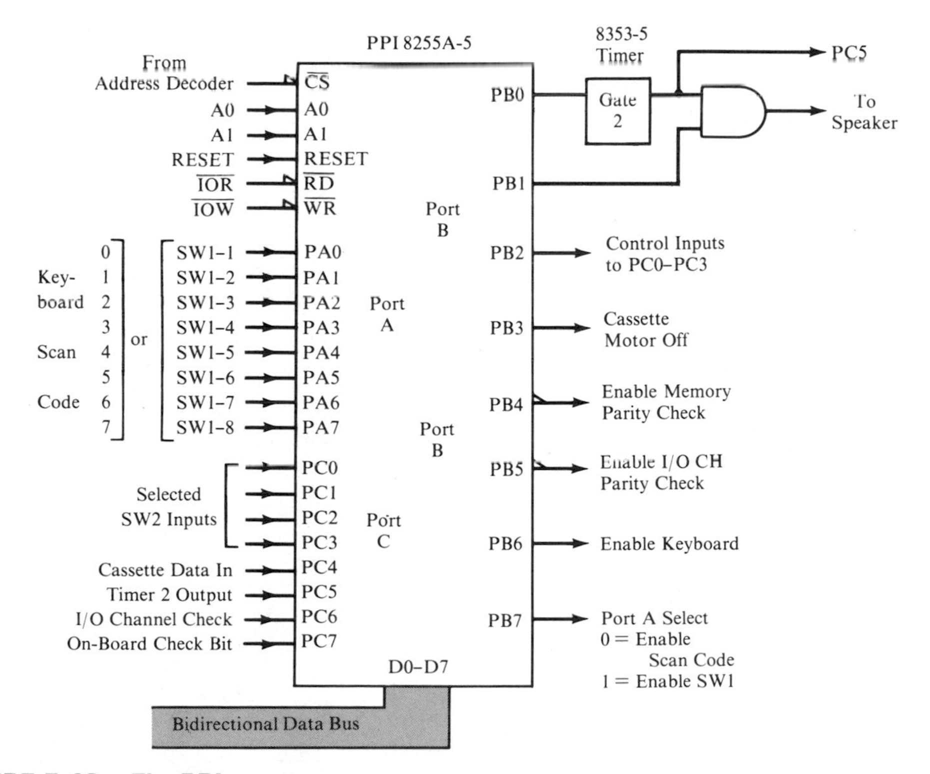

FIGURE 5-18. The PPI.

The three-state bidirectional data bus is used both for programming the chip and for communicating with the system.

Chip select $\overline{CS}$ is fed from a decoder output that is low-level active when address bits A9 through A5 are 00011, and bits A0 and A1 are connected directly. Thus, the chip address is $00011XXXA_1A_0$. Port addresses of 60H through 62H, used with IN and OUT instructions, address ports A, B, and C, respectively, and address 63H provides for writing a mode word into the control register. When $\overline{CS}$ is inactive or when both $\overline{RD}$ and $\overline{WR}$ are active, the pins of the data bus float in the three-state mode.

During a system reset, the RESET input from the clock generator clears the control register and sets all ports to the input mode. If the three 8-bit ports are to be used for input only, no initialization is required. However, by writing to the control register, the functions of the I/O pins can be changed. Programming can be done at any time, as often as desired.

There are three basic modes of operation. For mode 0 the control word is that of Fig. 5-19. The MSB is the mode set flag, which must be 1 for programming. The three 0s of the byte specify that mode 0 applies to all ports, and each remaining bit selects a group for input or output. Because the two nibbles of port C are individually programmable, 16 different combinations are possible.

During the power-on initialization of the PC, byte 99H is moved to AL, followed by the instruction OUT 63H,AL. As indicated in Fig. 5-19, this procedure programs ports A and C for input and port B for output. A byte sent to an output port is latched, and *the latched data can be read*. Thus, to read the last byte sent to output port B, instruction IN AL,61H is executed. A byte transferred through an input port is not latched.

Not used in the PC system are the other two basic modes of the PPI, but they are available, of course. One of these modes provides a means for transferring I/O data to or from a specified port using handshake signals for control. The other allows communication with a peripheral on a single 8-bit bus for both transmission and reception; the data flow of this mode is also controlled with handshake signals. Combinations of the different modes are allowed. Only mode 0 is considered here.

Output Port B

Port B provides eight outputs. Pin PB0 is connected directly to the GATE 2 input of the 8253 timer chip, which is described in the next section. *A logical 1 on pin PB0 enables the output of timer 2,* and a 0 forces the output high. The timer output connects to pin PC5 of port C and also to an input of an AND gate that feeds the speaker. The other AND gate input comes directly from bit PB1, which can supply the speaker with a CPU programmed signal. *If both PB0 and PB1 are held high, then the output of timer 2 is gated into the speaker.*

Bit PB2 controls the inputs to the low-order nibble of port C. With PB2 high, settings of switches 1, 2, 3, and 4 of DIP switch SW2 are fed into pins PC0-PC3,

1 = input
0 = output

				Upper Nibble			Lower Nibble
1	0	0	Port A	Port C	0	Port B	Port C

FIGURE 5-19. PPI control word for mode 0.

respectively, with an open switch supplying a logical 1. These switches are manually set according to the amount of memory installed in excess of 64 KB. For 64 KB or less, all are on, giving 0s. When PB2 is low, unused switches 5 through 8 are read. The next bit, PB3, turns the cassette motor off when high and on when low. With diskette drives, an audio cassette is not normally used, and PB3 is available for other purposes.

Bit PB4 enables the memory parity check when low, provided a memory read operation is implemented. Odd parity is employed. When a byte is written into memory, the parity generator/checker chip supplies a logical 1 to the ninth bit position if and only if the byte contains an even number of 1s. Thus, *every 9-bit word of memory should have an odd number of 1s at all times.* During a read operation, the checker verifies that the 9-bit word that is read does indeed have an odd number of 1s. Otherwise a memory error has occurred. The checker supplies an output parity bit of 1 for no error and 0 for an error. The parity check bit (PCK) is the complement of this, and therefore it should be 0. Otherwise the word read is not that which was written. This topic is examined in more detail later.

The I/O channel parity check enable bit (PB5) enables the circuit receiving the PCK from any memory that is present on an adapter card. When output enable bit PB5 is active, the PCK is supplied to input pin PC6 of port C of the PPI, allowing the bit to be read by the program. Also, if the nonmaskable interrupt circuits are enabled by the software, a parity error in the external memory produces an active signal on the nonmaskable interrupt (NMI) pin of the CPU. The interrupt routine generates an error message on the monitor.

In the high state, bit PB6 enables the keyboard. When a byte written to port B makes the bit 0, the keyboard clock is held low, completely disabling all keys. Bit 6 must be 1 and the scan code enable bit 7 must be 0 for keyboard operation. It is sometimes necessary to disable the keyboard while a sequence of instructions is processed; this can be accomplished by a write operation to port B. A second write operation can reenable it.

Input Port A

The high-order bit PB7 of port B programs port A for reading either the eight switch settings of DIP switch SW1 or the keyboard scan codes. When PB7 is 1, the eight parallel outputs of the scan code shift register are disabled, both the register and the keyboard interrupt request line IRQ1 are cleared, the keyboard serial input into the shift register is inhibited, and the operator has no keyboard control. At the same time, the outputs of the octal buffer fed by SW1 are enabled and supplied to port A. The effects are reversed when PB0 is 0, and port A then receives the octal keyboard scan code.

The settings of SW1 correspond to the installed hardware. Switch 1 is open, indicating that a diskette drive is attached. With no 8087 coprocessor, switch 2 is closed. Switches 3 and 4 are open, for 64 KB of on-board memory, and switches 5 and 6 are also open, indicating that an IBM monochrome display is present. These settings apply even though a color monitor is also connected. Switch 7 is open and 8 is closed, specifying two disk drives. Because switches 1 through 8 connect to port A inputs 0 through 7, respectively, with an open switch supplying a 1 and a closed switch supplying a 0, the word read by port A when PB7 is 1 is 7DH. When the processor needs information about these hardware options, an IN instruction to port 60H is executed. Newer versions of the PC contain 256 KB of on-board memory.

In a system having both monochrome and color monitors, it is necessary to change the switch settings as seen by a program, although the hardware switch positions must not be changed. During the initialization routine of a system reset, the settings of SW1 are read from port A and placed in the low byte of a word location with name EQUIP_FLAG. The high byte is used to indicate additional hardware options, such as a printer. The address of EQUIP_FLAG is 0040:0010. In the video routine in ROM BIOS, this location is examined to determine if the display is monochrome or color. Bits 5 and 4, corresponding to switches 6 and 5, can be changed with software to indicate a color monitor. Then the video routine automatically sends output to this unit.

To return to the monochrome monitor, it is necessary to restore the original data of EQUIP_FLAG and activate the video routine. It is not possible to operate both monitors simultaneously, but switching back and forth is done quickly. Routines that change the switch settings as indicated by EQUIP_FLAG are examined in Section 7-5, in both assembly language and BASIC.

Input Port C

The low-order nibble of port C is used to read the switches of SW2. When PB2 of port B is 1, the nibble read in from closed switches 1 through 4 is 0110, indicating a total read/write memory of 256 KB. When PB2 is 0, the nibble from open switches 5 through 8 is 1111. These nibbles have no significance in the system considered here.

Bit PC4 receives data directly from the audio cassette option. It was mentioned earlier that the gated output of timer 2 passes through an AND gate to the speaker. It is also fed into PC5 for use by the system as desired. Into pin PC6 is fed the I/O channel PCK, and the last bit PC7 is the on-board PCK. If it is 1, an on-board memory error is indicated. Fortunately, this is most unlikely.

Programming Examples

We now consider several examples illustrating the use of the PPI. These examples assume that PORT_A, PORT_B, and PORT_C are equated with their respective port numbers.

Parity Check. Suppose a check for an on-board parity error is desired after a memory write operation. The instructions of Fig. 5-20 provide this check.

In the sequence of Fig. 5-20, the first three instructions fetch the current byte

```
IN    AL, PORT_B     ;get the latched byte
AND   AL, 0EFH       ;clear bit 4
OUT   PORT_B, AL     ;enable parity bit PCK
IN    AL, PORT_C     ;get port C data
TEST  AL, 80H        ;isolate bit PCK
JNZ   addr           ;error if PCK = 1
```

FIGURE 5-20. Routine for on-board memory parity check.

```
    IN  AL, PORT_A   ;get SW1 settings
    AND AL, 0CH      ;clear all bits except 3 and 2
    ADD AL, 4        ;convert 0,4,8,12 to 4,8,12,16
    MOV CL, 12       ;shift count = 12 for multiplication
    SHL AX, CL       ;multiply by 4K, giving 16, 32, 48, 64K
    MOV CX, AX       ;number of locations to count reg CX
    XOR AX, AX       ;clear AX
    MOV ES, AX       ;clear ES for start of storage at 0
    XOR DI, DI       ;clear index
    MOV AL, 0ADH     ;storage byte to AL
    CLD              ;direction up
A:  STOSB            ;store a byte, decrement CX, inc DI
    LOOP A           ;loop until finished
```

FIGURE 5-21. Routine to determine the on-board memory size and write 0ADH into each location.

from the latch of port B, clear bit 4, and then return the byte to the latch. This procedure enables the parity check circuit that supplies PCK to pin PC7. The next three instructions read the byte from port C and test bit PCK. If it is 1, the program branches to an error routine. This branching should interrupt processing to send an error message to the display. The instruction TEST can be replaced with AND if desired.

Memory Size Determination and Data Fill. Next, suppose a program must determine from the switch settings the amount of on-board memory and then write byte ADH into each location. Assume that bit PB7 is set, so that the SW1 settings feed into port A. Switches 4 and 3 are set to give bits PA3 and PA2 that are 00 for 16 KB, 01 for 32 KB, 10 for 48 KB, and 11 for 64 KB of on-board memory. A similar routine is part of a memory test found in the system initialization following power-on. A sequence that accomplishes this task is given in Fig. 5-21.

In the routine of Fig. 5-21, if the on-board memory is 64K, the word of AX after the 12 left shifts is 0000H, with CF set. In the loop the first pass decrements CX to FFFFH. Thus, the total number of bytes transferred to memory is precisely 64K.

Keyboard Enable. Enabling of the keyboard is accomplished by loading AL with 48H and then executing the instruction OUT PORT_B, AL. Following this operation, receipt of a valid scan code to the keyboard shift register automatically initiates an interrupt, and the program branches to the keyboard service routine. In this routine the scan code is read into the accumulator with the instruction IN AL, PORT_A.

Now suppose AL is loaded with CCH. The OUT instruction to port B enables data from the equipment switches of SW1 onto the pins of port A, enables the keyboard and parity check, turns off the cassette motor if it was on, and places data from the hardware switches 1 through 4 of SW2 on pins 0 through 3 of port C. Although the keyboard circuits are enabled, operation is disabled because scan codes cannot be read when PB7 is 0.

Speaker Beep. Byte 4FH to port B connects the output of timer 2 to the speaker. This output is inhibited when PB0 is 0, which gives a high-level timer

```
        CLI                 ;Numbers given below are clocks:
        MOV   BX, 0C0H      ;           4
A:      MOV   AL, 4CH       ;           4
        OUT   61H, AL       ;          10
        MOV   CX, 100H      ;           4
B:      LOOP  B             ;          17
        OR    AL, 2         ;           4
        OUT   61H, AL       ;          10
        MOV   CX, 100H      ;           4
C:      LOOP  C             ;          17
        DEC   BX            ;           2
        JNZ   A             ;          16  (for jump)
        STI                 ;Interrupts on
```

FIGURE 5-22. Routine to beep the speaker.

output. In this case, bit PB1 is fed to the speaker. The program of Fig. 5-22 uses this bit to beep the speaker (see Prob. 5-18).

The routine repeatedly toggles bit PB1, with time delays inserted to give a desired frequency. The duration of the beep is determined by the initial count of BX. If the program is run with the interrupt system enabled, the beep has a warble sound. This warble is the result of the periodic timer ticks that interrupt the program and the beep once every 55 milliseconds. Execution of the routine with the interrupt system disabled eliminates the warble, giving a pure tone. Interrupts are enabled with instruction STI and are disabled with CLI. With interrupts disabled, the time-of-day clock receives no ticks, and of course, the keyboard is disabled.

BASIC Statements BEEP and SOUND. In BASIC, statement BEEP sounds the speaker at 800 Hz for 0.25 seconds. The statement executes a routine similar to that of Fig. 5-22. The SOUND statement is more flexible. Its format is

SOUND frequency, duration

The specified frequency in hertz is any numeric expression having a value from 37 to 32,767. The duration in clock ticks is a numeric expression with a value from 0 to 65,535. Accordingly, SOUND 1000,182 gives a 1000-Hz tone for 10 seconds.

These examples show some of the essential functions of the PPI chip. Another important peripheral found on the computer board of the PC is examined next.

5-6 TIMER/COUNTER

A binary counter can serve as a *timer,* and most computers have one or more timers. To illustrate the hardware and some of the applications of such peripherals, the timers and counters of the PC are examined here.

Timer Connections

The timer is the 24-pin 8253-5 NMOS chip containing three independent 16-bit down counters. Each counter is individually programmable, and accurate time

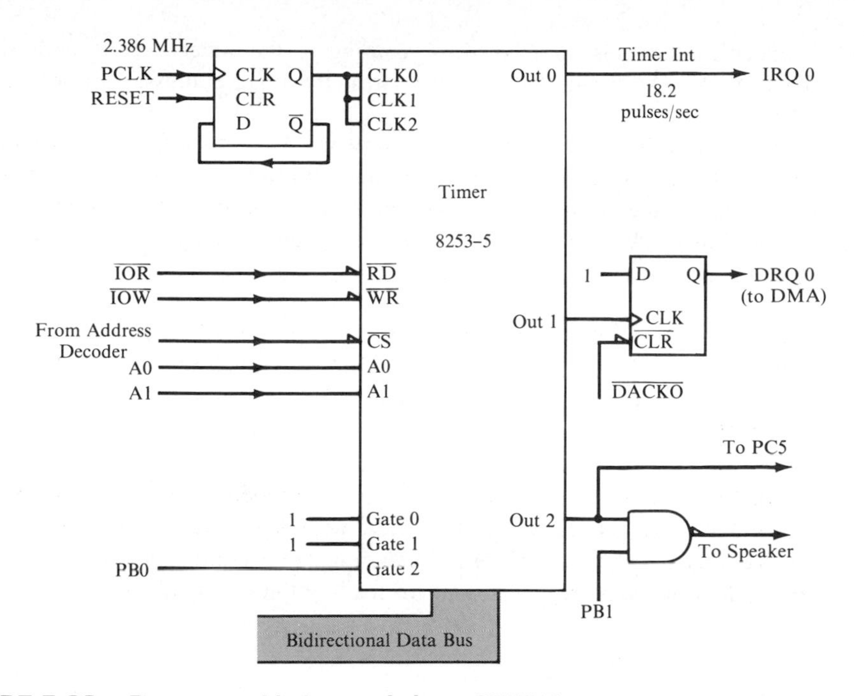

FIGURE 5-23. Programmable interval timer 8253-5.

delays can be generated under software control. Connections to the timer of the PC are illustrated in Fig. 5-23.

The Clocks. The 2.386-MHz PCLK line from the clock generator is divided by 2 by the 1-bit, D-type, positive edge-triggered register, supplying a 1.193-MHz square wave to each input CLK. This register is called a *D flip-flop*. On each low-to-high transition of the signal PCLK at the dynamic clock input of the flip-flop, the bit value of input D becomes the output Q, and output $\overline{Q}$ is its complement. Unlike the D latch, the input at D has no effect except during the active clock transition.

With the connections as shown in the figure, input D at the start of the active transition is always the complement of output Q. Accordingly, output Q switches state each time the dynamic clock input PCLK goes low to high. The waveforms of input PCLK, or Q, and output CLK are shown in Fig. 5-24. Because the frequency of CLK is exactly one-half that of PCLK, this is often called a *divide-by-2* circuit.

The three counters have individual CLK inputs. In the PC each is fed by the 1.193-MHz square wave from the output of the flip-flop.

Addresses. In the PC the chip select $\overline{CS}$ input comes from an address decoder, and $\overline{CS}$ is active when address bits A9 through A5 are 00010. Bits A0 and A1

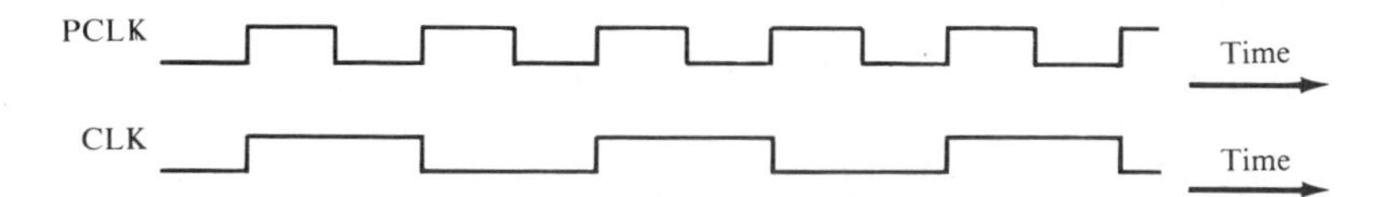

FIGURE 5-24. Waveforms illustrating frequency division.

connect to the pins so labeled. Accordingly, the timer port is addressed by $00010XXXA_1A_0$. Using 0s for don't-cares, the timer addresses are 40H through 43H.

With $\overline{CS}$ enabled, when the 2-bit input word A1,A0 is 00, the count register of counter 0 is accessed for reading or writing. Word 01 accesses the count register of counter 1, and word 10 is for counter 2. The control word register is accessed for writing by word 11. It cannot be read. Summarizing, the addresses are as follows:

Port 40H	Count register of counter 0	(read/write)
Port 41H	Count register of counter 1	(read/write)
Port 42H	Count register of counter 2	(read/write)
Port 43H	Control word register	(write only)

The bidirectional data bus is in the three-state mode if $\overline{CS}$ is inactive, or if both $\overline{RD}$ and $\overline{WR}$ are active, or if $\overline{RD}$ is active with A_1A_0 equal to 11. The three-state bidirectional data bus is used for programming the operating modes of the counters, for loading the count registers, and for reading the count values.

Programming

Before a counter can be used, it must be programmed. The counters can be programmed in any order and as often as desired. Programming a counter consists of writing a word to the control register and then writing the count value to its count register. The format of the control word is shown in Fig. 5-25.

With bit BCD equal to 0, the 16-bit down count is binary, and the initial count is presettable. When the bit is 1, the 16 bits give 4 BCD digits, and the count can go from 9999 to 0000. There are six different modes.

Counter 0 (Clock Ticks)

In the PC the counter of channel 0 is dedicated as a timer *that generates periodic clock ticks at a rate of 18.2 per second.* During a system reset, the initialization

SC1	SC0	RL1	RL0	M2	M1	M0	BCD

Select SC1, SC0	Read/load RL1, RL0
00 = counter 0	00 = Latch count
01 = counter 1	01 = Read/load LSB
10 = counter 2	10 = Read/load MSB
11 not allowed	11 = Read/load LSB, then MSB

Mode M2, M1, M0 = 000, 001, 010, 011, 100, or 101
for modes 0, 1, 2, 3, 4, and 5

BCD = 0 for binary count (16 bits)
BCD = 1 for decimal count (BCD, 4 decades)

FIGURE 5-25. Control word format.

```
MOV AL,   36H    ;select counter 0, LSB-MSB, mode 3
OUT 43H, AL      ;write the control register
MOV AL,   0      ;count 0
OUT 40H, AL      ;write LSB to timer 0 register
OUT 40H, AL      ;write MSB to timer 0 register
```

FIGURE 5-26. Counter 0 programming.

routine first checks the counter to verify that it operates satisfactorily. Then the counter is programmed with the sequence of Fig. 5-26.

Byte 36H to the control register selects *mode 3* for counter 0, specifying that *2 bytes* are to be assigned to the *binary* count. Mode 3 provides a *square wave generator*. The write operations to the count register of timer 0 load the 16-bit register with 0s, *giving the maximum count of 65,536*. The square wave output is high for 32,768 clocks and low for the same number. It follows that the frequency of the output square wave is the 1.19318-MHz signal of the clock divided by 65,536, or 18.2065 Hz.

Mode 3. A counter programmed for mode 3 operation has its output high for one-half of the programmed count and low for the other half. Then the count repeats. This operation is illustrated in Fig. 5-27. If the number loaded into the count register is odd, the output high level is one clock longer than that of the low level. Thus, for a count of 5, the output is high for three clocks, and low for two. In any event, the frequency of the output pulse train *is precisely equal to that of the clock input divided by the programmed count*. The gate input must be high; otherwise the output is constant at the high level.

In the PC, the gate of counter 0 is fed with a logical 1, or 5 V, as shown in Fig. 5-23. Thus, the counter is always enabled, and the count commences on the first high-to-low clock transition after the second write to the count register.

Timer Ticks. The output of counter 0 is connected to request level 0 of the 8259 interrupt controller. Each positive pulse of the square wave sends a request to the CPU to stop processing of the main program temporarily in order to provide a *tick* for the *time-of-day clock* maintained in memory. Normally the interrupt system is enabled, and the request is honored, giving a type 8 interrupt. Following initiation of the interrupt, the timer routine is executed, and then processing of the main program continues. Timer ticks occur at a rate of 18.2 per second. On the average, about 27 instructions are processed in the timer routine, taking about 0.1 millisecond. Since the interrupts are 55 milliseconds apart, they do not appreciably slow normal processing. Details of the interrupt procedure are examined in the next chapter.

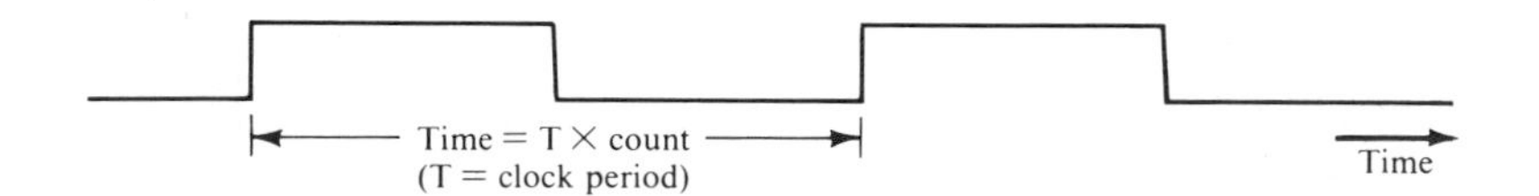

FIGURE 5-27. Mode 3 (square wave generator) output.

```
    INC   TIMER_LOW          ;increment low count
    JNZ   A                  ;jump if not 0
    INC   TIMER_HIGH         ;increment high count
A:  CMP   TIMER_HIGH, 24     ;is high count 24D?
    JNZ   B                  ;if not, jump to B
    CMP   TIMER_LOW, 176     ;is low count 176?
    JNZ   B                  ;if not, jump to B
    MOV   TIMER_HIGH, 0      ;24-hour settings
    MOV   TIMER_LOW, 0
    MOV   TIMER_OFL, 1
B:  (program continues)
```

FIGURE 5-28. Part 1 of the timer interrupt routine.

A portion of the timer interrupt routine is given in Fig. 5-28, with TIMER_LOW and TIMER_HIGH denoting defined words at respective locations 0040:006C and 0040:006E. TIMER_OFL is defined as a byte variable at 0040:0070.

Suppose the program sets the three memory locations to 0. Each interrupt increments TIMER_LOW, and during those interrupts that change the byte from 65,535 to 0, the byte of TIMER_HIGH is also incremented. When the high byte changes from 23 to 24, the low byte commences again from 0, but upon reaching 176, the two word locations are reset and a 1 is written into the overflow byte. The number of ticks in this time span is 176 plus the product of 24 and 65,536, for a total value of 1,573,040. Division by the number of ticks per second gives 86,400 seconds, or precisely 24 hours. A program can be designed that permits the operator to set the count to correspond to the real time and to read this time when desired, with the hours, minutes, and seconds displayed (Prob. 5-23).

A second function of the timer interrupt routine is *to time the drive motor of a diskette and to turn off the motor when the time expires*. With MOTOR_STATUS and MOTOR_COUNT denoting byte variables at location 0040:003F and 0040:0040, respectively, this portion of the program is shown in Fig. 5-29.

By means of a memory write, the motor count is set to a value between 0 and 255, as desired by the diskette program, but it is always decremented at periodic intervals of about 1/18th of a second by the timer interrupts, assuming the interrupts are enabled. As indicated in Fig. 5-29, during execution of the particular interrupt routine that decrements the count to 0, byte 0CH is written into the digital output register (DOR) of the floppy disk controller (FDC) using port address 3F2H. This operation turns off the motor. If the maximum count of 255 is moved to MOTOR_COUNT when the motor is turned on, the motor will run for 14 seconds before turning off, provided a subsequent program instruction does not change the count.

```
B:  DEC   MOTOR_COUNT            ;count for disk motor
    JNZ   C                      ;is count 0?
    AND   MOTOR_STATUS, 0F0H     ;yes, reset flag bits
    MOV   AL, 0CH
    MOV   DX, 3F2H               ;FDC output register
    OUT   DX, AL                 ;turn motor off
C:  (continue)
```

FIGURE 5-29. Part 2 of the timer routine.

```
MOV AL,  54H   ;select timer 1, LSB only, mode 2
OUT 43H, AL    ;write the control register
MOV AL,  18    ;count byte (LSB only)
OUT 41H, AL    ;write timer 1 count register
```

FIGURE 5-30. Initialization of counter 1.

Because most diskette read/write operations require only several seconds, programs normally change the count upon completion of the operation so as to turn off the motor within 2 seconds. Motor turn-off is always done by the timer interrupt. At the same time that the motor is turned off, the low-order nibble of MOTOR_STATUS is cleared. The bits here are flags that indicate whether a motor is running, with bits 0 through 3 corresponding to the motors of drives A, B, C, and D, respectively.

Although a few additional instructions follow those given, the major part of the timer interrupt routine has been examined. The few remaining instructions are related to the interrupt system and are discussed in the next chapter.

Counter 1 (Memory Refresh)

In the PC, counter 1 is used exclusively to supply pulses to the DMA controller for the purpose of activating memory refresh operations at regular intervals. Every 72 system clocks, the counter generates a pulse that latches a logical 1 into a D flip-flop, which feeds into channel 0 of the DMA chip. The flip-flop is cleared by the acknowledge signal $\overline{\text{DACK 0}}$, which is supplied by the DMA controller when refreshing begins. Input CLK1 to the counter is 1.193 MHz, and the counter is enabled at all times with pin GATE 1 connected to 5 V. The connections are shown in Fig. 5-23.

During the self-test of a system reset, counter 1 is tested to determine that it operates properly. Then it is initialized by the sequence of Fig. 5-30.

Mode 2. Byte 54H is written to the control register in the routine of Fig. 5-30, specifying the *rate generator* mode, and a count of 18 is written to the low byte of the count register. With the GATE input high, the count commences downward from the initial value. Upon reaching 0, the register is automatically reloaded with the initial value, and the count continues downward. Each count corresponds to one input clock. Except during the interval from count 1 to 0, the output of the counter is high, and from count 1 to 0 it is low. The periodic output has a frequency equal to that of the CLK input divided by the specified count. If the gate is low, the output is high at all times; thus, the gate input can be used to synchronize the counter. Figure 5-31 shows the input clock and the output of the counter for mode 2 with a count of 4.

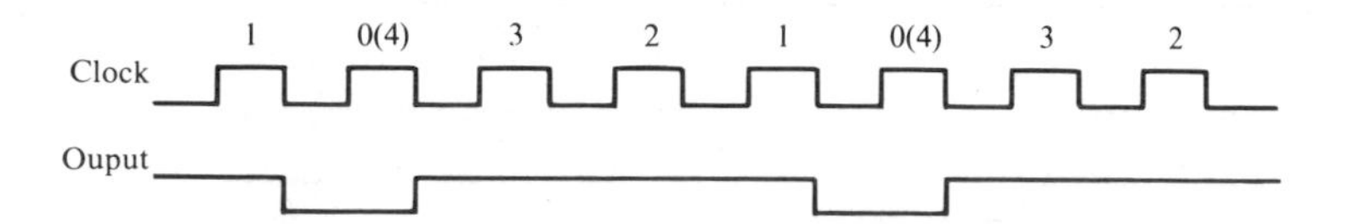

FIGURE 5-31. Mode 2 (rate generator) output for a count of 4.

Memory Refresh. With a count of 18, the frequency of the output pulse train is that of the input clock divided by 18. The frequency of the input clock is exactly one fourth that of the system clock (4.77 MHz). Therefore, the pulse train provides a low-to-high transition to the D flip-flop once every 72 system clocks, and each transition activates a DMA memory refresh. When refreshing occurs, the current instruction cycle of the processor has four wait states inserted, with timing controlled by handshake signals from the wait-state generator of the system board. Accordingly, bus operations by the CPU are inhibited for 4 cycles out of each block of 72 cycles, or for 5.6% of the time.

Only 1 of the 128 rows of every memory chip is refreshed during a DMA refresh operation. Thus, every cell gets refreshed once in a time span equal to the product of 72 clocks, 128 rows, and the 210 nanoseconds of a clock, which gives 2 milliseconds. This cycle is adequate.

Counter 2 (General Purpose)

Counter 2 is available to a program to be used as desired. Its CLK 2 input is the 1.19318-MHz square wave from the divide-by-two D flip-flop of Fig. 5-23, and the input to GATE 2 is from pin PB0 of output port B of the PPI. The timer output goes to pin PC5 of input port C and also to one input of the AND gate feeding the speaker.

At the end of the self-test routine of a system reset, just before the DOS system is booted, there is a beep on the speaker. It serves as an indicator that all tests were successful. The beep is from counter 2, which is programmed to supply a square wave to the speaker for 1.4 seconds. The specified count is 533H, which gives a frequency of 896 Hz. Accordingly, the counter is initially programmed as a square wave generator at 896 Hz. This is mode 3 with a 2-byte binary count.

Mode 1. Only modes 2 and 3 of the six available modes have been considered. Mode 1 gives a *one-shot* output. For a specified count *n*, the output is a single negative pulse with a duration of *n* clocks. After mode 1 is set by writing to the control register, the output is high. It is triggered by the signal at the gate input, and the negative pulse starts on the first high-to-low clock transition following the rising edge of the gate input. This operation is shown in Fig. 5-32 for a count of 3.

Other Modes. *Mode 0* is somewhat similar to mode 1. With the gate input high, the output is initially low following the write to the control register. The count is loaded, and on the next high-to-low input clock transition the downward count commences. The output remains low until the count reaches 0, but then it goes high and stays high. Because the output is often used to generate timed interrupts, mode 0 is called the *interrupt-on-terminal-count* mode.

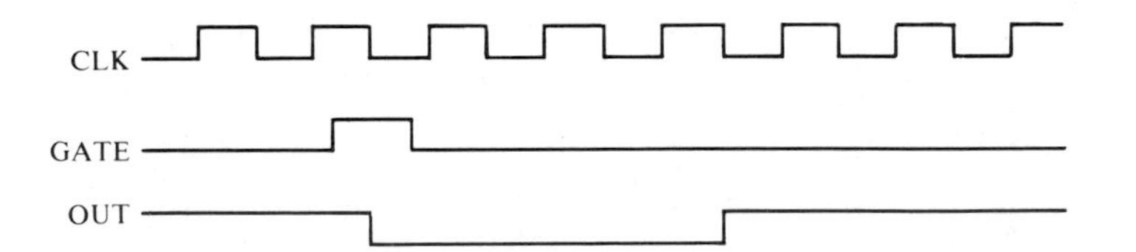

FIGURE 5-32. One-shot counter output for a count of 3.

```
MOV   AL,   80H
OUT   43H,  AL      ;latch the count of timer 2
IN    AL,   42H     ;read the count LSB
MOV   AH,   AL      ;move the LSB to AH
IN    AL,   42H     ;read the MSB
XCHG  AL,   AH      ;exchange the bytes
```

FIGURE 5-33. Routine to read a count register.

Modes 4 and 5 each generate a single pulse that can be used to strobe an input of a chip after a specified time delay. For each, the output is a negative pulse that begins when the downward count reaches 0, and the pulse duration is one input clock period. In mode 4 the count begins as soon as the count is written. Thus, the time at which the pulse occurs is *software triggered.* In mode 5 the count begins when the gate input goes low to high, and thus the pulse is *hardware triggered.*

Read Operations. The count of any of the counters can be read at any time. One way to do this is to inhibit the count by means of the gate or clock inputs and then to read the count register. There is a better way that has no effect on the count in progress. It consists of writing the control register with 00 for bits 5 and 4. Bits 7 and 6 select the counter, and the 4 low-order bits of the control word are don't-cares. When the byte is written, the count at that time is latched in a register for reading, but the count continues. The address for the read operation is the same as for the count register.

To illustrate, suppose it is desired to read the count of timer 2 at some point in a program. Assume that the timer is programmed for a 2-byte count. To put the count in AX, the sequence of Fig. 5-33 can be used.

Whenever a count is read, it is essential that the number of bytes read be the same as the number programmed for the count.

Example

To illustrate programming of a counter, a short routine is examined here that uses counter 2 as a square wave generator at a frequency of 1000 Hz. To verify the frequency, decimal digits 1, 2, 3, 4, and 5 are written to the monitor in a selected column at regular intervals timed by counter 2, and these intervals are measured with a digital watch. Also, the counter beeps the speaker at the end of the routine. Hexadecimal code is given in the comments section.

Initialize. In Fig. 5-34 is shown the initialization. Interrupts are turned on, because the timer interrupts do not affect the counters, which continue counting

```
STI                 ;turn on interrupts              (FB)
MOV DX, 5           ;5 digits to be displayed    (BA0500)
MOV AX, 0B000H      ;buffer segment address      (B800B0)
MOV ES, AX          ;buffer address to ES          (8EC0)
MOV DI, 1454        ;space 8, line 10 index    (C7C7AE05)
MOV BX, 0731H       ;ASCII 31H, attribute 7    (C7C33107)
SUB CX, CX          ;clear CX                      (2BC9)
```

FIGURE 5-34. Initialization module of a program.

```
MOV AL,   0B6H   ;counter 2, mode 3, LSB-MSB     (B0B6)
OUT 43H,  AL     ;write to control register      (E643)
MOV AX,   1193   ;count for 1 kHz              (B8A904)
OUT 42H,  AL     ;write LSB of count             (E642)
MOV AL,   AH     ;MSB to AL                      (8AC4)
OUT 42H,  AL     ;write MSB of count             (E642)
IN AL,    PORT_B ;get current byte               (E461)
OR AL,    1      ;set bit 0 (gate 2 enable)      (0C01)
AND AL, 0FDH     ;clear bit 1 (disable spkr)     (24FD)
OUT PORT_B, AL   ;return modified byte           (E661)
```

FIGURE 5-35. Counter 2 programming routine.

during interrupts. The initial count of 5 to DX fixes the number of displayed digits at 5. This choice is arbitrary. Segment register ES is loaded so as to point to the display buffer of the monochrome monitor, and offset DI points to space 8 of line 10. Recall that there are 160 bytes per line, and 2 bytes are needed per space.

The first character to be sent to the display is decimal 1, and its ASCII code 31H and the normal CRT attribute code of 7 are moved to register BX. When these codes are transferred to the display buffer, the ASCII code of BL goes to the low-byte location and the attribute code goes to the next one, as required. Register CX, which will maintain a count of the negative pulses of counter 2, is cleared.

Program the Counter and Disable the Speaker. Counter 2 is programmed by the routine of Fig. 5-35 to provide a square wave (mode 3) with a 2-byte binary count. Because division of the 1.193-MHz clock by 1193 gives the specified 1-kHz output, number 1193 is written to the count register of port 42H. The latch of output port B is read to get the current byte. Bit 0 is set to enable the counter, bit 1 is cleared to disable the speaker, and other bits remain unchanged. Then the modified byte is returned to the latch of port B.

Count the Negative Pulses. The next portion of the program, shown in Fig. 5-36, increments register CX each time the output of counter 2 goes low. Accordingly, *CX maintains a count of the negative pulses of the output of the counter,* up to a maximum value of 65,535. When an overflow occurs, as indicated by a new count of 0, the program exits the loop in order *to send the character of BX to the display.*

Transfer the Digit. Whenever the count changes from 65,535 to 0, a digit is sent to the monitor using the routine of Fig. 5-37. Memory-mapped I/O is used for addressing the display buffer. After the transfer, the ASCII code of BL is

```
A: IN AL, PORT_C       ;get timer output bit PC5   (E462)
   TEST AL, 00100000B  ;is output bit high?        (A820)
   JZ A                ;if not, go back (wait)     (74FA)
B: IN AL, PORT_C       ;get timer output PC5       (E462)
   TEST AL, 20H        ;is output bit low?         (A820)
   JNZ B               ;if not, go back (wait)     (75FA)
   INC CX              ;increment count              (41)
   JZ C                ;exit loop if count is 0    (7402)
   JMP A               ;repeat                     (EBEF)
```

FIGURE 5-36. Program module for counting pulses.

```
C: MOV ES:[DI], BX ;send character to CRT     (26891D)
   ADD DI, 160      ;pointer to next line    (81C7A000)
   INC BL           ;increment ASCII code        (FEC3)
   DEC DX           ;decrement character count     (4A)
   JNZ A            ;repeat if not 0             (75E3)
```

FIGURE 5-37. Routine for CRT output and housekeeping.

incremented so that the next character transferred to the display differs from the last one sent. Also, 160 is added to index DI so that characters are printed in a column. If this addition is omitted, the next digit overwrites the preceding one. The word of DX (initial value of 5) is decremented, and if it is 0, the program exits the routine. Otherwise, it goes back to count another 65,536 negative pulses of counter 2.

Beep the Speaker. The final part of the program uses the 1000-Hz counter to beep the speaker for about 2 seconds. Bits 0 and 1 of port B are set, thereby feeding the counter output to the speaker. A delay loop of 234 milliseconds is executed nine times to give the 2-second duration time, with the initial count of 9 stored in BL. Then the speaker is disabled, and the programs ends. The routine is shown in Fig. 5-38.

Results. The first character is printed after the counter goes through 65,536 periods, and subsequent characters are printed in the same time span. For the frequency of 1000 Hz, the interval is 65.5 seconds. After five digits are printed in a column, the speaker beeps. A digital watch was used to time the appearance of the digits, and the time from the start to the initial sound of the beep was 328 seconds. Successive digits appeared at precisely the expected intervals, indicating that the frequency was 1000 Hz, as programmed.

Without too much additional work, the program can include a display of the number of timer ticks alongside each printed digit, thereby eliminating the watch and increasing the accuracy of the timing. If this modification is made, the tick count should be initialized at 0. Timing resolution of the ticks is about 55 milliseconds, this being the interval between successive ticks.

Timing within the program is not appreciably affected by the timer interrupts. Interrupts occurring within the wait loops of the program do not affect the results at all, because the counter continues to count. If an interrupt occurs just before a digit is to be printed, the delay is a negligible 0.1 millisecond. During the 2-second beep, there are about 36 timer interrupts that add approximately 4 milliseconds to

```
   MOV BL, 9          ;delay count                 (B309)
   IN AL, PORT_B      ;get byte of latch           (E461)
   MOV AH, AL         ;save byte in AH             (8AE0)
   OR AL, 3           ;set bits 0 and 1            (0C03)
   OUT PORT_B, AL     ;turn on speaker             (E661)
   SUB CX, CX         ;clear loop counter          (2BC9)
D: LOOP D             ;234-ms delay loop           (E2FE)
   DEC BL             ;decrement repeat count      (FECB)
   JNZ D              ;repeat delay 9 times        (75FA)
   MOV AL, AH         ;get the saved byte          (8AC4)
   OUT PORT_B, AL     ;disable speaker             (E661)
```

FIGURE 5-38. Routine to beep the speaker.

the beep duration. This increase is insignificant. However, *the interrupts do noticeably affect the sound of the beep*. The effect can be eliminated, if desired, by turning off the interrupts at the start of the beep with instruction CLI, followed by STI at the end of the beep.

This program was assembled and loaded into memory, using the system DEBUG program described in Chapter 7. An assembler can be used, of course, and for long programs it is almost essential. Before running the program in the DEBUG mode, the screen should be cleared with repeated activation of the Enter key so that the digits displayed will stand out. A breakpoint must be set at the end of the program in order to stop execution. In Chapter 7, other ways to terminate programs are examined.

Remember that HLT does not work because of the interrupts. The instruction is present in the self-test routine of a system reset. Here it is used to halt operations if a malfunction is detected, but its use is restricted to the initial period of time prior to enabling of interrupts. When HLT is executed with interrupts disabled, only a power-off reset can restore control to an operator.

REFERENCES

Component Data Catalog, Intel Corp., Santa Clara, Calif., 1980.
Personal Computer Technical Manual, IBM Corp., Boca Raton, Fla., 1981.
The 8086 Family User's Manual, Intel Corp., Santa Clara, Calif., 1979.

PROBLEMS

SECTION 5-1

5-1. Define DOS, MS DOS, boot DOS, monitor, system reset, and DMA.

5-2. For the clock generator of Fig. 5-1, on identical time scales sketch the waveforms of outputs OSC, PCLK, and CLK, and give the period of each.

5-3. Following a reset, where is the first instruction that is fetched and executed? In the PC, what does this instruction do? Briefly describe what follows.

5-4. Explain in detail the function of each input and output signal of the clock generator of Fig. 5-1.

SECTION 5-2

5-5. For the 8288 bus controller, state the function of each input and output shown in Fig. 5-1.

5-6. Sketch the timing diagram for the memory write cycle of the 8088 of the PC, which is similar to that of Fig. 5-5. Include ALE and DT/$\overline{\text{R}}$.

5-7. (a) Sketch the complete logic diagram of the 74LS373 chip. Include eight latches, each with an active-low enable $\overline{\text{G}}$ and an output three-state buffer with active-high enable. Inputs OE and $\overline{\text{G}}$ should be buffered. (b) Also, sketch the complete logic diagram of the 74LS245 octal transceiver chip.

5-8. Explain the action of signals $\overline{\text{DMA WAIT}}$ and $\overline{\text{AEN}}$, including the effect on the status bits to the bus controller, when a DMA operation is implemented.

5-9. For an 8088 in the minimum mode, prepare a table relating the bus transactions to the different combinations of the bits of $IO/\overline{M}$, $DT/\overline{R}$, and $\overline{SS0}$.

SECTION 5-3

5-10. With standard I/O, give the minimum number of mnemonic instructions that will (a) transfer a word from port 150H to the TOS and (b) move the contents of SS to ES.

5-11. With a minimum number, give the instructions for reading from PORT_E if PORT_E has been equated to (a) 42H and (b) 3F0H. Then repeat for a write operation to the port.

5-12. Suppose a peripheral chip of the PC has three-state data bus pins, along with inputs $\overline{CS}$, $\overline{RD}$, $\overline{WR}$, A1, and A0. The chip addresses are 60H through 63H. Sketch the circuit showing the address lines and the controls from the bus controller. Use an AND gate decoder. Recall that the PC uses 10 address bits for ports.

5-13. In the positive logic TTL circuit of Fig. 5-39, an open circuit at a gate input is a logical 1 and ground is 0. Determine the state of Y when the switch (a) contacts A, (b) is in the position shown after leaving A, (c) contacts B, and (d) is in the position shown after leaving B. If the switch is moved to A and bounces 50 times before coming to rest, what is the effect of the bounces on Y?

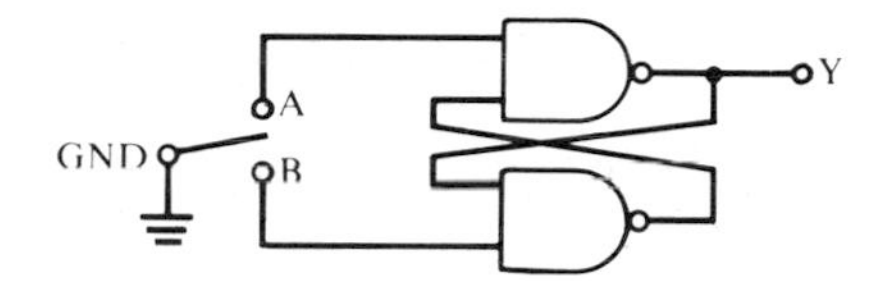

FIGURE 5-39. TTL circuit.

SECTION 5-4

5-14. Rewrite the assembly language routine of Fig. 5-11 so that only 40 characters are put on one line with a space between each, and with a blank line between adjacent character lines. Start the character display at the left side of line 8. Save all registers on the stack and restore them at the end.

5-15. Rewrite the BASIC routine of Fig. 5-11 so that only 40 characters appear on a line with a space between each, and with a blank line between adjacent character lines. Start the character display at the left side of line 8.

5-16. For the PC, provide an assembly language routine with no more than five instructions that give a delay of 1 second. Include the hexadecimal machine code. (b) Repeat (a) for a 1-millisecond delay, but use only two instructions.

SECTION 5-5

5-17. Sketch a chip tester circuit that simultaneously tests a 7400 quadruple of two-input NAND gates, a 7402 quadruple of two-input NOR gates, and a 7404 hex inverter. Use the PPI, with port A for output and ports B and C for input. Include a routine that programs the PPI and performs the tests. All four combinations of bits at the NAND and NOR gate inputs should be included in the tests.

5-18. For the beep-the-speaker program in Fig. 5-22, determine the time duration and frequency of the beep.

5-19. For the sequence in Fig. 5-40, suppose the first input from PORT_C of the PC is byte x, which changes 0.5 millisecond later to byte y and then remains constant. Find the delay within the loop having label A1 if (a) x = 85H, y = C5H; (b) x = 85H, y = 95H; (c) x = 96H, y = 94H; (d) x = 96H, y = 86H. Note that the loop gives a delay that allows a change in input bit 4, if any, to be detected.

```
    MOV CX, 60H         ;  4 clocks
    IN  AL, PORT_C      ; 10 clocks
    AND AL, 10H         ;  4 clocks
    MOV AH, AL          ;  2 clocks
A1: IN  AL, PORT_C      ; 10 clocks
    AND AL, 10H         ;  4 clocks
    CMP AL, AH          ;  3 clocks
    LOOPZ A1            ; 18 for loop, 6 no loop
```

FIGURE 5-40. Routine for Problem 5-19.

5-20. When byte 010011A0 is fed to PORT_B of the PPI of the IBM PC, bit A goes directly to the speaker. Logic is positive with 0 and 5-V levels. (a) Sketch and dimension two cycles of the waveform fed to the speaker by the program in Fig. 5-41.

```
   MOV BX, 2000            OUT PORT_B, AL
A: MOV AL, 4EH             MOV CX, 150
   OUT PORT_B, AL       C: LOOP C
   MOV CX, 100             DEC BX
B: LOOP B                  JNZ A
   MOV AL, 4CH             HLT
```

FIGURE 5-41. Routine for Problem 5-20.

(b) Determine the frequency, duty cycle, and time duration of the signal. (c) Give the hexadecimal object code for the program.

5-21. With *r* denoting the word of register BX, write a program that beeps the speaker at about 1 kHz for *r* milliseconds. Refer to the beep-the-speaker routine of the text. Give both assembly and hexadecimal machine language programs.

5-22. With *m* denoting the byte of register DH and *n* being that of DL, write an assembly language program for sounding the speaker with *m* three-second beeps followed by *n* one-second beeps. The delay between beeps should be about 0.25 second.

SECTION 5-6

5-23. Determine the calendar month and day, as well as the reading in hours, minutes, and seconds (H:M:S), of a precise 24-hour digital clock at the instant when locations TIMER_LOW, TIMER_HIGH, and TIMER_OFL store F0D8, 000E, and 03, respectively. Assume that the locations are initially cleared precisely at midnight on December 31.

5-24. If TIMER_LOW and TIMER_HIGH are initialized at 12 minutes and 25 seconds past 10 A.M., so that the contents of each location become 0 at precisely midnight, determine the initial hexadecimal values written into the locations. Also, what is the time later that day when the words are ABCDH and 0014H?

5-25. At the beginning of a program to write a portion of a diskette, the instruction MOV MOTOR_COUNT,0FFH is executed. At the end of the program is the instruction MOV MOTOR_COUNT,20H. Approximately how long does the diskette spin if the write operation requires (a) 1 second, and (b) 1 minute?

5-26. In the Example in Section 5-6, determine the count to be written to the count register for digits to be displayed exactly 60 seconds apart in time. Also, if the count is changed to give a frequency of 100 Hz, and if count register CL replaces CX in the program, determine the count to be written to the count register and the rate at which characters are shown.

5-27. Revise the program in the Example of Section 5-6 so that all 256 ASCII characters are displayed on the monitor, with characters appearing at precisely 1-second intervals. The program should put 80 characters on a line, starting at the left side of line 6. List all instructions and code. Omit the beep.

CHAPTER 6

Subroutines and Interrupts

In 8088 assembly language, a subroutine is referred to as a *procedure* and is so labeled. It consists of a sequence of instructions that can be called at any point of a program. When "CALL *procedure-name*" is encountered in a program, the address of the next instruction, referred to as the *return address,* is pushed onto the current stack. The program then branches to the routine of the procedure. After processing, the return address is popped from the stack, and processing of the main program resumes.

An obvious advantage of a subroutine is that it can be called as often as needed. For example, at various points in a program, it may be necessary to beep the speaker. Including a subroutine for this purpose is simpler than inserting the routine at the various points of the main program. Also, the program is reduced in length and requires less memory for storage. A second major advantage is the division of the program into modules, which can be developed and tested independently of one another. This modular approach greatly simplifies the design and testing of long programs.

A *software interrupt* is quite similar to a CALL instruction. With n denoting an interrupt type number from 0 through 255, INT n pushes the doubleword return address CS:IP onto the stack, and the program automatically branches to a new logical address, CS:IP. This new address is fetched from the doubleword memory location at absolute address $4 * n$. Flexibility is increased, because the programmer *may change the stored address* so as to give a branch to any desired location.

Here, *additional code can be inserted,* followed by a jump to the original subroutine, or an entirely different subroutine can be used. In each case, the original routine is preserved, and it can be called in the normal manner at any point of the program.

There are 256 different interrupt types. A disadvantage of the software interrupt relative to CALL is that memory space must be reserved for storage of the doubleword pointers for each of the possible interrupts. Also, the number of interrupt types is limited to 256, whereas the number of procedures that can be called is unlimited.

We shall also investigate *hardware interrupts*. These are similar to software interrupts, except that the interrupt of the running program occurs when an external event, such as a key closure or a timer pulse, activates appropriate hardware. The hardware is designed so that the external event effectively inserts an interrupt instruction into the running program at the time of the event.

6-1 THE SUBROUTINE

Each procedure of an 8088 assembly language program is assigned a name and is defined as either a NEAR or FAR type. The instruction CALL must be followed by an operand that refers either directly or indirectly to a named procedure. The name and type definition precede the assembly language routine, with termination indicated by the assembler directive ENDP. Procedures are usually located within a program so as to follow the activating code of the main program, but this is not a requirement. They may be called from any point of the program, and nesting is allowed, with one procedure wholly contained within another. Two examples are presented in Fig. 6-1 to illustrate the format.

The code of a *NEAR procedure* must be stored in memory within the current code segment CS; hence, a NEAR procedure is referred to as an *intrasegment procedure*. Activation requires that a new value be assigned to IP, with no change in CS. On the other hand, a *FAR procedure* may be placed anywhere within the memory space, including the current code segment. Because access to a FAR procedure requires that a CS address be included in the instruction in addition to the new IP, it is known as an *intersegment procedure*. Code supplied by the assembler for CALL *target* depends on whether the procedure has been defined as NEAR or FAR. A FAR procedure must precede its call.

The procedure name is similar to a label in that it can be used as the target of a jump instruction. Instructions of a procedure placed within a program without a bypassing jump are executed just like those of any other sequence. Thus, care must be taken not to enter a procedure inadvertently. Normally, CALL is used for this purpose, with the instruction RET placed at the end of the procedure to provide the return to the main program. Both CALL and RET give a program branch, and hence are examples of program transfer instructions.

`ERR_BEEP  PROC  NEAR`	`A2  PROC  FAR`
Instruction Sequence	*Instruction Sequence*
`ERR_BEEP  ENDP`	`A2  ENDP`

FIGURE 6-1. Formats of NEAR and FAR procedures.

Program Transfer

Instructions are always fetched from the 64K code segment CS, using the instruction pointer IP for the offset. When an instruction is fetched, IP is automatically incremented so as to point to the next one in sequence. Usually, the next instruction to be executed has already been prefetched and put in the queue. However, *program transfer* instructions, which change either CS or IP or both, empty the queue; fetching then commences from the new transfer target, CS:IP. The status flags are not affected.

There are four basic groups, which are called *iteration control, conditional transfer, unconditional transfer,* and *software interrupt*. The iteration control, or loop, instructions were presented in Section 5-4, and the conditional transfer group was summarized in Section 4-4. Here we examine the three unconditional transfer instructions, shown in Fig. 6-2, with the software interrupts treated in the next section.

CALL. When a CALL is executed, the program branches to a procedure, and the return address is saved on the stack. After the procedure has been processed, the return address is popped from the stack, and the main program continues. The procedure may be called repeatedly by the same program. There are four types of CALLs:

1. Intrasegment (NEAR) direct CALL
2. Intrasegment (NEAR) indirect CALL
3. Intersegment (FAR) direct CALL
4. Intersegment (FAR) indirect CALL

These CALLs are now examined in order.

1. NEAR CALL (Direct). For an *intrasegment direct call,* the instruction includes a 16-bit displacement that gives the offset to the target when added to IP. Prior to the addition, the contents of IP are pushed onto the stack, thereby saving the return address. Then the displacement is added to IP, and the new IP is the pointer to the target. Because the CALL instruction and its target are in the same segment, the relative displacement is not affected by changing the word of CS. Thus, this form of CALL is appropriate for position-independent, or dynamically relocatable, code. It is the most common of the four types of CALLs.

The displacement can be calculated and treated as either a signed or an unsigned number. *It makes no difference,* because the same location is accessed either way. To illustrate, suppose IP is 6K and the displacement is F800H. This is -2K for a negative number and $+62$K for a positive number. These values give 4K and 68K when added to 6K, but the 68K becomes 4K when bit 17 is eliminated, as it is in the processor. An address that passes through one end of a 64K segment

Unconditional Transfer	
CALL target RET pop-value (if any) JMP target	Call procedure: CALL S2 Return from procedure Jump; JMP BETA[BX]

FIGURE 6-2. Unconditional transfer instructions, with comments and examples at the right.

wraps around to the other end. For the NEAR procedure ERR_BEEP, the instruction is CALL ERR_BEEP, and the code is byte E8H followed by the 16-bit displacement.

2. NEAR CALL (Indirect). For an *intrasegment indirect call,* IP is pushed onto the stack, and the new IP is obtained from a register or from memory, using any of the addressing modes. To illustrate, CALL AX gets the new IP from AX, and both CALL BETA [BX][SI] and CALL WORD PTR [BX] get IP from memory.

3. FAR CALL (Direct). An *intersegment direct call* must push both CS and IP onto the stack, and the instruction must give new values. First, CS is pushed, followed by IP. For the FAR procedure A2 the instruction is CALL A2, and the assembler recognizes a FAR procedure call. The code is 9AH, followed by the new 16-bit IP and then by the new 16-bit value of CS. Low-order bytes precede high-order bytes. For example, for machine code 9A1234ABCD, the call is to a procedure at CDAB:3412.

4. FAR CALL (Indirect). Next is the *intersegment indirect call*. The new values of IP and CS are obtained from memory. The doubleword pointer stored at the addressed memory location contains, in order of ascending addresses, the low byte of IP, the high byte of IP, the low byte of CS, and the high byte of CS. As before, the return address is saved on the stack, with CS pushed first. Examples are CALL DWORD PTR [BX] and CALL NUM_PTR, with NUM_PTR defined in the program as a doubleword pointer.

Parameter Passing. Most procedures require parameters to be supplied by the main program and results to be delivered to it. A common method is to *place necessary data in selected registers*. For example, if a procedure performs certain operations on two words that must come from the running program, these can be placed in two of the general registers, as specified by the procedure. Results can be returned to the program by means of the registers. A disadvantage of this method is the limited number of registers available.

Data can, of course, be passed through memory; this operation is usually done with the stack. *Parameters are pushed on the stack prior to the CALL,* with each PUSH causing SP to be decremented by 2. The procedure can address these locations, use the parameters, and perhaps replace them with results for transfer to the main program. After the return, the SP must be decremented by the number of parameter entries multiplied by 2, so as to restore SP to its proper value. Although this can be done with POP operations, the return RET *n* instruction can do it more efficiently provided the parameters are no longer needed. The method is illustrated in Example 1.

RET. The last instruction of a procedure activated by CALL should be RET, which returns control to the next instruction of the main program. The word pointed to by SP is popped into IP, and SP is incremented by 2. In a FAR procedure, the POP of IP is followed by a second POP, with the word placed in CS. The assembler generates code CB for the intersegment return, but otherwise the code is C3. This latter code applies if RET is used either without an associated procedure or with a NEAR procedure.

RET may be followed by a *pop* value *n,* if desired. In this case, after the return address has been popped, the value *n* is added to SP. For example, RET 6 causes SP to jump over three stack words after popping the return address. The optional value is normally used *to discard parameters* no longer needed, which were pushed on the stack ahead of the execution of CALL. An example using this feature

```
PUSH AX          ;move address of AX to stack
PUSH BX          ;move address of BX to stack
PUSH CX          ;move address of CX to stack
CALL SUB_PROC    ;branch to the subroutine
```

FIGURE 6-3. Sequence of the main program of Example 1.

follows. Although the example is trivial, it illustrates the method of parameter passing through memory, which is especially useful with programs having a large number of parameters to be passed.

EXAMPLE 1

Design a FAR procedure to subtract 16-bit number B from A, giving the result. Assume that A, B, and C denote memory locations reserved with DW directives, *having offset addresses in segment DS that have been placed in AX, BX, and CX,* respectively. Locations A and B store binary numbers, and C is initially clear. Use the stack to pass all addresses.

SOLUTION

In the main program is placed the sequence of Fig. 6-3. The three PUSH operations move the *addresses* of the operands to the stack. Because SUB_PROC is defined as FAR, the CALL transfers CS and then IP to the stack. At this point in the program, SP points to word IP, with CS and the three addresses stored at higher addresses.

The subroutine is shown in Fig. 6-4. The initial contents of registers AX and SI are destroyed by the subroutine, but the base pointer BP is saved. The word of SP is moved to BP, which is then used as a pointer to the addresses of the stack. As stated in the "Based Addressing" subsection of Section 3-4, register BP is a pointer to a location of the current stack segment unless a segment override prefix is specified. The stack pointer SP cannot be used for this purpose, as indicated by Table A-2 of Appendix A. The routine is similar to one given in Appendix C of the IBM *BASIC manual.*

Words of the stack following the PUSH of BP are shown in Fig. 6-5. After the POP of BP, the SP points to the return offset IP. When RET 6 is executed, IP is popped, followed by a POP of CS. Then SP is incremented by six byte locations so that it now points to the entry just above the address of A. Thus, the three addresses below SP are effectively discarded. ■

```
SUB_PROC  PROC  FAR
    PUSH BP               ;save BP
    MOV  BP, SP           ;initialize BP with SP
    MOV  SI, [BP + 10]    ;move address of A to SI
    MOV  AX, [SI]         ;transfer number A to AX
    MOV  SI, [BP + 8]     ;move address of B to SI
    SUB  AX, [SI]         ;subtract B from A, store AX
    MOV  SI, [BP + 6]     ;move address of C to SI
    MOV  [SI], AX         ;transfer difference to memory
    POP  BP               ;restore
    RET  6                ;FAR return, discard 3 addresses
SUB_PROC ENDP
```

FIGURE 6-4. FAR procedure of Example 1.

Address of A	SP + 10
Address of B	SP + 8
Address of C	SP + 6
Return CS	SP + 4
Return IP	SP + 2
BP	← SP

FIGURE 6-5. Stack of Example 1 after PUSH BP.

JMP. The instruction JMP is unconditional, and no return address is saved, as in CALL. In addition to CALL-activated procedures, a program may have JMP-activated procedures, in which case the last instruction of the procedure is an unconditional jump back to the main program rather than RET. For example, suppose an interrupt program has 10 procedures, *only one of which is activated each time the interrupt is executed*. The selection may depend on data present in AX or some other register when the interrupt is initiated, and processing of the procedure might terminate the service program. An efficient method is to use JMP-activated procedures, with the last instruction of each being an unconditional jump to instruction IRET, which provides the interrupt return. Actually, this method is essentially the same as that using labels in place of the procedure names. However, the listing program is easier to read, because the separate routines appear as modules with start and end directives.

As in the case of CALL, jumps may be direct or indirect, either intrasegment or intersegment. For a direct jump within the current code segment, the target of the jump is specified by the label that follows JMP. The relative displacement of the target from the jump instruction is added to IP. The instruction is position independent in that its code does not depend on the value assigned to CS.

If the assembler can determine that the target is within 127 bytes of JMP, it automatically generates a 2-byte code, and the jump is said to be *SHORT*. The second byte of a SHORT jump is the 8-bit signed displacement. In the instruction, if the jump is both forward and short, the optional word SHORT can be placed immediately after JMP, which simply saves 1 byte of code. This procedure is not recommended unless memory is very limited, because of the extra typing and because of the error that is generated if the jump is not short. An example is JMP SHORT A1, with A1 denoting the label of the target having a higher address. Direct intrasegment jumps that are not short require 3 bytes. These are called *NEAR jumps*.

For an intrasegment indirect jump, the new IP is obtained from either a register or memory. An example is JMP BX, and the word of BX is moved to IP. With a memory operand as in JMP BETA, the word BETA of memory replaces IP.

A direct jump to a target in another segment replaces IP and CS with values indicated by the instruction. An example is "JMP FAR PTR *label*," with the FAR PTR directive *indicating to the assembler that the direct jump is intersegment*. If the target identified by the label has logical address B123:0100, the machine code is EA000123B1.

An intersegment indirect jump gets IP and CS from memory. The addressing mode must indicate to the assembler that the memory location stores a doubleword. For example, if PARM_PTR is defined in the program as a variable of a doubleword

```
      (running program)
A1   PROC FAR                    ;Must be specified as FAR
        PUSH DS                  ;   so that assembler will
        MOV AX, 0                ;   generate a FAR return.
        PUSH AX
        RET                      ;FAR return
A1   ENDP                        ;ENDP always required
      (continue)                 ;continue at new CS:IP
```

FIGURE 6-6. Routine for inserting (DS) into CS and 0000H into IP.

location, JMP PARM_PTR is an intersegment indirect jump. We note that the instruction operands used for indirect jumps are variables stored in memory, whereas those for direct jumps are labels, or perhaps names of procedures.

Although conditional jumps are always SHORT, a conditional jump to any location of memory can be implemented with a combination of a conditional jump and JMP. The sequence that follows illustrates a conditional jump to a FAR location, with the jump made only if flag ZF is set:

```
     JNZ A1
     JMP DWORD PTR [BX]
A1:  (continue)
```

Directive DWORD PTR identifies the jump as FAR.

Loading CS. Seldom is it necessary to change the contents of CS with a program, but on occasion this procedure may be necessary. Two examples with this requirement are given in the next chapter. Register CS is not allowed as the destination of either a MOV or POP instruction. However, CALLs and JMPs to FAR targets and RETs from FAR procedures change CS, as do the interrupt INT instructions. Any of these instructions can be used to change CS. The example that follows illustrates a method using RET.

EXAMPLE 2

Develop a routine based on the return from a FAR procedure that transfers the word of DS to CS and clears IP.

SOLUTION

The specified changes in CS and IP can be accomplished by inserting a FAR procedure directly into the main program at the point where the changes are desired. There is no CALL, and the procedure is *not* used as a subroutine. When program execution reaches the procedure, its instructions are processed in order, just like the instructions of any other sequence. The routine is shown in Fig. 6-6.

Hexadecimal code generated by the assembler for the sequence of Fig. 6-6 is 1E B80000 50 CB. If the procedure start and end directives were omitted, the code for RET would be C3, and only IP would be popped from the stack. ■

6-2

SOFTWARE INTERRUPTS

Another way to call subroutines, and one extensively used in the PC, is through the software interrupt instruction INT *n*. It may be placed anywhere within a

Interrupts	
INT *n*	Activate interrupt routine (type *n*)
IRET	Interrupt return
INTO	Executes INT 4, but only if OF = 1

FIGURE 6-7. Interrupt instructions and comments.

program, and when executed, it causes the program *to branch automatically to a new logical address obtained from the doubleword location at absolute location $4 * n$*. For each interrupt type used by a program, the program must store the appropriate values of CS and IP in the doubleword location prior to execution of the interrupt, and a service routine must be included at the new logical address CS:IP. The return address is saved on the stack, as in the case of CALL. In addition, the flag word is automatically saved on the stack.

Normally, at the end of each interrupt service routine is the instruction IRET, which pops the return values of IP and CS from the stack, as well as the flag word. Interrupts can be nested, with interrupt instructions included within interrupt routines. Only the size of the stack limits the degree of nesting. Shown in Fig. 6-7 are the interrupt instructions. Each one is described in detail in this section.

Dummy Return

Frequently, an INT *n* instruction is included in a program with the "service routine" consisting only of the instruction IRET, which is called a *dummy return*. The purpose is *to provide the user of the program with the opportunity to add a service routine* if desired. If no service routine is added, the interrupt essentially does nothing except to waste a small amount of time.

Interrupt Flags

Bits 8 and 9 of the flag word of Fig. 3-3 are the *trap flag TF* and the *interrupt-enable flag IF*, respectively. First, we examine the trap flag.

Trap Flag TF. When TF is set, the processor automatically generates a type 1 interrupt after each instruction. If the routine consists of a dummy return, the set TF does nothing of value. However, because an interrupt is generated after each instruction, the service routine for INT 1 is usually designed to display register contents, flags, the address and code of the next instruction, and other information. Also, the routine puts the program in a WAIT loop until some appropriate action is taken by the operator, such as closure of a specified key. Then the next instruction is executed, and the process is repeated. This operation is referred to as *single-stepping*. Normally, INT 1 is dedicated to single-stepping, and when TF is set, the processor is said to be in single-step mode. This mode is very useful in debugging programs.

Each time INT 1 is activated with TF set, the flag word is pushed onto the stack and TF is cleared. In fact, whenever any interrupt is processed, both TF and IF are automatically cleared after the PUSH of the flag word. Therefore, the service routine *executes at normal speed,* without single-stepping. However, when IRET

```
PUSHF                              PUSHF
MOV BP, SP                         MOV BP, SP
OR  WORD PTR [BP], 0100H           AND WORD PTR [BP], 0FEFFH
POPF                               POPF
```

FIGURE 6-8. Routines for setting and clearing TF.

is executed at the end of the routine, the flag word with TF set is restored, and the single-step mode again applies.

There is no instruction for setting or clearing TF. The sequences of Fig. 6-8 can be used for these purposes. The sequence at the left sets TF, and the one at the right resets it.

Interrupt-Enable Flag IF. Flag IF affects only interrupt requests supplied by external hardware to the CPU on input pin INTR. If flag IF is set, the CPU is allowed to recognize an external request at this pin. However, if flag IF is clear, the external requests are ignored. The flag has no effect whatsoever on software interrupts, on an interrupt request from external hardware fed to input pin NMI, or on interrupts generated internally, such as INT 1 when TF is set. Flag IF is automatically cleared whenever an interrupt service routine is entered.

A summary of flag instructions is presented in Fig. 6-9. As shown, instructions STI and CLI, respectively, set and clear flag IF.

Interrupt Instructions

Excluding STI and CLI, there are only three interrupt instructions. INT *n* and IRET correspond closely to CALL FAR_PROC and RET. The other interrupt instruction (INTO) has relatively limited use. We now examine each of these instructions.

INT *n*. When INT *n* is executed, with type number *n* from 0 to 255, the following actions take place in the order listed:

1. PUSHF
2. Reset IF and TF
3. PUSH CS
4. Move word from location (4**n* + 2) to CS
5. PUSH IP
6. Move word from 4**n* to IP

Flag Operations	
PUSHF	Push flag word to stack
POPF	Pop stack word to flags
STC	Set CF
CLC	Clear CF
CMC	Complement CF
STD	Set DF
CLD	Clear DF
STI	Set IF
CLI	Clear IF

FIGURE 6-9. Flag instructions.

Except for the type 3 interrupt, the hexadecimal code is CD *type-number*, and 71 clocks are required. For INT 3, the code is CC with no type number. The doubleword pointer at location 4 * *n*, with CS at the higher word location and IP at the lower word location, is called the interrupt *vector address*. After execution, the words on the stack are the flag word, return CS, and return IP, given in order of descending addresses. All words are stored with the higher byte at the higher byte address. The next instruction executed, which is the first instruction of the interrupt service routine, is the one at the new logical address CS:IP.

IRET. Normally, the last instruction of an interrupt service routine is IRET. When IRET is executed, the following actions take place in the order listed:

1. POP IP
2. POP CS
3. POPF

The hexadecimal code for IRET is CF, and 36 clocks are required. The next instruction executed is the one of the main program that follows INT *n*. In special cases, it may be necessary to keep the flags of the service routine rather than restore the old flags. This is done with intersegment RET 2 in place of IRET, which pops IP and CS; then the pop value 2 discards the flag word image of the stack.

INTO. Whenever interrupt-if-overflow INTO is executed, flag OF is examined. If it is set, a type 4 interrupt is generated. Activation follows the procedure described for INT *n*. The difference between INTO and INT 4 is that INTO gives INT 4 *only if OF is set*. Otherwise, control passes to the next instruction in sequence.

INTO is often placed in a program immediately after an arithmetic or logical operation involving signed numbers, for the purpose of activating a service routine if overflow occurs. The routine may be designed to display an error message and perhaps to halt program execution. The processor was not designed to activate INT 4 after every operation giving a set overflow flag because such a condition denotes an error only when the numbers are signed. The CPU does not know if a signed operation was intended.

Service Routines

Interrupt service routines can be located in RAM and ROM, and each can be used by a program as often as desired. Up to 256 are allowed, with their vector addresses stored in locations having absolute addresses from 0 through 1023. The vector address of INT 0 is located in the doubleword location from 0 through 3, and that of INT 0FFH occupies locations 1020 through 1023. The low 1 KB of memory should be RAM, so as to allow a program to establish and perhaps change at times the vector addresses. Although this 1 KB memory space is normally reserved for vector addresses, portions of it not used for interrupts may be used for other purposes. Because interrupt service routines are often developed as modules that are linked with the main program after the design is done, they are frequently labeled as assembly language procedures. This arrangement improves the readability of the listing program.

Most routines use a number of processor registers. If the original contents of these registers are important to the main program, they should first be saved.

Therefore, interrupt routines typically have several PUSH instructions at the start and an equal number of POP instructions at the end. Usually, it is desirable to permit the interrupt program itself to be interrupted by a peripheral signal at pin INTR. Because flag IF is cleared by the interrupt operation, instruction STI is required. It is normally placed ahead of the PUSH instructions. In general, *hardware interrupts should not be disabled except when absolutely necessary*. When pin INTR is disabled in the PC, the operator loses control of the keyboard and timer ticks do not go to the time-of-day clock. After IRET is executed, the original flags are restored, and IF has its original state.

Quite common are interrupt routines designed to service a peripheral such as a keyboard, with activation by hardware. Hardware interrupts are discussed in the next section. Each of these routines can, of course, be called by the appropriate software interrupt instruction. This procedure provides a way of testing interrupt programs that service external devices.

Software interrupts are extensively used by operating system programs that control computers. A number of such routines are examined later. In addition, the interrupt instruction often serves as a communications link between two different programs. A request by a program for service from an operating system, whether by a software interrupt or some other method, is sometimes referred to as a *supervisor call*. Because each interrupt vector address is at a fixed location in memory, different programs of a system can use the same interrupt routines. It makes no difference where the programs are located or even relocated. Clearly, interrupt routines can be designed so that two programs may communicate with each other. The routines themselves may be shifted in position provided the interrupt pointer table is updated.

Internal Interrupts

External, or hardware, interrupts are those activated by signals on input pins NMI and INTR. All other interrupts are classified as *internal*. These interrupts are generated by the processor in the absence of an external signal. The 256 software interrupts (INT *n*) are internal, and there are 3 internally generated interrupts, as shown in Fig. 6-10. Two of them have been examined.

The three internally generated interrupts of Fig. 6-10 are named *divide error, single step,* and *interrupt-if-overflow (INTO)*. Whenever the unsigned divide instruction DIV or the integer divide instruction IDIV is executed, if the calculated quotient is too large for the specified destination, a type 0 interrupt is automatically

Name	Type	Activation Condition
Divide error	INT 0	After DIV or IDIV, if quotient is too large for the specified destination
Single step	INT 1	After each instruction if TF is set
INTO	INT 4	In response to INTO if flag OF is set

FIGURE 6-10. Internally generated interrupts.

generated by the processor. Thus, the type 0 interrupt is dedicated to the service of a divide overflow, often referred to as *divide-by-0*.

Interrupts that are generated internally are often called *traps*. When they are activated by an error, the error is said to be *trapped,* and the routine takes corrective action. This action may include an error message. In debug programs, breakpoints are set by software interrupt instructions that are inserted into the program at locations specified by the user; these breakpoints are often referred to as traps. At each trap, the service routine normally halts execution temporarily while the operator examines results.

INT 1 is dedicated to single-stepping, and INT 4 is dedicated to the servicing of an overflow, which activates INT 4 only when INTO is executed. In contrast, INT 0 is activated whenever a divide-by-0 error occurs. For these dedicated interrupts, as for all others, the vector addresses and service routines must be provided by the programmer and placed in memory prior to the interrupts.

Internal interrupts have higher priority than external ones, except for the single-step interrupt, which has the lowest of all priorities. For example, if a divide-by-0 error occurs when an active signal is present on one of the two input interrupt pins, the divide error INT 0 is executed first. On the other hand, if a program is in the single-step mode, with trap flag TF set, an interrupt request from a divide error or from an external signal is processed ahead of INT 1.

Occasionally, a program that is to be run on a computer has a 2-byte INT *n* instruction that is not absolutely essential, and perhaps the user does not care to add an interrupt routine. One option is to replace the 2-byte code of the interrupt instruction with two NOP instructions, each of which has code 90H. NOP simply causes no operation, while using three clocks for execution. Another method, which is usually more practical because it does not involve changing the program, is to call a dummy return. In the PC, at address F000:FF53 in ROM is byte CFH for IRET. The logical address of this location is used by a number of programs as the vector address of an interrupt with a dummy return, which allows the user to insert a service routine, if desired.

Breakpoint INT 3. INT 3 is reserved for use as a *breakpoint* interrupt. The CPU was designed so that INT 3 has only a *1-byte code* (CCH), which is essential for a breakpoint interrupt. A breakpoint is a trap within a program where execution is temporarily stopped so that some kind of special processing, such as the display of registers, can be performed. It is important to be able to set breakpoints when debugging programs. Single-stepping is slow and not always necessary.

A breakpoint is set by a debugger program by *replacing temporarily* the first part of an instruction with code that transfers control to a breakpoint routine. Upon completion of the routine, the original byte is replaced. Suppose INT 6 is used for this purpose. Then 2 bytes must be inserted, which might replace a 1-byte instruction and the first byte of a second instruction. Now if the program happens to jump to the second of the replaced instructions, it tries to process the second byte of INT 6. *This problem is avoided with the 1-byte instruction INT 3.*

There is one other interrupt type that is reserved for special use. This is number 2, which is automatically activated by an external signal on pin NMI. It is examined in the next section. Thus, interrupt types 0, 1, 2, 3, and 4 are reserved, and they should not be used for other purposes unless the dedicated functions are never going to be used.

Example Program

A short program is presented to illustrate the use of a software interrupt. In the PC the routine of INT 10H is stored in the BIOS ROM. It provides for video I/O, and this interrupt is employed in the program for the purpose of displaying a message. Initially, the cursor is assumed to be at the left side of the screen. The requirement is to write an assembly language program that will print on the monitor of the PC the following message:

```
Insert your work diskette in drive B.
    Press any key when ready.
```

"Press" should be indented four spaces from the left, with a blank line between the two lines. Storage of the message is in segment DS. The message can be displayed by writing directly to the display buffer, as done in the program of Fig. 5-11. Here the BIOS interrupt type 10H is used. It is described in more detail in Section 6-5.

In order to write a byte in ASCII to the monitor, register AH must store byte 14D and the character byte must be in AL prior to execution of INT 10H. Also, BH should be cleared for normal mode operation of the CRT. Thus, the passing of parameters to the subroutine is accomplished through processor registers. The contents of all registers are destroyed by the routine except those of the segment registers and BX, CX, and DX. This information is given in the BIOS program of the IBM *Technical Reference Manual*.

Stored at absolute location 40H during a reset is the interrupt doubleword pointer F000:F065. Only one character is transferred.

A table is set up in a program segment named BSTORE, and the segment with the mnemonic instructions is assigned the name CODSEG. Following an ASSUME directive, register DS is initialized by the program to segment address BSTORE. Register BH is cleared for the proper CRT mode, and CX is initialized to the byte count. Then a pointer to the message is established. After the interrupt routine has transferred a character, a LOOP instruction provides the required repetition, with CX determining the number of repeats. The program is presented in Fig. 6-11.

When assembled and loaded on the PC by the DEBUG program, the hexadecimal values initially assigned to DS, ES, CS, and SS were 90B, 90B, 920, and 922, respectively, with SP = 400H and IP = 0. The CODSEG address was that of CS, STACKS was that of SS, and the segment address of BSTORE was 91B. Hexadecimal codes for the 12 instructions were FB, B8C504, 8ED8, B700, B94500, BE0000, 8A04, 46, B40E, CD10, E2F7, and F4. The 24 bytes were located at absolute addresses 09200H through 09217H, and the preceding 69 ASCII bytes of data were stored at 091B0H through 091F4H. The 1 KB stack starts at 09220H.

Execution requires a DEBUG breakpoint to be set after HLT. In Section 5-6, we learned that about 18 timer interrupts occur every second. As explained in Section 5-1, such external interrupts cause processing to continue beyond HLT; hence, *the use of HLT in the program is symbolic only*. Suitable ways to terminate a program without setting a breakpoint are described in the next and later chapters. When run, the specified message appears on the monitor.

The EQU statement at the end of the table equates the number of bytes of the table named MSG to the name MSG_C. As shown, following EQU is $-*table-name*, which was described in Section 4-4. The assembler counts the number of

```
  BSTORE SEGMENT
     MSG DB  'Insert your work diskette in drive B'
         DB   0DH, 0AH, 0AH                      ;CR and two LF's
         DB  '    Press any key when ready'  ;four blanks
         DB   0DH, 0AH                           ;CR, LF
     MSG_C  EQU   $-MSG                          ;number of bytes
  BSTORE ENDS                                    ;segment end
;
  STACKS SEGMENT STACK
     DW  512 DUP (?)             ;this is the stack
  STACKS ENDS
;
  CODSEG SEGMENT
     ASSUME  CS:CODSEG, DS:BSTORE, SS:STACKS
     START: STI                  ;enable interrupts
        MOV AX, BSTORE
        MOV DS, AX               ;establish DS
        MOV BH, 0                ;CRT normal mode
        MOV CX, MSG_C            ;byte count to CX
        MOV SI, OFFSET MSG       ;establish pointer
;
; initialization OK, now for the transfer from memory
;
     A: MOV  AL, [SI]            ;ASCII byte to AL
        INC  SI                  ;point to next byte
        MOV  AH, 0EH             ;code for write
        INT  10H                 ;print character
        LOOP A                   ;print all bytes
        HLT                      ;idle computer
  CODSEG ENDS                    ;segment end
  END START                      ;end of program
```

FIGURE 6-11. Program for message display.

bytes and assigns this number to the name at the left of EQU. The table has 69 bytes.

The ASSUME Directive

The *ASSUME* directive is a promise to the assembler that each segment register will store the address of the named segment at execution time. Assembly is based on this information, which allows the assembler to develop proper code for instructions with labels or named variables. Unless ASSUME properly identifies the segment *of each label and variable prior to its use in an instruction,* the assembler cannot generate the code, and an error message is produced. ASSUME directives are placed in the code segment wherever needed. A routine with several ASSUME directives is given in Fig. 6-22 of Section 6-5.

Suppose a program has two code segments named CODE1 and CODE2. When instructions are to be fetched from CODE1, the statement ASSUME CODE1 is required ahead of the code. Later in the program, if code is to be fetched from CODE2, ASSUME CODE2 must precede the sequence of segment CODE2. Code is assembled on the assumption that the word of the segment register will be the base segment address of the specific one identified in the ASSUME statement. Note that the directive does not cause the assembler to generate code for initializing a segment register. This operation is done by the program or by a loader.

The STACK Attribute

The *STACK* attribute at the right side of the STACKS SEGMENT statement of the preceding program is required by the LINK program. It signifies that the stack defined by the SEGMENT directive is a LIFO stack *to be used by the PUSH, POP, CALL, and INT instructions.* A stack segment is essential. Without it, the program goes "into left field."

The END Directive

At the end of the program is the *program termination directive END.* This directive notifies the assembler that there is no more code. Unlike the BASIC statement END, the assembly language END must always be present. The label name START after END indicates the starting address for program execution. If it is not included in the program of Fig. 6-11, CS:IP is 91B:0, which is the start address of the message, not that of the code. Thus, *the label name should always be present.*

6-3 EXTERNAL INTERRUPTS

Data transfer between the CPU and peripheral devices is accomplished in various ways. Frequently, both software and hardware interrupt techniques are involved. Here we examine hardware, or *external,* interrupt techniques. An example is the usual input from the keyboard. If the interrupt-enable flag IF is set when a key is depressed, the running program is interrupted and branched to the INT 9 keyboard routine in the BIOS monitor. This program examines the scan code, performs service depending on the function of the key, and then returns control to the main program. The processor ignores the keyboard until a key is pressed. Without an external interrupt system, the CPU would have the check the keyboard and other I/O devices periodically, a method known as *polling.* Communication links between the processor and peripherals are provided by interface chips, or controllers.

Polling

Polling is a simple way to handle asynchronous servicing of external devices. Each device has an associated bit, or flag, that indicates whether or not service is required. At periodic intervals, the processor interrogates the flag bits in sequence to determine if attention is required. Elements are polled in order of priority, with the higher priorities normally assigned to those having the higher speeds. The procedure is wasteful of computer time.

An alternate and more efficient method, which requires a minimum of additional hardware, consists of feeding the device flags into the input of an OR gate, with the output connected to the interrupt-request INTR terminal of the CPU. This operation is shown in Fig. 6-12 for three I/O units. The connecting buses are omitted.

When a device needs attention, it activates the output of the OR gate, providing the CPU with an interrupt request. An interrupt-acknowledge handshake signal ($\overline{\text{INTA}}$) is transmitted from the processor along an output line to which the I/O

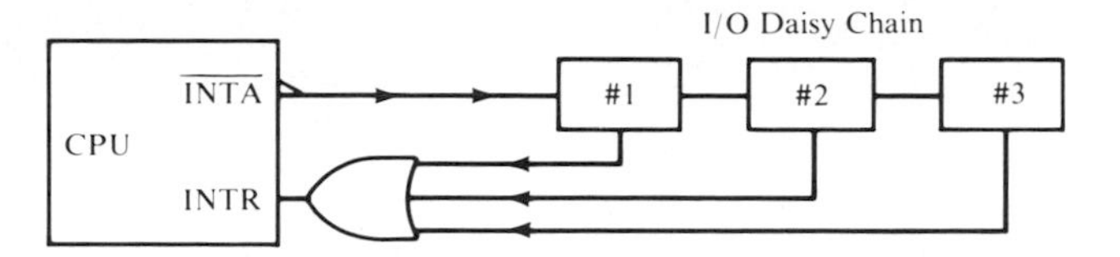

FIGURE 6-12. Polled interrupt with three I/O units connected in a daisy chain.

devices are connected in chain like fashion, arranged in order of priority. This arrangement is called a *daisy chain.* The first unit of the daisy chain, which has the highest priority, receives the $\overline{\text{INTA}}$ signal and passes it on only if there is no need for service. The signal stops at the highest-priority device needing attention, and the program branches to the routine designed for servicing this element. When execution is completed, processing continues with the next instruction of the main program. This procedure is known as *polled interrupt.*

The interrupt-enable flag IF within the processor can be set or cleared by software. This operation allows the programmer either to permit or to reject external interrupts during execution of any segment of the program.

Vectored Interrupts

An even better way to implement interrupts is provided by the *interrupt controller.* This chip receives interrupt requests from peripherals, examines their priorities, and sends an INTR signal to the processor. When an interrupt acknowledge $\overline{\text{INTA}}$ is received from the CPU, the controller transmits to the processor via the data bus a code that gives, either directly or indirectly, a vector address mapping to a routine in memory. The routine is designed to perform the service required by the device. Except for activation by an external event rather than by a program instruction, the effect is the same as that of instruction INT *n*.

To illustrate an advantage of the interrupt method, suppose a computer with an interrupt controller sends ASCII-coded characters to a slow teletypewriter. After the teletype interrupts the processor, the program jumps to the service procedure. The routine fetches a character from memory and feeds it through the processor to the teletype machine. During the next 0.1 second that may be required for printing the character, the computer fetches and executes 20,000 or more instructions. Then it is again interrupted, and the process is repeated. No time is lost polling devices or waiting for a slow peripheral, and throughput is maximized.

In the PC, some service routines are stored in the BIOS nonvolatile ROM and others are placed in RAM by the operating program in use, such as the BASIC interpreter, the DOS, or a word processor program. This system is very flexible, and all interrupt routines can be accessed by program instructions.

8088 External Interrupts

Two input pins of the 8088 are available for interrupt requests from external sources. One is the nonmaskable interrupt NMI pin with the higher priority; the other is the interrupt request INTR pin, which is usually fed from the 8259 programmable interrupt controller.

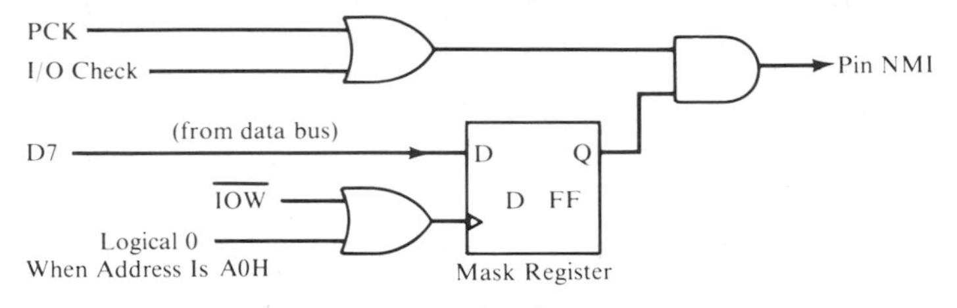

FIGURE 6-13. Pin NMI-associated circuits.

Input NMI. Pin NMI is *positive edge-triggered,* which means that it is activated when its input voltage makes a low-to-high transition. Once it is activated, the interrupt request is *latched* by the chip and held until serviced, even if the input signal changes. Flag IF has no effect, as indicated by the nonmaskable descriptor. Because of edge triggering, a request signal that remains high after being serviced *does not give another interrupt.*

Because input NMI cannot be masked, it is normally used only for catastrophic events that justify interrupting a program no matter how important its present task. One common use of the pin is to detect a power failure. A signal from an external device might indicate that the line voltage has dropped below 100 V and is falling. During the remaining few milliseconds of operation, the CPU can possibly move important data to nonvolatile storage, if available.

However, in the PC the NMI pin is used for parity error, and provision is included for masking the error signal before it reaches this pin. This is done by feeding the pin from an AND gate, with the software-controlled mask signal fed to one input. The basic circuit is shown in Fig. 6-13. The on-board memory parity check bit PCK and the I/O channel parity bit I/O CHCK are both 0 when there is no error or when they are disabled. They are enabled by writing 0s to respective output pins PB4 and PB5 of the PPI. Although not shown in Fig. 6-13, the two check bits can be read from respective pins PC7 and PC6 of the PPI (see Prob. 6-11). Recall that they were enabled and read in the routine of Fig. 5-20. Following a system reset, the parity system is active.

The 1-bit parity mask register of Fig. 6-13 is positive edge-triggered at the dynamic input. The output Q becomes equal to data bus bit D7 when $\overline{\text{IOW}}$ goes low to high during execution of OUT 0A0,AL. If AL is loaded with 80H ahead of OUT, pin NMI is enabled, but if AL stores 0, it is disabled. Assuming both pin NMI and the parity bits are enabled, NMI interrupt type 2 is activated when a memory parity error occurs. All internal interrupts except single-step have priority, however.

INT 2. In the PC, the service routine of INT 2 consists of 37 bytes of code at F000:E2C3 in the ROM BIOS, plus 32 bytes in an error message table (see Prob. 6-12). When activated by either software or by a signal on pin NMI, it reads PCK and I/O CHCK from the PPI, and if either bit is 1, an error message is displayed with the aid of INT 10H. Then the processor is put into the halt state, with execution of instruction CLI followed by HLT. This operation disables the periodic timer interrupts and the keyboard. Power must be turned off and then on to regain keyboard control.

Input INTR. Pin INTR is *level triggered* by a high signal. This term signifies that the interrupt is activated by a signal that must be high and remain high until the request is serviced. There is no latching of the input, and pin INTR is disabled when flag IF is clear. Interrupt requests at INTR have the lowest priority of all interrupts, except for single step. An interrupt procedure begins only when execution of the current instruction is completed. This rule applies to both internal and external interrupts. Unlike the type 2 NMI, an INTR request must send the CPU the 1-byte type number via the data bus. The 8259 interrupt controller is used for this purpose. Because INTR is level triggered, *a request that remains high after servicing gives another interrupt.*

Suppose a request is received on pin INTR with IF set. The interrupt is acknowledged by executing two consecutive interrupt-acknowledge INTA bus cycles, with the 3 status bits to the bus controller in their low states, indicating the INTA cycle. During states T2 and T3 of each cycle, the 8288 bus controller emits an active-low $\overline{\text{INTA}}$ signal. It is shown in Figs. 5-1 and 6-15. The $\overline{\text{INTA}}$ output from the bus controller goes to the $\overline{\text{INTA}}$ input of the 8259 controller. The 8259 chip then places a byte on the data bus, and this byte is the type number. It is read by the CPU during the active $\overline{\text{INTA}}$ signal of the second bus cycle, and a 2-bit left shift multiplies the type number by 4. The result is the absolute address of the doubleword pointer to the service routine. Processing is now precisely the same as for a software interrupt.

Let us compare the operations of an INTR interrupt with those of the same software interrupt. The hardware interrupt is activated by an external event, whereas the software interrupt is activated by a program instruction. For the INTR interrupt, the first two bus cycles are INTA cycles that send out an $\overline{\text{INTA}}$ signal and read the type number from the controller. For the software interrupt, the first two bus cycles read the 2 bytes of the instruction from memory. The rest of the operations are identical. Pushes save the flag word, CS, and IP, and the new values of CS and IP are obtained from memory at the absolute address $4 * n$. After the flags are saved, IF and TF are reset.

Interrupt Pointer Table

Doubleword pointers to the 256 possible interrupt routines are stored in the lowest 1 KB of memory space, from location 0 through 3FFH. Within the 4-byte space of one pointer, the low byte of IP is at the lowest address, followed in order by the high byte of IP, the low byte of CS, and the high byte of CS. Figure 6-14 shows the *interrupt pointer table* for the 8088 microprocessor.

Types 5 through FFH are not dedicated. All are internal software interrupts available as needed, and any of them can also be used for hardware interrupts provided suitable external control is available. Those of the table with names are reserved by the PC for special use, and those with 8259 at the right side are activated by external events.

Except for types 0 through 4, the interrupts of the table having descriptive names are initialized by the BIOS program of the monitor during power-on and reset. The doubleword pointers placed in memory are vector addresses to service routines in ROM. Those not initialized by BIOS are set up as needed by the program being executed. Initialization of the pointers by BIOS and any service routines stored in RAM can be changed by a program at any time. Although a service routine in ROM is fixed, the pointer can be changed to point to an alternate routine set up in RAM.

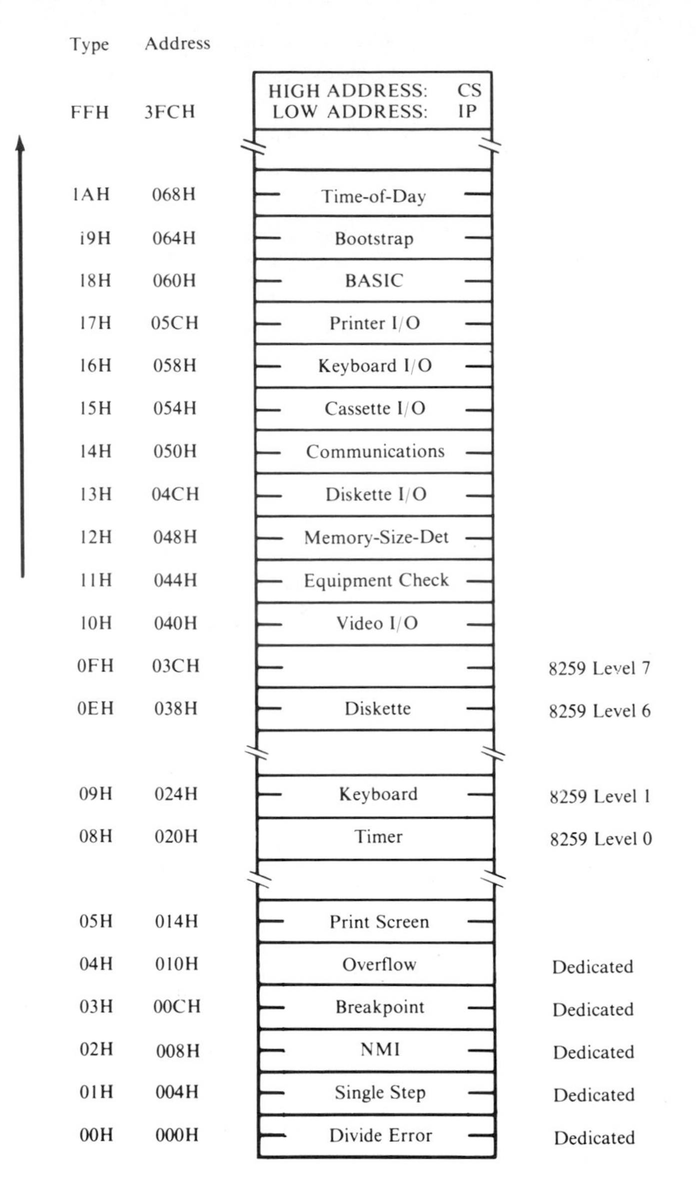

FIGURE 6-14. Interrupt pointer table.

Types 8, 9, EH, and FH of the table are external interrupts fed to pin INTR of the processor from the 8259 interrupt controller. Type FH, which is not used, is reserved for a printer. Level 0, corresponding to INT 8, has the highest priority of this group, and level 7, corresponding to INT 0FH, has the lowest. It is possible for requests to be present simultaneously on both pins INTR and NMI, along with a request generated internally. The internal request has top priority. Next is the latched NMI request and then INTR.

6-4 INTERRUPT CONTROLLER

In the PC, eight prioritized interrupt levels are provided by the 8259A programmable controller, with pin connections as shown in Fig. 6-15. The NMOS circuit

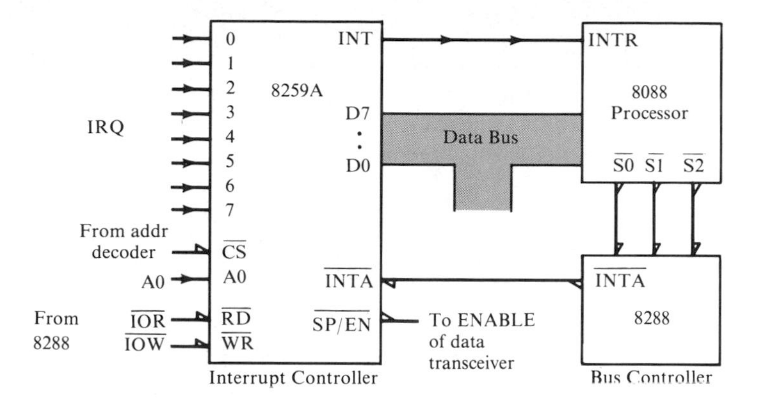

FIGURE 6-15. The 8259A controller and connections.

is contained in a 28-pin DIP. Two of the five pins not indicated in the figure provide the 5 V supply, and the other three are not connected. The controller output INT is connected to pin INTR of the processor.

Interrupt Levels

There are eight input *request levels*. They are labeled IRQ and are assigned level numbers 0 through 7. Level 0, with the highest priority, provides the periodic interrupt from the general-purpose timer. It activates INT 8 at a rate of 18.2 per second. Next in priority is level 1, which receives interrupt requests from the keyboard control circuit whenever a scan code is sent and activates INT 9.

The other six request levels come from the I/O channel, providing for interrupts from devices connected to the expansion slots. Level 6, corresponding to INT 0EH, is used by the diskette system, and there is a request line from the printer to level 7. The asynchronous communications adapter supplies a request line to level 4. Programs can be included for these and also for levels 2, 3, and 5, which are available for use by I/O devices that might be added. In Fig. 6-16 are shown the request levels in order of priority top to bottom, along with the locations of the vector addresses, the interrupt type numbers, and their functions.

Interrupt Level	Vector Address Location	Interrupt Type	Interrupt Function
IRQ0	20–23H	INT 08H	Timer
IRQ1	24–27H	INT 09H	Keyboard
IRQ2	28–2BH	INT 0AH	
IRQ3	2C–2FH	INT 0BH	
IRQ4	30–33H	INT 0CH	Communications
IRQ5	34–37H	INT 0DH	
IRQ6	38–3BH	INT 0EH	Diskette
IRQ7	3C–3FH	INT 0FH	Printer

FIGURE 6-16. Hardware interrupt levels.

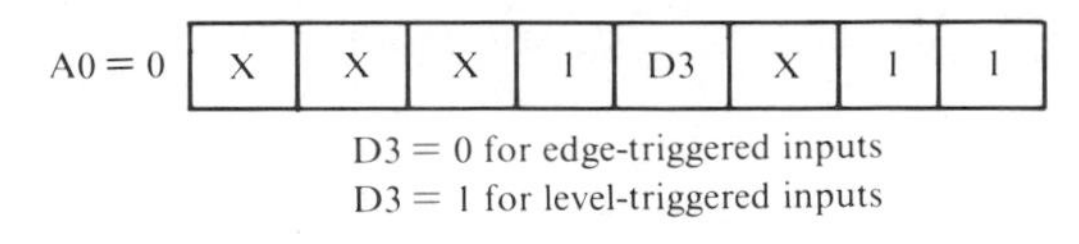

FIGURE 6-17. ICW1 for a single 8259 controller.

Addressing

In the PC, chip input $\overline{CS}$ is supplied from an address decoder. It is active when the binary address is 00001XXXXA, with the low-order bit A denoting A0. This bit feeds pin A0 of the chip. Selecting 0s for the don't-care X's, byte addresses 20H and 21H are used for programming the controller and reading its registers. Although only two addresses are applicable from the address bus, bits D4 and D3 of the data bus provide some additional internal selectivity. When either $\overline{CS}$ or both $\overline{RD}$ and $\overline{WR}$ are high level, the data bus floats. Inputs $\overline{IOR}$ and $\overline{IOW}$ to pins $\overline{RD}$ and $\overline{WR}$ are outputs of the bus controller passed through a buffer not shown in Fig. 6-15.

Initialization

Before normal operation can begin, 3 bytes must be written to the controller. Each is an *initialization command word ICW*. In order, the bytes written are ICW1, ICW2, and ICW4. Byte ICW3 applies only when there is more than one 8259 controller chip in the system. Thus, it is ignored for the PC.

ICW1. An OUT instruction with A0 clear and with data bus bit D4 set is interpreted by the controller as ICW1. For the 8088 processor, the byte is that of Fig. 6-17.

Selecting 0s for don't-cares, as we will always do, for edge-triggered inputs byte 13H is written to port 20H. An advantage of edge triggering is that a request remaining high level does not activate a second interrupt after the first one has been processed. The value of D4 identifies ICW1, that of D1 indicates a single controller chip, and D0 is always 1 when the processor is the 8086/8088.

ICW2. Each request level provides an interrupt type number *n*. Although consecutive values of *n* are required for levels 0 through 7, these values may be selected within the allowed type number range 0 - FFH. The selection is accomplished by ICW2, shown in Fig. 6-18.

The address is 21H, and data bus bits are not involved in addressing. For the PC, level 0 corresponds to INT 8. Accordingly, ICW2 is 08H. With this choice, the controller automatically generates interrupt type numbers from 08H through

A0 = 1	D7	D6	D5	D4	D3	X	X	X

Type numbers for levels 0–7 are determined by D7–D3, with D2, D1, and D0 automatically assigned values 000–111.

FIGURE 6-18. Initialization command word ICW2.

```
MOV   AL,   13H     ;               (B013)
OUT   20H,  AL      ;   ICW1,       (E620)
MOV   AL,   8       ;               (B008)
OUT   21H,  AL      ;   ICW2,       (E621)
MOV   AL,   9       ;               (B009)
OUT   21H,  AL      ;   ICW4,       (E621)
```

FIGURE 6-19. Controller initialization with code.

0FH for respective interrupts IRQ0 through IRQ7. If byte ICW2 is changed to 30H, the type numbers of the eight hardware interrupts become 30H through 37H.

ICW4. For the PC, the binary byte of ICW4 is 00001001, and the write address is 21H. The 3 high-order bits are always 0s. For a system with only one controller, D4 must be 0. The values selected for bits D3 and D2 cause an output signal to be generated at pin $\overline{SP}/\overline{EN}$. This signal goes low when the interrupt type number is placed on the data bus. It is used in the PC to disable the data transceiver, thereby isolating the CPU from peripherals other than the interrupt controller. A logical 0 for D1 specifies the normal end-of-interrupt (EOI) procedure, and the 8088 processor is indicated by the logical 1 assigned to D0.

The controller is programmed during the initialization procedure that follows a system reset. The instructions and their machine codes are shown in Fig. 6-19. After the sequence is executed, the controller is ready to accept external interrupts.

End-of-Interrupt (EOI)

When a hardware interrupt is accepted for processing, the *in-service* bit *IS* of a latch of the controller is set, which inhibits all further interrupts of the same or lower priority. Interrupts of higher priority are allowed, provided software of the current service routine has set flag IF. At the end of the service routine, an instruction must be included that resets the IS bit. This is done by writing operational command word OCW2. The address is 20H, with bits D4 and D3 both 0, and the byte that must be written is 20H. Otherwise, there can be no further interrupts of the same or lower priority.

Figures 5-28 and 5-29 contain the greater part of the service routine of the timer interrupt (INT 8) of level 0. Three instructions that should be added at label C of Fig. 5-29 are as follows:

```
INT   1CH
MOV   AL,   20H
OUT   20H,  AL
```

This sequence completes the actual routine in the ROM BIOS, except for PUSH, POP, and IRET instructions. The last two instructions of the preceding sequence reset the IS bit.

The timer tick INT 1CH gets its vector address from absolute location 112. During a reset, the vector is initialized to F000:FF53. The variable of this location points to an IRET instruction in ROM, constituting a dummy return. *The purpose is to allow a programmer easy access to the periodic timer ticks.* Unless the vector address is changed to point to a suitable program in RAM, the software interrupt does nothing useful. An example that changes the address is given in Section 7-6.

Interrupt Mask Register

By writing OCW1 to the mask register of the 8259 chip, one can mask any of the 8 levels. A set mask bit of a level inhibits the interrupt, and a reset bit enables it. Normally, all bits are clear. An OUT instruction to port 21H following the initialization writes to the mask register, and data bus bits D7 through D0 correspond to the respective mask bits of levels 7 through 0. There are additional registers and features not examined here. For example, it is possible to read the active interrupt level, the interrupt-request register, the mask register, and the in-service register. A detailed description of these is given in Intel's *Component Data Catalog*.

Interrupt Sequence

Suppose that interrupt requests IRQ4 and IRQ6 are active when the interrupt system is enabled with instruction STI. Signal INT from the 8259A provides a logical 1 to input INTR of the processor. The current instruction, including all WAIT states, is completed. Then the three status lines from the processor to the 8288 bus controller are activated, and two interrupt acknowledge $\overline{\text{INTA}}$ output signals are generated by two successive INTA bus cycles.

When the second $\overline{\text{INTA}}$ signal is received by the 8259A, byte 0CH corresponding to IRQ4 of higher priority is transmitted from the 8259A to the CPU. A 2-bit left shift converts byte 0CH to 30H, and the program branches to the INT 0CH routine having the vector address that is stored in absolute locations 30–33H. The flags and the contents of CS and IP are pushed onto the stack, and the trap and interrupt flags TF and IF are cleared. The first instruction in the service routine is usually STI, so that higher-priority interrupts can interrupt the interrupt. On completion of the service routine for IRQ4, request IRQ6 is serviced.

Although an interrupt request is normally acknowledged at the end of execution of the current instruction, there are exceptions. When a POP or MOV instruction loads a segment register, an interrupt is not allowed until after the instruction that immediately follows has been executed. This procedure prevents an interrupt occurring between establishment of a new segment address, such as SS, and an associated pointer, such as SP.

6-5 BIOS INTERRUPTS

In the BIOS ROM are 15 interrupt service routines, excluding tables and dummy returns. Most of them provide interfacing between the processor and I/O units. Listings of each, with comments alongside the instructions, are given in the IBM *Technical Reference Manual*. They are identified by name in the table of Fig. 6-14, with types 0, 1, 3, 4, and 18H (BASIC) excluded. The vector addresses of these interrupts are initialized during a system reset, and parameters are passed to and from the routines through the 8088 registers. By means of both external and internal interrupts, I/O control is provided for devices normally connected to the PC.

We investigate these and other routines here, beginning with those that are normally activated externally. A major objective is to show the kind of service support provided in a typical monitor of a microcomputer. Those who write assembly language programs for execution by a computer should become familiar with the services available.

External Interrupts

Four of the 15 routines in the BIOS ROM are activated externally. The first one is requested by a signal on the nonmaskable interrupt pin NMI, and the other three requests come through the 8259 controller to pin INTR. All can be activated by software.

INT 2 (NMI). As described in Section 6-3, INT 2 from pin NMI is for *parity error*. With the parity system enabled, an NMI interrupt prints an error message and halts operation.

INT 8 (Timer Interrupt). INT 8, a level 0 external interrupt, is fed from counter 0 of the 8253 timer; the routine was described in Sections 5-6 and 6-4. The periodic interrupts *supply 18.2 ticks per second to the internal clock*. A count of the interrupts is maintained in five byte locations at address 0040:006C in RAM, starting from 0 at power-on. The clock variables of the locations are TIMER_LOW, TIMER_HIGH, and TIMER_OFL.

The routine also decrements byte variable MOTOR_COUNT of location 0040:0040 on each interrupt. Then, if a diskette motor is running, the routine *turns off the motor* when the count reaches 0 by writing byte 0CH to the digital output register (DOR) of the diskette adapter card. In addition, for possible use by a program, it *executes INT 1CH with a dummy return*. The 78-byte routine starts at address F000:FEA5.

INT 9 (Keyboard Interrupt). A level 1 request from the 8259 controller activates INT 9. With interrupts enabled, *the scan code is processed,* and both the scan code and the ASCII code, if there is one, are put in KB_BUFFER. In Section 1-5 is a BASIC program that examines the buffer contents after entries are made. The vector address is F000:E987, and there are 887 bytes of code and data. The keyboard buffer and the interrupt were described in Section 4-4.

INT 0EH (Diskette Interrupt). INT 0EH, a level 6 interrupt type 14, has only 20 bytes of code at F000:EF57. It simply *sets an interrupt flag* of a certain memory location, which is investigated in detail in Section 8-5.

Internal Interrupts

Although the 11 routines described here are activated only by software, any of them can be placed within a routine activated externally.

INT 5 (Print Screen). The INT 5 routine contains 119 bytes of code at F000:FF54 and is activated by instruction INT 5. The instruction is present in the keyboard routine of INT 9 and is triggered by the key combination Left-PrtSc with interrupts enabled. *Execution sends the entire contents of the display buffer to the printer,* giving a hard copy of the CRT display. It utilizes INT 10H (video I/O) and INT 17H (printer I/O) to accomplish the task.

INT 10H (Video I/O). Software INT 10H was used in the program of Section 6-2 to write to the monochrome display monitor. It consists of 2045 bytes of code, starting at F000:F045 in the BIOS ROM. There are 16 subroutines, corresponding to values 0 through 15 in AH at entry. Other parameters are passed to the routines through registers AL, BX, CX, and DX. Among the functions are those that *set the mode, set the cursor type and position, scroll the display, read a character, write a character, and set the color pallette for a color monitor.*

INT 11H (Equipment Check). The 12-byte code of INT 11H at F000:F84D simply *moves the word of EQUIP_FLAG to register AX,* where it can be examined or changed by a program. Variable EQUIP_FLAG is the word at location 0040:0010 (see Prob. 6-17).

During a system reset, the eight switches of SW1 are read from port A of the PPI, and data are read from communications and printer ports. Based on the data, EQUIP_FLAG is loaded during a system reset so as to indicate optional devices attached to the system. In particular, the stored word gives the number of diskette drives, printers, and communications cards, the amount of on-board memory, whether or not a game adapter is present, and the initial video mode to use, whether monochrome or color. The word put in EQUIP_FLAG can, of course, be modified by software. This is necessary when switching between monochrome and color monitors, as shown in the first example of Section 7-5.

INT 12H (Memory Size Determination). It is often necessary for a program to know the amount of RAM available, and the memory size determination INT 12H is available for this purpose. The 12-byte code of the routine at F000:F841 *reads word variable MEMORY_SIZE at location 0040:0013H and puts the result in AX.* The word of MEMORY_SIZE is encoded to indicate the amount of memory available, including on-board and I/O channel RAM. During the power-on diagnostics, the memory variable is loaded with data found by reading switches SW1 and SW2 from ports A and C of the PPI. The 16-bit binary number of the variable represents the number of 1-KB blocks of memory. An example using this interrupt is given in Section 7-1.

INT 13H (Diskette I/O). INT 13H, a 863-byte software interrupt at F000:EC59, *provides access to the diskette drives.* There are six basic operations, with the one selected depending on the number in AH at entry. These operations reset the diskette, read the diskette status, read a specified number of sectors into memory, write the desired sectors from memory, verify an operation, and format a track. The read and write operations are examined in detail in Section 8-5.

INT 14H (Communications). At F000:E729 are the 261 bytes of the communications routine. *It transfers serial byte streams to and from the communications port* in accordance with parameters encoded and stored in AL on entry. After initialization, AL is used for storage of the character byte to be transmitted or received. Registers AH and DX also pass parameters. The interrupt is investigated in Section 10-4.

INT 15H (Cassette I/O). The cassette I/O routine at F000:F859 has 533 bytes that interface the processor with an *audio cassette.* This device is not normally used when a diskette drive is available because it is relatively very slow.

INT 16H (Keyboard I/O). Software INT 16H *implements keyboard functions* depending on the bit placed in AH before INT 16H is executed. If AH stores 0, the routine reads a word from KB_BUFFER and transfers it to AX. However, first the pointers BUFFER_HEAD and BUFFER_TAIL are compared. If they are equal, indicating an empty buffer, the routine loops until a hardware keyboard INT 9 places a word in the buffer. When one or more words are present, the one to which BUFFER_HEAD points is moved to AX, and then the pointer is incremented by 2. The scan code is placed in AH, and the extended ASCII code goes to AL. Then IRET is executed.

The main program must process the data of AX. When a program needs input from the keyboard, such as the current date or time, a request message is printed on the display, followed by execution of INT 16H. With the buffer empty, the routine waits in the loop until a keyboard entry activates hardware INT 9, which places a character in the buffer. The process is repeated until all required characters have been entered. There are 84 bytes in the routine stored at F000:E82E.

The WAIT loop is of special interest because the computer normally spends a lot of time there. Whenever the computer is waiting for the operator to press a key, it is probably in this loop. The routine of the loop is given in Fig. 6-20.

STI sets IF, which enables external interrupts after execution of the next instruction, and CLI resets IF, disabling interrupts immediately. Clearly, the no-operation instruction NOP between STI and CLI is essential. *Without it, there could be no interrupts.* When an external interrupt occurs with IF reset, the request is held until IF is set. Then it is processed. Note that an interrupt is not allowed between the MOV and CMP instructions, thereby preventing a routine from changing BUFFER_TAIL at an inopportune time.

Programs commonly use INT 16H when there is nothing to do until a key is pressed. The sequence of the loop takes 48 clocks, or 10 microseconds. Thus, the loop is executed 100,000 times per second. Of course, about every 55 milliseconds, or 5500 loops, there is a timer interrupt of the loop from level 0 of the 8259 controller, execution of which takes about 100 microseconds. When a key is pressed, the level-1 INT 9 is processed, and the routine of INT 16H moves the word of the buffer to AX, followed by IRET. The program processes the code in AX and then may very likely execute INT 16H again, returning almost immediately to the WAIT loop. Even when a key is held down, there is about 0.1 second between the scan code interrupts, corresponding to 480,000 clocks. Many instructions can be executed in this interval.

INT 17H (Printer I/O). There are 115 bytes of code at F000:EFD2 that *provide communication with the printer.* If AH stores 0 on entry, the character in AL is sent to the printer. For a value of 1 in AH, the printer port is initialized, and for 2 the status of the printer is read into AH.

```
A: STI                      ; 2 clocks
   NOP                      ; 3 clocks
   CLI                      ; 2 clocks
   MOV BX, BUFFER_HEAD      ; 6 clocks
   CMP BX, BUFFER_TAIL      ; 19 clocks
   JZ  A                    ; 16 (4 for no jump)
```

FIGURE 6-20. The wait-for-interrupt loop.

INT 19H (Bootstrap Loader). The 55-byte routine at F000:E6F2 is normally entered by a direct intrasegment JMP instruction located at the end of the self-test routines of a system reset. After the jump to label BOOT_STRAP, the program examines variable EQUIP_FLAG. If a diskette drive is found to be present, the sense switch of the latch door of drive A is tested. With the latch closed, the *boot record* of sector 1 of diskette track 0 is read into RAM at 0000:7C00, using INT 13H (diskette I/O), with this initial location assigned label BOOT_LOCN. Then a FAR jump is made to the boot record routine at BOOT_LOCN, thereby passing control to the disk operating system. This operation is known as *booting the diskette system*.

If a diskette drive is not available, the program executes INT 18H (BASIC). The vector address of the interrupt points to F600:0000, which is the start address of the 32-KB BASIC interpreter in ROM. The complete bootstrap loader is shown in Fig. 6-21, with the initial ASSUME directive omitted.

Execution of the *bootstrap loader* of any computer is the initial step in bringing up the system. It is a short routine that is, in some systems, manually entered from the keyboard. In other systems it is stored in ROM as part of the resident monitor. The usual function of the routine is *to load a more powerful loader from an external unit*. In this case, the bootstrap loader either gets the boot record from a diskette or, in the absence of a diskette drive, passes control to the BASIC interpreter.

Exit from INT 19H is either by INT 18H to the cassette BASIC interpreter or by a FAR jump to the boot record in RAM at 0000:7C00. In the latter event, the boot record is first written from the DOS diskette to RAM. There is no IRET instruction.

The boot record consists of 400 bytes of code, including about 114 bytes of ASCII messages. It reads the DOS system from the disk into memory and then transfers control to DOS. Afterward, the boot record of memory has no use, being referred to as *garbage*. Its memory locations can be used by other programs.

```
     STI                             ;set IF
     MOV  AX, 40H                    ;
     MOV  DS, AX                     ;establish DS
     MOV  AX, EQUIP_FLAG             ;get hardware data
     TEST AL, 1                      ;examine drive sense switch
     JZ   A3                         ;if no drive, go to BASIC
     MOV  CX, 4                      ;count for up to four retrys
A1:  PUSH CX                         ;save count
     MOV  AH, 0                      ;AH = 0 for diskette reset
     INT  13H                        ;DISKETTE_IO, reset of FDC
     JC   A2                         ;reset failure if CF = 1, retry
     MOV  AH, 2                      ;a read operation is specified
     MOV  BX, 0                      ;move 0 to BX for move to ES
     MOV  ES, BX                     ;BOOT_LOCN seg-addr is 0
     MOV  BX, OFFSET BOOT_LOCN       ;BOOT_LOCN offset is 7C00H
     MOV  DX, 0                      ;head and drive IDs to DX
     MOV  CX, 1                      ;track and sector IDs to CX
     MOV  AL, 1                      ;number of sectors (1) to AL
     INT  13H                        ;read record (disk to RAM)
A2:  POP  CX                         ;get retry count
     JNC  A4                         ;no carry signifies success
     LOOP A1                         ;retry up to four times
A3:  INT  18H                        ;go to BASIC interpreter
A4:  JMP  BOOT_LOCN                  ;jump to boot record at 07C00
```

FIGURE 6-21. Bootstrap loader of the BIOS ROM.

INT 1AH (Time of Day). The 55-byte time-of-day routine at F000:FE6E is designed to provide *for reading and setting the clock of TIMER_LOW, TIMER_HIGH, and TIMER_OFL*. For reading the clock, register AH must store 0 before the call. The routine moves the word of TIMER_LOW to DX, that of TIMER_HIGH to CX, and the byte of TIMER_OFL to AL. For the set operation, register AH must initially store 1, and the count to be set must be present in registers DX and CX. Then the routine moves the word of DX to TIMER_LOW, that of CX to TIMER_HIGH, and TIMER_OFL is cleared (see Prob. 6-19). The precise interrupt rate is 18.206482 per second. A program that reads and sets the clock is given in Section 7-5. Although INT 1AH is not included explicitly in the program, it is called by higher-level DOS functions that are present.

Dummy Return Interrupts

There are two interrupts initialized during a reset with vector addresses that point to F000:FF53, at which location is stored byte CFH for IRET. The interrupts allow the programmer to write service routines if desired.

INT 1BH (Keyboard Break). Within the keyboard interrupt routine (INT 9) is software instruction INT 1BH, which is executed only if the key combination Ctrl-Break is activated. Unless a program has provided a new vector pointer to a service routine of RAM, the key combination does nothing.

For example, in the BASIC mode of operation the interpreter replaces the vector address at location 6CH with F600:4D34, which points to a routine within the ROM of the interpreter. In the DOS mode the pointer is 0070:0140, and at this RAM location is a service routine loaded from the DOS diskette.

INT 1CH (Timer Tick). The timer tick INT 1CH instruction is *included in the routine of timer INT 8*. It is initialized to give a dummy return (see Prob. 6-20). When using the cassette BASIC interpreter, the doubleword pointer is F600:5744, which addresses a routine within the interpreter. The TICKS program of Section 7-6 uses this interrupt.

Interrupt Tables

Three interrupts of the PC are dedicated exclusively to *storage of parameters*. None has instructions in a service routine, and software instruction INT must not be used. The stored data constitute a table in ROM with an assigned label, and parameters are accessed in the usual manner. A base address and an index are used, but the base address is the doubleword pointer stored in the low region of RAM.

The reason for designating a table as an interrupt is simply *to provide a fixed location in RAM for the vector address*. This action provides an easy way to change one or more parameters. In fact, a new table can be established in RAM. Then only the doubleword pointer is changed, not the instructions that use the table. The method is illustrated in the routine of Fig. 6-22.

```
ABSO   SEGMENT   AT 0             DATA   SEGMENT   AT 40H
   ORG  30 * 4                    (variables defined)
TABLE_PTR LABEL   DWORD           DATA   ENDS
ABSO   ENDS                       STACK SEGMENT STACK
                                     DW 100H DUP(?)
                                  STACK ENDS

        CODE SEGMENT
        ASSUME   CS:CODE,   DS:DATA, SS:STACK
        START:   (initialize DS)
           (continue)
           PUSH DS
           MOV   AX, 0
           MOV   DS, AX
        ASSUME   DS:ABSO
           LDS   SI, TABLE_PTR
           MOV   AL, [SI + 5]
           POP   DS
        ASSUME   DS:DATA
           (continue)
        CODE     ENDS
        END     START
```

FIGURE 6-22. Parameter fetch from table to AL.

In the routine of Fig. 6-22, segments ABSO, DATA, and STACK must be entered on separate lines, of course. Note the *AT directive* to locate segments ABSO and DATA. *This directive cannot, however, be used to locate segments containing code or initialized data.* The label TABLE_PTR is defined by the ORG and DWORD directives to identify the vector address of INT 1EH (diskette parameters). Variables defined in DATA are not shown, and only part of an actual program is included.

Instruction LDS moves the segment address of the 32-bit vector address to segment register DS and the offset to register SI. Note the ASSUME statement preceding LDS. *This is essential,* because TABLE_PTR is in segment ABSO. If it is omitted, the assembler reports that there is no ASSUME statement that makes the variable reachable, and the line with TABLE_PTR is displayed. The MOV instruction that follows LDS transfers to AL the sixth byte of the diskette parameter table, which has index 5.

INT 1DH (Video Parameters). The vector address of INT 1DH *points to a table of parameters used by the video I/O routine of INT 10H.* The table has 88 bytes at F000:F0A4.

INT 1EH (Diskette Parameters). There are 11 bytes at F000:EFC7 *intended for use by the diskette I/O routine of INT 13H. These bytes specify parameters* such as the head settle time, the motor start time, the gap length, and the number of bytes per sector. In the updated DOS used by the author's system, this table is replaced with one in RAM that has more suitable parameters. The revised vector address is 0:0522, and this address and the table in RAM are established automatically by DOS when the system is booted. *This change would not have been possible had the table been accessed by direct calls.*

INT 1FH (Character Generator). The INT 1FH interrupt *is reserved to serve as a pointer to a RAM character generator table for ASCII characters 128 - 255*. However, the table is not provided, and there is no initialization of the vector address. In the *graphics mode* of the color/graphics display, the read/write character interface of the video I/O INT 10H routine forms characters from a character generator in ROM. This operation was examined in Section 3-3. Each of the 128 ASCII characters from 0 through 127 is defined by an 8-byte dot pattern that produces the character when sent to the CRT. Although the display handles 256 characters, only the first 128 are given in the ROM.

A 1-KB table for the remaining 128 character images can be set up in RAM by a program. INT 1FH is reserved for addressing this table through a doubleword pointer at the fixed location 1FH * 4. In addition to establishing the table, the program must load the pointer into the fixed location.

INT 18H (BASIC)

The service routine for the cassette BASIC is not in the BIOS ROM. However, when INT 18H is executed, the vector address F600:0000 at location 0:60 *transfers control to the BASIC interpreter* in ROM, and the BASIC prompt appears on the screen.

Not all programs use the routines of BIOS without modification. Programs frequently change the doubleword pointer and insert in RAM a service routine that either adds to or replaces the one in BIOS. Additions are often made to the keyboard program of INT 9. In cassette BASIC there is neither a vector address nor a routine provided for the divide-by-0 interrupt type 0. Internal generation of this interrupt is avoided by examination of operands prior to division. In addition to the service routines of the BIOS ROM, there are those of DOS. These routines are at a *higher level,* in that they use the routines of BIOS to perform more complicated tasks. *They are placed in RAM when the diskette system is booted.* We examine them next.

6-6 DOS INTERRUPTS

The DOS interrupts use the software interrupts of BIOS within their own programs, thereby providing services to user programs at a higher level. There are many different functions accessible to user programs through software interrupt instructions, with parameters passed through the registers of the 8088. Only a few of these functions are described in detail here, but others are examined later.

Thirty-two interrupt types, from INT 20H through INT 3FH, are reserved for use by DOS. The corresponding transfer address locations are 80H through FFH. However, many of them are undefined, having no routines. When the system is booted, programs IBMBIO, IBMDOS, and COMMAND are moved from the diskette to memory. They provide service routines for the eight interrupts INT 20–27H available for use by a programmer. These interrupts are given in Fig. 6-23.

Number	Name	Purpose
20H	Program end	Terminate a program
21H	Function request	88 function calls
22H	Terminate address	Normally used only by DOS
23H	Ctrl-Break address	Normally used only by DOS
24H	Critical error	Normally used only by DOS
25H	Absolute disk read	Read scetors to memory
26H	Absolute disk write	Write sectors from memory
27H	End but stay resident	End program, keep resident

FIGURE 6-23. The eight available DOS interrupts.

Interrupts 20H and 27H

Program-end INT 20H is the traditional way to end a program. However, before it is executed, *the CS register must contain the segment address of the program's program segment prefix* (see Prob. 6-21). This topic is discussed in the next chapter. Normally, control is returned to the COMMAND processor of DOS, which issues a system prompt and waits for the next command from the keyboard. A replacement for INT 20H is described in Section 11-2.

If the program changes the vector address of either INT 22H (terminate address) or INT 23H (Ctrl-Break address), execution of INT 20H *restores the values to those they had on entry*. Also, if file buffers are present in memory, they are flushed to the diskette, but files must be closed prior to execution of INT 20H. File management is described in Chapter 11. When another program is loaded, it overlays the one terminated with INT 20H.

Terminate-but-stay-resident INT 27H has almost the same requirements as INT 20H and accomplishes the same tasks, but the program is not overlaid by the next one loaded into memory. Prior to execution, register DX must be written with the offset at which the next program is to be loaded. This should be the address of the last byte of the program plus one. INT 27H allows a utility program to be placed in memory and to *be write-protected from overlay*. This program might be an interrupt service routine that replaces one in BIOS, perhaps to service a serial printer instead of the usual parallel one. An example is given in Section 7-6. A replacement for INT 27H is described in Section 11-6.

Interrupts 22H, 23H, and 24H

Each of these three interrupts is normally used only by the DOS programs. However, *the user's program can, if desired, change the vector address and insert a new service routine* in RAM. In fact, the routines are contained within the COMMAND processor, and this entire program can be replaced with one designed by the user.

Terminate Address. The vector address of *INT 22H* is the *terminate address*, which points to the location within the COMMAND processor of DOS to which control transfers when the program ends. The terminate address is moved into the *program segment prefix (PSP)* when it is created, which occurs before a program is run, and it is returned to the interrupt table by INT 20H or 27H at the end of a running program. Then when the program terminates, *control is transferred to the* *original terminate address even if the vector address of INT 22H was changed during execution.* Instruction INT 22H is present in DOS routines, but it should not be placed directly in a user program.

A program that changes the vector address of INT 22H is given in Section 11-6. After doing so, it transfers control to a second program, and when this second one has completed its run, the newly assigned terminate address shifts control back to the first program, which ends normally.

Key Combination Ctrl-Break and INT 23H. Activation of key combination Ctrl-Break does several things of importance. First, it clears KB_BUFFER. Then it sets a flag bit of memory, which is the MSB of the byte at 0040:0071, referred to as BIOS_BREAK. Software instruction INT 1BH is called, and on return, the word 0000H is moved to KB_BUFFER. The key combination has no effect, of course, if interrupts are disabled.

Although the vector address of INT 1BH is initialized during power-on to point to a dummy return in the BIOS ROM, the DOS changes the vector address to point to a service routine within IBMDOS at address 0070:0140. Among other things, this service routine puts byte 03H in a reserved location at 0070:015A in IBMBIO. A program can check either the flag bit of BIOS_BREAK or the reserved byte location of IBMBIO. The check for byte 03H is done routinely by many functions of DOS, and if the check indicates that Ctrl-Break has been activated, the program is usually terminated by calling INT 23H. The vector address 053A:0299 of INT 23H transfers control to the COMMAND processor, and the processor stops the running program, displays ^C followed by the normal DOS prompt, and waits for the next command from the operator. Thus, *Ctrl-Break is often used to stop a running program.*

The vector address of *INT 23H* is called the *Ctrl-Break exit address.* The service routine is normally entered only when the key combination is activated during standard keyboard input, display output, printer operations, and communications. All registers are set to the value they had at the start of the function that initiated the interrupt.

A user program may change the vector address and insert a new service routine. BASIC does this, giving a jump to a service routine within the BASIC interpreter. Prior to execution of each BASIC statement, a check for Ctrl-Break is made; if it is found, the program aborts with ^C displayed. Many other programs have their own Ctrl-Break routines. There are no restrictions on what can be done. The Ctrl-Break exit address is stored in the program segment prefix and returned to the interrupt table by INT 20H or INT 27H, which terminates the program, just as was done for the vector address of INT 22H. *This procedure saves the normal DOS vector address in case it was changed by the running program.* An example using a different Ctrl-Break function, with the exit address of INT 23H changed, is given in Section 11-6.

Critical Error Handler INT 24H. When a *critical error* occurs in DOS, control is transferred to a DOS routine that services the error and displays an error message. If the error was a diskette operation, the operator is given the option of retrying, aborting, or ignoring the operation causing the error. The transfer to the error handler is by means of *INT 24H.* Its vector address points to the service routine. If desired, the user program may change the vector address and insert a new service routine.

Interrupts 25H and 26H

INT 25H is used for reading and INT 26H for writing absolute sectors of a diskette, each of which contains 512 bytes. On entry, register AL must store the drive number, which is 0 for drive A. CX has the number of sectors to be transferred, and DX has the beginning logical record number. For double-sided diskettes, logical record numbers of track 0 go from 0 through 17, with the first nine numbers applicable to side 0 and the next nine to side 1. Similarly, logical record numbers of track 1 are from 18 through 35, and so on. For single-sided diskettes, the numbers of the sectors of side 0 are sequential. The memory transfer address is DS:BX (see Prob. 6-22).

The contents of all registers except the segment registers are destroyed. An unsuccessful operation gives a set CF and puts an error code in AX. Exit from the routine is with RET rather than IRET, which leaves the original flag word on the stack. To prevent stack growth, the flag should be popped after examination of CF. Control is transferred to INT 13H of BIOS for implementation of the diskette read/write operations.

INT 21H (Function Request)

Associated with INT 21H are 88 different functions, each assigned a number from 0 through 57H. Thirteen of these numbers are for internal use only by DOS, and all others are described in Chapter 5 of the *DOS Technical Reference Manual.* They are easily accessed by user programs. The function that is implemented when the interrupt is called is identified by the number in register AH at entry. Each has its own routine stored in IBMDOS, and for device operations, calls are made to IBMBIO, which interfaces with the BIOS ROM interrupts. Figure 6-24 lists 18 of the 88 functions.

A number of the functions omitted in Fig. 6-24 are related to the operation of the diskette system. Included in this group are functions that reset the diskette controller, select the drive, return the number of the current default drive, and get the amount of free space on a diskette. Operation of the diskette system is examined

AH	Function
1	Wait for a character input to AL and display
2	Transfer character in DL to display
3	Wait for character from communications adapter
4	Move character in DL to communications adapter
5	Transfer character in DL to printer
6	Key input to AL if DL has FF; if not, CRT←(DL)
7	Same as 1, but no echo and no Ctrl-Break
8	Same as 1, but no echo to display
9	Display message; pointer is DS:DX; terminator is $
AH	Create a buffer for keyboard input
BH	Check keyboard status (INT 23H if Ctrl-Break)
CH	Clear keyboard buffer, execute keyboard function
25H	Set interrupt vector (type in AL; address in DS:DX)
2AH	Get the date and put it in encoded form in CX:DX
2BH	Set the date from encoded words of CX:DX
2CH	Get the time and put it in encoded form in CX:DX
2DH	Set the time from encoded words of CX:DX
35H	Move interrupt vector to ES:BX for type in AL

FIGURE 6-24. Selected functions of INT 21H.

in Chapter 8. Many of those functions omitted are dedicated to file controls, which create, open, close, read, write, rename, delete, and search files. Some operate on file directories. Many of these functions are examined in Chapter 11. Not included in the table is function 0, which is exactly the same as INT 20H. We now examine the listed functions.

Keyboard Functions

With 1 in AH when INT 21H is executed, the program waits for a character to be entered from the keyboard, unless one is already in KB_BUFFER ready to be processed. When the character is typed, it is moved from KB_BUFFER to AL and echoed to the display. However, for key combination Ctrl-Break, the only action is execution of INT 23H. As explained in Section 1-5, if the current character of KB_BUFFER has 00H in the low-byte position, such as 4700H for the Home key, the character is said to have an extended 2-byte code. The function returns the low byte, 00H, when first called, indicating a special character. A second call of the function returns the high byte. This procedure applies also to functions 6, 7, and 8.

Execution of *function 6* sends the character of DL to the display, provided it is not FFH. For byte FFH, the keyboard buffer is checked. If a character is available, it is moved to AL and flag ZF is cleared. Otherwise, ZF is set. The program should put the desired byte in DL prior to the function call.

Function 7 is the same as 1, except that there is no echo to the display and no execution of INT 23H in response to Ctrl-Break. For Ctrl-Break, byte 03H is put in AL. *Function 8* is identical to 1, except for omission of the echo.

It is often necessary to set up a special keyboard buffer, perhaps for transfer of a record (a number of bytes) to a diskette file. *Function 0AH* performs this operation. On entry, DS:DX points to the location of the buffer. At this address must be a byte that specifies the number of characters that can be held. Then when the interrupt is called, characters are read from KB_BUFFER and put into the new buffer, beginning at the third byte location. The second byte of the buffer is set to the number of characters received, excluding the carriage return 0DH that terminates the operation. The last byte of the buffer is always 0DH.

An attempt to enter valid characters in excess of one less than the value of the first byte causes a beep on the speaker, and the character is rejected. Thus, if the first byte is 100, up to 99 characters plus 0DH are allowed. Key closures are displayed, and characters placed in the buffer remain there until displaced by a memory write (see Prob. 6-23).

Function 0BH is used to check the status of the normal keyboard. If a character is available, FFH is put in AL. Otherwise, AL has 00H. For code of Ctrl-Break, INT 23H is executed.

KB_BUFFER is cleared by *function 0CH,* which then executes the function number that was placed in AL before entry. This number must be either 1, 6, 7, 8, or 0AH.

Display and Printer Output

Function 2 transfers the character of DL to the display and places it in AL. Following the output, if a Ctrl-Break is detected, INT 23H is executed. Quite similar is function 5, but the character of DL is sent to the printer.

Several routines have been used previously for sending a message to the display. However, *function 9* provides an easy way. Prior to entry, place the pointer to the message in DS:DX and put the dollar symbol $ (24H) at the end of the message. When the interrupt is executed, the message is sent to the display. If the terminator is carelessly overlooked, characters are sent until and if byte 24H is encountered. Line feed, carriage return, and backspace codes are accepted. For display of a single character, function 2 is more appropriate (see Prob. 6-24).

Set and Get Interrupt Vector

Function 25H provides a convenient way for a program to set the vector address of INT *n* in the interrupt table. Prior to entry, type *n* is placed in AL, and the vector is placed in DS:DX (see Prob. 6-25).

To read the vector address of an interrupt, the type number must be in AL when *function 35H* is called. The vector address is returned as ES:BX.

Date and Time

Functions 2A through 2D are for setting and getting the date and time, using the system time-of-day clock. *Number 2A* places the date in CX:DX, with the number of the year stored in CX, the month in DH, and the day in DL. The date is adjusted when the time-of-day clock rolls over to the next day, with consideration given to the number of days in the month and the effect of leap years. *Function 2B* is for setting the date, which is accomplished by placing the correct date in CX:DX prior to calling the interrupt.

The *time operations of 2C and 2D* are similar, except that CH has the hours from 0 through 23, register CL has minutes, DH has the seconds, and DL has the number of hundredths of a second. All data are stored in the registers in binary form (see Prob. 6-26). Selected formats for both the date and the time are readily converted by a program to printable form, yet they are suitable for calculations.

6-7 SUBROUTINES IN BASIC

As in assembly language, both subroutines and interrupts are quite useful in BASIC. Subroutines are examined here, including calls to machine language programs; such programs are often interfaced with those written in high-level languages. Also included are discussions of the cursor control statements, the BASIC functions, and the KEY statement. BASIC interrupts are investigated in Section 7-7.

Subroutines may be placed anywhere in a program, and a branch to one is accomplished with the GOSUB statement. There may be more than one RETURN statement, with logic dictating which one is implemented. GOSUB can be used in conjunction with ON to direct the branch to one of several subroutines. With each *line* denoting a statement number, the formats for both GOSUB and ON-GOSUB are as follows:

GOSUB *line*
ON *n* GOSUB *line, line, . . .*

If *n* is 1 in the second statement, the branch is to the subroutine identified by the first line number. The branch for $n = 2$ is to the one identified by the next line number, and so on. An expression that returns a number can be used for *n*, and if the result is not an integer, then *n* is rounded to one. In case *n* is either 0 or greater than the number of items in the list, there is no branch.

A related instruction is ON *n* GOTO followed by line numbers. For this instruction, the unconditional branch has no return. In the first illustrative program of this section are several additional instructions, some of which are now examined.

Locate

The LOCATE statement is used to position the blinking cursor on the screen and, optionally, to turn it on or off and to specify its size. The format is as follows:

LOCATE *row, column, cursor, start, stop*

Each of the parameters following LOCATE is optional. If one is omitted, the current value applies, except that if *start* is given but not *stop*, the value of *stop* becomes that specified for *start*. Parameter *row* can be from 1 to 25 and *column* can be from 1 to 80. At *home* both values are 1, and they are 25 and 80 at the lower-right-hand corner of the screen. A *cursor* value of 0 turns it off and 1 turns it on. At the start of program execution, the cursor is normally off. It can be turned on with LOCATE ,,1 placed in the program.

The size of the cursor is fixed by *start* and *stop*. A character block of the monochrome screen is 14 dots high and 9 dots wide. The cursor of a block is a horizontal bar 9 dots wide, but any or all of the possible 14 lines can be active. The lines are numbered from 0 at the top to 13 at the bottom. Values of *start* and *stop* can be from 0 through 13. For example, LOCATE ,,,6,13 activates the bottom 8 bars from 6 through 13, and LOCATE ,,,13,2 activates bar 13 at the bottom and bars 0 through 2 at the top, giving a two-part cursor. Normal values for *start* and *stop* are 12 and 13. For the color display, there are only eight lines, numbered 0 through 7.

Arithmetic Functions

Unlike commands and statements, *BASIC functions* return a value from one or more arguments, and the value may be a number, a string, or code. There are 15 arithmetic functions intrinsic in the system, and they return numeric values. Each function is listed in Fig. 6-25. Variable *x* denotes a numeric expression. Programs can use these functions as often as desired.

In addition, the user can define new functions by means of the DEF FN statement. Its format is as follows:

DEF FN*name (arg1, arg2, . . .)* = *definition*

Name is any valid numeric or string variable name, and it must be preceded by FN. The name of the defined function becomes *FNname*. Do not use reserved names, such as SIN. Each *arg* is a variable name that is replaced with a value when the function is called. However, if no arguments are supplied, current values

Name	Function
ABS (x)	Absolute value
ATN (x)	Arctangent
CDBL (x)	Convert to double precision
CINT (x)	Round to an integer
COS (x)	Cosine
CSNG (x)	Convert to single precision
EXP (x)	Raise e to power x
FIX (x)	Truncate x to an integer
INT (x)	Largest integer <= x
LOG (x)	Natural logarithm
RND (x)	Random number
SGN (x)	Sign of x
SIN (x)	Sine
SQR (x)	Square root
TAN (x)	Tangent

FIGURE 6-25. Intrinsic arithmetic functions.

are used. Recall that numeric variables are initialized to 0 by the system. The *definition* is an expression selected as desired. Only a single logical line is allowed, and the statement cannot be used in the direct mode.

Subroutine Example

Many programs require the use of a menu that is presented on the screen, with several choices for the user. After a selection is made, a routine is processed, with an eventual return to the menu. Because the menu and its selections may be called repeatedly, it is convenient to put their related instructions into subroutines. An elementary program illustrating the way this concept is commonly implemented is given in Fig. 6-26. There are three subroutines. One of them is for servicing the call to the menu, and the other two are service calls to two of the choices presented by the menu.

The three subroutines are placed at the beginning of the program, following a GOTO statement that prevents direct entry. Execution is usually fastest when they are so placed, but this is not a requirement. If these subroutines are put at the end of a program, the statement END normally precedes them, terminating execution. In the example, the operator either selects from the menu one of two user-defined functions for computation or decides to terminate the program. The main routine starts at line 400.

Functions FNA and FNB are defined on lines 400 and 410. They were rather arbitrarily selected, with assigned names that are simple, though nondescriptive. Note that FNA uses the intrinsic SIN function, and FNB uses both TAN and FNA.

Following the function definitions, the KEY OFF statement of line 420 eliminates the display of the string values assigned to the soft keys (10 function keys), which gives the display a nicer appearance. When it is on, the display occupies line 25 of the screen. The KEY statement is described in detail later in this section.

After this procedure, the WHILE statement is executed, and variable NUM is compared with 3. Because its initial value is 0, execution proceeds sequentially. Only if NUM were 3 would the program branch out of the loop to the statement END at line 470. The display is cleared by CLS, and the ON NUM GOSUB statement is ignored because NUM is neither 1 nor 2. At line 450, after another

```
10   REM  ***  Use of Subroutines  ***
20   GOTO 400
30   '
100  REM  ***  SUBROUTINE 1  ***  FNA(A) Computation
110     LOCATE  6, 4:  INPUT  "The value of A is ", A
120     LOCATE  9, 4:  PRINT  "FNA(A, B) is "; FNA(A)
130     LOCATE  14, 4
140     INPUT "Press Enter to return to menu", X$
150  RETURN
160  '
200  REM  ***  SUBROUTINE 2  ***  FNB(A, B) Computation
210     LOCATE  6, 4:  INPUT  "Enter A, B:  ", A, B
220     LOCATE  9, 4:  PRINT  "FNB(A, B) is "; FNB(A, B)
230     LOCATE  14, 4
240     INPUT  "Press Enter to return to menu", X$
250  RETURN
260  '
300  REM  ***  SUBROUTINE 3  ***  Menu Display and Number Check
305     LOCATE  3, 26:  PRINT  "Menu"
310     LOCATE  6, 15
315     PRINT  "FNA(A) = (sin A)/A + (sin 2A)/(2A)"
320     LOCATE  8, 15
325     PRINT  "FNA(A, B) = B * TAN( (FNA(A)) ^ B)"
330     LOCATE  11, 18:  PRINT  "1. Compute FNA(A)"
335     LOCATE  13, 18:  PRINT  "2. Compute FNB(A, B)"
340     LOCATE  15, 18:  PRINT  "3. Terminate program"
345     LOCATE  18, 15:  INPUT  "Enter 1, 2, or 3: ", NUM
350     '    ——Verify proper entry——
355     WHILE  NUM <> FIX(NUM)  OR  NUM < 1  OR  NUM > 3
360        BEEP
365        LOCATE  18, 15:  PRINT  SPC(30)
370        LOCATE  18, 15
375        INPUT  "Erroneous entry.  Enter 1, 2, or 3:  ", NUM
380     WEND
385  RETURN
390  '
400  DEF FNA(A)   = SIN(A)/A + SIN(2*A)/(2*A)
410  DEF FNB(A,B) = B * TAN( (FNA(A))^B )
420  KEY OFF
430  WHILE  NUM <> 3
440     CLS:  ON  NUM  GOSUB 100, 200
450     CLS:  GOSUB  300
460  WEND
470  END
```

FIGURE 6-26. A program based on subroutines that develop a menu.

CLS that does nothing on this first pass, the jump to subroutine 3 at line 300 occurs. The target can be changed to line 305. There is no reason to jump to a REM statement, but it is permissible.

The Menu. The menu of subroutine 3 is designed to display the two functions that can be computed, followed by three numbered statements. The first of these statements is selection of FNA, the next one is for FNB, and the third choice is for program termination. The INPUT statement of line 345 *requests the operator to enter choice 1, 2, or 3*. LOCATE statements of the subroutine position the start of each printed line. The numbers were chosen to give a nice appearance on the screen.

Entry Check. Lines 355 through 380 consist of a WHILE loop that examines the keyboard entry for choice 1, 2, or 3 and *rejects any other values*. Consider the WHILE statement of line 355. If the entry assigned to the variable NUM is less than 1 or greater than 3, statements of the loop are executed. They produce a warning beep and erase the last line of the menu, replacing it with another entry request. Suppose the entry is within the range but fractional, such as 1.2. Then noninteger NUM and integer FIX(NUM) are unequal, and this condition also executes the loop.

The SPC(x) function of line 365 prints x blanks from the current cursor position. Thirty blanks are sufficient to erase the previous input request and its response. With a proper entry, the program returns to WEND of the WHILE loop of the main program and checks the entry for 3. If it is 3, the program ends. Otherwise the screen is cleared, and a branch occurs to either subroutine 1 or 2.

Subroutines 1 and 2 (Computation). Subroutines 1 and 2 are identical, except for the particular function processed. An INPUT statement displays a request message and *waits for the operator to enter the necessary data from the keyboard.* Then the PRINT statement that follows gives the result. Note that the numerical value of a function is the one based on the current values of its arguments. The second INPUT statement of each of the two subroutines is used *to give a delay,* so that the result remains displayed until the operator presses Enter. A dummy variable X$ is used. The RETURN is to line 450, and the screen is cleared and the menu is again displayed.

CALL Statement in BASIC

Machine language programs are easily run on the PC in the DOS mode. However, for short routines, it is sometimes easier to execute a machine language routine in the BASIC mode. Furthermore, high-level languages often include machine language code in special routines that must be processed at higher speed. The CALL statement can be used for these purposes. The format is as follows:

CALL *numeric-variable (var1, var2, . . .)*

The value assigned to the integer *numeric-variable* becomes the 16-bit *offset into the current defined segment* of the subroutine that is called, and this segment is that of the last DEF SEG statement. Optional integer variables *var* are passed as arguments into the machine language routine by pushing onto the stack their *offsets* into BASIC's data area. The call is always to a machine language procedure, which must have an intersegment (FAR) return. *Variables must be designated as integers,* because the processor cannot work directly with BASIC's floating point variables. An elementary example that assumes n bytes of machine language code is given in Fig. 6-27.

Statement 10 defines the user segment address as 2000H, giving an absolute address of 20000H, or 128K. Then all variables are defined as integers, which is essential. The next three statements read n bytes of machine code into memory, starting in segment 2000H at offset 0. Because POKE puts bytes into memory, *the data items must be bytes.* In line 70 the 16-bit 0 offset is assigned to variable SUBR, which is the name selected for the subroutine. Integer variables A, B, and

```
10  DEF SEG = &H2000
20  DEFINT  A-Z
30  FOR I = 0  TO n - 1 'numerical value replaces n - 1
40    READ  J
50    POKE  I, J
60  NEXT
70  SUBR = 0
80  A = 6: B = 8; C = 0
90  CALL  SUBR(A, B, C)
100 PRINT  A, B, C
110 DATA  (n bytes of machine code)
```

FIGURE 6-27. BASIC program for execution of a machine language routine.

C in statement 80 are assigned word values, which are stored in BASIC's data space in the lower end of memory, along with SUBR.

The CALL statement defines SUBR as a FAR procedure. When it is executed, the *offsets* of arguments A, B, and C are pushed in order onto the stack, followed by CS and then IP, with CS:IP being the return address to the next BASIC statement. The program branches to the machine code at location 2000:0000. Upon entry to the procedure, *the DS, ES, and SS registers hold the segment address of BASIC's data space, and CS stores the address specified by DEF SEG.*

In Fig. 6-4 is a sequence of 10 assembly language instructions that add integers A and B and store the result in word location C. The routine assumed that the stack was used for parameter passing, with offsets of A, B, and C pushed in order. Thus, the code of this procedure can be used for the DATA in line 110 of the program of Fig. 6-27. Hexadecimal code for the 10 instructions consists of 22 bytes, which are as follows:

55 8BEC 8B760A 8B04 8B7608 2B04 8B7606 8904 5D CA0600

With the bytes identified as hexadecimal and entered in order into the DATA statement, using 22 for *n*, execution of the program gives a display of the values of A, B, and their difference C, or 6, 8, and -2.

Machine Code Storage. To keep BASIC from writing over a subroutine, the machine language code should be located in memory above the 64-KB region used by BASIC for program storage and work space. With disk BASIC present in memory the BASIC data segment starts at about 40 KB, which is above the interrupt table (1 KB), the BIOS ROM data area (0.25 KB), the DOS programs and workspace (23 KB), and the disk BASIC extension (16 KB) loaded from the diskette. The maximum memory space used by BASIC is the 64 KB of its data segment, or less if there is not that much memory available. Thus code should be above about 105 KB, requiring a minimum memory of 128 KB. If memory is less than this, there are ways to get around the problem.

The command CLEAR can be used to limit space for BASIC so that memory can be set aside for machine language subroutines. This operation was examined in Section 3-5. Another method is *to place the machine code in defined arrays in the BASIC data area.* The program of Fig. 6-28 is similar to that of Fig. 6-27, except that the *n* bytes of machine language code are put into an array.

```
10  DEF SEG
20  DEFINT A-Z
30  DIM  A(m)                       'Replace m with the
40  FOR  I = 1 TO m                 '   number of words.
50    READ  A(I)
60  NEXT
70  A = 6:  B = 8:  C = 0
80  SUBR = VARPTR(A(1))
90  CALL SUBR(A,B,C)
100 PRINT A, B, C
110 DATA (m words)                  'Insert machine code.
```

Figure 6-28. Program with machine code stored in an array.

The first statement in the program of Fig. 6-28 establishes the BASIC data segment as the current segment for use by the CALL statement. The FOR loop reads the code and assigns the words to the 16-bit variables of the array. Variable A(0) is not used. The assignments of line 70 define A and B and reserve a location for variable C. In 80, SUBR is assigned the 16-bit offset of the first element of the array, which stores the first *2-byte code*. Note that each data item *must be a word* containing 2 bytes of machine code, *with the low-order byte being the one to be stored at the lower address*. To illustrate, the 11 data entries for the previous code, with &H omitted, become

```
8B55, 8BEC, 0A76, 048B, 768B, 2B08, 8B04, 0676,
                0489, CA5D, 0006
```

These should be compared with the entries given previously for the program of Fig. 6-27. For an odd number of bytes, the high-order byte of the last entry is a don't-care value, because it is never executed.

A word of caution with respect to the use of VARPTR is in order. If lines 70 and 80 are interchanged, the program does not work. In this case, SUBR is assigned the value of the offset to array element A(1) at the time that the assignment statement is executed. However, when word variables A, B, and C are assigned in the next statement, *the array variables are moved 6 bytes upward in memory to make room for the simple variables,* which are always stored ahead of array variables. Thus, *the value of SUBR errs by 6 bytes*. If the program is run with this error, it probably will "go into left field," requiring a power-off.

There are alternate ways to run machine language routines in the BASIC mode. The USR function, which is somewhat similar to CALL, can be used. String variables can be passed to subroutines. Programs that have been assembled, linked, and stored on a diskette can be loaded into memory using the BLOAD command and then called. These and other methods are described in Appendix C of the IBM *BASIC Manual*.

The KEY Statement

The KEY statement allows one to *set, display,* or *turn off* the soft keys. Initially, the 10 soft keys, or function keys, are assigned string values by BASIC. Up to six characters of each assignment are displayed on the bottom row of the display. For example, key F2 is assigned RUN←, which indicates that pressing F2 is equivalent to entering RUN. Thus, to run a program, the operator can either press key F2 or type RUN followed by Enter. The display is removed, but the function keys are not disabled, by executing KEY OFF. This action makes line 25 available for program use. However, whether off or on, the line is not scrolled, as are the others. Values are returned to the display by KEY ON.

With KEY *n,x$*, the initial assignments can be changed by the operator. The function key number from 1 to 10 is *n*, and the string *x$* is the assignment. Up to 15 characters are allowed. When a function key is pressed, the string is input to BASIC just as though it were typed from the keyboard. Assignment of a null string (length 0) disables the key. All assigned values are listed by KEY LIST.

REFERENCES

BASIC Manual, IBM Corp., Boca Raton, Fla., 1981.

Component Data Catalog, Intel Corp., Santa Clara, Calif., 1980.

Disk Operating System Technical Reference Manual, version 2.10, IBM Corp., Boca Raton, Fla., 1982.

N. GRAHAM, *Programming the IBM Personal Computer: BASIC,* Holt, Rinehart and Winston, New York, 1982.

Personal Computer Technical Manual, IBM Corp., Boca Raton, Fla., 1981.

The 8086 Family User's Manual, Intel Corp., Santa Clara, Calif., 1979.

PROBLEMS

SECTION 6-1

6-1. Suppose instruction CALL A1 at offset A566H in code segment 200H is executed. The code is byte E8H followed by a 16-bit displacement. Assume that the NEAR procedure A1 begins at offset 5B43H. Determine the complete code for the CALL instruction and give the logical address of each byte. Also, identify each byte pushed on the stack and its byte address relative to pointer SP.

6-2. Executed is CALL A2 at offset 1234H in code segment A00H. The first byte of the 5-byte code is 9A. The FAR procedure A2 begins at logical address 400:5000. Determine the complete code for the CALL instruction, and identify each byte pushed on the stack and its byte address relative to pointer SP.

6-3. Twenty-five binary numbers stored in the data segment of memory are to be added. The offset addresses are in consecutive locations of DS, starting at offset 100H. The word location just above the last number has the address for storage of the sum. Design a routine that uses a FAR procedure to add the 25 numbers and return the 16-bit result. Employ the stack to pass all addresses, following the method of Figs. 6-3 and 6-4.

6-4. With AX = 7CA7H, BX = B24EH, SP = 100H, SS = 800H, flag IF set, and ERR_BEEP PROC NEAR at offset 1234H, suppose the statement ORG 500H is followed by the following instruction sequence: SUB AX,BX; PUSHF; POP DX; AND DX,0AD5H; PUSH DX; and CALL ERR_BEEP. After these instructions are executed, determine AX, IP, SP, and the contents and logical addresses of the words pushed on the stack. Also, give the machine code for the instruction sequence.

SECTION 6-2

6-5. With DS, SS, CS, and SP equal to hexadecimal 0500, 0600, 0700, and A000, respectively, determine the absolute 20-bit address of the pointer to the service routine and the absolute addresses of the bytes pushed onto the stack just after execution of the instruction INT 0B1H at offset 0300H. The service routine is at 0800:01F2, and only flags DF and SF are set. Also, give the binary content of each of these locations and the new IP.

6-6. Give a sequence of instructions, ending with IRET, that gives a far jump to address DS:0. Repeat for address F00:ABC.

6-7. Design a routine that displays the message of the example program of Section 6-2 without using INT 10H. Refer to Fig. 5-11.

6-8. Instruction INT 11110111B is executed at address 01AB:2B39, with SS = 0FEB and SP = 00F4 immediately before execution. The routine is at absolute address 0E4AB. Assume that all flags are clear and that undefined flag bits are 1s. In hexadecimal, determine (a) the interrupt type, (b) each pointer byte, using 0BAC for CS, and its absolute location, and (c) the absolute locations and contents of each byte pushed on the stack.

SECTION 6-3

6-9. A peek into absolute memory locations 0024H through 0027H, using the keyboard of the IBM PC, reveals that the respective stored bytes are 87H, E9H, 00H, and F0H. Interpret the significance of these bytes. If byte 00H is poked into location 0027H, what would you expect this to do to the computer?

6-10. INT 0FBH is used in a program, with the start address of the routine at absolute address 08000H. With CS = 04BAH, determine both the absolute addresses of the locations that store the pointer to the routine and the contents of each location in hexadecimal.

6-11. Redraw the circuit of Fig. 6-13 with the following additions: AND gate enable circuits for PCK and I/O CHCK, connections to ports B and C of the PPI, and a line from RESET of the 8284 clock generator to a CLR input added to the parity mask register.

6-12. Design a service routine that implements the functions of INT 2 as described in Section 6-3. Include IRET in case no error is read from the PPI.

SECTION 6-4

6-13. For each of the eight interrupt requests feeding into the 8259A, identify the interrupt type n and give for each the absolute address of the first of the four locations that store the pointer to the routine. Use hexadecimal numbers.

6-14. Suppose the interrupt request of level 6 of the 8259 is active, with the service routine located at absolute address FEF57 in ROM. Determine the byte supplied by the interrupt controller to the data bus. Also, identify the memory locations that store the vector address and give the hexadecimal contents of each, with CS = F000. How is the routine accessed with software?

6-15. List all instructions of timer interrupt type 8. Turn on interrupts at the start of the routine, save all registers used, and establish DS at 40H.

6-16. Suppose bytes 1BH, 1BH, 09H, and F0H are written in order to respective addresses 20H, 21H, 21H, and 21H of the 8259 controller of the PC. Determine the type numbers of the interrupts of levels 0–7 and determine the type of triggering.

SECTION 6-5

6-17. Design service routines for interrupts 11H and 12H. At the start, turn on the interrupts, save DS, and move 40H to DS. Also, give the hexadecimal code.

6-18. Suppose an operator takes a 30-minute break with the PC running. Estimate the number of passes of the WAIT loop of Fig. 6-20 during this interval. Also, estimate the number of times the loop is interrupted. What would be the effect of eliminating NOP?

6-19. Design the routine for INT 1AH. At the start, turn on the interrupts, save DS, and move 40H to DS. Disable the interrupts immediately before reading the variables and enable them immediately thereafter. The routine should take no action if AH stores neither 0 nor 1.

6-20. Design a service routine for interrupt INT 1BH that is initiated by Ctrl-Break, after which register DL is decremented with each tick. Each time DL becomes 0, a letter is sent to the display, and the displayed letters should start with A and proceed in sequence to Z. Then the process should terminate. Use INT 10H for output, and include hexadecimal code.

SECTION 6-6

6-21. Suppose code for INT 20H is stored at DS:0, and CS:0 is the same absolute address as is DS:100. Give a sequence using a FAR procedure RET that can be used to terminate a program properly.

6-22. With 1 in AX, 1000H in BX, 0AH in CX, 9 in DX, and 40H in DS, suppose INT 25H is executed. What is done? Repeat with the same values for INT 26H.

6-23. With 0A00H in AX, 1000H in DX, 0040H in DS, and byte 0DH stored at 40:1000, suppose INT 21H is executed. (a) During the wait, key *x* is pressed 15 times, followed by Enter. State the observed results and list all bytes of the buffer. (b) Repeat, assuming key *x* is pressed only six times before Enter.

6-24. Using function 9 of INT 21H, give a sequence that displays a message of 50 characters followed by byte 24H. The message starts at 0:2000H, and the program's data segment address is 0300H, which must be restored after the operation. Also, give instructions that display ABC using function 2 as needed.

6-25. Suppose 2540H is in AX, 0500H is in DX, and 0A00H is in DS when INT 21H is executed. What is done?

6-26. (a) With 2B55H in AX, 5000H in BX, 07C1H in CX, and 0B17H in DX, INT 21H is executed. What date is set? (b) With 2D00H in AX, 0F3AH in CX, and 160AH in DX, what time is set?

SECTION 6-7

6-27. Sketch the display when the program of Fig. 6-29 is run, with R = 8 and C = 40.

```
100  LOCATE  R, C:   PRINT "*"
110  LOCATE  R + 1, C - 1:  PRINT  "***"
120  LOCATE  R + 2, C - 2:  PRINT  "*****"
130  LOCATE  R + 3, C - 3:  PRINT  "*******"
```

FIGURE 6-29. A short routine.

Using the statements of Fig. 6-29 as a subroutine, design a program that displays five of the patterns in a symmetrical arrangement of your choice. The patterns may overlap, if desired. First, clear the screen and turn the soft key display off.

6-28. Repeat Prob. 6-27, but use extended ASCII characters with code 219 in place of the asterisks. Then try ASCII 221 and perhaps 254. To enter the code, press the Alt key and type the three digits from the numeric pad. The character appears when Alt is released.

6-29. Repeat Prob. 6-27, but use a subroutine that generates a diamond-shaped pattern. Locate one pattern in the middle of the display, and locate the other four immediately above, below, to the left, and to the right, with the patterns meeting but not overlapping.

6-30. For a 64-KB system, design a BASIC program with a machine language subroutine in which variables A, B, C, D, and E are added, with the sum stored as variable S. Assume that each variable and the sum qualify as integers, and initialize the variables as desired. Give machine code in the data statements. Use CLEAR and put code in memory with POKE. Then repeat with data read into an array.

6-31. In the program of Fig. 6-26, give IF-THEN-ELSE statements that can replace the two WHILE-WEND loops. Also, explain what happens if the program is run with NUM initialized to a value of 1 and also to a value of 3. Are both CLS statements needed? Explain.

6-32. For x = $-4/3$, evaluate each function of Fig. 6-25 that gives a fixed value.

Program Creation, Assembly, Debugging, and Execution

We have studied assembly language and a number of assembler directives. Here we investigate *how a prepared program is edited, put on a diskette, assembled, debugged, and executed.* The utility programs of the PC are used, along with the IBM PC MACRO Assembler. Also considered is the design of (EXE) execution and (COM) command programs. The chapter concludes with a look at debugging techniques and interrupts in BASIC.

7-1 EXAMPLE: PROGRAM CREATION

Shown in Figs. 7-1 and 7-2 is a short routine that is given primarily for the purpose of illustrating procedures for processing a prepared program. It does nothing very useful, simply reading the word of MEMORY_SIZE at 0040:0013, converting the value read into ASCII decimal digits, and displaying the result in a one-line message at the bottom of the display. The binary word of MEMORY_SIZE equals the total number of kilobytes of RAM, including both on-board RAM and that connected through an I/O slot. Its value is obtained from switch settings that are read during a power-on reset. For a 256-KB memory, when the assembled program is executed, line 25 of the display reads

```
Total RAM is 256 kilobytes
```

The 26 characters of the message start in column 1.

Data and Stack Segments

Four of the five segments of the program are shown in Fig. 7-1, along with the title *RAM*. The code segment is given in Fig. 7-2. With up to 60 characters, optional TITLE *name* specifies a title to be put on the first line of each page of the listing program that is generated by the assembler. Because the fixed location of MEMORY_SIZE has to be read, the segment named DATA must include an AT statement. Also, the segment named VIDEO requires AT in order to identify the desired region of the video buffer. Being more than 64K apart, the two segments cannot be combined into one.

The video buffer of the monochrome display starts at B000:0000, and the offset of the first location of a row is $(r - 1) * 160$, with r denoting the row number from 1 to 25. Thus, the initial word of row 25, which is assigned the name CRT, has offset $24 * 160$. Each word should have attribute 7 in the high byte position and the ASCII code of the desired character in the low byte position.

In segment MSTORE, the message has three underlines for the overwrite of a number from 016 to 640, depending on the memory size. It is not possible to combine MSTORE with either DATA or VIDEO. MSTORE includes data in the form of ASCII code, and neither data nor code is allowed in a segment with an AT statement. However, there is no need to put the message in a separate segment, as is done here. It can be placed at the front end of the code segment, if desired, provided it is preceded by an unconditional jump. This position is usually preferred, since it gives a more compact program; examples are shown later. The STACK segment provides for 50 entries, which is adequate for the two pushes and for the stack requirements of terminating interrupt 20H.

```
TITLE  RAM
;
DATA  SEGMENT AT 40H
   ORG  13H
   MEMORY_SIZE DW ?
DATA  ENDS
;
VIDEO  SEGMENT  AT 0B000H
   ORG  24 * 160
   CRT  LABEL  WORD
VIDEO  ENDS
;
MSTORE  SEGMENT
   MSG  LABEL  BYTE
      DB  'Total RAM is ___ kilobytes'
   NUM  EQU $-MSG
MSTORE  ENDS
;
STACK  SEGMENT STACK
   DW  50 DUP(?)
STACK  ENDS
```

FIGURE 7-1. Four segments of the RAM.ASM program.

The Code Segment

In the code segment of Fig. 7-2, only essential ASSUME statements are shown. Although programs often include ASSUME SS:STACK, this statement is not required. The attribute *STACK* that follows STACK SEGMENT adequately identifies the segment for PUSH, POP, and other normal stack operations.

```
CODE   SEGMENT
  ASSUME   CS:CODE
;
START:  PUSH DS                          ;save DS for program terminate
        MOV AX, DATA
        MOV DS, AX
    ASSUME DS:DATA
        MOV AX, MEMORY_SIZE
        MOV BL, 100
        DIV BL
        ADD AL, 30H
        MOV BH, AL                       ;BH stores high-order ASCII digit
        MOV AL, AH
        SUB AH, AH
        MOV BL, 10
        DIV BL
        ADD AL, 30H
        MOV BL, AL                       ;BL stores next-order ASCII digit
        ADD AH, 30H
        MOV DH, AH                       ;DH stores low-order ASCII digit
;
        CLD
        MOV AX, MSTORE                   ;message-store segment
        MOV DS, AX
        MOV AX, VIDEO                    ;video buffer starts at 0B000H
        MOV ES, AX
        MOV SI, OFFSET MSG
        MOV DI, 24 * 160                 ;pointer to CRT row 25, col. 1
        MOV CX, NUM
        MOV AH, 7                        ;CRT mode 7 to AH
;
   A1: LODSB                             ;move byte of MSG to AL
          STOSW                          ;move mode and byte to CRT
          DEC CX
          JNZ A1                         ;repeat until MSG is displayed
;
    ASSUME ES: VIDEO
        MOV AL, BH
        MOV CRT + 26, AX                 ;high-order digit to display
        MOV AL, BL
        MOV CRT + 28, AX                 ;next-order digit to display
        MOV AL, DH
        MOV CRT + 30, AX                 ;low-order digit to display
;
        MOV AX, 0                        ;begin program terminate
        PUSH AX
   DUMMY  PROC  FAR
        RET                              ;FAR return to INT 20H
   DUMMY ENDP
 CODE   ENDS
 END   START
```

FIGURE 7-2. Code segment of the RAM.ASM program.

In the legend of Fig. 7-2 is the program filename RAM.ASM. The extension ASM identifies the program as one written in assembly language. When the program is assembled and linked, *the load module is given the filename RAM.EXE,* and when RAM.EXE is loaded into memory for execution, the loader assigns addresses to the segment registers. The segment addresses of both DS and ES are set to point to the 256-KB *program segment prefix (PSP)* control block, which is discussed later. At offsets 0 and 1 of this block are stored the respective bytes CDH and 20H. These bytes constitute the code for instruction INT 20H. Recall that this software interrupt is the normal way to terminate a program. *The first instruction saves the address on the stack for later use by the program terminate routine at the end of the program.*

The address assigned to CS is the one that follows the PSP. To illustrate, when the program is loaded by DEBUG, registers DS and ES store 8F7H, and register CS has 907H. Segment address 907H is greater than 8F7H by 256 byte locations, or 100H.

In order to fetch the word of MEMORY_SIZE, it is necessary to load DS with the address of DATA, or 40H. In addition, before the assembler can assemble MOV AX, MEMORY_SIZE, it must be told that DS will contain DATA; this task is done with the ASSUME directive. An alternate method is to omit the ASSUME directive and place the segment prefix DS: ahead of MEMORY_SIZE in the MOV instruction. *One or the other is required,* and both are acceptable. In either case, no code is added for the segment prefix, because DS is the default segment. However, *failure to include either the ASSUME directive or the segment prefix causes the assembler to report a serious error.*

The word of AX is a binary number from 16 through 640. It is necessary to convert the number to decimal digits in ASCII code. The high-order ASCII digit is found by dividing the word of AX by 100D and adding 30H to the integral part of the result. Then the remainder is divided by 10D, and 30H is added to the integral part of the result to give the next-order digit. The addition of 30H to the new remainder gives the low-order digit (see Prob. 7-2). These operations are implemented by the 12 instructions that follow the fetch of MEMORY_SIZE, with the digits put in registers BH, BL, and DH. We now examine the divide instructions.

Divide Instructions. The instruction DIV *source* treats both the source divisor and the dividend as unsigned numbers, with the source being that of a register or memory. Any of the register and memory addressing modes can be used. If the source is a byte, it is divided into the 16-bit dividend of AX. The integral byte quotient is placed in AL, and the remainder goes to AH.

For a word divisor, the 32-bit dividend is that of DX and AX, with the high half in DX. *Register DX is often treated as an extension of AX when 32-bit words are encountered.* The integral quotient is returned to AX, and the remainder goes to DX.

For division of signed integers, IDIV is available. It is similar to DIV, *except that the numbers are regarded as signed.* For both DIV and IDIV, the quotient is truncated toward 0, and the six status flags are undefined following the operation. If the quotient exceeds the capacity of its destination register, the quotient and the remainder are undefined, and a type 0 interrupt (divide-by-0) is automatically generated. The algorithms that implement DIV and IDIV are similar to those described in Section 2-3.

Two instructions sometimes used in conjunction with the divide instructions are *convert-byte-to-word CBW* and *convert-word-to-doubleword CWD*. Each is a one-byte instruction with no operand. CBW extends the sign of the byte in AL throughout register AH. Thus it can convert a byte dividend into a word prior to performing byte division. For word division, CWD is useful for converting a word dividend into a doubleword. It does this by extending the sign of the word in AX throughout register DX. Neither instruction affects any flags.

Multiply Instructions. Because the multiply instructions are somewhat similar to those of division, it is appropriate to examine them here. The two instructions are MUL *source* and IMUL *source*, which refer to unsigned and signed multiplication, respectively. The source multiplier may be either a byte or a word of a register or a memory location. If it is a byte, the multiplicand is the byte of AL, and the 16-bit product is stored in AX. Both flags CF and OF are set if and only if AH contains significant bits of the product. For signed multiplication, this operation implies that CF and OF are clear if the byte of AH is a sign extension of AL.

When the source is a word, the multiplicand is that of AX, and the 32-bit product is stored in DX and AX, with the high half in DX. For unsigned multiplication, flags CF and OF are set if and only if the word of DX is not 0, and for signed multiplication, they are set if the word of DX is other than the sign extension of AX. Thus, a program can examine either of these flags to determine if the upper half of the product has significant bits, and this is true for both byte and word multiplication. Flags AF, PF, SF, and ZF are undefined after a multiply operation. The algorithms that implement MUL and IMUL are similar to those described in Section 2-3.

The Display

The next group of instructions send the message to the display, using memory-mapped I/O. Beginning with CLD are nine instructions that initialize the registers for the transfer. Then the message is moved to the video buffer by means of the four instructions of the loop at label A1, but only underlines are sent to locations that are to store the digits. However, these underlines are almost immediately replaced.

The six-instruction sequence following ASSUME ES:VIDEO transfers the digits. The ASSUME directive notifies the assembler that segment register ES will store the address of VIDEO, or B000H. When MOV CRT + 26, AX is encountered, the assembler recognizes that CRT is a location of the VIDEO segment, and the ASSUME directive has notified the assembler that the address of VIDEO is in ES. Because the default segment of the destination of MOV is DS, *the assembler automatically generates code 26H for the program segment prefix ES,* placing it ahead of the code for the MOV instruction. The same operation applies to the other two instructions containing CRT. Without the ASSUME directive, the assembler reports three serious errors, one for each instruction with CRT.

As before, there is an alternative to ASSUME. With the directive omitted and with the segment override prefix ES: included in each of the three instructions containing CRT, the program assembles without error. The prefix should immediately precede CRT, and the code 26H is generated as before. If one desires, both the ASSUME directive and the segment override prefixes can be used. In case of a conflict, the segment override prefix dominates.

It is important that CRT be specified as a word label, not a byte label. The instruction MOV CRT + 28, AX will generate an assembler error if CRT is defined as a byte label; the error message reports that a word cannot be moved to a byte location. Of course, the directive WORD PTR can be used to override a byte label whenever necessary.

Program Terminate

Upon entry to the DUMMY procedure, the two words that have been pushed onto the stack are the initial value of DS and 0000H. *The only purpose of the FAR procedure is to inform the assembler to generate code for a FAR return.* When RET is executed, 0000H is popped into IP and the initial value of DS is popped into CS. Thus, CS:IP points to the first location of the PSP, and the code stored here is CD20H for INT 20H. *This is the normal way to terminate an EXE program* produced by the assembler and linker programs. Usually the procedure name is selected as the program name and placed at the start of the program. However, *the only requirement is that it precede RET.* Without the procedure definition, the code for RET would be that of a NEAR return.

It is not permissible to replace RET with INT 20H. This instruction works only if the CS register contains the segment address of its PSP when INT 20H is executed. The method of the routine will be used for terminating EXE programs until the replacement for INT 20H is described in Chapter 11.

Although program storage and execution are discussed later, brief comments on execution are appropriate here. When the assembled and linked program, which has the filename RAM.EXE is stored on the diskette of drive A, *execution is accomplished by typing "RAM",* followed by the Enter key. The diskette spins, the internal processor program COMMAND loads RAM.EXE, and RAM is immediately executed. The appropriate message appears on the bottom line of the display. RAM can be put on the DOS diskette and treated as another utility program, such as the program FORMAT, which formats a diskette. Then whenever RAM is entered, the operator receives a report on the amount of memory present. This is not a very useful utility, but it illustrates the way utility programs are developed.

Program Creation

Let us suppose that the program of Figs. 7-1 and 7-2 has been developed on paper. The problem now is to put it in memory and then to store it on a diskette for eventual assembly, linking, debugging, and execution. Here we shall examine the steps required for entering the assembly language program into memory, followed by transfer to a diskette. Assembly is described in the next section.

EDLIN. The DOS diskette includes a 4608-byte program with the filename EDLIN.COM. It is an *editor* that is designed to create, alter, and display ASCII source files, either in assembly language or in a high-level language. Because cursor movement is restricted to a single line, editing is done line by line. For this reason, EDLIN is called a *line editor*. It can also be used for editing any text file in ASCII. Examples other than programs are letters, memos, and data files designed for serial communications.

With the DOS diskette in default drive A and a formatted work diskette in drive B, in order to create RAM.ASM we enter EDLIN B:RAM.ASM. The filespec of the new file to be created and stored on the diskette of drive B is B:RAM.ASM. The actual name, including the extension, is of no special significance to EDLIN. After the entry, the diskette of drive A spins, and EDLIN is loaded into memory. Then the work diskette of drive B spins, and *its directory is loaded into memory and searched for the filename RAM.ASM*. Not finding a file with this name, the editor prints on the display "New file", followed by the asterisk prompt (*). EDLIN is now ready to receive the characters of the file to be created.

Pressing key I followed by ENTER gives the *insert mode,* and line 1 appears on the display. The first line of the assembly language program is typed and entered. Line 2 is automatically displayed, and the second program line is typed. This procedure continues until completion. Ctrl-Break is activated to exit the insert mode. Up to 253 characters can be placed on a line, and lines are numbered consecutively, with the numbers used only for reference while in EDLIN. They are not part of the program being created. The complete RAM.ASM file has 1748 bytes.

If an error occurs while typing a line in the insert mode, it can be corrected with the Backspace, DEL, and INS keys. However, if the error is not noticed until after ENTER is pressed, it is usually best to ignore it until all of the lines are entered. Then the error is easily corrected.

After entry of all lines, *followed by Ctrl-Break to exit the insert mode,* the next step is to edit the program. The first 23 lines can be displayed and examined by entering the list command L, preceded by 1. Thus, we enter 1L, with an optional space allowed between 1 and L. Suppose an error is observed in line 5 of the listing. To edit the line, enter 5; the line is displayed with a blank line under it, also numbered 5 and containing the cursor at the left side. A new line can be typed on the blank line.

Usually it is not necessary to type an entire new line. If the cursor is moved to the right with the cursor movement key, the characters of the displayed line 5 at the top are automatically entered as the cursor passes corresponding locations of the bottom line. Moving the cursor to the left deletes characters. At any point the cursor can be stopped, and characters can be inserted by first pressing the INS key; alternatively, characters can be deleted with the DEL key.

Perhaps an entire line was carelessly omitted. If the line is to be inserted just ahead of line 9, we enter 9I, which puts us in the insert mode. As many new lines as desired can be inserted ahead of the previous line 9, with exit from insert by means of Ctrl-Break. Lines are also easily deleted. To delete line 8, we enter 8D. After an insertion or deletion, all line numbers are automatically updated. At any time, a list of 23 lines can be obtained by entering a start line number followed by L. Thus, 36L gives a listing of lines 36 through 58. The prompt (*) always appears at the beginning of the current line. This is the line to which the location counter points, and it is the default line for an insert, delete, list, or other command when a line number is not specified.

There are numerous variations of the commands mentioned that add to the power of the line editor. Also, there are various other commands. For example, one command replaces all occurrences of a character string with another string, and another searches for a specified string. The user should refer to the EDLIN section of the IBM *DOS Manual* for details. With a little practice, creation and editing of text files become easy.

```
10  CLS: KEY OFF
20  DEF SEG = 0
30  A = 256 * PEEK(&H414) + PEEK(&H413)
40  LOCATE 25, 1
50  PRINT  "Total RAM is";  A;  "kilobytes"
```

FIGURE 7-3. The example program in BASIC.

Program Storage. After the complete program has been entered and edited in EDLIN, the *end-edit* command *E* is entered. The diskette of drive B spins while RAM.ASM is automatically written to it. Now suppose the program is assembled and one or more errors are found. It is necessary to invoke EDLIN again. We type EDLIN B:RAM.ASM, and as before, EDLIN is loaded from the default drive A into memory, followed by the directory of drive B. This directory is examined, RAM.ASM is found, and *then it is loaded into memory*. The display reads "End of input file", indicating that the file can now be edited.

Upon completion of editing, the command E is entered, and the edited file is stored on the diskette of drive B. The old file is renamed RAM.BAK, with extension *BAK* denoting a *backup file*. If it is not needed, it can be erased with the DOS command ERASE B:RAM.BAK. We will not erase it.

Now suppose an attempt to assemble, link, and execute the program indicates that more revision is necessary. We enter EDLIN B:RAM.ASM, and both EDLIN and RAM.ASM are moved to memory. Also, the backup file RAM.BAK on the diskette of drive B is automatically erased. However, when the edited file is returned to the diskette with command E, the previous file RAM.ASM is renamed RAM.BAK. *Thus, the backup file is always the one that existed just prior to the last update*. Occasionally, it may be desirable to terminate an edit without saving the file. The quit command Q performs this operation.

The line editor is not a *word processor*. It processes lines, one at a time. Although it can be used to edit typed text material, such as letters and papers, it is not nearly as convenient and powerful as a word processor program. It is designed mainly for creating and editing source programs consisting of lines of instructions. For this purpose, it is excellent.

The Program in BASIC

The program RAM is easily implemented in BASIC, as shown in Fig. 7-3 (see Prob. 7-1). Note the simplicity of the high-level language in comparison with assembly language.

7-2 ASSEMBLY

We are now ready to examine the assembly of the diskette file with filespec B:RAM.ASM. The IBM PC MACRO Assembler will be used. It consists of 14 programs on a single diskette, accompanied by a manual with operating instructions. Initially, the work diskette is assumed to be in drive B, and the DOS diskette is in default drive A.

Included in the programs of the assembler are three of special importance to us here. One is the MACRO Assembler MASM.EXE with 67,584 bytes, requiring a

```
IBM Personal Computer Assembler
Version 1.00 (C)Copyright IBM Corp 1981
Source filename [.ASM]:
```

FIGURE 7-4. Initial assembler message.

memory of at least 96 KB for its use. Another is the 52,736-byte program ASM.EXE, referred to as the *Small Assembler,* which can be used in a machine with only 64 KB of memory. The third program is CREF.EXE, with 13,824 bytes. We will use it later to generate the file RAM.REF, which is useful but not essential.

Assembly Procedure

We replace the DOS diskette of drive A with the assembler diskette and enter MASM. The assembler is loaded into memory. If the machine has insufficient memory, one can enter ASM in place of MASM. The MASM program is preferable, because it allows the use of *macros,* (user-defined instruction sequences) and provides error messages rather than error codes that have to be looked up in the assembler manual. A few seconds after entry of MASM, the diskette stops spinning, and the message of Fig. 7-4 appears. At this point, the assembler program is stored in memory. Because the assembler diskette is no longer needed, it can be replaced with the DOS diskette. If this operation is not done now, a message after assembly instructs the operator to do so.

In the third line of the message of Fig. 7-4, the prompt in brackets indicates that the file to be assembled must have extension ASM, which is used only for assembly language programs. The operator now enters the leading part of the filespec, or B:RAM in this example, and the extension is added by the assembler. After entry of the source filename, a message appears requesting the ID of the object file. Its entry is followed by a request for the ID of the source listing, and after this one appears asking for the name of the cross reference file. Beginning with the source filename of Fig. 7-4, the display is that of Fig. 7-5, with all entries shown at the right side in lowercase letters.

Let us consider the responses shown in Fig. 7-5. That of the first line specifies the filespec of the source program to be B:RAM.ASM, and the response on the second line gives B:RAM.OBJ as the filespec of the *object file* to be generated. The object file has machine language code, which is the output of the assembly operation. In the last two lines of Fig. 7-5 are prompts with NUL. NUL indicates that there will be no generation of listing or cross-reference files if the operator responds only with Enter, without typing the filespecs shown. Although these two *optional* files are not used in generating the executable file, they are useful, and the illustrated responses cause the files to be generated.

In a one-drive system, B: is omitted in the responses, and the work diskette is in drive A. A two-drive system is assumed here. Following the entry to the cross

```
Source filename [.ASM]:      b:ram
Object filename [RAM.OBJ]:   b:
Source listing  [NUL.LST]:   b:ram
Cross reference [NUL.CRF]:   b:ram
```

FIGURE 7-5. Assembler requests and responses.

```
Warning   Severe
Errors    Errors
0         0
```

FIGURE 7-6. Display after assembly.

reference request, *the source file is loaded from drive B, the program is assembled, and the object, listing, and cross reference files are moved to the work diskette.* For our short program, the process takes only a few seconds. The display now gives the message of Fig. 7-6, which is followed by the DOS prompt.

The number of both small errors and severe errors is reported, as indicated in Fig. 7-6. A report of no errors is encouraging but does not, of course, indicate that the program will accomplish its intended task. If errors are detected, the lines involved are displayed, along with error messages.

The work diskette now has five files related to the RAM program. These files are listed in Fig. 7-7, along with comments and byte size. *Only the text files (ASCII) are printable,* these being the assembly language and listing files.

```
RAM.ASM   ;current assembly language file (1748 bytes)
RAM.BAK   ;assembly language backup file
RAM.OBJ   ;object file in machine language (293 bytes)
RAM.LST   ;listing file in ASCII (4369 bytes)
RAM.CRF   ;cross reference file (267 bytes)
```

FIGURE 7-7. Diskette files after assembly.

Suppose one or more errors are reported. All errors should be corrected in the source file using EDLIN, and the program should be reassembled. *Generated new files replace the old ones automatically.* With no errors encountered in assembly, we are ready to proceed to the linker that generates the load module. This operation is described in the next section. We now examine the object file.

The Object File

The 293-byte *object file* includes 27 bytes for the ASCII code of the title and the segment names, 26 bytes of ASCII for the message, and 84 bytes of machine language code. Most of the remaining 156 bytes *provide information that enables the linker to perform its functions.* Neither the object file nor its 84 bytes of machine language code are in a form suitable for execution. If the file is printed, the binary bytes are interpreted as ASCII code, and the result is meaningless. Many of the bytes of the machine language code correspond to graphics and nonprintable control characters. *Both assemblers and compilers convert source programs into object files.* The machine code of the object file RAM.OBJ is given in the listing program that follows.

The Listing

The complete *listing file RAM.LST* is shown in Figs. 7-8, 7-9, and 7-10. It can be displayed on the screen by entering TYPE B:RAM.LST and *using key combination*

```
The IBM Personal Computer Assembler 09-27-84 PAGE 1-1 RAM
1                                     TITLE  RAM
2                                     ;
3  0000                               DATA SEGMENT AT 40H
4  0013                                 ORG 13H
5  0013 ????                            MEMORY_SIZE  DW  ?
6  0015                               DATA ENDS
7                                     ;
8  0000                               VIDEO SEGMENT AT 0B000H
9  0F00                                 ORG 24*160
10 0F00                                 CRT LABEL WORD
11 0F00                               VIDEO ENDS
12                                    ;
13 0000                               MSTORE SEGMENT
14 0000                                 MSG LABEL BYTE
15 0000 54 6F 74 61 6C 20                DB 'Total RAM is ___ kilobytes'
16      52 41 4D 20 69 73
17      20 5F 5F 5F 20 6B
18      69 6C 6F 62 79 74
19      65 73
20 = 001A                               NUM EQU $-MSG
21 001A                               MSTORE ENDS
22                                    ;
23 0000                               STACK SEGMENT STACK
24 0000   32 [                            DW   50 DUP(?)
25               ????
26                                 ]
27
28 0064                               STACK ENDS
29                                    ;
30 0000                               CODE SEGMENT
31                                      ASSUME CS:CODE
32                                    ;
33 0000 1E                            START: PUSH DS                 ;Save DS for program terminate
34 0001 B8 ---- R                            MOV AX, DATA
35 0004 8E D8                                MOV DS, AX
36                                      ASSUME DS:DATA
37 0006 A1 0013 R                            MOV AX, MEMORY_SIZE
38 0009 B3 64                                MOV BL, 100
39 000B F6 F3                                DIV BL
40 000D 04 30                                ADD AL, 30H
41 000F 8A F8                                MOV BH, AL              ;BH stores high-order ASCII digit
42 0011 8A C4                                MOV AL, AH
43 0013 2A E4                                SUB AH, AH
44 0015 B3 0A                                MOV BL, 10
45 0017 F6 F3                                DIV BL
46 0019 04 30                                ADD AL, 30H
47 001B 8A D8                                MOV BL, AL              ;BL stores next-order ASCII digit
48 001D 80 C4 30                             ADD AH, 30H
49 0020 8A F4                                MOV DH, AH              ;DH stores low-order ASCII digit
```

FIGURE 7-8. Page 1 of the listing program.

Ctrl-NumLock to stop scrolling when the screen becomes full. Scrolling is resumed by pressing the space bar or any other key. If a hard copy is desired, *prior to entry of the TYPE command, press and release key combination Ctrl-PrtSc* with the printer turned on. After this is done, any text sent to the display is automatically sent also to the printer, at least until *toggle switch Ctrl-PrtSc* is pressed a second time.

At the beginning of the listing is the first page number 1-1, followed by the title RAM, which also appears below the page number of each of the other two pages. Then there are 80 numbered lines having offset addresses that start at 0 in each segment. *Line numbers are generated only if the cross reference file was requested* in the last response shown in Fig. 7-5. The lines contain both object code and assembly language statements, including the comments. The last part of the program consists of two tables.

Object Code of the Listing. Although most of the listing is self-explanatory, a few comments may clarify some of the items. All offsets and code at the left side are given in hexadecimal. The term = 001A in line 20 is the equate for NUM.

```
The IBM Personal Computer Assembler 09-27-84 PAGE 1-2 RAM
50                         ;
51 0022 FC                        CLD
52 0023 B8 ---- R                 MOV AX, MSTORE       ;message-store segment
53 0026 8E D8                     MOV DS, AX
54 0028 B8 ---- R                 MOV AX, VIDEO        ;video buffer starts at 0B000H
55 002B 8E C0                     MOV ES, AX
56 002D BE 0000 R                 MOV SI,OFFSET MSG
57 0030 BF 0F00                   MOV DI, 24*160       ;pointer to CRT row 25, col. 1
58 0033 B9 001A                   MOV CX, NUM
59 0036 B4 07                     MOV AH, 7            ;CRT mode 7 to AH
60                         ;
61 0038 AC                    A1: LODSB                ;move byte of MSG to AL
62 0039 AB                         STOSW               ;move mode and byte to CRT
63 003A 49                         DEC CX
64 003B 75 FB                      JNZ A1              ;repeat until MSG is displayed
65                         ;
66                            ASSUME ES:VIDEO
67 003D 8A C7                      MOV AL, BH
68 003F 26: A3 0F1A R              MOV CRT+26, AX      ;high-order digit to display
69 0043 8A C3                      MOV AL, BL
70 0045 26: A3 0F1C R              MOV CRT+28, AX      ;next-order digit to display
71 0049 8A C6                      MOV AL, DH
72 004B 26: A3 0F1E R              MOV CRT+30, AX      ;low-order digit to display
73                         ;
74 004F B8 0000                    MOV AX,0            ;begin program-terminate
75 0052 50                         PUSH AX
76 0053                       DUMMY PROC FAR
77 0053 CB                         RET                 ;FAR return to INT 20H
78 0054                       DUMMY ENDP
79 0054                     CODE ENDS
80                          END  START
```

FIGURE 7-9. Page 2 of the listing program.

In lines 5 and 25, ???? is used to indicate an uninitialized word location, and the 32 ahead of the bracket in line 24 denotes the hexadecimal number of such locations. Code for MOV AX,DATA is given in line 34 as B8 ---- R. The four hyphens replace the word represented by DATA, with the value 0400H specified by AT to be filled in when the linker creates the executable module. Not all segment addresses are known at the time of assembly. An example is the segment address of MSTORE in line 52, which is not assigned until the executable program is placed in position by the loader of the COMMAND processor program in memory.

In general, an R after a word indicates a value that may be modified by the linker or loader; *all segment addresses and offsets are so marked.* The assembler makes two passes of the program. During the *first pass,* the offset for each line of the program is determined, and a symbol table is formed. The table associates an offset with each label and name identified in the program. If an undefined item is found, an assumption is made so that the number of bytes generated by instructions related to this item can be determined. No object code is produced.

Labels are identified by segment, offset, and type NEAR or FAR. *One followed by a colon is always type NEAR,* and thus can be used as an operand for a jump or call only from within the segment. Variables are also identified by segment, offset, and type, with the type defining the number of bytes reserved for the variable.

During the *second pass,* the listing, object, and cross reference files are generated, using the relative offsets and the values contained in the symbol table formed in pass 1. The program has only one *relocatable address* that cannot be determined until it is actually loaded into memory. This is the segment address of MSTORE, which is assigned by the loader of the COMMAND program that resides in memory.

```
The IBM Personal Computer Assembler 09-27-84 PAGE Symbols-1 RAM

Segments and groups:
        Name                Size      align     combine     class
CODE . . . . . . . . . . . . . .  0054      PARA      NONE
DATA . . . . . . . . . . . . . .  0015      AT        0040
MSTORE . . . . . . . . . . . . .  001A      PARA      NONE
STACK  . . . . . . . . . . . . .  0064      PARA      STACK
VIDEO  . . . . . . . . . . . . .  0F00      AT        B000

Symbols:
        Name                Type      Value     Attr
A1 . . . . . . . . . . . . . . .  L NEAR    0038      CODE
CRT  . . . . . . . . . . . . . .  L WORD    0F00      VIDEO
DUMMY  . . . . . . . . . . . . .  F PROC    0053      CODE        Length =0001
MEMORY_SIZE  . . . . . . . . . .  L WORD    0013      DATA
MSG  . . . . . . . . . . . . . .  L BYTE    0000      MSTORE
NUM  . . . . . . . . . . . . . .  Number    001A
START  . . . . . . . . . . . . .  L NEAR    0000      CODE

Warning Severe
Errors Errors
0       0
```

FIGURE 7-10. Symbol tables of the listing program.

Note the segment override prefixes of lines 68, 70, and 72. *These codes were generated by the assembler as a consequence of the preceding ASSUME directive.*

Each item of code is shown as either a byte or a word. Actually, *the 2 bytes of each word are reversed in order when the object code is converted into the machine code of the run program.* For example, the code of line 68 is given as 26: A3 0F1A, but the corresponding code of the load module is 26 A3 1A 0F, with 26 at the lowest address and 0F at the highest. This arrangement places the low-order byte of a word at the lower address of the 2-byte location.

Listing Tables. The two tables of the listing are shown in Fig. 7-10. The first one gives each segment name, the segment size in bytes from the initial location through the location of the last identified byte, and the *align, combine,* and *class* entries. For segments CODE, MSTORE, and STACK, no align-type entries, which specify the start of a segment, were given in the assembly language program. As indicated in the table, the default is PARA, for paragraph. A *paragraph* is defined as *a 16-byte region of the address space with a starting address divisible by 16.* PARA specifies that the segment should begin on a paragraph boundary. This means that the absolute address will have its least significant hexadecimal digit equal to 0. For DATA and VIDEO, the align entry was specified as AT.

The combine-type entries of the table are self-explanatory, with STACK having previously been described. There are no class entries, which are described in the next section. The information in both this table and the one following it is passed on to the linker.

The second table lists all labels, names, and variables of the program. The L in the "Type" column refers to labels and variable names. Both A1 and START are shown to be NEAR labels, whereas DUMMY is identified as the name of a FAR procedure. CRT and MEMORY_SIZE are words, MSG is a byte location, and NUM is a number. Entries in the "Value" and "Attr" columns should be apparent. Note that the DUMMY procedure is reported to have only 1 byte.

```
Symbol Cross Reference    (# is definition)   Cref-1
A1 ........................... 61#   64
CODE ......................... 30#   31    79
CRT .......................... 10#   68    70    72
DATA .........................  3#    6    34    36
DUMMY ........................ 76#   78
MEMORY_SIZE ..................  5#   37
MSG .......................... 14#   20    56
MSTORE ....................... 13#   21    52
NUM .......................... 20#   58
STACK ........................ 23#   28
START ........................ 33#   80
VIDEO ........................  8#   11    54    66
```

FIGURE 7-11. The cross reference file RAM.REF.

Although the optional listing is not essential, it contains information useful to the programmer, and it is an aid in debugging. In most assemblies, a listing program is requested.

The Cross Reference File

The optional RAM.CRF file of the diskette contains a compact representation of the variable references and definitions of the program. However, it is not in printable form. To obtain a useful cross reference file it is necessary to call again on the assembler diskette. We replace the DOS diskette of drive A with the assembler and enter CREF. A 13,824 byte program is loaded from the diskette into memory, and the following message with the appropriate response at the right is displayed:

```
Cref filename [.CRF]: B:RAM
```

As soon as the message appears, either before or after the response is entered, the assembler diskette can be replaced with the DOS diskette. Entry of the response causes a new file named RAM.REF to be created from RAM.CRF and placed automatically on the diskette of drive B. This is the printable *cross reference* file RAM.REF, which has 717 bytes.

To display the file, we enter TYPE B:RAM.REF, and the file appears on the screen. Key combination Ctrl-PrtSc can be activated, as previously explained, to send the file to the printer if a hard copy is desired. A copy of RAM.REF is shown in Fig. 7-11. The file lists all named labels, variables, and constants, and gives the line numbers of the listing program that contain these names. For the short program treated here, it is of little value, but in long ones it can be helpful in debugging. The line number with # denotes the line in which the quantity is defined. We are now ready to proceed to the linker.

7-3 THE LINKER

Assembly has given us the object file RAM.OBJ, but it is not ready for execution. In order to obtain a *run file,* we enter LINK with the DOS diskette in drive A. A 41,856-byte linker program named LINK.EXE is loaded into memory, and a message appears on the display, along with a request for a response. When the

```
A> link

IBM Personal Computer Linker
Version 1.10 (C)Copyright IBM Corp 1982

Object Modules [.OBJ]:  b:ram
Run File  [A:RAM.EXE]:  b:
List File   [NUL.MAP]:  b:ram
Libraries      [.LIB]:
```

FIGURE 7-12. Message, requests, and responses of the linker.

entry is made, another response is requested, and then a third and a fourth. After the last response is entered, the linking process is implemented, and a run file and a listing file are created. These files are automatically loaded onto the diskette of drive B.

Link, Responses, and Files

The command LINK and the message that follows, along with the four requests and the responses, are shown in Fig. 7-12. The operator's entries are given in lowercase letters, and each is followed by Enter, which is not indicated on the figure. The last of the four responses is simply Enter.

The first response shown in Fig. 7-12 identifies the object module as that of drive B, which has filename RAM.OBJ. As indicated by the term in brackets, the extension is assumed to be OBJ if none is specified. The second response tells the linker that the run file RAM.EXE that is to be created should be transferred to the diskette of drive B. This command overrides the indicated drive A in brackets. Next is the response requesting that the optional file RAM.MAP be created and sent to drive B, with NUL in brackets signifying that there is no MAP file created if the only response is Enter. Because the program being processed has no libraries, the final response is Enter, and the linker then takes about 2 seconds to create and load on the diskette the two new files RAM.EXE and RAM.MAP.

Including the assembly language backup file, there are now eight RAM files on the diskette of drive B. These files are listed in Fig. 7-13, with the optional ones on the left.

Linker Functions

RAM was designed as a single assembly language program, and assembly generated only one object module. However, longer programs are often designed as a

Optional Files	Essential Files
RAM.BAK (1748) RAM.LST (4369) RAM.CRF (267) RAM.REF (717) RAM.MAP (256)	RAM.ASM (1748) RAM.OBJ (293) RAM.EXE (640)

FIGURE 7-13. RAM files of the work diskette, with the byte size of each in parentheses.

number of modules, perhaps involving different programmers. Procedures are used extensively. After the various assembly language routines are assembled, the separately produced object modules must be linked into a single executable load module, or run file. External cross references from one file to another must be resolved. The *linker* accomplishes these tasks. For example, suppose the main program has a CALL to a procedure that is assembled separately. Clearly, the assembler cannot determine the transfer address. This is the type of address resolution handled by the linker.

Source programs written in high-level languages are compiled into object modules. These source programs often utilize numerous routines found in the system library of the high-level language. The *library routines* consist of compiled procedures and functions stored on a diskette in object code; they become part of the executable program after the linkage. For example, a program library might contain a procedure for handling floating point or decimal arithmetic. The linker is designed so that it can search library files and link both the object modules and the specified library routines into a single loadable and executable EXE file.

In the response to the request "Object Modules [.OBJ]:" in Fig. 7-12, only one filename was entered. For programs consisting of more than one object module, each file specification must be separated from the next by either a plus (+) sign or a blank space. If the extension is omitted from any filename, LINK assumes OBJ. The object files do not have to be on the same diskette. If LINK cannot find a filename on the diskette of the specified drive, a message instructs the operator to change the diskette and hit Enter. For a filespec with no specified drive, the default drive is assumed. A linker session can be terminated prior to its normal end with Ctrl-Break.

The second request of the linker, "Run File [A:RAM.EXE]:", provides a default filespec that can be changed if desired. However, the extension EXE is assigned even if another is specified. *All run files receive this extension.* The response to the third request of Fig. 7-12 provides a list file called the *linker map,* and the response to the last request is either Enter or a list of specified library files separated by plus signs or spaces.

There are various options that can be typed at the end of each response in Fig. 7-12. Each included option is preceded by a slash (/). Normally, programs are loaded at the lowest vacant area of memory, but one option allows the operator to specify placement at the high end. Another provides a pause for a change of diskette before the run file is loaded. A third can be used to specify a stack size that differs from that specified by the assembler or compiler. There are also others, all of which are described in detail in the chapter on LINK in the DOS manual, along with additional features of LINK not examined here.

Unless at least one object module contains a SEGMENT statement with the combine-type STACK, the linker returns the message "Warning: No Stack statement" and one error is flagged. However, the program is properly linked. As we shall see in Section 7-5, assembly language programs that are to be converted into command-type COM programs must *not* have a segment with the STACK statement.

The Linker Map

Figure 7-14 shows the optional *linker map* for the RAM program. It consists of 256 bytes of ASCII code, more than half of which is made up of spaces, CRs, and

```
  Start      Stop       Length    Name       Class
  00000H     00053H     0054H     CODE
  00060H     00079H     001AH     MSTORE
  00080H     000E3H     0064H     STACK
 Program entry point at 0000:0000
```

FIGURE 7-14. The linker map RAM.MAP.

LFs. The 20-bit addresses in the "Start" and "Stop" columns are relative to the actual initial location of the load module. This initial location is found by subtracting the entry shown at the end of the map listing from address CS:IP, assigned by the loader of the COMMAND processor when the run file is placed in memory. The locations and lengths of the various segments of a program are useful when examining a program that is loaded in memory along with the DEBUG program.

The "Class" column of the linker map is blank. In SEGMENT statements of the assembly program, a *class* entry can be used for the purpose of grouping segments at linker time. The entry is a selected name enclosed in single quotation marks, placed at the end of the SEGMENT statement. The linker examines class names and creates the executable program so that all segments with the same class name will be loaded into memory contiguously. No class names were specified in the RAM program.

The Run Program

The RAM.EXE program is run simply by entering the filespec, and it can be run as often as desired. With RAM.EXE on the diskette of drive B, entry of B:RAM causes the program to be loaded into memory and immediately executed. The memory-size message appears on the display within 2 seconds after pressing Enter. Note that *the filename extension can be omitted* if desired.

Although there are 640 bytes of code, the actual sequence of machine instructions consists of only 84 bytes and the message contains only 26 bytes. When DOS is booted, the COMMAND.COM program is transferred to memory; it includes a routine called a *loader* that supervises the loading of other programs. Within the 530 remaining bytes of RAM.EXE is information needed by the loader for locating the segments in memory and for calculating any relocatable addresses.

The RAM.EXE program has a *header* of 512 bytes, followed by a block of 128 bytes containing the segments CODE and MSTORE. Although the diskette file has two sectors with 1024 bytes, only the 640 bytes of the header and segments are significant. However, all 1024 bytes are read from the diskette when the program is activated. The header provides the information needed by the loader. The segment CODE, with 84 bytes, will be loaded at the module's entry address of relative 0, extending through offset 83; MSTORE comes next, with its 26 bytes of ASCII code extending from offset 96 through 121. Recall that segments start on paragraph boundaries. The 12 bytes between CODE and MSTORE, from offset 84 through 95, store null codes (00H). Then there are six null codes, from offset 122 through 127, that complete the last paragraph of MSTORE. This arrangement accounts for all 128 bytes.

Header Word	Significance
5A4D	Code indicating that file is linked EXE type
007A	7AH (122D) bytes in code and data block
0002	Two blocks (512 bytes each) in EXE disk file
0001	One relocation item in the header
0020	Number of paragraphs in header (512 bytes)
0000	No paragraphs required above the load module
FFFF	Load the program as low as possible in RAM
0008	Relative stack segment address is 0008H
0064	Value for the stack pointer SP
DCD8	A checksum; recalculated after loading
0000	Value for IP
0000	Relative segment address for CS
001C	First relocation item of header at offset 1CH
0000	Header refers to the resident portion of EXE
0024	Offset of load module word needing relocation

FIGURE 7-15. RAM.EXE header words and comments.

In the 512-byte header of RAM.EXE are only 15 significant words, which are shown in hexadecimal form in Fig. 7-15. All additional words are 0000H. *The minimum size of a header is 512 bytes,* and programs with many relocatable addresses may have even longer ones.

The header words in Fig. 7-15 are stored with the most significant byte at the higher byte location. The second word specifies 122 bytes in the code and data block. These are the 84 bytes of code, the 26 bytes of the message, and the 12 null codes between them. The relative stack segment address of 0008H indicates that the stack starts at offset 80H above the beginning of the load module. This start address is also given in the linker map of Fig. 7-14. The checksum word DCD8H is the negative of the sum of all words of the EXE program. After the program is loaded, a new checksum is calculated by the loader and compared with the original value. A difference indicates a loading error, in which case the program is not executed.

For RAM.EXE the loader has to assign only one relocatable address, which is the segment address of MSTORE. It must be supplied to complete the code for MOV AX,MSTORE. As shown on line 52 of Fig. 7-9, the segment address word is located at offset 0024H, and this is the value of the last header word of Fig. 7-15. In the 122-byte code block at offset 24H is word 0006H, indicating that *the loader must add relative segment address 0006H to the start segment address of the load module to obtain code for MSTORE in the instruction.* The relative segment address of CS is given in Fig. 7-15 as 0000H. Thus, the segment address of MSTORE is that of CS plus 6H. The start of the message is 60H bytes above the entry point of the program, which agrees with the offset of 96 previously stated and with the MSTORE start address of the linker map.

In longer programs with more relocatable addresses, the last word of the header of Fig. 7-15 is followed by additional pointers to such addresses. In each case, *the offset relative to the start address of the load module is found from the code of the blocks following the header blocks, and this information is used by the loader to calculate actual addresses to be inserted into the code.* These addresses can change whenever the program is loaded, depending on other programs in memory. For example, when RAM.EXE is loaded above a previously loaded DEBUG program, the address of MSTORE differs from its value without DEBUG.

DEBUGGING

Because the RAM program has been debugged, it can be assembled, linked, and run with no problems. However, it is normally not good practice to run a new program just after it has been linked. There are likely to be *bugs* that cause trouble, and if a defective program is run, there is a good chance that it will disable interrupts and never stop. In this event, the only way to regain control is to switch power off and on, losing all programs stored in RAM.

On the DOS diskette is a 11,904-byte program named DEBUG.COM that is designed to aid in debugging. Let us assume that RAM.EXE has just been processed by the linker and may quite likely have errors. We wish to debug it prior to execution. With the DOS diskette in drive A and the work diskette in drive B, we enter DEBUG B:RAM.EXE. First, the diskette of drive A spins while DEBUG is loaded into the lowest available memory space, just after DEBUG's 256-byte PSP. Then RAM.EXE is read from the diskette of drive B. The *loader of DEBUG* examines the header, evaluates the one relocation address, and loads the segments in the memory space following the PSP for RAM.EXE, which is located above DEBUG. The DEBUG prompt is the hyphen (-).

The Register Command R

Suppose the *register* command *R* is entered. The response is that of Fig. 7-16. Registers AX, DX, BP, SI, and DI are cleared, and BX and CX store the byte length of the program, excluding the header, with the high count in BX. The stack pointer SP is set to the stack size, which is 64H bytes, corresponding to the specified 50 words. *DEBUG uses only hexadecimal numbers,* and there is no indicator H.

In the second row of Fig. 7-16 are the initial contents of the segment registers, the instruction pointer, and the flags. In order, the flag indicators are *no overflow, direction up, disabled interrupts, positive sign, not zero, no auxiliary carry, parity odd,* and *no carry*. The respective opposite indicators are OV, DN, EI, NG, ZR, AC, PE, and CY.

Registers DS and ES point to the start of the PSP, and CS, SS, and IP are set according to information passed to the loader by the header of RAM.EXE. For this program, CS points to the first location above the 256-byte PSP, offset IP is 0, and SS is 90FH, which is above the entry point 907:0 of the load module by 80H bytes. This arrangement is in accordance with the stack segment offset value of the header, as shown in Fig. 7-15. On the bottom row of the display is the location, code, and assembly language of the first instruction to be executed. Whenever a running program stops at a breakpoint, the registers, flags, and the next instruction are automatically displayed in the format shown in Fig. 7-16.

The register command can also be used to set the contents of any of the registers or the setting of any of the flags. For example, the command R AX displays the

```
AX=0000  BX=0000  CX=007A  DX=0000  SP=0064  BP=0000  SI=0000  DI=0000
DS=08F7  ES=08F7  SS=090F  CS=0907  IP=0000  NV UP DI PL NZ NA PO NC
04C5:0000 1E            PUSH    DS
```

FIGURE 7-16. Display after the register command R.

```
0907:0000 1E           PUSH DS
0907:0001 B84000       MOV  AX,0040
0907:0004 8ED8         MOV  DS,AX
0907:0006 A11300       MOV  AX,[0013]
0907:0009 B364         MOV  BL,64
```

FIGURE 7-17. Display example for the unassemble command.

current contents of AX and allows the operator to change the contents, if desired, by typing new hexadecimal digits. To change one or more flags, enter R F and the current flag settings are displayed. To change a flag, type its opposite code. This operation can be done for more than one flag, and the order in which the opposite flags are typed is unimportant. Spaces between the entries are optional. For example, the entry CYEIZR PE will set flags CF, IF, ZF, and PF.

Command Unassemble U

Now suppose the *unassemble* command *U* is entered. This command translates approximately 32 bytes of code into statements that closely resemble assembly language; addresses and code are also displayed. To illustrate, with U entered after loading the RAM program, or following the R command, 15 assembly language instructions are displayed. The first five of these instructions are shown in Fig. 7-17. Note that there are no H indicators after the hexadecimal numbers, and the variable MEMORY_SIZE is replaced with its offset [0013] in DS. This format is not supported by the macroassembler, of course.

By entering the U command repeatedly, successive sequences of instructions are unassembled. Also, it is possible to enter a start address CS:IP after U, or both a start and stop address can be entered. For example, U 907:100 130 will unassemble and display instructions with bytes from offset 100H to 130H in segment 907H. If the code does not represent machine language instructions, the result will be meaningless.

Command Go G

To execute the program, simply enter the command *G* for *go*. With RAM.EXE in memory, the program runs, the memory-size message is displayed, and all registers and flags are restored to their initial values, allowing the program to be run again, if desired. The message "Program terminated normally" is also displayed. The starting point for the execution is the current value of CS:IP. If G = 11 is entered, execution starts at offset 11H in CS.

Command G can be run with breakpoints. Enter G13 and the program goes until it reaches the instruction starting at offset 13H. *Then it halts and all registers and flags are displayed, along with the next instruction.* Because programs frequently branch, a specified breakpoint may be bypassed, in which case it has no effect. Multiple breakpoints are allowed. The command G=100 140 120 160 starts execution at offset 100H, and the program stops if one of the three breakpoints is reached. If none are encountered, execution proceeds to the normal end. DEBUG sets breakpoints by replacing the first byte of an instruction with the code CCH, which is the 1-byte code for breakpoint INT 3. When any one breakpoint is reached, all breakpoint codes are replaced with the original instruction bytes. New breakpoints can then be set.

Breakpoints provide a convenient way to trace a program. The displayed registers and flags can be examined to see if the results are reasonable. However, there are two precautions to observe. First, each breakpoint must be set *to the offset of the first byte of an instruction*. Failure to do this may cause the computer to go into left field, possibly requiring a power-off reset. This nuisance should definitely be avoided. A second precaution is to avoid rerunning a program that executed from a breakpoint through INT 20H *without first using the register command to restore all registers and flags to their initial conditions*. Although INT 20H is the normal terminator, it restores the registers to their values at the last breakpoint. A rerun from this point will probably not have the initial value of DS on the stack, and the FAR return may start a runaway program.

Command Trace T

Trace T is the easiest way to trace execution of a program. Each time T is entered, one instruction is executed, and registers, flags, and the next instruction are displayed in the format of Fig. 7-16. The command T = 5A gives a trace starting at offset 5AH, and T commands proceed from this point. Instead of single-stepping, a number of instructions to be executed can be specified. For example, the command T11 executes 17 instructions in sequence, but the registers are displayed after execution of each instruction. A space between T and the hexadecimal value is optional. Scrolling of the screen is stopped with Ctrl-NumLock and resumed by pressing any key.

If a CALL or INT instruction is encountered in a trace, *the trace branches into the routine*. If branching is not desired, the command G can be used, with a breakpoint set at the start of the instruction following the CALL or INT. Similarly, G with a breakpoint can provide quick passage through a loop having many repetitions.

The trace command T provides a convenient way to examine the program of a procedure or interrupt, including those of the ROM BIOS. However, single-stepping through DOS interrupts sometimes fails because both DEBUG and the interrupt may store bytes in identical memory locations.

Other Commands

Only 4 of the 18 commands of DEBUG have been examined. Figure 7-18 summarizes all the commands. Each consists of a single uppercase or lowercase letter, usually followed by an address, a range of addresses, or another value. The range can be specified *either as a logical address followed by an offset or as an address followed by L and the byte length*. To illustrate, ranges 0:A000 AFFF and 0:A000 L 1000 are equivalent. If the segment address is omitted, a default of DS or CS is provided, depending on the instruction. All numbers are understood to be hexadecimal, *including those entered by the operator*. A space between the letter of a command and a parameter that follows is optional. Thus, D CS:100 and dcS:100 are equivalent. Brief descriptions of most commands not yet considered are given here.

Command	Function	Example
Assemble	Assemble routines	A907:100
Compare	Compare two memory blocks	C40:100 200 9
Dump	Display a memory block	d AB:5000
Enter	Change memory bytes	E B9:ABC
Fill	Fill a memory region	F5AC:0 100 "AB"
Go	Run program	g
Hexarithmetic	Add and subtract numbers	H 0F6 1C
Input	Display a byte from a port	I2FB
Load	Load from disk to memory	L CA5:0
Move	Move a memory block	M50:0 10 60:0
Name	Name a file for loading	N filespec
Output	Send a byte to a port	o 1F0 B5
Quit	Return to DOS	Q
Register	Display registers; modify	rCS
Search	Search memory for code	S40:0 A6 "AB"
Trace	Trace program execution	T = 0:A00 5
Unassemble	Unassemble code	u8F7:100
Write	Write data to disk	w

FIGURE 7-18. Summary of DEBUG commands.

Assemble A

Within DEBUG is a *mini-assembler* that can assemble statements of an assembly language program, *with the generated code placed directly in memory*. The mini-assembler is activated by entering "A *address*", and the initial logical address is displayed. An assembly language statement can then be typed and entered. It is immediately converted into code, which is stored in memory at the specified address, and the start address of the next instruction is displayed. All numerics are hexadecimal without the identifier H. To terminate the operation and return to the DEBUG prompt, the Enter key is pressed in lieu of a statement.

If an invalid statement is entered, it is rejected and "^Error" is displayed, with the current assemble address redisplayed. Prefix mnemonics such as REP and ES:, must be entered ahead of the opcode to which they refer, either on the same line or on a separate preceding line. Offset values replace labels; for example, JMP A68 gives a jump to offset address 0A68H in CS and CALL A0B0:55 calls a procedure at the specified logical address. All addressing forms are supported. A short example showing three lines of a program follows:

```
907:200  neg byte ptr [155]
907:204  sub bx, 45 [bp + 5] [si - 2]
907:207  rep movsw
```

The first instruction uses the BYTE PTR directive, because otherwise the mini-assembler would be unable to determine whether the operand pointed to a byte or a word. Execution of the 4-byte instruction converts the byte at offset 155H into its 2s complement.

The second instruction, with 3 bytes, uses based indexed addressing, described in Section 3-4. The address of the source word is (BP) + (SI) + 48H. Note the use of the numeric 45 in place of a variable name, as required by the macroassembler.

After code is generated with the A command, it can be unassembled with the U command, which displays the hexadecimal code alongside the assembly language statements, as shown in Fig. 7-17. *This method of obtaining code is easier than applying the rules of Appendix A*. Even so, there is much to be learned from familiarity with the rules.

In the next section, we will learn how to create programs that can be assembled with the mini-assembler, stored on a diskette, and run. For *short* programs the method is easier than using the macroassembler. However, because the assembly language program is not generated, *editing with EDLIN is not possible*. Thus, lines cannot easily be inserted, deleted, or modified.

Dump D

Dump D is used to display the contents of a portion of memory, with the bytes of the specified region displayed in rows. Each row starts on a paragraph boundary (offset divisible by 16) and contains one paragraph (16 bytes). At the left side is the logical address of the paragraph boundary, and at the right are the ASCII characters that correspond to the bytes, with each unprintable character indicated by a period. The example that follows is that of a single displayed row, printed here on two lines:

```
08F7:0140 A3 1A 0F 8A C3 26 A3 1C-0F
                   8A C6 26 A3 1E 0F B8   #...C&#...F&#..8
```

The first # sign at the right is ASCII for the first byte, A3, at the left, and the last byte, B8, is interpreted by DEBUG as 38, giving the digit 8.

If command D or D *address* is entered, eight rows with 128 bytes are displayed from an initial address. For an unspecified address, the start address is the location following the last one displayed by a previous D command, or DS:100 if there was no previous command. A second option allows specification of an address range. Thus, D CS:100 10C will display lines starting at offset 100H in CS and continuing through 10CH. By entering D repeatedly, consecutive 128-byte blocks of memory are displayed. *Dump is useful for rapid examination of a region of memory*.

Enter E

Enter E provides a way to change the contents of memory. To illustrate, suppose a short program has been hand-assembled, and we want to load it into memory at DS:100 for execution. The command E100, or EDS:100, is entered, and the byte at this location appears. When the new byte is typed and the space bar is pressed, the next one in sequence is displayed. Again, the replacement is typed followed by the space bar, and so on. Pressing the space bar without typing a new byte leaves the original one unchanged. When all bytes of the program have been keyed in, Enter is pressed and the data are converted to the new values.

A program with known code, including characters in quotation marks, is easily and quickly put into memory with the command E. It can then be unassembled to determine if the code correctly represents the intended instructions. If the program is assigned a name with the command N, it can be stored on a diskette using *write W*. Other forms of Enter are available, one of which allows for insertion of multiple bytes of a list into consecutive locations.

Input and Output Commands

The command *I port* reads and displays a byte from the specified input port, and *O port byte* sends the byte to the output port. In each case, *port* denotes one of the input or output hexadecimal port addresses of the PC. For example, I60 displays the byte of port A of the PPI, and O 61 7C sends byte 7CH to output port B of the same chip.

Move M

The contents of specified memory locations are transferred with *move M*. To illustrate, M0:A000 BFFF 0:B000 moves the 8K bytes from 0:A000 through 0:BFFF to the region of memory starting at 0:B000. Note that the first two addresses after M specify the range to be from 0:A000 through offset BFFF, and the third address specifies the destination. As in this example, the source and destination regions may overlap, but there is no loss of data during the transfer. The default segment address is DS.

Name N and Load L

Initially, program RAM.EXE was loaded into memory by DEBUG by entering DEBUG B:RAM.EXE. An alternate way is to enter DEBUG alone, followed by the *name* command *N B:RAM.EXE,* and then *L,* for *load*. There are other uses of the name and load commands.

Quit Q

Entering *Q* terminates DEBUG and returns control to DOS. It is a normal way to end a debugging session, but the file in memory is not saved. However, because of the header, *it is not possible to write EXE programs from the debugger to the diskette*. DEBUG uses the header for relocating addresses and loading the program, and changes in the program may affect the discarded header. If errors are found in an EXE program, it is usually necessary to correct the assembly language with EDLIN, reassemble it, and process it through the linker.

Write W

The *write* command *W* will write data being debugged to a diskette provided the data are not those of an EXE program. An attempt to use W with an EXE program generates an error message. For most other types of programs, the write operation transfers data either to absolute sectors of a diskette or to a named diskette file. *Registers BX and CX must first be set to the number of bytes to be written,* and the file must have an assigned name. The command is often used with COM programs, which are examined in the next section.

For simple replacement of a few bytes of erroneous code of an EXE file, one method sometimes used is to rename the file, with extension EXE changed or omitted. Then the renamed file is loaded with DEBUG. *This process leaves intact the header,* which is not used here for relocating addresses or for loading. Of course, the program is *not* executable. However, errors can be corrected by means of the various commands, after which the program and its preceding header are transferred back to the diskette. Now the filename is changed to the original one, and the corrected EXE file is ready for execution.

Although commands *compare C, fill F, hexarithmetic H,* and *search S* have not been described, they are quite useful and important. These and various additional features of the commands examined here are presented in the DOS manual. With a little practice, all are easily mastered.

7-5 COM PROGRAMS

There are two kinds of machine language programs, known as EXE and COM. As we have learned, an EXE program is the normal output of the assembler and linker. However, in most cases *COM programs are easier to create and debug, and they require less storage space.* We shall examine the design and execution of COM programs, starting with a short example program named COLOR.COM.

Program COLOR.COM

The program presented here is used to switch control from the monochrome monitor to the color monitor while in the DOS mode. Although not one of the original DOS programs, it was created with the mini-assembler of DEBUG and was then loaded on the DOS diskette as a utility of considerable importance. There are only 20 bytes in the complete program, *which has no header.* The assembly language routine is shown in Fig. 7-19, along with the hexadecimal code at the right.

In the BIOS ROM, variable EQUIP_FLAG is defined as a word location of the BIOS data area at address 0:410, and bits 4 and 5 of the low byte of EQUIP_FLAG correspond to the settings of respective switch numbers 5 and 6 of switch SW1. Whenever an IBM monochrome monitor is installed, these switches *must be manually set in the off position.* During the power-on self-test routine, the settings of SW1 are read, and the bits of EQUIP_FLAG are set or reset to serve as flags indicating switch settings, with a set bit denoting an off switch. Thus, bits 5 and 4 are both set during power-on, *giving initial control to the monochrome monitor.*

In order to transfer control to the IBM color monitor, it is necessary *to clear bit 4 of the low byte of EQUIP_FLAG, followed by execution of the set mode function of the video I/O interrupt type 10H.* Resetting bit 4 gives flag indicators corre-

```
XOR AX, AX                          ;clear AX               (33C0)
MOV DS, AX                          ;clear DS               (8ED8)
MOV AL, BYTE PTR EQUIP_FLAG         ;get switch flags     (A01004)
AND AL, 0EFH                        ;for reset of bit 4     (24EF)
MOV BYTE PTR EQUIP_FLAG, AL         ;reset bit 4          (A21004)
MOV AL, 3                           ;color mode value       (B003)
MOV AH, 0                           ;set mode number        (B400)
INT 10H                             ;video I/O              (CD10)
INT 20H                             ;terminate program      (CD20)
```

FIGURE 7-19. COLOR.COM program.

sponding to installation of an 80 × 25 high-resolution color monitor, and these flags are read by the set mode routine of INT 10H. The bit is cleared by the program of Fig. 7-19, which also places mode value 3 in AL and code 0 in AH for set-mode. Then INT 10H is executed, and control passes to the color monitor. Mode value 3 is interpreted by the interrupt as that of an 80 × 25 color monitor.

With only two changes, the program will return control to the monochrome monitor. The instruction AND AL,0EFH is replaced with OR AL,10H so that the next instruction will set bit 4. The new code is 0C10H. Then the color mode value 3 must be changed to the monochrome value 7. With these two changes, the new program, named MONO.COM, is also placed on the DOS diskette. Entering COLOR gives control to the color monitor, and MONO transfers operation to the monochrome set. In each case, the target screen is cleared. The same two programs can be used in the BASIC mode for switching monitors by placing the code in machine language subroutines (see Prob. 7-15). Alternate programs are given in Appendix I of the IBM *BASIC Manual*.

Program Creation. One way to create a COM program such as this one is to use the mini-assembler to generate the code or to hand-assemble it, using Appendix A. Then with the DOS diskette in drive A we enter DEBUG COLOR.COM, which loads DEBUG into memory, searches the diskette directory for COLOR.COM, and prints the message "File not found".

At the end of the DEBUG program in memory, a 256-byte PSP is set up for the new file. By entering R, we see that the four segment registers and IP store the following values:

DS = ES = SS = CS = 08F7H   IP = 100H

Values assigned to the segment registers are pointers to the bottom of free memory, just above DEBUG. Starting at offset 0 is a PSP with 100H bytes. *The instruction pointer IP points to the first location above it*. The stack pointer SP is set to the end of the segment or to the bottom of a portion of the COMMAND program placed at the high end of memory, whichever is lower. Note that *the stack builds downward from high memory, whereas a COM program builds upward from low memory*.

To create COLOR.COM, we now use the command E100 and enter the 20 bytes of hexadecimal machine code. If the code has not been determined, the assemble command A can be activated to place the code in memory. For this operation, the statements of Fig. 7-19 should be typed, with [410] replacing EQUIP_FLAG. If DEBUG is entered without the filespec of the new file, command *name N* must be used. Assuming the write-protect tab has been removed and byte 14H placed in CX, *the 20-byte program is written to the DOS diskette of default drive A with command W*, followed by quit Q. Back in the DOS mode, the program is ready for use. *Execution consists of entering COLOR*, with the COM extension ignored.

If B:COLOR.COM is specified as the filespec, the program is transferred from DEBUG is to the diskette of drive B. In this case, program execution is implemented with B:COLOR. *Only machine language programs having extensions EXE and COM are executed simply by entering the filespec*, and these two types have quite different forms. If both types are present with the same name, *the COM program is chosen*.

When COLOR is entered, the COM program is loaded into the low region of memory *immediately above its PSP*. *All segment registers point to the start of the PSP, and IP points to the first byte of code at offset 100H*. Because address CS is at the beginning of the PSP, *INT 20H can be used as the last instruction,* giving the normal program end. Although the initialized stack is confined to the segment that also has the data and code, it is made as large as possible.

Creation with the Assembler. For longer programs, especially those that are modular and may require considerable debugging and editing, the macroassembler is more convenient. The steps of the procedure are as follows:

1. Prepare the assembly language *without a stack segment.* The loader initializes a stack segment, as previously stated. *Begin the code segment with ORG 100H,* which allows code to be loaded at the end of the PSP. For the COLOR program, the assembly language program is shown in Fig. 7-20.
2. Because relocatable addresses are not allowed in COM programs, which have no header, *all initialized data must be put in the code segment.* A good procedure is to start the code segment with a jump instruction followed by the initialized data, with the jump used to branch around the data. In COLOR there are no initialized data.
3. Assemble and link the program, *ignoring the one serious error reported by the linker and accompanied with a warning that there is no stack segment.*
4. For the COLOR program, we now have COLOR.EXE on the diskette of drive B, along with others generated by the assembler and linker. After the DOS prompt, *we enter the DOS command*

```
EXE2BIN   B:COLOR   COLOR.COM
```

Execution of the preceding DOS command *creates COLOR.COM from COLOR.EXE,* which completes the process. Both the EXE and COM files are now present on the diskette of drive B, but the COM file is selected for execution when B:COLOR is entered. It has only 20 bytes, whereas the EXE file has 896 bytes, including its header and the 256 bytes that result from the ORG 100H directive. With COLOR.COM on the same diskette as COLOR.EXE, the latter program can be run only by first changing its name. However, this should not be done, because there is no stack segment and because the termination is faulty. The file should be erased, using the DOS command ERASE B:COLOR.EXE.

```
DATA   SEGMENT   AT 0
   ORG   410H
   EQUIP_FLAG   DW   ?
DATA   ENDS
CODE   SEGMENT
   ASSUME CS:CODE, DS:DATA
   ORG 100H
START: XOR AX, AX
       MOV DS, AX
       MOV AL, BYTE PTR EQUIP_FLAG
       AND AL, 0EFH
       MOV BYTE PTR EQUIP_FLAG, AL
       MOV AL, 3
       MOV AH, 0
       INT 10H
       INT 20H
CODE ENDS
END  START
```

FIGURE 7-20. The COLOR program for the assembler.

As shown by this example, COM files are smaller than EXE files, having no header. Thus, *they conserve diskette space, and they load and execute faster*. Also, *errors are easily corrected with the DEBUG program*. For these reasons, COM programs are strongly preferred. However, the EXE2BIN command works only for programs assembled without a stack segment, and all code, procedures, and initialized data should be placed in one 64K segment to avoid relocatable addresses. If these requirements cannot be satisfied, *an EXE file must be used*.

Program SET_TIME

To illustrate the development of a COM program containing initialized data and procedures, along with several DOS interrupts not previously used, the SET_TIME program of Figs. 7-21 and 7-22 is examined. It is designed to display the current time of the internal clock and to allow the operator to set this time to any desired value.

```
01   TITLE SET_TIME
02   ;
03   CODE SEGMENT
04      ASSUME  CS:CODE, DS:CODE
05      ORG 100H
06   ;
07    START: JMP A1
08   ;
09      MSG1     DB  'Current time is $'
10      MSG2     DB   0DH, 0AH
11               DB  'Enter new time: $'
12      BUFFER   DB   09, 10 DUP (?)
13   ;
14   ;DISPLAY TWO DECIMAL DIGITS FROM THE BINARY BYTE OF AL
15   ;
16      DISPLAY   PROC   NEAR
17           AAM                      ;convert byte to decimal
18           OR   AX, 3030H           ;decimal digits to ASCII
19           MOV  DX, AX
20           XCHG DL, DH
21           MOV  AH, 2
22           INT  21H                 ;display digit
23           XCHG DL, DH
24           INT  21H                 ;display digit
25           RET
26      DISPLAY   ENDP
27   ;
28   ;LOAD TWO DIGITS FROM THE KEYBOARD BUFFER INTO AL BINARY
29   ;
30      LOAD     PROC    NEAR
31           LODSB
32           CMP AL, 0DH              ;check for CR
33           JNZ B1                   ;if not CR, continue at B1
34              POP AX
35              JMP C1                ;CR, go to end
36      B1:  MOV AH, AL
37           LODSB
38           SUB AX, 3030H            ;ASCII to decimal digits
39           AAD                      ;digits to binary byte
40           RET
41      LOAD     ENDP
```

FIGURE 7-21. Part 1 of the SET_TIME program.

The program is a simplified version of that of the DOS command TIME, with the much larger TIME routine reduced to one with only 169 bytes. SET_TIME performs the same service as TIME, but without refinements. Figures 7-21 and 7-22 show the assembly language in a format suitable for conversion into a COM file. For reference, line numbers are included. At the beginning is the required ORG 100H directive, and the program has only one segment, which contains the data of two messages, a keyboard buffer, two NEAR procedures, and instructions. All data and both procedures are located at the beginning, just after a jump instruction that branches the program around them. If desired, the procedures may be placed in the region after INT 20H, which terminates the program.

EDLIN can be used to develop the program and store it on the diskette of drive B. Then it is assembled and linked. One error is reported by the linker, along with a warning that there is no stack segment. The COM file is generated from the EXE file by entering EXE2BIN B:SET_TIME SET_TIME.COM.

```
42    ;DISPLAY THE CURRENT TIME OF THE INTERNAL CLOCK
43    ;
44         A1:  STI
45              MOV   DX, OFFSET MSG1
46              MOV   AH, 9
47              INT   21H                                 ;display MSG1
48              MOV   AH, 2CH
49              INT   21H                                 ;clock time to CX:DX
50              MOV   BX, DX
51              MOV   AL, CH
52              CALL  DISPLAY                             ;display the hours
53              MOV   DL, 3AH
54              MOV   AH, 2
55              INT   21H                                 ;display ":"
56              MOV   AL, CL
57              CALL  DISPLAY                             ;display the minutes
58              MOV   DL, 3AH
59              MOV   AH, 2
60              INT   21H                                 ;display ":"
61              MOV   AL, BH
62              CALL  DISPLAY                             ;display the seconds
63    ;
64    ;ENTER THE NEW TIME INTO THE INTERNAL CLOCK
65    ;
66              MOV   DX, OFFSET MSG2
67              MOV   AH, 9
68              INT   21H                                 ;display MSG2
69              MOV   DX, OFFSET BUFFER
70              MOV   AH, 0AH                             ;wait for key
71              INT   21H                                 ;  entries
72              MOV   SI, OFFSET BUFFER + 2
73              CALL  LOAD                                ;hours to AL
74              MOV   CH, AL                              ;  and CH
75              INC   SI
76              CALL  LOAD                                ;minutes to AL
77              MOV   CL ,AL                              ;  and CL
78              INC   SI
79              CALL  LOAD                                ;seconds to AL
80              MOV   DH, AL                              ;  and DH
81              MOV   DL, 0
82              MOV   AH, 2DH
83              INT   21H                                 ;set time (Hr:Min:Sec)
84         C1:  INT   20H
85    CODE   ENDS
86    END    START
```

FIGURE 7-22. Part 2 of the SET_TIME program.

When the program is executed by entering B:SET_TIME, the following message appears, with the actual clock time replacing that shown:

```
The current time is 15:27:56
Enter new time:
```

At this point, the operator can either accept the reported time by pressing Enter or enter a new time. All six digits are required; otherwise there is no change. The command can be repeated to verify that the entry was accepted. In the TIME program of the DOS diskette, an entry of one digit for the hours, minutes, or seconds is accepted, and if the entry does not adhere to the correct format, the request is repeated following display of "Invalid time". Also, the time display of DOS includes hundredths of a second.

Procedure DISPLAY. The DISPLAY subroutine is called three times from the main program. When it is called, register AL stores a binary number representing the hours, minutes, or seconds of the internal clock. The subroutine *converts the binary number into two decimal digits in ASCII code and then sends them to the display in the proper order.*

Conversion of the binary byte of AL into two decimal digits is done by the AAM instruction, which divides the number by 10 and puts the quotient in AH and the remainder in AL. Thus, AH stores the high-order digit and AL stores the low-order digit, each from 0 to 9. Detailed descriptions of this and the other ASCII-adjust instructions are given in the next section. Next, 3030H is ORed with AX, converting the decimals into ASCII.

In preparation for the display, the digits are moved to DX, with the high-order digit placed in DL and the low-order one in DH. Function 2 of INT 21H sends the digit in DL to the display. Then the low-order digit is moved to DL, and INT 21H is again executed, displaying the next digit alongside the first one, at its right. Parameter passing from the main program to the procedure is through register AL, and registers AH and DL pass parameters to the interrupt routine. No parameters are returned to the main program.

Procedure LOAD. On entry to the LOAD subroutine, *index register SI points to a location of the keyboard buffer named* BUFFER, *which stores either a sequence of two decimal digits in ASCII code or byte 0DH for CR (Enter).* If the byte is CR, the program jumps to INT 20H and ends. POP AX is placed ahead of the jump simply to restore the stack pointer to its original value, thereby preventing stack shrinkage. On the other hand, if the byte is not CR, *the subroutine transfers the two decimal ASCII bytes from BUFFER and converts them into a binary byte stored in AL.*

The LOAD procedure is called three times from the main program. The LODSB instruction is used twice for loading the 2 ASCII bytes, and each time it is executed, pointer SI is updated. After the ASCII codes are transferred in order to AH and AL, word 3030H is subtracted from AX, thereby converting the ASCII digits to unpacked BCD. Then AAD multiplies the digit of AH by 10 and adds the result to AL, thereby transforming the two decimal digits into a binary number stored in AL. This instruction is described in detail later. Parameters are passed to the procedure through SI and the keyboard buffer, and the result is returned in AL.

Display the Current Time. In Fig. 7-22 the main program is divided into two sections. The first one displays the current time of the internal clock. When a program is run, *hardware interrupts are enabled,* but not when run with DEBUG. Therefore, it is good practice to enable interrupts by placing STI at the start of the program. The three instructions of the main routine following STI activate function 9 of INT 21H, which displays the message "The current time is ". Note that the message contained in the program terminates with $, as required by the function.

After this operation, function 2CH of INT 21H returns the time of day. In binary form, CH stores the hours, CL the minutes, DH the seconds, and DL the number of 1/100th seconds. First, the binary hours are transferred to AL and DISPLAY is called to show the decimal digits. Then a colon is sent to the display using function 2 of INT 21H. Using the same method, the minutes, another colon, and the seconds are displayed in order.

Enter the New Time into the Clock. The second section of the main program of Fig. 7-22 *allows the operator to type new entries that reset the internal clock.* Function 9 of INT 21H is used to send the next message to the display. The first 2 bytes of the message are CR and LF, placing the cursor at the left side of the next line. The displayed message is "Enter new time: ".

The offset of BUFFER is placed in DX, and function 0AH of INT 21H is called. As required at entry, DS:DX points to an input keyboard buffer, which is set up in the data section of the code segment. As shown in Fig. 7-21, the first byte of this buffer is 9, *indicating that up to nine characters can be entered, including CR.* Line 12 specifies 10 reserved locations with uninitialized bytes. In addition to the first byte, with value 9, these 10 locations provide space for the byte denoting the number of entries, six decimal digits, two colons, and CR.

When the keyboard input function 0AH of INT 21H is executed, the computer *waits for entries from the operator.* Pressing Enter terminates the program during the first call to LOAD. However, suppose the keyboard entry is the new time, assumed to be 12:34:56. This entry fills the space of the keyboard buffer that is available for data, and when Enter is pressed, the program continues. The first entry of 1 is at offset 2 of the buffer, and index register SI is loaded with this offset by the instruction of line 72.

The LOAD procedure is called, and the ASCII digits of the entry for hours are converted to binary and put in AL. After transfer to CH, pointer SI is incremented to bypass the colon character of the buffer. LOAD is again called, and the binary byte of minutes is put in CL. After pointer SI is incremented to jump over the next colon entry, the seconds are converted to binary and loaded in DH. Because the program does not accept an entry for hundredths of a second, byte 0 is put in DL. Function 2DH of INT 21H moves the new time to the internal time-of-day clock, provided the data for HR:MIN:SEC are within allowed ranges, and INT 20H terminates the program normally, giving the DOS prompt.

The program uses DOS interrupt 20H and functions 2, 9, 0AH, 2CH, and 2DH of INT 21H. As in this example, *the functions of INT 21H are commonly used in programs written for the PC and compatible machines.* The COM format is generally preferred, except for large programs that require more than one segment for code and initialized data. These must be EXE programs, with headers. In the next section, we examine another COM program and the arithmetic ASCII-adjust instructions, two of which were present in SET_TIME.

A RESIDENT PROGRAM

The TICKS program presented here is designed *to reside in memory and to perform its function while other computer operations proceed in a normal manner*. Usually, when a program is loaded into memory for execution, it is placed in the lowest available free space, overwriting any user programs present. However, by using INT 27H, referred to as terminate-and-stay-resident, a program in memory will not be overlaid. When TICKS is executed, it remains in memory, protected from overwrite, and sends changing code to the display even while most other programs are loaded and executed. As we shall see, the EXE format is almost essential. Following the discussion is a description of the ASCII-adjust instructions.

Earlier, it was mentioned that interrupts with dummy returns provide a way for a user to add to the interrupt routine. This procedure is shown by the example used here. Also, a shortcoming of COM routines relative to EXE routines is illustrated.

Shown in Figs. 7-23 and 7-24 is the complete assembly language program. Note the termination with INT 27H. At the start of the program, DS is pushed onto the stack. Then, just ahead of the FAR return, the sum of OFFSET A4 and 100H is pushed. Accordingly, the FAR return branches the program to INT 27H, with CS storing the segment address of its PSP as required. In addition, the sum of OFFSET A4 and 102H is placed into DX by the instruction with label A3, providing a pointer in CS to the first empty location following TICKS. The method used here

```
TITLE  TICKS
;
STACK  SEGMENT  STACK
   DW  100H  DUP(?)
STACK  ENDS
;
VIDEO   SEGMENT  AT  0B000H
   ORG  (24 * 160) + 150
   CRT  LABEL  WORD
VIDEO   ENDS
;
CODE   SEGMENT
  ASSUME  CS:CODE, DS:CODE, ES:VIDEO
   START: PUSH DS
          MOV  AX, CODE
          MOV  DS, AX
          MOV  DX, OFFSET A1
          MOV  AX, 251CH
          INT  21H
          JMP  A3
;
      A1: NOP                   ;can be replaced with CF for IRET
          JMP  A2
;
      W1  DW   0, 10, 182, 1000, 100
;
     DISPLAY   PROC   NEAR
          MOV  AH, 7
          OR   AL, 30H
          MOV  ES:[SI], AX
          ADD  SI, 2
          RET
     DISPLAY   ENDP
```

FIGURE 7-23. Part 1 of the TICKS program.

illustrates the proper way to terminate an EXE program with the terminate-and-stay-resident INT 27H.

Analysis of the program and its function is left as an exercise for the reader. After assembly and linking, it can be executed by entering B:TICKS, assuming drive B. Then DEBUG can be entered and used to examine the program. NOP in the line with label A1 *allows for insertion of code CF for IRET while the program is running,* using the DEBUG enter command E. This procedure inactivates exe-

```
     A2: STI
         PUSH AX                 ;all pushes are essential
         PUSH DX
         PUSH SI
         PUSH DS
         PUSH ES
         MOV  AX, 0B000H
         MOV  ES, AX
         MOV  AX, CODE
         MOV  DS, AX
;
         INC  W1                 ;increment the word each run
         MOV  AX, W1             ;multiply by 10
         MUL  W1 + 2             ;divide by 182
         DIV  W1 + 4
         XOR  DX, DX
         DIV  W1 + 6
         MOV  SI, OFFSET CRT
       CALL   DISPLAY
         MOV  AX, DX
         XOR  DX, DX
         DIV  W1 + 8
       CALL   DISPLAY
         MOV  AX, DX
         XOR  DX, DX
         DIV  W1 + 2
       CALL   DISPLAY
         MOV  AX, DX
       CALL   DISPLAY
         POP  ES
         POP  DS
         POP  SI
         POP  DX
         POP  AX
         IRET
;
     A3: MOV  DX,   OFFSET A4 + 102H
         MOV  AX,   OFFSET A4 + 100H
         PUSH AX
     DUMMY  PROC  FAR
        RET
     DUMMY  ENDP
;
     A4: INT  27H
;
CODE  ENDS
END   START
```

FIGURE 7-24. Part 2 of the TICKS program.

cution of instructions that follow IRET and *allows for modification of the routine,* if desired. Then code CF can be replaced with 90H for NOP, and execution continues.

To convert to a COM program, a few changes are needed. First, the stack segment must be eliminated and the directive ORG 100H added. Then delete PUSH DS at the start, moving the label START to the next instruction. In the line with label A3, change 102H to 2, and delete the next five program lines. With these changes, the program assembles and links, but the DOS command EXE2BIN is unable to convert the EXE program to COM. The reason for this failure is the *relocatable address of CODE,* which appears twice in the routine. CODE denotes the start address of segment CS, which is not known until the program is loaded. Thus, EXE2BIN cannot possibly generate the necessary code, and it reports this fact in a message.

For generating a COM program, it is necessary to determine the start address of CS and to replace CODE with the value at both places where it appears in the program. When this operation is done, command EXE2BIN can make the conversion. The result is not very satisfactory, because the start address depends on whether the program is loaded by COMMAND or DEBUG and also on the options present. *The EXE form is more appropriate.* In the author's machine, the start address CODE of a COM program loaded by COMMAND equals address CS displayed by the *register* command of DEBUG less segment address 2F1H. At this location is code CD 20 for the INT 20H instruction that begins the PSP.

AAA, AAS, AAM, AAD

Instructions AAA, AAS, AAM, and AAD adjust arithmetic results of operations with *unpacked* decimals, which use a byte for a digit. The first two A's denote *ASCII-adjust* and the last letter symbolizes either *add, subtract, multiply,* or *divide.*

Although ASCII is the most common form of unpacked BCD, other forms can be used. CPU binary operations are adjusted for BCD arithmetic by means of the instructions. When a BCD arithmetic operation is performed with unpacked numbers, the appropriate adjust instruction must be used. AAA, AAS, and AAM follow add, subtract, and multiply operations, respectively, whereas AAD *precedes* a decimal division. Figure 7-25 shows the six arithmetic ASCII-adjust instructions, including DAA and DAS, which apply to packed BCD addition and subtraction, respectively.

ASCII-Adjust-Add AAA. Instruction AAA examines the low nibble of AL. If it is greater than binary 1001, or if AF is set, binary 0110 is added to this nibble, 1 is added to the byte of AH, and flags AF and CF are set. Otherwise the low

Instruction	Comments
AAA	Adjusts AL after adding unpacked BCD numbers
AAS	Adjusts AL after subtracting unpacked BCD numbers
AAM	Adjusts AX after multiplying unpacked BCD numbers
AAD	Adjusts unpacked digits of AX before division
DAA	Adjusts AL after adding packed BCD numbers
DAS	Adjusts AL after subtracting packed BCD numbers

FIGURE 7-25. Arithmetic ASCII-adjust instructions.

nibble is not changed, and flags AF and CF are 0. In any event, the upper nibble of AL is cleared, so that after execution of AAA, register AL stores a number from 0 through 9. Flags OF, PF, SF, and ZF are undefined after the operation.

To illustrate, suppose instructions ADD AL,7 and AAA are executed in sequence, with initial values of 5 in AL and 0 in AH. The word of AX following the ADD instruction is 000CH, and after AAA it is 0102H, corresponding to BCD 12.

ASCII-Adjust-Subtract AAS. For a subtraction of unpacked BCD numbers, instruction AAS normally follows the operation. If the low nibble of AL exceeds 9, or if AF is set, binary 0110 is subtracted from the nibble, 1 is subtracted from AH, and flags AF and CF are set. Otherwise the low nibble is not changed, and flags AF and CF are clear. In any event, the upper nibble of AL is cleared and the other four status flags are undefined. After the operation, AL stores a number from 0 through 9. *Subtraction of 6 is equivalent to the addition of its 2s complement 1010B*.

For example, with 0105H stored in AX, corresponding to BCD 15, suppose SUB AL,7 and AAS are executed. After execution of SUB, the word of AX is 01FEH, and after AAS it is 0008H, which is decimal 15 minus 7. Flags AF and CF are set.

ASCII-Adjust-Multiply AAM. Instruction AAM corrects the product of a multiplication of two unpacked decimal operands, each of which must have a high nibble of 0. The two valid decimal digits are stored in AH and AL in binary form. Flags PF, SF, and ZF are updated, and the other status flags are undefined. Specifically, AAM divides the number of AL by 10D, and the integer result is stored in AH with the remainder placed in AL.

Consider MUL BL followed by AAM, with initial bytes of 15D in both AL and BL. After the multiplication, the word of AX is 225D, or 00E1H, and after AAM it is 1605H. The byte of AH is 22D and that of AL is 5, corresponding to 225. There are various applications unrelated to the multiply operation, such as that of the DISPLAY procedure of SET_TIME.

ASCII-Adjust-Divide AAD. Unlike the other adjust operations, AAD precedes the divide instruction. *It converts the dividend of AL into valid unpacked BCD form*. Specifically, AAD multiplies the high byte AH of AX by 10D, adds the result to the low byte AL, and stores the result in AL; then the byte of AH is replaced with 0s. For example, if AX stores 0608H when AAD is executed, the word of AX becomes 0044H. The instruction requires 60 clocks. Flags PF, SF, and ZF are updated, and the other status flags are undefined. There are numerous specialized applications of this instruction.

To illustrate, suppose AAD and DIV CL are executed in order, with initial BCD values of 0509H and 16H in AX and CL, respectively. The BCD dividend is 59D and the divisor is 22D. After AAD, the word of AX is 003BH, or 59D, and following the division it is 0F02H, corresponding to a quotient of 2 and a remainder of 15. Note that AAD changes the double-digit decimal of AX into an equivalent binary number. This is its only function in the LOAD procedure of SET_TIME.

DAA and DAS

DAA and DAS are decimal adjust instructions for addition and subtraction of *packed* decimals. Packing permits storage of two decimal digits in a byte location, thereby conserving memory. Their normal use is for correcting the result of an addition or subtraction of packed decimal operands, giving a packed decimal result.

Decimal-Adjust-Add DAA. DAA examines the low nibble of AL. If it is greater than 9 or if flag AF is set, 6 is added to AL and AF is set. Then if the result in AL exceeds 9FH, or if flag CF is set, 60H is added to AL and CF is set. All flags are affected except OF, which is undefined.

With BCD arithmetic assumed, suppose we add 86H to AL, with AL containing 59H. This operation represents the addition of 86 and 59, which gives 145. The binary addition of the computer yields DFH, with both AF and CF clear. Adding 6 to AL gives E5H, and then adding 60H converts this result to 45H. With the set carry flag CF included, the BCD result is 145D.

Decimal-Adjust-Subtract DAS. DAS is the corresponding decimal adjust for subtraction of packed BCD numbers. The operation is identical to that of DAA, except for subtraction of 6 and 60H in place of addition. Again, OF is undefined. To illustrate, suppose 54H is subtracted from 81H of AL, assuming BCD arithmetic. The correct packed BCD result is 27H, but the computer gives 2DH. Subtracting 6 converts this result to the correct value, 27H.

7-7 BASIC DEBUGGING

There are similarities between debugging in BASIC and in assembly language. In both, programs can be traced and breakpoints can be set, but the techniques are different. Although there is no equivalent DEBUG program in BASIC, there are some special commands, statements, and variables in the interpreter designed to help the programmer find and eliminate bugs. After investigation of these elements, we will look at some BASIC interrupts.

Regardless of the language of the source program, perhaps the most important way to obtain programs that run properly is to use good programming techniques. For large programs, some kind of structure chart is usually prepared, with the main program at the top feeding downward into smaller and simpler modules. Each of these modules should have a limited and clearly defined function, so that individual units can be designed and tested independently of one another. This procedure is done in BASIC with either subroutines or the more restricted functions. Care must be exercised in passing parameters between the main program and the subroutines.

Testing the main program is simplified by replacing each subroutine with a dummy routine, or *stub,* that simply prints any variables passed to it, passes back a set of dummy variables, and then gives a return. *The routine of the subroutine is omitted.* By this method, the main program can be tested to determine if it calls each subroutine at the proper time, passes it appropriate data, and processes returned values in a satisfactory way. After the routines of the modules have been designed and tested, they can be added one at a time, with a test done following the addition of each new one. An alternate way is to include all modules following the test of the main program. Then they are replaced one by one with stubs, and tests are performed after each change. Either method should help to pinpoint errors. Even when considerable care is used during the design process, bugs do occur, and they must be found and eliminated.

Tracing and Setting Breakpoints

Programs in BASIC can be traced, and breakpoints can be set. *Tracing simply shows the order in which instructions are executed,* allowing the operator to relate each set of output data to the corresponding path of the program. In the trace mode, no way is provided for examination of registers and results at each step, as is possible with DEBUG. However, at breakpoints fixed by STOP statements, program variables can be examined.

Commands TRON and TROFF. The *trace-on* command *TRON* turns the trace on by enabling an internal *trace flag*. Once executed, it remains in effect until the trace flag is disabled. Disabling is normally done with the *trace-off* command *TROFF*. The command NEW, which deletes a program from memory and clears all variables, also disables the trace flag. Each of the commands may be entered in either the direct or indirect modes.

When the trace flag is enabled, each line number of the program is printed as the line is executed. These numbers are in *square brackets,* which distinguish them from data. Output generated by an instruction is displayed next to the line number, as shown in the elementary example of Fig. 7-26. The unstructured routine of the example is presented simply to illustrate the result of a trace.

In the program of Fig. 7-26, TRON is entered in the direct mode, followed by the prompt Ok. There are two jumps, as shown by the trace. Note the different values generated by INT and FIX.

Statement STOP. There are several ways to stop program execution. One is the key combination Ctrl-Break, which displays ^C and the message "Break in 30", assuming 30 is the line number at which the break occurred. However, to stop execution at a specified point, the statement STOP is used; it may be placed anywhere in a program. Suppose a running program has STOP at line 70. The

```
10  A = &HFFF6 + 4.4
20  B = ABS(A): C = INT(A):  D = FIX(A)
30  GOTO 60
40  PRINT C, D
50  END
60  PRINT A; B
70  GOTO 40
TRON
Ok
RUN
[10][20][30][60] -5.6  5.6
[70][40]-6    -5
[50]
Ok
```

FIGURE 7-26. Program and results with a trace.

program halts and the message "Break in 70" is displayed, followed by the command-level prompt Ok. *It is now possible to examine variables with direct-mode PRINT commands.* Execution may be resumed from the break with the continue command CONT.

In order to read from or write to a diskette file, thereby transferring data between the file and memory, it is first necessary to *open* the file. This operation can be performed with function 0FH of INT 21H and also with the OPEN statement of BASIC, both of which are examined in Chapter 11. Some commands, such as LOAD and SAVE, open a file automatically. When read/write operations are completed, the file should be *closed.* In addition to the CLOSE statement, various BASIC commands perform this operation automatically.

The END statement is similar to STOP, but there are two differences. END does not cause a BREAK message to be printed, and END closes all files, which STOP does not do. However, for programs that do not involve reading and writing diskette files, END can be used in place of STOP to provide breakpoints. This procedure is not recommended.

Command CONT. To resume execution of a program after a break, the command CONT is normally used. The break may have been caused by either Ctrl-Break, STOP, or END. In any case, *execution proceeds from the point of the break.* CONT is normally used in conjunction with STOP. For debugging, *STOP provides a breakpoint at which variables can be examined. Then execution is resumed, with CONT entered in the direct mode.* Another way to continue is to execute a *direct-mode* GOTO statement, with GOTO followed by the line number at which execution is to resume. The continue command is invalid if the program is changed during the break, and if entered, the message "Can't continue" is displayed.

Error Trapping

When an error is detected by BASIC, *a message is printed and the command or program is aborted.* In Appendix A of the IBM *BASIC Manual* are 49 error messages with numbers in the range 1-73. Error 2, or "Syntax error", is probably most common. The message is printed whenever the interpreter encounters a line in either the direct or the indirect mode that has an incorrect sequence of characters, including erroneous punctuation. Other common error messages are *Overflow, FOR without NEXT, Out of data,* and *File not found.*

Statement ON ERROR GOTO. It is possible *to trap errors* so that BASIC will not halt program execution and print a message when an error is detected. The trap is a subroutine designed to handle the situation as well as possible; with error trapping enabled, all errors cause the program to branch to the subroutine. Thus, *error trapping is a BASIC interrupt.* It is enabled by the statement ON ERROR GOTO *line,* with line 0 excluded. The line number must exist; otherwise the message "Undefined line number" is printed. Error trapping remains enabled until disabled by the statement ON ERROR GOTO 0. Both are executable in either the direct or the indirect mode.

Assume that ON ERROR GOTO 0 is present in an error-handling routine. If it is executed, it immediately disables error trapping, stops execution, and prints the error message for the one that caused the trap. Error-trapping routines often include ON ERROR GOTO 0 to stop processing if the routine is unable to handle the error properly and resume normal program execution.

Suppose error trapping is enabled with ON ERROR GOTO 100. At any time thereafter, an error detected by BASIC in either the direct or the indirect mode gives a branch to line 100, and there is no error message. The service subroutine at line 100 is executed, and it must be designed to handle the situation properly. After the error has been processed, execution can be continued by means of RESUME. Error trapping does not occur in an error-handling routine. If one occurs here, the program halts and the appropriate error message is printed.

Statement RESUME. *RESUME provides the return from the subroutine of the error trap.* Three choices are available:

RESUME RESUME NEXT RESUME *line*

The statement RESUME transfers *to the line that generated the error,* RESUME NEXT goes to the one immediately following the one causing the error, and the third form branches to the specified line number. RESUME can be used only in error-trapping routines.

Statement ERROR *n*. *An error can be simulated by means of the statement ERROR n,* with *n* denoting an integer expression having a value from 0 through 255. If *n* is one of the error codes of Appendix A of the BASIC manual, the error simulated is the one that corresponds. For values of *n* not used by BASIC, the code is defined by the user as desired.

To illustrate, suppose ERROR 15 is placed in a program. This number corresponds to the BASIC error "String too long", which signifies a string with more than 255 characters. With error trapping disabled when the statement is encountered, the message is printed and processing stops. On the other hand, with trapping enabled by prior execution of ON ERROR GOTO 80, the program jumps to line 80, and there is no message.

Next, assume that ERROR 200 is executed with trapping disabled. Because there is no defined error message, the response is "Unprintable error", and execution stops. With trapping enabled, there is no message, and a branch to the error subroutine occurs.

Variables ERR and ERL. Provided by BASIC is variable ERR, which normally has *the code of the most recent error,* including an error simulated with ERROR *n*. *The corresponding line number is stored as variable ERL.* To illustrate, if a syntax error occurs on line 50, the direct-mode statement PRINT ERR gives 2 and PRINT ERL gives 50. The same results are obtained if line 50 consists only of the statement ERROR 2.

Next, suppose the statement ERROR 200 appears on line 80. After its execution, ERR is 200 and ERL is 80. Variables ERR and ERL are assigned values only by actual errors or by the statement ERROR *n*. In case of a direct-mode error, there is no line number, and ERL becomes 65535, or FFFFH.

Variables ERR and ERL are usually used in IF-THEN statements to direct pro-

```
10  CLS: LOCATE 5,1
20  ON ERROR GOTO 200
30  RANDOMIZE
40  A = INT(100 * RND: PRINT A
50  INPUT "Enter a number from 1 through 10: "; B
60  IF B < 1 OR B > 10 THEN ERROR 150
70  END
200 PRINT "ERR is"; ERR; "and ERL is"; ERL
210 IF ERL = 30   THEN  PRINT:  RESUME
220 IF ERR = 150  THEN  PRINT:  RESUME 50
230 ON ERROR GOTO 0
```

FIGURE 7-27. Error-trapping program with an error in line 40.

gram flow in the error-handling routine. An example illustrating their use is presented in the short routine of Fig. 7-27.

Assume the program of Fig. 7-27 is entered, followed by entry of TRON in the direct mode. When the program is run, suppose the response is that of Fig. 7-28.

The display shown in Fig. 7-28 indicates that syntax error 2 occurred in line 40, and following the prompt, the line is displayed for correction. A branch from line 40 to line 200 is indicated by the trace. Because the conditions of both IF statements were false, execution continued to ON ERROR GOTO 0, which printed the error message and stopped processing. *No statements of an IF-THEN logical line are executed when the condition is false.* Neither of the RESUME statements were executed.

Now suppose the missing parenthesis is entered into the displayed line. Then the program is run again, giving the display of Fig. 7-29, which includes user responses. In the second line of the display is a report of error 6 in line 30. This is the *overflow* error, which occurs when a number is too large for storage. With the statement ON ERROR GOTO 200 omitted, the error would have halted processing with display of the message "Overflow". Following the first error is another one, which is error 150 in line 60. ERROR 150 was defined in the program *to exist if the entry of line 50 was outside the specified range*. It has no other meaning. If ERROR 150 is entered in the direct mode, the response is "Unprintable error". *All errors not defined by BASIC have this message.*

In the error-trapping program of Fig. 7-27, note the two forms of RESUME that are used. The first one can be replaced with the equivalent statement RESUME 30. In lines 210 and 220, the statement PRINT: can be omitted. In each case, it gives a blank line in the display. The trace of Fig. 7-29 shows that both RESUME statements were executed.

```
[20][30]Random number seed (-32768 to 32767)? 4
[40][200]ERR is 2 and ERL is 40
[210][220][230]
Syntax error in 40
Ok
40  A = INT(100 * RND: PRINT A
```

FIGURE 7-28. Display after execution of an error-trapping program with a syntax error.

```
[20][30]Random number seed (-32768 to 32767)? -33333
[200]ERR is 6 and ERL is 30
[210]
[30]Random number seed (-32768 to 32767)? 0
[40] 54
[50]Enter a number from 1 through 10: 12
[60][200]ERR is 150 and ERL is 60
[210][220]
[50]Enter a number from 1 through 10: 9
[60][70]
Ok
```

FIGURE 7-29. Display after rerun of an error-trapping program.

Event Trapping

In addition to error trapping, with BASICA it is possible to trap events, allowing the computer to respond to external events. *Event trapping is the BASIC interrupt,* and it is especially useful in video games. Just as error trapping causes a program to branch to a specified service routine when an error occurs, event trapping gives a jump to a specified subroutine when an external event occurs.

Four types of events can be trapped, and each is activated with an ON statement. ON COM is used to trap a message arriving through the communications adapter, ON PEN traps an activated light pen, and ON STRIG traps a joystick button, or trigger. *Only ON KEY, which traps a key closure, is examined here.*

Statement KEY(*n*). To activate or to disable the ON KEY statement, it is necessary to execute KEY(*n*), which has the following three forms:

KEY(*n*) ON KEY(*n*) OFF KEY(*n*) STOP

Value *n* must be from 1 to 14, indicating the key to be trapped. Numbers 1-10 denote the function keys F1 to F10, number 11 is the Cursor Up key, 12 is the Cursor Down key, 13 is the Cursor Right key, and 14 is the Cursor Left key. The KEY(*n*) statement is available only in advanced BASIC, and it can be used only in programs. Do not confuse KEY(*n*) with the statement KEY used to set or display the soft keys. The latter has no effect on the Soft Key values displayed at the bottom of the screen.

To illustrate, when KEY(5) ON is executed in a program, a previous or subsequent ON KEY(*n*) GOTO 200 statement is enabled. After execution of both statements, key F5 can be used to trap to line 200. *Each time the key is pressed, there is an interrupt of the program, with the service routine starting at line 200.* Execution of KEY(5) STOP disables trapping of F5, but *a closure of the key is remembered,* so that if KEY(5) ON is again executed, an immediate trap takes place. The statement KEY(5) OFF also disables the trap, but subsequent closures of key F5 are *not remembered.* It is often necessary to disable trapping for brief intervals, such as when variables used by the subroutine are being changed by the main program.

Statement ON KEY(*n*) GOSUB *line*. The ON KEY(*n*) GOSUB *line* statement *sets up a line number for the trap of key(n).* If the line number is 0, the trapping of the specified key is disabled. To illustrate its use, suppose the statement contains values 9 for the key number *n* and 50 for the line number when it is executed in

```
10   RANDOMIZE
20   CLS: TIME$ = "0"
30   ON KEY(10) GOSUB 200: KEY(10) ON
40   ON KEY(9) GOSUB 300: KEY(9) ON
50   KEY(10) STOP
60   LOCATE 23,70: PRINT TIME$
70   KEY(10) ON
80   GOTO 50
90   '
200  LOCATE INT(23 * RND + 1), INT(79 * RND + 1)
210  PRINT "*"
220  RETURN
230  '
300  RETURN 20
```

FIGURE 7-30. Key trapping program.

a program. Assume that the statement KEY(9) ON is processed either before or after the ON KEY instruction. Following execution of both instructions, *each time BASIC starts to process a new line, it first checks to see if key F9 has been pressed; if so, it performs a GOSUB to line 50.* When a trap occurs, an automatic KEY(9) STOP instruction is executed in order to prevent a recursive trap. The return from the subroutine automatically processes a KEY(9) ON instruction unless a KEY(9) OFF was encountered within the trap routine.

All trapping is automatically disabled when an error trap occurs. This includes ERROR, KEY, COM, PEN, and STRIG trapping. The return from the subroutine is with RETURN, as for the ON *n* GOSUB statement examined in Section 6-7. However, in advanced BASICA, *a line number can follow RETURN,* in which case the return is to the line specified. This nonlocal return is useful in event trapping. A short program is given in Fig. 7-30 to illustrate event trapping.

In the key-trapping program of Fig. 7-30, the TIME$ statement of line 20 sets the hours, minutes, and seconds of the internal clock to 0. Note that key F10 interrupts are inhibited while the statements of line 60 are locating and printing the time. Try running the program with the statement of line 50 removed and key F10 held down. A convenient way to do this is *to insert an apostrophe at its left side,* converting it temporarily into a comment. The interrupt service routine at line 300 has only a RETURN, but *the included line number provides key F9 with a useful function.* Ctrl-Break will stop program execution.

REFERENCES

MCS-86 Assembly Language Reference Manual, Intel Corp., Santa Clara, Calif., 1978.

D. J. BRADLEY, *Assembly Language Programming for the IBM Personal Computer,* Prentice-Hall, Inc., Englewood Cliffs, N.J., 1984.

P. NORTON, *Inside the IBM PC-Access to Advanced Features and Programming,* Robert J. Brady Corp., Bowie, Md, 1982.

PC *DOS, DOS Technical Reference, BASIC, and MACRO Assembler* manuals, IBM Corp., Boca Raton, Fla.,1981.

L. J. SCANLON, *IBM PC Assembly Language-A Guide for Programmers,* Robert J. Brady Corp., Bowie, Md., 1982.

D. C. WILLEN AND J. I. KRANTZ, *8088 Assembler Language Programming: The IBM PC,* Howard W. Sams, Inc., New York, 1982.

PROBLEMS

SECTION 7-1

7-1. Word location TIMER_LOW at 40:6C is incremented 18.206 times per second by timer interrupt type 8 from level 0 of of the interrupt controller. Design a BASIC program that reads TIMER_LOW and repeatedly displays the count in a one-line message on the IBM monochrome monitor. Start the message at the left side of row 15. (The program can be stopped with Ctrl-Break.) Also, calculate the time required to increment the variable from 0 through 65,535.

7-2. Design an assembly language program similar to the one of RAM in Figs. 7-1 and 7-2 that does the same task as the BASIC program of Prob. 7-1. If possible, develop the program with EDLIN and store it on a diskette for assembly.

7-3. (a) Indicate all changes that must be made in the program RAM if the only ASSUME directive is ASSUME CS:CODE. (b) Identify memory variables that have no specified address. (c) What is the effect of changing FAR after DUMMY PROC to NEAR? (d) What happens if WORD after CRT LABEL of segment VIDEO is changed to BYTE? (e) With only two PUSH instructions in RAM, is a two-word stack adequate? Explain.

7-4. Suppose registers AX, CX, and DX initially store F32AH, 7BCDH, and 4321H, respectively. (a) Give the hexadecimal contents of these registers after execution of MUL CX. (b) Repeat (a) for IMUL CX.

SECTION 7-2

7-5. (a) With reference to Figs. 7-8 and 7-9, describe the differences between the object code and the machine language code of the load module. (b) Which of the three blank addresses of the object code are known by the assembler? (c) Explain why MSG LABEL BYTE cannot be replaced with MSG:, located on the same line with the message. (d) Explain why RET cannot be replaced with INT 20H.

7-6. Prepare a detailed flowchart for the program RAM.ASM.

SECTION 7-3

7-7. Identify the meaning of the following filename extensions: ASM, BAK, BAS, CRF, EXE, LST, MAP, OBJ, and REF.

7-8. Suppose RAM.EXE is loaded at CS:0000. Relative to CS, give the beginning offset addresses of the instruction code, the message of MSTORE, and the stack. Identify all unused locations between offset 0 and the stack.

SECTION 7-4

7-9. For each example at the right side of Fig. 7-18, state precisely what is done and give the results, if appropriate. Also, determine flags after the response entry of cypoovdnac following the DEBUG command rf. Precisely what does the DEBUG command M F000:E000 FFFF 0:A000 accomplish?

SECTION 7-5

7-10. Using DEBUG with all segment registers normal, the machine code for the following nine instructions is placed in memory at offset 100H:

```
 AAM      OR AX,3030H       MOV DX,AX       XCHG DL,DH
MOV AH,2      INT 21H       XCHG DL,DH        INT 21H
INT 20H
```

With IP at 100H, command T is entered repeatedly, which single-steps the routine. Each interrupt routine is jumped with G and a breakpoint. Give a trace of the execution, showing changes in AX and DX at each step, if the initial value of AX is (a) FC0EH, (b) AB3AH, and (c) 2D17H. Explain exactly what the program accomplishes.

7-11. If $ at the end of MSG1 of SET_TIME is carelessly omitted, determine the display at the keyboard-entry wait. What happens if $ is omitted at the end of MSG2?

7-12. In SET_TIME, suppose the keyboard entry is 23:57:46. (a) After this, give in hexadecimal the contents of CX:DX and all 11 locations of the keyboard buffer when INT 21H of line 83 is reached. (b) Repeat (a) if the first digit 2 is omitted, recalling that the last entry is 0DH. Comment on the set time.

7-13. Determine the setting of the internal clock if the keyboard entry in SET_TIME is 3$,h5/4A. Repeat if the entry is $3;a6-XY.

7-14. The display has the form HH:MM:SS.ZZ when hundredths are included, and the entry of the latter can be made along with the others during the wait. Revise the program to accommodate this addition. Give the new assembly language starting at line 64 of Fig. 7-22. Include comments. No changes are needed in the first part of the program or in the procedures, but the buffer must be enlarged.

7-15. Write a BASIC program COLOR.BAS containing machine language code in an array that executes the COLOR program of Fig. 7-19 when COLOR is loaded and run in the BASIC mode.

SECTION 7-6

7-16. Prepare an assembly language program that, when assembled and run, will stop execution of TICKS, transferring control to the dummy IRET of the ROM BIOS.

7-17. List each line of TICKS, with a comment beside each stating the purpose of the instruction or directive.

7-18. Explain exactly what TICKS accomplishes. Precisely where is the display, and what is the reading if W1 stores FC4AH when the routine at offset A2 is executed? What is the maximum value displayed if the program is run for hours?

7-19. Revise TICKS so that the display has the form of HR:MIN:SEC and run it on a PC.

7-20. Revise TICKS in a form suitable for generating a COM file, assuming a start segment address CODE of 606H.

7-21. (a) Explain why all PUSH instructions of TICKS are essential, and comment on the consequences of omitting one and any associated POP. (b) List each MUL, XOR, and DIV instruction in the proper sequence, and state what each does.

(c) Is operation affected if the MUL and DIV instructions are replaced with IMUL and IDIV? Why is TICKS dormant in the BASIC mode?

7-22. Decimal 58,479 is stored at DS:0100H. Give the bytes in binary form, and their hexadecimal offsets, for storage as (a) a decimal in unpacked ASCII, (b) a decimal in packed ASCII, and (c) a binary number.

7-23. If AX stores C9ABH, determine the hexadecimal word of AX and all status flags after execution of (a) AAA; (b) DAA; (c) AAS; (d) DAS; (e) AAM; (f) AAD; and (g) CBW. Tabulate the results.

7-24. Prepare a COM program that transfers with immediate MOV instructions three signed 8-bit binary numbers X, Y, and Z to memory at 0:8000. The program should calculate the fourth powers of these numbers, designated X4, Y4, and Z4, and store them in consecutive doubleword locations starting at 0:9000H. Design and use procedure FOUR, which places the fourth power of the signed number of AL in register AX and its high-word extension in register DX. (b) Run the program with X, Y, and Z equal to 80H, 7FH, and B7H, respectively, and examine the results with DEBUG.

SECTION 7-7

7-25. In the BASIC program of Fig. 7-26, give the displayed values of the variables if A of line 10 is changed to (a) &HFD53 + 96.3; (b) (&HFD53) * 96.3; and (c) &HFD53 / 96.3.

7-26. Give the direct-mode response to each entry that follows: ERROR 2; ERROR 6; ERROR 15; ERROR 88.

7-27. With TRON active, sketch the display of the error-trapping program of Fig. 7-27 if the response to the INPUT request of line 50 is &H9ABCD.

7-28. With reference to the key-trapping program of Fig. 7-30, what are the functions of keys F9 and F10? What is the effect of deleting the two +1 terms of line 200?

7-29. Design a program with key trapping that does something of interest to you.

7-30. Design a program with a useful error trap.

CHAPTER 

Direct Memory Access and the Diskette System

Magnetic disks and tapes provide nonvolatile storage for programs and data. Here we investigate the characteristics and functions of a typical disk system, which is that of the PC. Included is a study of the floppy disk controller and the way it transfers data between a diskette and memory. Presented is a program that illustrates the precise internal mechanisms controlling the transfer. The DOS commands for managing diskette operations are described in the next chapter.

In the PC, the transfer of bytes between a diskette and memory is done by means of a technique called *direct memory access (DMA)*. This method is extensively used in most computing systems. It also provides the required refreshing of memory cells. Accordingly, we begin our study with DMA.

8-1 DIRECT MEMORY ACCESS

One way to move data between a peripheral and memory consists of first moving the word into the accumulator and then transferring it to the memory location or the I/O device. If a large block of words is to be transferred, the method is usually wasteful of processor time. A much faster procedure is DMA; it floats the address, data, and appropriate control pins of the CPU and uses an external controller chip to transfer the words directly between the peripheral and memory via the data bus.

DMA is sometimes used for high-speed transfer of data from one block of memory to another, which is a common operation, especially in large systems with many concurrent users. Transfers can be made from a peripheral to memory or from memory to a peripheral. No processor registers are involved in the actual movement of data, but the CPU must initially program the controller with appropriate information. Although operating speed is increased and software is reduced, additional hardware is needed. This hardware is present in the PC.

DMA Controller 8237A

Direct memory access is controlled by a DMA interface chip, which has a data-bus buffer and a unit with control logic, plus a transfer channel for each peripheral served. Each channel has its own *memory address register* and a *word count register* that stores the number of words to be transferred. The controller is initialized by the processor. Employed by the PC is the 8237A chip, which is a 40-pin programmable DMA controller with four independent channels.

Wait-State Generator

When a peripheral requests DMA service on one of the four prioritized request lines (DREQ) to the DMA controller, the controller issues a hold request (HRQ) to an input of the *wait-state generator*, which provides a set of timed control signals. Other inputs to this generator that are related closely to DMA operation are status lines $\overline{S2}$, $\overline{S1}$, and $\overline{S0}$ from the CPU. Outputs include control line $\overline{AEN}$, a hold-acknowledge HLDA line that goes to the DMA controller, and a $\overline{DMA\ WAIT}$ line to the 8284A clock generator, as shown in Fig. 5-1. A block diagram showing the DMA-related signals of the wait generator is given in Fig. 8-1.

At one input to the wait-state generator is *the hold-request HRQ handshake signal from the DMA controller*. When HRQ is inactive low, each of the three outputs is inactive. Suppose a DMA request from a peripheral to the DMA controller causes the chip to issue an active HRQ. The three outputs remain inactive until the 3 status bits from the processor become 1s. As mentioned in Section 5-1, this occurs during state T3 of each bus cycle.

Then the wait-state generator, which consists of logic gates and flip-flops, activates in time sequence the signals HLDA, $\overline{AEN}$, and $\overline{DMA\ WAIT}$. The precise

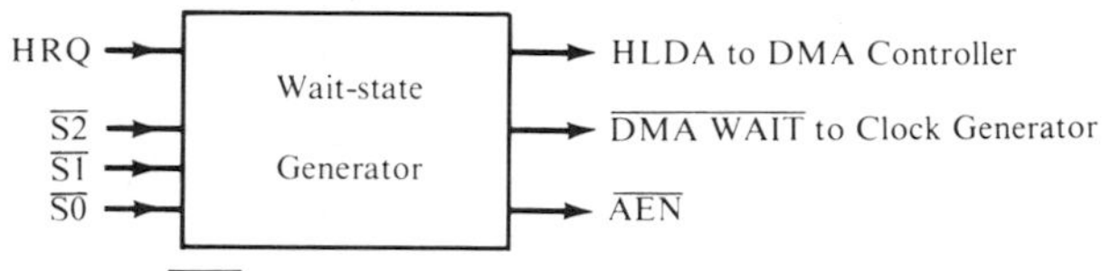

Output $\overline{AEN}$ supplies signals to:

Address bus and data bus transceivers
Peripheral-address decoder
8288 bus controller
Diskette and communications adapters (inverted)

FIGURE 8-1. DMA control signals of the wait-state generator.

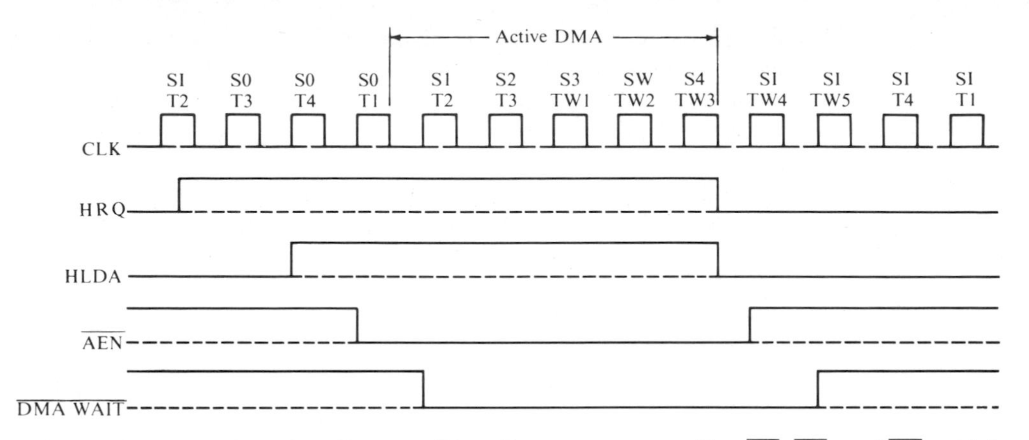

FIGURE 8-2. DMA timing diagram. Not shown are status bits $\overline{\text{S2}}$, $\overline{\text{S1}}$, and $\overline{\text{S0}}$, which go inactive high during T3 and remain high until the end of the next T4 state.

timing is shown in Fig. 8-2. Recall that each bus cycle of the 8088 has four T states in addition to any TW wait states inserted between T3 and T4. The processor T states during DMA are indicated on the timing diagram. Above these are S states that refer to the DMA controller, with SI denoting an idle state. State S0 occurs after HRQ is received but prior to HLDA, and the other S states denote active DMA operation.

As shown in the timing diagram, thc hold-acknowledge HLDA signal becomes active on the positive clock transition of state T4, provided HDR is active when the 3 status bits go high during the preceding T3 state. *This handshake signal to the DMA controller initiates the DMA process.*

On the positive clock transition of the following T1 state, control $\overline{\text{AEN}}$ becomes active. It floats the address and data buses at the transceivers shown in Fig. 5-1, but the processor can continue to use the local buses. $\overline{\text{AEN}}$ also disables outputs of the peripheral address decoder, thereby inactivating the PPI ports and all read/write operations for the interrupt controller, the timer/counter, and the DMA controller. In addition, it floats output pins of the 8288 bus controller that control read/write operations. During transfers, read/write controls and required addresses are supplied by the DMA chip.

Finally, on the positive clock transition of T2, signal $\overline{\text{DMA WAIT}}$ goes active. It feeds into the 8284A clock generator of Fig. 5-1, which then inactivates the READY input of the CPU, causing wait states to follow state T3.

At the end of DMA state S4, HRQ is driven low by the DMA controller. This action causes the wait-state generator to inactivate HLDA, $\overline{\text{AEN}}$, and $\overline{\text{DMA WAIT}}$ in sequence, as shown in the timing diagram. Processor wait states end only after $\overline{\text{DMA WAIT}}$ becomes inactive. The timing is such that there are no bus conflicts between the CPU and the DMA controller, which does not use the buses until after they have been floated by $\overline{\text{AEN}}$. Although there are five CPU wait states and five active DMA states, they do not coincide.

Channels

Channel 0, with the highest priority of the four channels, is programmed to refresh the dynamic memory of both the system board and that added through the expansion slots. The diskette drives use channel 2. These drives exercise control via DMA request, terminal count, and acknowledge lines, which connect the diskette computer card with the controller. The other two channels are available for use by peripherals that may later be added to the system through the I/O slots.

Diskette I/O is implemented with the instruction INT 13H, and the interrupt program is stored in the ROM BIOS. The routine receives from the user program appropriate parameters that specify the source and destination of data, the number of bytes to be moved, and other information. Data transfers between the diskette and memory employ DMA, with the DMA controller initialized by the interrupt program.

Basic Operation

When a peripheral requests DMA service on one of the four prioritized request lines (DREQ) to the DMA controller, the controller issues a hold request (HRQ) for control of the system buses, assuming the channel mask bit is not set. The handshake signal is sent to the wait-state generator, and the delays there allow the current bus cycle to be completed. After receiving the return handshake signal HLDA, the IC notifies the individual peripheral by means of the DMA acknowledge signal $\overline{\text{DACK}}$ that the transfer may begin. The cycle changes from idle to active. Transfer of a byte requires 1.05 microseconds, or five clock periods. When the transfer is done, the DMA controller inactivates line HRQ, and the processor regains control of the buses. Input DREQ must be held active at least until $\overline{\text{DACK}}$ becomes active.

Block Diagram

The controller is shown in Fig. 8-3, with some control lines omitted. The chip has 15 addressable registers, a timing and control unit, a command control unit, a priority encoder, and buffers. It allows external devices to transfer data to or from system memory *without processor intervention;* transfer of data from one region of memory to another is also possible. The three-state I/O read and write lines $\overline{\text{IOR}}$ and $\overline{\text{IOW}}$ are bidirectional, as are address pins A3-A0.

Chip-Select Address. Chip select $\overline{\text{CS}}$ is fed from an output of the address decoder that is active only when signal $\overline{\text{AEN}}$ is inactive high and address bits A9 through A5 are 00000. An active input allows the CPU to read/write the DMA registers. At other times, including the intervals between DMA transfers, $\overline{\text{CS}}$ is inactive. Address bits A3 through A0 are supplied directly to the chip. With don't-care bit A4 taken as 0, the peripheral addresses are 00H through 0FH.

Controls RESET and RDY. Connected to the chip is the system RESET line. When the signal to the input pin is active, it clears the command, status, request, and temporary registers, clears the first/last flip-flop, idles the controller, and sets the 4 bits of the mask register, which disables all DMA. The DMA chip must be programmed by the CPU, using OUT instructions. The OUT instruction enables pin $\overline{\text{CS}}$ and activates line $\overline{\text{IOW}}$ at the proper time. For reading the internal registers, IN instructions are used, which activate $\overline{\text{CS}}$ and control $\overline{\text{IOR}}$. For both IN and OUT, the input/output pins $\overline{\text{IOW}}$ and $\overline{\text{IOR}}$ function as inputs.

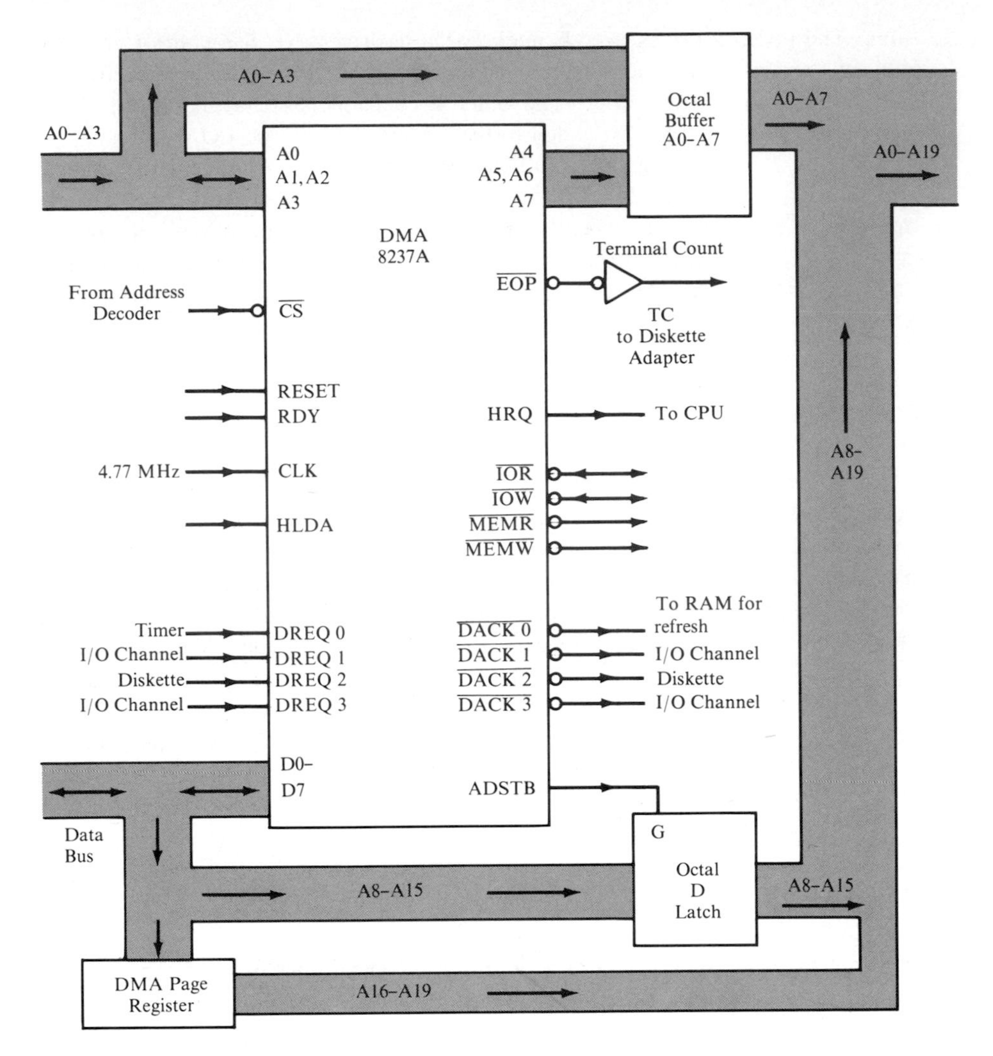

FIGURE 8-3. DMA controller 8237A and associated circuits.

The chip design is such that the associated processor is normally idled for four clock periods during each byte transfer. This design can be changed by supplying a signal to input pin RDY from a wait-state generator. If RDY is driven inactive low at the end of bus cycle state T2 and held low, the DMA controller extends the active duration of its output read and write pulses, thereby causing additional wait states to be inserted between states S3 and S4. The number depends on the duration of the inactive low state of RDY. Although not shown in Fig. 8-1, the wait-state generator of the PC uses this method to insert one additional wait state in all byte transfers; this is done to accommodate slower peripherals. The inserted wait state SW is shown in Fig. 8-2. However, DMA memory refresh operations do not use SW, and the CPU has only four TW wait states.

The Idle Cycle. In the *idle cycle,* with states defined as SI, the chip samples the DREQ request lines during each clock pulse, searching for a request. Also, chip select $\overline{CS}$ is sampled to determine if the CPU has signaled for a write or read of the internal registers. Reading and writing of these registers is done as necessary, with address lines A3-A0 being inputs during the idle cycle.

The First/Last Flip-Flop. The low byte of a 16-bit register is addressed if the logic state of the *first/last* flip-flop is 0, and the high byte is addressed if the state is 1. A write operation with instruction OUT 0CH,AL clears this flip-flop regardless of the byte stored in AL, and each read or write of a 16-bit register switches the state. In the idle cycle, lines $\overline{IOR}$ and $\overline{IOW}$ are inputs. However, in the active cycle in which DMA transfers occur, controls $\overline{IOR}$ and $\overline{IOW}$ and addresses A3-A0 are programmed for output.

Operating Modes

Prior to DMA transfers, the channel to be used must be initialized. There are four modes of operation: single transfer, block transfer, demand transfer, and cascade mode. The *single transfer* mode transfers a single data byte, decrements the word count, changes the address by one, and then relinquishes the buses. A block of data can be moved in this mode by successive single transfer operations. This is the mode most often used.

Especially suitable for high-speed peripherals is *block transfer,* or *burst,* which moves a block of data without relinquishing the buses. However, block transfer does not allow other devices to use DMA until the entire block has been moved, and this procedure is unacceptable in most cases. If used in the PC, it would delay memory refresh much too long. Some DMA units allow a higher-priority request to override a burst that is in progress. *Demand transfer* can move data until the I/O device has exhausted its capacity. The *cascade* mode applies to multiple controllers, providing additional DMA channels.

Single Transfer. Because it is very common and is the method used in the PC, single transfer merits special attention. It is often called *cycle stealing,* with DMA transfer cycles regarded as stolen from the processor. Between transfers there is always at least one full machine cycle execution. In many cases, the stolen cycle occurs when the processor is not using the buses, in which case the DMA operation is transparent to the CPU. However, it is not usually possible to make a device wait for such a spare cycle. Most peripherals are unable to wait for more than a very short time between each byte transfer without causing a system error. An example is a spinning diskette.

Only one byte is moved during a single transfer. If DREQ is held active throughout the single transfer, the request line HRQ still goes inactive after the transfer, releasing the buses to the CPU. However, HRQ again becomes active, and upon receipt of a new HLDA, another single transfer occurs. The operation is repeated over and over until the specified number of bytes has been moved. The procedure ensures that the CPU goes through a full machine cycle between byte transfers. Many peripherals, including diskettes, are sufficiently slow to require multiple bus cycles between transfers.

Other devices can be serviced with DMA between the single transfers. If two or more peripherals request DMA before an operation has started, the one with the higher priority is served first. Clearly, the maximum delay for a peripheral having the highest priority is one machine cycle. Interrupts can occur between active DMA transfers, and these transfers may be implemented during interrupt routines.

The handshake protocol previously described is followed for each byte movement, with the processor regaining control of the the buses between transfers. Each channel has a current word-count register that is decremented. When the count becomes 0, indicating that all bytes have been moved, a *terminal-count (TC)* pulse is generated. This pulse causes transfers to cease, and the TC pulse pulls the state of pin $\overline{\text{EOP}}$ (end-of-process) low. In addition to the current count register, each channel has a nonaddressable *base count register* that holds the programmed initial count.

As indicated in Fig. 8-3, signal $\overline{\text{EOP}}$ is inverted and fed to the diskette adapter as control TC, notifying the floppy disk controller (FDC) that the operation has ended. The FDC then performs some housekeeping chores, such as putting certain data into some of its registers for reading by the processor, followed by an interrupt request (INT 0EH) through level 6 of the interrupt controller. *The interrupt simply notifies the CPU that the operation has ended and that certain results and status information are ready for reading*. An example is treated in detail in Section 8-5, illustrating the way DMA serves diskette I/O.

Transfers can be classified as *read, write,* or *verify*. A verify operation applied to a diskette performs a normal read, except that data read are not placed in memory. Thus, verify provides a test of the system by generating addresses, sending and receiving control bits, and verifying successful operation.

Memory Addresses

For all transfers, initial memory addresses are sent from the CPU to the DMA controller via the data bus. Because the DMA address registers hold only 16 bits, an external 4-bit register file is needed for storage of bits A19-A16. This is the *page register* shown in Fig. 8-3. Outputs float when $\overline{\text{AEN}}$ is inactive. The file has read/write controls and can store four 4-bit words, with write addresses 80H through 83H. For channel 2, address 81H writes the 4-bit word from data lines D3-D0 into the register that feeds into lines A19-A16 of the address bus when $\overline{\text{DACK2}}$ goes active.

Each channel has a *current address register,* which contains the 16 low bits of the address used in DMA transfers, and also a nonaddressable *base address register,* which holds the initial value. The word of the current address register is automatically incremented or decremented, as programmed, after each transfer. If *autoinitialize* is specified, the current address register is reinitialized to its original value following an end-of-operation signal on pin $\overline{\text{EOP}}$.

Reading and writing of DMA registers by the CPU are done with 2 successive bytes transferred along the data bus. Output address pins A4-A7 have the three-state mode during the idle cycle. When addressing memory for DMA transfers, pins A0-A7 supply the low byte of the 20-bit address bus through the octal buffer; pins D0-D7 supply bits A8-A15 to the bus through the octal D latch, which is strobed at the proper time by address strobe ADSTB from the controller; and the page register provides bits A16-A19 directly. The chips are identified in Fig. 8-3.

8-2 PROGRAMMING THE 8237A

All valid port numbers are listed in Fig. 8-4. For the 16-bit address and word count registers, the low byte is accessed when the first/last flip-flip is reset, and the high

Address	Register	Number of Bits	Read/Write
00H	Ch 0 current address	16	Read/Write
01H	Ch 0 current word count	16	Read/Write
02H	Ch 1 current address	16	Read/Write
03H	Ch 1 current word count	16	Read/Write
04H	Ch 2 current address	16	Read/Write
05H	Ch 2 current word count	16	Read/Write
06H	Ch 3 current address	16	Read/Write
07H	Ch 3 current word count	16	Read/Write
08H	Status register	8	Read only
08H	Command register	8	Write only
09H	Request register	8	Write only
0AH	Mask register (1 bit)	8	Write only
0BH	Mode register	8	Write only
0CH	Clear first/last FF	1	Write only
0DH	Temporary register	8	Read only
0DH	Master clear	0	Write only
0FH	Mask register (all bits)	8	Write only

FIGURE 8-4. Valid port numbers of the 8237 DMA controller.

byte is accessed when the flip-flop is set. Each access of one of these registers automatically switches the state of the flip-flop. The instruction OUT 0CH,AL clears the flip-flop, with the value of AL having no significance. Both read and write of the address and word count registers are allowed. Initial values must be written before commencing DMA.

The instruction OUT 0DH,AL gives a master clear. It is a pseudo write, with the contents of AL unimportant. This is the same as the hardware reset previously described. Ports 0AH and 0FH refer to the same mask register. OUT instructions to these two port numbers have a different meaning, however. The temporary register, with port 0DH, is used only for temporary storage of bytes of a memory-to-memory transfer.

Status Register. Bits 7-4 of the read-only *status* register are set if a DMA request is present on respective channels 3-0, and bits 3-0 are set each time a terminal count TC is reached on the corresponding channel. All bits are cleared by a read operation or by a reset.

Command Register. The encoding of the *command* register is given in Fig. 8-5. In the examples considered later, we will write 00H to this register, which specifies active low-level $\overline{\text{DACK}}$ signals, active high-level DREQ signals, normal-width write pulses, fixed priority, normal timing, controller enable, channel 0 address-hold disable, and memory-to-memory disable.

Bit 7 - ($\overline{\text{DACK}}$ state)	0 low;	1 high
Bit 6 - (DREQ state)	0 high;	1 low
Bit 5 - (Write pulse width)	0 normal;	1 extended
Bit 4 - (Priority)	0 fixed;	1 rotating
Bit 3 - (Timing)	0 normal;	1 compressed
Bit 2 - (Pin $\overline{\text{CS}}$)	0 enable;	1 disable
Bit 1 - (Ch 0 address-hold)	0 disable;	1 enable
Bit 0 - (Mem-to-mem transfer)	0 disable;	1 enable

FIGURE 8-5. Encoding of the 8237 command register.

Bits 7,6 - (Transfer) 00 demand; 01 single;	10 block;	11 cascade	
Bit 5 - (Address)	0 increment;	1 decrement	
Bit 4 - (Autoinitialize)	0 disable;	1 enable	
Bits 3,2 - 00 verify; 01 write;	10 read;	11 not allowed	
Bits 1,0 - (Channel select)	00,0; 01,1;	10,2; 11,3	

FIGURE 8-6. Encoding of the 8237 mode register.

Bit 4 of the command register can be set to establish a rotating priority. The order is always 0123, with channel 0 initially having the highest priority. However, after a DMA operation, the channel with the number following the one processed is given top priority. Normally, fixed priority is used.

When set, bit 1 of the command register holds the channel 0 address, rather than incrementing or decrementing it as usual. This procedure allows the same byte to be written to a block of memory locations in a memory-to-memory transfer. Such a transfer always uses channel 0 for the source and channel 1 for the destination. Bit 0 determines if memory-to-memory transfer is to be used.

Mode Register. Encoding of the write-only *mode* register is given in Fig. 8-6. For memory refresh we write 58H to this register, specifying single-transfer, address-increment, enable-autoinitialize, memory-read, and channel 0. The operation is pseudo-read, with no data actually transferred. Autoinitialization is necessary so that the count continues endlessly. For a diskette read operation, byte 46H should be written. It selects channel 2 for single transfer, with address-increment and autoinitialization disabled.

Request Register. Bits 1 and 0 of the *request* register are encoded to select one of the four DMA channels, and bit 3 gives an active DMA request if set. All other bits are don't-cares. The request register can be used to initiate software DMA operations. It is cleared by a reset and also by an active terminal count TC.

Mask Register. The write-only *mask* register has two port numbers. Port 0AH is used for setting or resetting the mask bit of a single channel, with bits 1 and 0 selecting the channel. A logical 0 bit 3 resets the mask bit, and a logical 1 sets it. Other bits are don't-cares. A set mask bit disables the DMA request line DREQ, and all bits are set following a reset operation. For memory refresh, 00H should be written to port 0AH after the initialization. This operation clears the mask bit of channel 0, allowing refresh operations to commence. For a diskette operation, byte 02H should be written to the port.

By using port 0FH, all mask bits can be written at once. Bits 3, 2, 1, and 0 apply to respective channels of the same number, and the other bits are don't-cares.

Memory Refresh

Channel 0 is dedicated to *memory refresh*. Once every 72 clock periods, counter 1 of the 8253 chip examined in Section 5-6 supplies a DMA request to channel 0. During the reset initialization, after completion of the tests on the DMA system, five OUT instructions transfer bytes FFH, FFH, 58H, 0, and 0 to respective port addresses 1, 1, 0BH, 8, and 0AH.

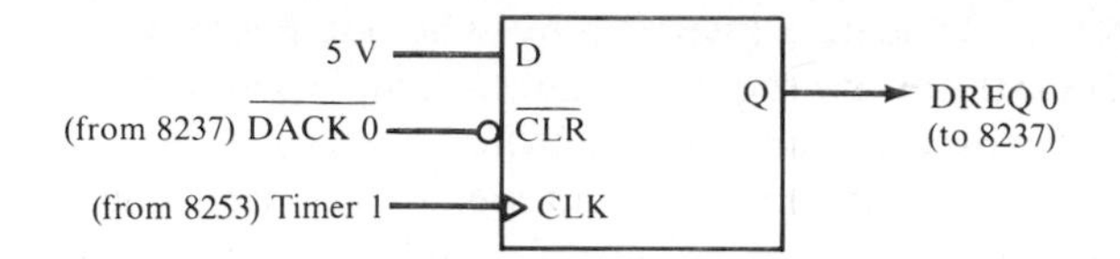

FIGURE 8-7. DMA request 0 from the 8253 timer 1.

The first two of these instructions load the word-count register of channel 0 with FFFFH. Port 0BH is the mode register, and byte 58H written to it selects channel 0 for READ transfer with address increment and single transfer. The byte also specifies that the word count is to bc automatically restored to FFFFH after becoming 0. Accordingly, the count never terminates.

Port 8 is the command register, which is cleared by the specified write operation. The 0 bits fix the desired active levels and give normal timing and priority. The last of the 5 bytes clears the mask bit of channel 0, thereby enabling the input request line DREQ 0. After this operation, a periodic timer request leads to transmission of active output signals on lines $\overline{\text{DACK0}}$ and $\overline{\text{MEMR}}$ of the DMA controller. Each request activates the row address strobc (RAS) of each RAM chip.

All RAM chips are strobed, causing refreshing of the cells of the row that is addressed by the 7-bit word A6-A0. The DMA addrcss register is incremented, the count register is decremented, and DMA terminates until the next timer pulse. There are numerous on-board memory chips, plus additional chips on a memory card, and one row of each is refreshed. Both 16-KB and 64-KB chips are refreshed simultaneously if both types are present.

After 72 clock pulses following the previous request, the timer sends another request, and the next row is refreshed. There are 128 rows per chip, and the 7-bit address word is cycled from 0 through 127 and repeat. Each bit of memory is refreshed during an interval of 128 DMA operations, occurring in a time span of less than 2 milliseconds. Although the refresh operation implements a read transfer, this is a dummy procedure in which no data are moved to the data bus. The memory design is such that a READ of a cell is automatically followed by a WRITE, with the bit written into the cell restoring the original charge to its full strength. During the pseudo READ, the CPU idles through four wait states, perhaps doing some internal processing.

Recall that the DREQ must be held active until $\overline{\text{DACK}}$ becomes active. This requirement is satisfied by the circuit of Fig. 8-7. Note that the DREQ 0 line is activated when the timer pulse goes low to high, and it remains high until the D type flip-flop is cleared by $\overline{\text{DACK 0}}$ at the asynchronous $\overline{\text{CLR}}$ input.

The I/O Channel

In Section 8-4 the diskette adapter is examined. It is connected to the system through the *I/O channel,* and its interaction with the DMA controller and other chips is through the contacts of the slot. A few comments are in order before we proceed to the study of the diskette system.

The 62 conductors that constitute the I/O channel of the system unit are connected to the 62 contacts of a slot, half on each side. An adapter card inserted into a socket becomes part of the computer. Several such cards in their slots are shown in Fig. 5-10.

In the I/O channel are 20 address lines, fed by either the microprocessor or the DMA controller. Those from the processor are demultiplexed, with the address obtained from latches. Most peripherals are I/O mapped, using the 10 lower address bits, and input and output are accomplished with IN and OUT instructions. However, we have learned that the display buffers of the adapters for the monochrome and color monitors are memory mapped, allowing the use of the powerful string instructions. The eight data lines serve the microprocessor, memory, and I/O devices. There are eight power lines, consisting of three GND conductors, two lines with +5 V, and single lines with −5, +12, and −12 V. All signal lines have TTL logic levels of 0 and 5 V.

Six interrupt request lines go from the channel to levels 2 through 7 of the 8259 interrupt controller. The one of level 4 is from the communications adapter, and that of level 6 is from the diskette adapter. The others are available for use as desired.

A number of the channel lines connect to the DMA controller. Three DMA channels are controlled by prioritized request lines. One of these lines is DREQ 2 from the diskette drive, and the other two are not currently in use. A terminal count TC line from the DMA controller of the system unit through the slot to the diskette adapter provides a pulse when the terminal count for a channel is reached. Then there are four DMA acknowledge lines from the 8237 to the channel. Included are $\overline{\text{DACK2}}$ to the diskette adapter and $\overline{\text{DACK0}}$ for refresh of 192 KB of dynamic memory present on an adapter.

Also serving DMA is an address-enable (AEN) line to the channel from the wait-state generator. There are four lines that control reading and writing of I/O devices and memory, and those to the RAM of an adapter are controlled by either the processor or the DMA controller.

Two clock lines are present. One is fed by the system clock, and the other has a pulse train with a 50% duty cycle at the crystal frequency of 14.32 MHz. A reset-drive line serves to reset or initialize external logic when power is turned on. Available to the I/O channel is the address-latch-enable (ALE) signal from the bus controller. It is used on both the system board and the memory card to latch valid addresses from the CPU.

Parity errors in memory expansion options or elements in the I/O channel are provided by a channel check line. An I/O channel ready (I/O CH RDY) line can be used by a slow device to extend processor machine cycles by an integral number of clock cycles. One of the 62 lines is not used.

8-3 DISKETTE FORMAT

Examined here are the diskettes and their drive circuits. Figure 8-8 is a photograph showing the DOS manual, two diskette drives, and the diskette adapter. The two drives are located at the right side of the system unit, which is without its protective metallic jacket, and the adapter has been removed from its channel slot.

Diskettes are also called *floppy disks*. When the electric power to the computer is switched off, all data bits of the volatile memory are lost. However, those of a diskette remain stored. Accordingly, the diskette is convenient for permanent storage of programs and data. A *one-sided* diskette has a formatted capacity of 180 KB, and a *double-sided* one stores 360 KB. Both single-sided and double-sided drives are available. The computer can read from or write to a diskette, with writing

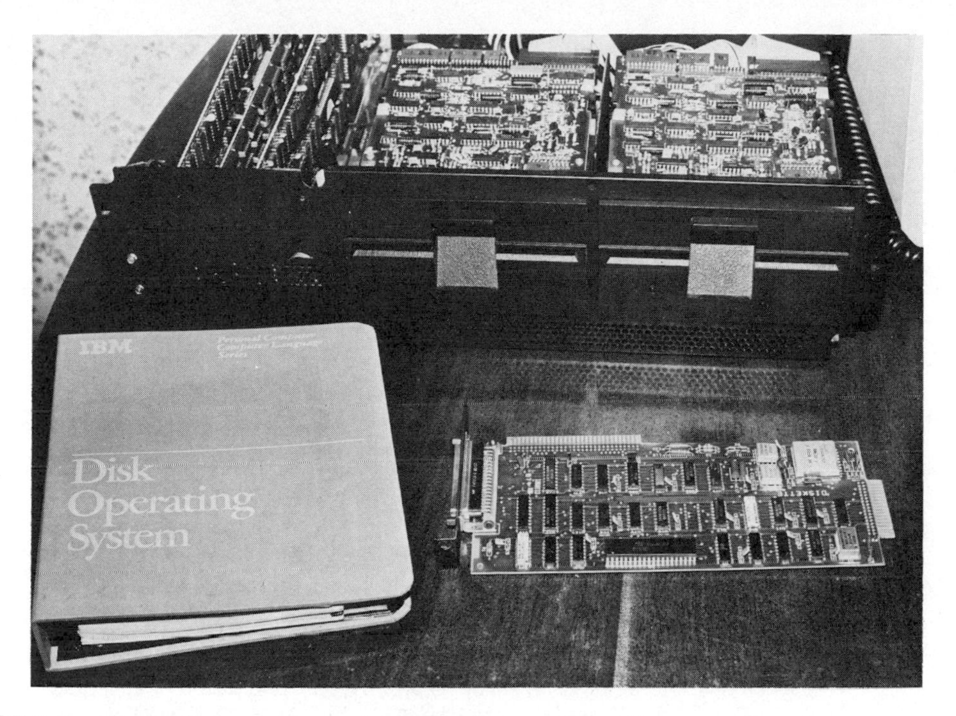

FIGURE 8-8. The DOS manual, two diskette drives, and the diskette adapter.

automatically replacing old data. Although the ROM is nonvolatile, as is the diskette, it cannot store new data.

Reading from or writing to a diskette involves a system that utilizes the CPU, the timer/counter chip, the interrupt controller, the DMA controller, the floppy disk controller, and other circuits. Some LSI circuits contain most of these functions on a single chip. Interaction between the various functions is presented in detail in the programming example of Section 8-5. Another especially interesting feature is the interfacing between computer electronics and the mechanical drives of the diskettes, including control of the *magnetic head* that reads and writes the flux patterns.

The Diskette

The 5.25-inch circular diskette is thin, flexible, and coated with a metallic oxide that can store digital data serially in the form of magnetic flux patterns. A top view of a diskette within its protective cardboard jacket is shown in Fig. 8-9. At the center is a circular opening in the envelope, within which is a smaller hole in the diskette, providing access for the spindle system that rotates the diskette. With the diskette inserted into one of the two slots at the front of the system unit, closing the latch at the opening of the slot positions the diskette and clamps it to the drive hub that projects through its center opening. For a single-sided diskette, recording head #0, which can read and write magnetized spots, is pressed against the bottom side, or *side 0,* of the disk at the outer edge of the elongated head aperture of the envelope. A double-sided diskette has head #1 pressed against the aperture of the top side, called *side 1*.

When operating, the diskette spins within its jacket at a rate of five revolutions

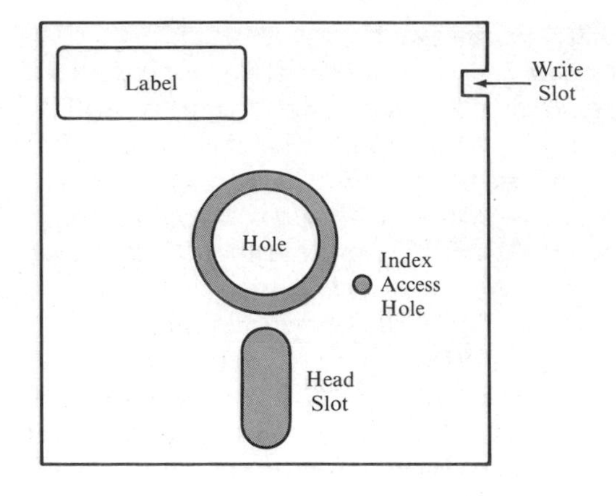

FIGURE 8-9. Circular diskette in its square envelope.

per second, driven by a servo-controlled DC motor. A soft fabric lining continuously cleans it. In the envelope near the center region is the small circular *index access hole*. Once each revolution, an *index hole* in the diskette appears within this aperture, and light from a light-emitting diode (LED) passes through the two aligned holes to a phototransistor. One pulse per revolution is generated and transmitted to the index pin of the floppy disk controller chip.

The presence of the rectangular slot at the right side of the illustrated diskette of Fig. 8-9 indicates that writing is allowed. To protect a diskette from accidental erasure, a plastic tab is used to close the indentation. This tab depresses a mechanical switch when the diskette is in position, which instructs the diskette controller to disable the write circuitry. Diskettes with purchased programs or those with important data should be *write-protected*. Accidental overwrites of programs do occur. When a program with a specified filespec is copied from memory to a diskette that is not write-protected, *it will overwrite only one having the same filename*. Otherwise, it is written into empty space. If there is insufficient space available, no writing takes place, and an appropriate message is printed.

Tracks and Sectors. Each storage side of a diskette is divided into circular *tracks*, or *cylinders*. There are 48 tracks per inch, but only 40 tracks of a side are used. Each stores 4.5 KB of data. Commencing at the outer edge of the head slot, the tracks are numbered 0 through 39, and the track address is its number along with the side number. Number 0 refers to both the bottom side and the head, and number 1 addresses the top side and the head. The physical index hole identifies the start of a track, and a track is completely scanned by its head in one revolution. Head movement from one track to another is implemented with a step motor.

The tracks are subdivided into *sectors*, numbered in order from 1, and a sector number serves as an internal address. On a single-sided diskette, data transfer is always by an integral number of sectors. Thus, if sectors store 512 bytes and a file has 513 bytes, the file requires two sectors for storage. The vacant space of the second one is not available for storage of other files. On double sided diskettes, the minimum allocation unit is a pair of sectors, and a 1-byte file uses two 512-byte sectors. Because the diskette is *soft-sectored*, the number of sectors in a track and the size of each sector are programmable, limited to a maximum of nine sectors. In *hard-sectored* diskettes, these parameters are fixed by equally spaced index holes in the jacket, each of which signals the start of a sector.

In addition to being soft sectored, the one- or two-sided diskette is classified as double-density. *Single-density* storage employs *frequency modulation (FM)* encoding, in which each data bit is preceded by a clock pulse. In contrast, a *double-density* diskette uses a special bit-encoding scheme, referred to as *modified frequency modulation (MFM),* which places clock pulses only between successive 0s. The procedure effectively doubles the density of stored bits relative to that of single-density diskettes used in many systems. Encoding is said to be *frequency modulated* because the clock and logical-1 pulses are irregularly separated by low-level 0s.

Standard Format

The standard format of a track is shown in Fig. 8-10, along with numerical values indicating typical numbers of bytes. Identification (ID) and data fields of sectors 2 through 9 are identical to those of sector 1. Each data field is preceded by a 7-byte *ID field*, with a *gap* between the fields. The first byte of identification is a unique

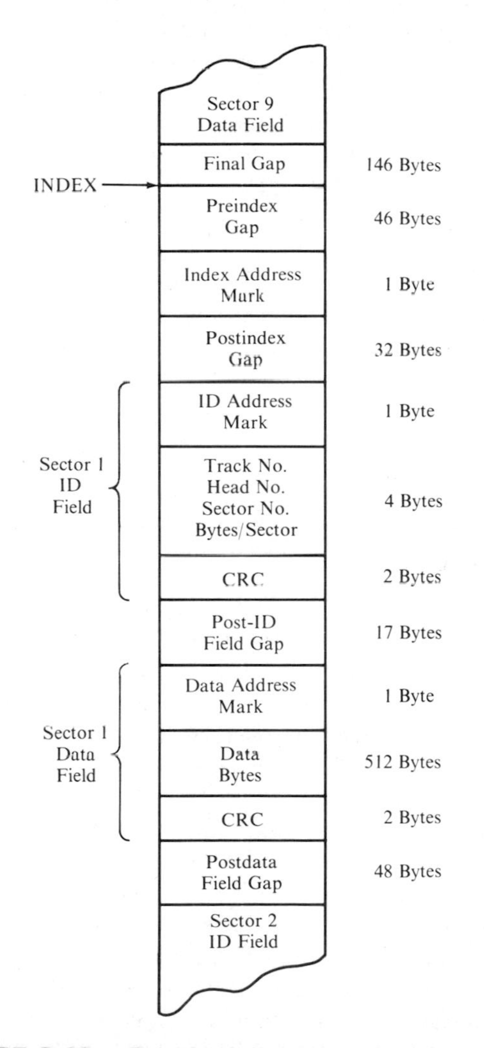

FIGURE 8-10. Standard diskette track format, with each numerical value indicating a typical number of bytes.

address mark. The next 3 bytes provide track, head, and sector addresses, followed by one indicating the sector size in bytes. Two *cyclic redundancy check (CRC)* bytes terminate the field. These bytes are generated when writing, and they are examined when reading to verify that data are being read correctly. Read and write operations use the ID field to locate the proper track and sector and to determine the number of bytes to be transmitted to or from a sector.

The data field begins with a data address mark. This information is followed by the user data, and the field terminates with two CRC bytes. The physical index mark, along with the sectors and their ID fields, are separated by gaps; these gaps compensate for mechanical and electrical tolerances of the head position and provide spacing so that drive speed changes and drive switching will not produce errors. The lengths and bytes of the gaps are programmable, but the last 6 bytes of most gaps are 0s used for synchronization.

The diskette controller is initialized during reset with essential format parameters loaded into RAM. Included are time-delay parameters for motor start and head movement, along with gap lengths and *filler bytes* of hexadecimal F6. The format specifies nine sectors per track with 512 data bytes each. During the read operation, all bits are transmitted to the data separator, including the bits of the gaps and ID fields. There are more than 40,000 bits per track, and the transfer rate is 256,000 bits per second.

Although the operator cannot change the initialization of the controller, a program can change the parameters, and diskettes can then be formatted with the new parameters. The pointer to the table is the INT 1EH vector address 0:522, which is placed in the interrupt table by the DOS program when it is booted. The parameter table at 0:522 has 11 bytes placed there initially by DOS. The original DOS system 1.0 used a table of parameters in ROM, but the updated DOS 2.1 has the more satisfactory table in RAM. Using a software interrupt to call the table allowed the parameters to be changed without replacing the BIOS ROM. The ROM table is used only during start-up.

8-4 A DISKETTE CONTROLLER

Interface between the software of the processor and the mechanical system of the dual disk drives is provided by the *diskette drive adapter* card. This card is designed to interface as many as four diskette drives, designated A, B, C, and D. Only physical drives A and B are present in the system here. The adapter and its components are examined, including the diskette controller, its registers, and its commands. The use of a simulated disk drive that can be set up in memory by a software program is also described.

Card Connections

Shown at the left side of Fig. 8-11 are connections to the adapter card. The $\overline{\text{read}}$, write, and reset signals are controlled by the CPU; address-enable AEN is $\overline{\text{AEN}}$ inverted from the wait-state generator; and lines terminal count TC and $\overline{\text{DACK2}}$ supply handshake signals from the DMA controller. These six controls are fed to the adapter from the I/O channel through the slot connector.

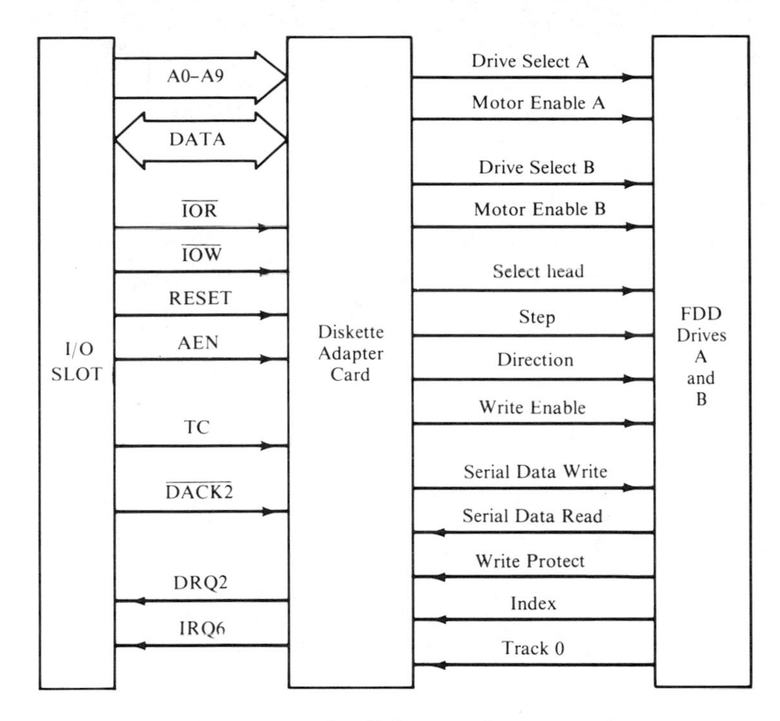

FIGURE 8-11. Connections to the diskette adapter card.

Also at the left side of Fig. 8-11 are the two outputs from the adapter to the I/O channel. They consist of a DMA request line (DRQ2) to channel 2 of the DMA controller and an interrupt request line (IRQ6) to level 6 of the interrupt controller. The level 6 interrupt is INT 0EH.

Supplied to disk drives A and B are drive-select and motor-enable lines for each of the motors, three signals used for selecting head 0 or 1 and for stepping the head inward or outward, a write-enable line, and a line for writing serial data to the selected diskette. From the diskette drive circuits to the adapter are lines for the serial data, the write-protect signal, the index pulses, and a logic state indicating whether or not track 0 is accessed. These 13 lines are shown at the right side of Fig. 8-11.

Block Diagram

A simplified block diagram of the drive adapter is shown in Fig. 8-12. Input and output lines at the left side are those of the I/O channel, and those at the right connect to the disk drives. Drives A and B are mounted within the enclosure of the system unit, and when present, physical drives C and D are external. The bidirectional data bus passes through a transceiver for buffering. Whenever the read pin RD of the *floppy disk controller (FDC)* is inactive, the direction of transfer through the transceiver is left to right, as is appropriate for write operations. An active read signal reverses this direction.

Addresses. The eight-input NAND gate at the top left of Fig. 8-12 decodes the address, with an active low-level output. To show this operation, let us suppose that the output is inactive high. Clearly, the clock input of the *digital output register (DOR)* is fixed at a high level, which disables this dynamic input. Consequently,

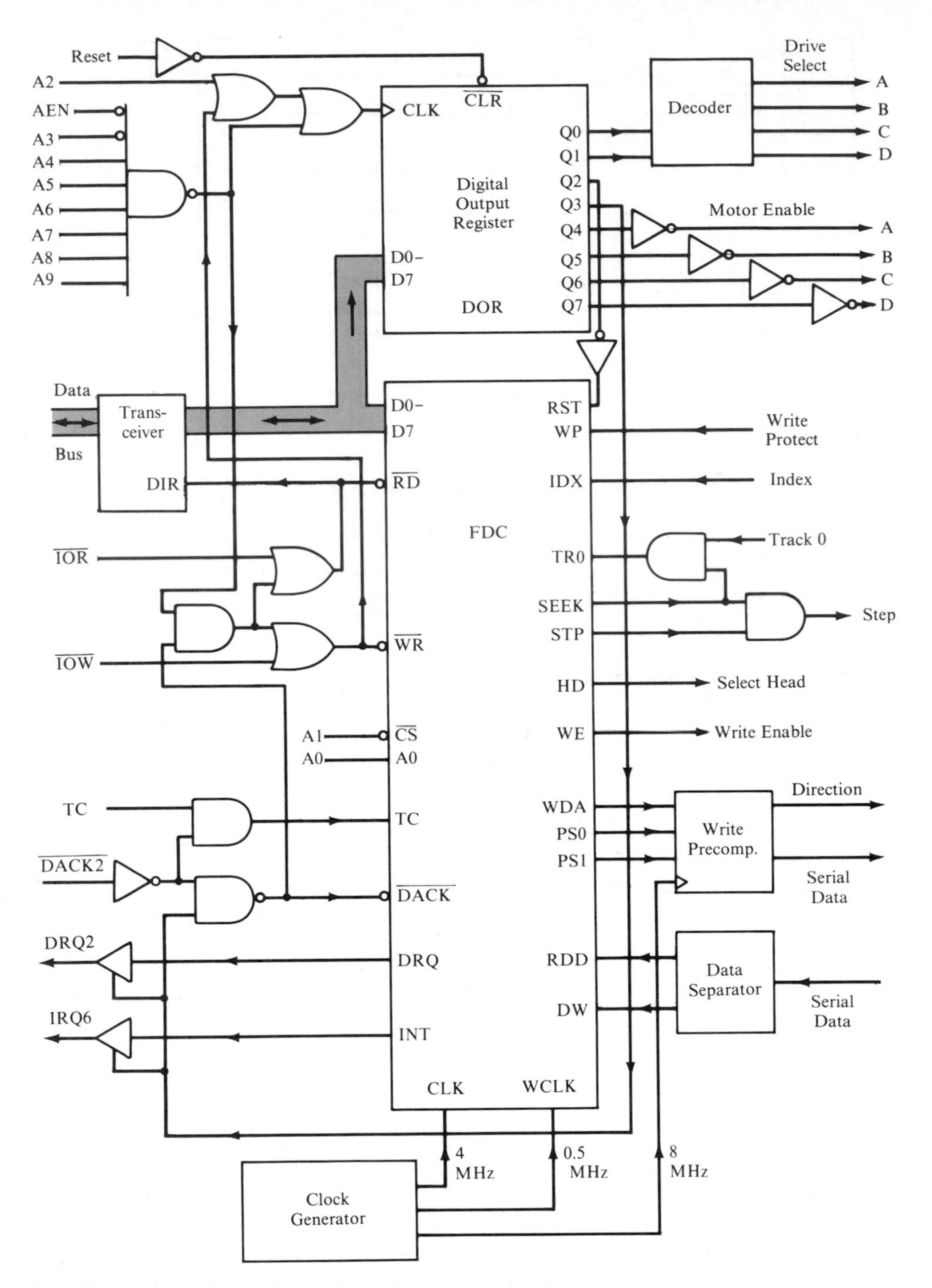

FIGURE 8-12. Simplified block diagram of the diskette drive adapter. The FDC is a NEC uPD765 chip.

the eight positive edge-triggered flip-flops of DOR are in the latch state. Furthermore, with $\overline{\text{DACK2}}$ inactive along with a high-level NAND output, both inputs to the two-input AND gate located in Fig. 8-12 between the $\overline{\text{IOR}}$ and $\overline{\text{IOW}}$ inputs are high level. The output of this AND gate supplies logical 1s to the FDC inputs $\overline{\text{RD}}$ and $\overline{\text{WR}}$, which are therefore inactive.

Accordingly, in order to write into the DOR, or to read from or write into the FDC, the output of the eight-input NAND gate must be low. Thus, the address-enable AEN must be 0 and the address bits A9-A3 must be 1111110. It follows that the addresses of the diskette adapter card are 3F0H through 3F7H.

For writing to the adapter card, the instruction OUT DX,AL is used, with the address stored in DX and the byte in AL. For the DOR, the address is 3F2H. The 0 state of bit A2, along with the 0 state at the output of the eight-input NAND gate, enable the two OR gates at the CLK input of DOR so that a low-to-high transition of $\overline{\text{IOW}}$ will activate this dynamic CLK input, causing the byte of the data bus to be written into DOR at the end of the write pulse. Furthermore, the high state of A1 feeding into the chip-select pin of the FDC floats its data bus, read and write pins, and pin A0.

Within the FDC are two 8-bit registers accessible to the data bus: the data register and the status register. For data register reading or writing, the address is 3F5H, and the read address of the status register is 3F4H. These addresses inhibit the DOR clock, enable $\overline{\text{CS}}$, and select the register determined by bit A0.

Digital Output Register DOR. Output bits 0 and 1 of DOR select drive A, B, C, or D. In order for a motor to be on, its drive select must be active and its motor-enable line must be low level. The low state of bit Q2 places the FDC in the idle state and resets its control outputs. The FDC interrupt and DMA requests are enabled by bit Q3, and the high nibble provides the motor-enable signals. All DOR outputs are cleared by a system reset.

FDC Registers. An FDC status byte essential for effective processor control is stored in the read-only *status register* of port 3F4H. Its encoding is shown in Fig. 8-13. Especially important are bits 7 and 6. Bit 7 indicates whether or not the data register is ready to receive or send data, and *the bit must be examined by a program prior to a read or write of this register*. Then bit 6 indicates whether the next operation on the data register must be a read or a write. Although not done in the program of the next section, it is good practice to check this bit also before reading or writing.

If the FDC is instructed by the SPECIFY command to operate in the non-DMA mode, then bit 5 of the status register is set and no DMA requests are issued by the FDC. In the PC, diskette transfers use DMA, and bit 5 is reset. Bit 4 indicates if a read/write operation is in progress. A bit of the low nibble is set if the corresponding floppy disk drive (FDD) is busy moving the head to a specified track in response to the SEEK command.

The read/write *data register* of port 3F5H is actually one register of an array arranged as a stack. It is defined as *the one currently accessible to the data bus*.

Bit 7	(Data register ready)	0 no;	1 yes
Bit 6	(Data register R/W)	0 write;	1 read
Bit 5	(non-DMA mode)	0 no;	1 yes
Bit 4	(R/W in progress)	0 no;	1 yes
Bit 3	(Drive 3 seek mode)	0 no;	1 yes
Bit 2	(Drive 2 seek mode)	0 no;	1 yes
Bit 1	(Drive 1 seek mode)	0 no;	1 yes
Bit 0	(Drive 0 seek mode)	0 no;	1 yes

FIGURE 8-13. Encoding of the FDC status register of port 3F4H.

Some registers of the stack store status information supplementing that of the main status register of port 3F4H, and others store parameters, commands, and results. Also, the data register is the source or destination of the data bus during DMA transfers. The controller can execute 15 different commands. For most of them, the results depend on certain parameters stored in the stack, which must be written into the data register prior to execution of the command. Parameters specify sector ID and other information.

Write Precompensation. Figure 8-12 shows a *write precompensation* block. Write precompensation is a technique used to reduce the detrimental effect of the interference of magnetic flux reversals on one another. It is accomplished by introducing bit shifts into the write data stream that are equivalent but opposite to those of the diskette.

Data Separator. Bits read from a diskette include not only the data bytes but also synchronizing and clock bits, the specified bit patterns of gaps, and ID information in the form of bytes representing the cylinder number, the head number, the sector number, and the physical length. An entire sector is read at a time. The *data separator* chip provides a data window that separates the data from the composite mixture of bits. Separated data bits enter the FDC controller at the data window pin DW, whereas all bits that were read enter pin RDD. An 8-bit shift register within the FDC transforms each serial byte into parallel form.

FDC Commands

Some commands, such as READ DATA, READ ID, and READ A TRACK, transfer data from a diskette to the main system, and there are various WRITE commands. Included in the set are SCAN commands that compare on a byte-by-byte basis data stored in memory with data that are read from a diskette. The SEEK command positions the head over the proper track of the diskette. Others are RECALIBRATE, which returns the head to track 0, and SENSE DRIVE STATUS, which provides status information about the FDD. The FORMAT A TRACK command formats one complete track into specified sectors containing a prescribed number of bytes. Before a new diskette can be used, all tracks must be formatted.

Initialization of the diskette system is implemented by a diskette system reset. This operation causes appropriate data to be written to the DOR and the SPECIFY command to be executed. SPECIFY sends to the data register the 3 bytes 03H, DFH, and 02H. The first of these is the command, the next one specifies a step-rate time of 26 milliseconds and a head-unload time of 480 milliseconds, and the last one prescribes a 4-millisecond head-load time and DMA operation. The command RECALIBRATE must be executed for each drive prior to the first read or write operation. It initializes the position of the head, moving it to the start of track 0.

Results Phase. Following execution, most commands have a *results phase,* initiated either by successful execution of the command or by an error occurring during execution. In this phase the CPU must fetch the results from the data register of the FDC. They must be read without delay, at the time the head is at the sector ID field following the last one read; otherwise the diskette system must be reinitialized before a new command will be accepted. On completion of an operation, the next-sector ID information is stored, along with 4 status bytes, in status registers of the data stack.

The Floppy Disk Drive

Each floppy disk drive (FDD) has spindle and step motors, along with associated analog and digital electronics, on a circuit board mounted above the drive. A pulse on the STEP input line of Fig. 8-11 moves the magnetic head one track in or out, with the direction depending on the state of the DIRECTION input line. When WRITE ENABLE is active, a low-to-high transition of the WRITE DATA input causes a flux change to be placed by the head on the diskette. In the read mode, flux changes encountered by the head generate pulses that are supplied to the READ DATA output. Included in the electronics are a read amplifier, a differentiator, a zero-crossing detector, and digitizing circuits for pulse generation. Present are discrete circuit components, transistors, and diodes, as well as IC operational amplifiers and logic gates.

Simulated Disks

Inexpensive software is available that allows the designation of a portion of memory for use as a *simulated disk*. When this operation is performed, the "memory disk" is designated as diskette drive A, B, C, or D as desired, and the number of bytes reserved for storage is specified in the command setting up the memory disk. The reserved space can no longer be used for loading and running programs, but a new memory disk is available. Because the RAM disk has certain superior properties, it is sometimes called an *emulated* disk.

To illustrate the advantages of such a disk, consider the preparation of the manuscript of this book. It was typed using the copy-protected word processor program Easywriter, which is stored on a diskette. When booted on drive A, Easywriter requests the user to insert the work diskette in drive B and press any key when ready. Creating and editing files on the work diskette involve frequent transfers of data between memory and the diskette of drive B. Each transfer requires a spin of the diskette, accompanied by a short time delay, wear on the diskette, and some noise. These problems are eliminated by creating a RAM disk and designating it as drive B.

To accomplish this operation, a command is given following the boot of DOS to set up a simulated disk in RAM, designated as drive B, with physical drive B changed to drive C. Next, any files already present on the work diskette of physical drive C are copied to memory drive B, using the DOS copy command. Then the word processor is booted from drive A.

There are now three effective drives. Easywriter is present in drive A, with its major operating routines stored in normal user memory. The simulated work diskette resides in the RAM disk, and the actual work diskette is located in physical drive C. All file editing, creation, moving, printing, and so forth are done using the memory disk. Operation is fast and quiet. In fact, data transfers between the user memory and the simulated diskette appear to be instantaneous. At the end of the session, the updated files of memory disk B are copied to physical disk C for nonvolatile storage.

There are many other types of programs that require frequent transfers of data between a diskette and memory, and the simulated diskette can be used to advantage. More than one RAM disk can be set up if desired, with the byte storage of each specified. However, the system supports only four drives with designations A, B, C, and D. Thus, with a two-drive system, the number of simulated drives is limited to two.

8-5 PROGRAM DISK_READ

A specific example involving data transfer between a diskette and memory is examined in this section. Suppose it is desired to read the 1024 bytes of sectors 3 and 4, track 9, side 0, of diskette B, with the bytes to be stored in memory starting at absolute location 08000H. The 205-byte program is named DISK_READ.COM, and for reference, the statements are numbered.

The desired transfer is easily accomplished with INT 25H (absolute disk read), which uses INT 13H (diskette I/O) of the BIOS ROM (see Prob. 8-15). However, these interrupts will not be used, because the objective here is *to describe the handshake signals and interaction between the processor, the FDC, the DMA controller, the interrupt controller, and timer 0*. Initialization of these chips as required for the specific program is included, and steps of the program are followed and examined in detail.

Execution of the SPECIFY command is required prior to the read or write operation of a diskette, but once executed, it does not have to be repeated. Within the initialization that follows a system reset, SPECIFY is executed. The specified data apply to all drives. The command is not needed and is not included in the program developed here.

During the first read or write operation of a diskette after a reset, DOS automatically executes the command RECALIBRATE. The routine is within INT 13H of the BIOS ROM. It must be executed for each drive before the drive can be used for reading or writing, but once executed for a drive, it does not have to be repeated. RECALIBRATE is included in the program, which allows it to run properly even if drive B has not been used.

The Data Segment

In Fig. 8-14 is the title and data segment. The symbolic names are similar to those used in the listing program of the BIOS ROM, given in the IBM *Technical Reference* manual.

```
1    Title DISK_READ
2    ;
3    DATA SEGMENT AT 40H
4      ORG 3EH
5        SEEK_STATUS    DB    ?
6      ORG 40H
7        MOTOR_COUNT    DB    ?
8      ORG 42H
9        FDC_STATUS     DB    7 DUP(?)
10   DATA ENDS
```

FIGURE 8-14. Title and data segment of DISK_READ.

The MSB of SEEK_STATUS is used as an interrupt flag, which is designated INT_FLAG. Whenever INT 0EH is activated by the FDC through level 6 of the interrupt controller, this bit is set. In fact, INT 0EH does nothing else. INT_FLAG can be reset by the instruction *AND SEEK_STATUS, 7FH*. The other bits of SEEK_STATUS have no significance to DISK_READ.

MOTOR_COUNT was encountered in the timer interrupt (INT 8) routine of Fig. 5-29. As shown in this routine, each tick of the internal clock decrements the byte of MOTOR_COUNT, and when MOTOR_COUNT becomes 0, byte 0CH is sent to the DOR of the FDC. Examination of Fig. 8-12 reveals that writing 0CH to DOR disables all drive motors and enables FDC outputs IRQ6 and DRQ2 and input $\overline{\text{DACK2}}$. Whenever interrupts are enabled, MOTOR_COUNT is decremented at a rate of 18.2 per second.

After a diskette read or write operation, 7 bytes of results must be read from the data register of the FDC. The 7 byte locations of FDC_STATUS are reserved for their storage.

Procedure FDC_OUT

Following the DATA segment is the CODE segment. It begins with an ORG 100H statement, an ASSUME statement, and a jump over two procedures to label A4 of the main program. These statements and the procedure FDC_OUT are shown in Fig. 8-15.

```
11   ;
12   CODE   SEGMENT
13      ORG   100H
14      ASSUME   CS:CODE, DS:DATA
15       A1: JMP A4                        ;jump to main routine
16   ;
17      FDC_OUT   PROC   NEAR
18           MOV   DX, 3F4H                ;FDC status-reg addr to DX
19       A2: IN    AL, DX                  ;get status
20           TEST AL, 80H                  ;is data register ready?
21           JZ    A2                      ;if not, recheck status
22           MOV   AL, AH                  ;move byte to AL
23           INC   DX                      ;data-reg addr to DX
24           OUT   DX, AL                  ;send byte to FDC
25           RET
26      FDC_OUT   ENDP
```

FIGURE 8-15. The start of code and the FDC_OUT procedure.

The NEAR procedure FDC_OUT provides for *writing the byte of register AH into the data register of the FDC,* which indicates when it is ready by setting bit 7 of the status register. This bit is tested, and the program stays in a loop until it is set. Then the byte of AH is transferred to AL and sent to the data register. It is called 14 times.

Procedure WAIT_FOR_INT

Immediately after execution of the RECALIBRATE, SEEK, and READ commands, the program must wait until the FDC signals that it is ready. This is done

```
27   ;
28   WAIT_FOR_INT PROC NEAR
29    A3: TEST SEEK_STATUS, 80H
30        JZ  A3
31        AND SEEK_STATUS, 7FH
32        RET
33   WAIT_FOR_INT ENDP
```

FIGURE 8-16. The wait-for-the-interrupt procedure.

by setting the INT_FLAG by means of INT 0EH, which is activated through level 6 of the interrupt controller. The WAIT_FOR_INT procedure of Fig. 8-16 examines the INT_FLAG *and loops until it is set.* Then it clears the flag and returns to the main program.

Program Start and DMA Setup

The main routine starts at label A4 with STI. Without this instruction there would be no interrupts from levels 0 and 6, and the program would not function properly. Registers are saved, and DS is initialized to value 40H of the DATA segment. The eight OUT instructions to the DMA clear the first/last flip-flop, specify the single-transfer mode, write the 20-bit address 08000H to the page and base address registers, supply the count 5FFH to the count register, and clear the channel 2 mask bit. The routine is shown in Fig. 8-17.

```
34  ;
35  ; Set Interrupts, Save Registers, and Initialize DS
36    A4: STI
37        PUSH AX
38        PUSH BX
39        PUSH CX
40        PUSH DX
41        PUSH DI
42        PUSH DS
43        MOV  AX, 40H
44        MOV  DS, AX         ;initialize DS to 40H
45  ;
46  ; Program the 8237 DMA Controller
47        MOV  AL, 46H
48        OUT  0CH, AL        ;clear first/last flip-flop
49        OUT  0BH, AL        ;single-transfer mode
50        MOV  AL, 0
51        OUT  81H, AL        ;four 0s to page reg (Ch 2)
52        OUT  4, AL          ;low byte 00H to addr reg
53        MOV  AL, 80H
54        OUT  4, AL          ;high byte 80H to addr register
55        MOV  AL, 0FFH
56        OUT  5, AL          ;low byte FFH to count register
57        MOV  AL, 3
58        OUT  5, AL          ;high byte 05H to count register
59        MOV  AL, 2
60        OUT  0AH, AL        ;clear channel 2 mask bit
```

FIGURE 8-17. Program start and DMA setup.

In the routine, the line 48 OUT instruction to port 0CH clears the first/last flip-flop with a source operand that can be any value whatsoever. Byte 46H is used. It is also written to the mode register of port 0BH, specifying the *single-transfer* mode with a write to memory via DMA channel 2. In addition, it specifies that the address is to be incremented after each byte is moved.

The 20-bit destination address 08000H is placed in the page and base address registers having respective port addresses of 81H (for channel 2) and 04H. Because the two sectors to be read contain 1024 data bytes, the count register of port 05H is loaded with the initial count of 03FFH, or 1023 decimal, which is decremented after a transfer. The last OUT instruction enables channel 2.

Single Transfer Mode

During a read or write operation, the diskette spins at the rate of five revolutions per second. Specifications indicate that bits are read at a rate of 32,000 bytes, or 256,000 bits, per second. Thus, the time required to read a byte is about 31 microseconds. Only five clocks, or 1 microsecond, are needed to move a byte to memory, leaving a time interval of about 30 microseconds between DMA transfers. The interval is longer when the head is not actually over a data field, but instead, over an ID or gap field. An interval of 30 microseconds corresponds to 143 clocks.

By using the single-transfer mode, *the processor controls the buses during the intervals between the transfers of bytes*. Also, other DMA operations can be implemented. In particular, memory refresh on DMA channel 0, which has the highest priority, must be allowed to occur every 72 clocks, or every 15 microseconds. *These memory refreshes take place between diskette transfers.*

After the READ command for the FDC has been executed, the processor polls the INT_FLAG until it indicates that all bytes have been transferred. During this time interval, bytes are being read from the diskette and moved to memory under the control of the DMA chip. *Only at the time of an individual transfer does the processor become momentarily idle.*

Let us consider an individual transfer. When a byte from the spinning diskette is placed in the data register, the DMA request line DRQ2 from the FDC to the DMA controller is driven active, which causes the DMA controller to send a hold-request HRQ to the wait-state generator. The transceivers and the bus controller float the buses and control lines, the peripheral address decoder is disabled, and a hold-acknowledge $\overline{\text{HLDA}}$ signal is returned to the DMA chip. During DMA state S2, active signal $\overline{\text{DACK2}}$ is sent to the FDC. Simultaneously, the DMA controller puts the memory address on the address bus, as shown in Fig. 8-3, and issues an active $\overline{\text{MEMW}}$ signal.

The DMA acknowledgment $\overline{\text{DACK2}}$ to the FDC degates the original request DRQ2 and indirectly gates the terminal count input TC to the interrupt output pin INT, which feeds output line IRQ6. It also places the byte of the data register on the data bus, and when the active $\overline{\text{MEMW}}$ pulse from the DMA controller goes high, the byte is latched into the addressed memory location. Latching occurs during DMA state S4, which is shown in the DMA timing diagram of Fig. 8-2. During this operation, the CPU has five wait states.

When the next byte is ready, the procedure is repeated. Finally, the last of the 1024 bytes is sent, and terminal count TC from the DMA chip to the FDC goes active. This action generates a high signal on pin INT, which activates INT 0EH through level 6 of the interrupt controller.

```
61  ;
62  ; Clear the INT_FLAG and Turn On the Motor
63         AND  SEEK_STATUS, 7FH     ;clear INT_FLAG
64         MOV  MOTOR_COUNT, 5 * 18 ;5-sec maximum on-time
65         MOV  DX, 3F2H             ;DOR address
66         MOV  AL, 2DH              ;byte 2DH for DOR
67         OUT  DX, AL               ;motor starts now
```

FIGURE 8-18. The routine to clear INT_FLAG and start the motor.

INT_FLAG and Motor Start

It is important to clear the INT_FLAG before execution of the RECALIBRATE, SEEK, and READ sequence, so that it can be set at the appropriate time by INT 0EH. The reset is done by the instruction of line 63. Then the motor of drive B is turned on, with MOTOR_COUNT initialized to 90D. This value allows about 5 seconds before normal turn off, which is considerably more time than needed for the READ operation. In addition to turning on the motor, byte 2DH to DOR enables FDC outputs IRQ6 and DRQ2 and input $\overline{\text{DACK2}}$ from the DMA controller, as shown in Fig. 8-12. The routine is presented in Fig. 8-18.

Recalibrate

The RECALIBRATE command is implemented by writing bytes 07H and 01H in sequence to the data register of the FDC. The first of these bytes is code for RECALIBRATE, and the second one identifies the drive. *After the head has been located at the start of track 0,* the FDC requests IRQ6, and INT 0EH is executed. This operation sets bit 7 of SEEK_STATUS, which is INT_FLAG. The call to the WAIT_FOR_INTERRUPT procedure of line 74 puts the program into a WAIT loop until RECALIBRATE is done. The command is essential only when there has been no prior read or write operation on drive B since the last reset. The routine is that of Fig. 8-19.

SEEK

Next is the SEEK command, which moves the head to track 9. After the required 3 bytes are written to the FDC data register, the program must wait until command execution has been completed. This action is indicated by a level 6 interrupt. Following the interrupt is a DELAY loop, which allows 2 milliseconds for the head to settle into place. Shown in Fig. 8-20 is the routine.

```
68  ;
69  ; Recalibrate Drive B
70          MOV   AH, 7
71         CALL   FDC_OUT
72          MOV   AH, 1
73         CALL   FDC_OUT
74         CALL   WAIT_FOR_INT
```

FIGURE 8-19. The RECALIBRATE routine.

```
75   ;
76   ; Execute the SEEK Command
77          MOV   AH, 0FH              ;SEEK command code
78         CALL   FDC_OUT
79          MOV   AH, 1                ;drive B
80         CALL   FDC_OUT
81          MOV   AH, 9                ;track 9
82         CALL   FDC_OUT
83         CALL   WAIT_FOR_INT         ;wait for INT 0EH
84      A5: MOV   CX, 550
85      A6: LOOP  A6                   ;2 millisecond delay
```

FIGURE 8-20. The SEEK command and head-settle delay.

The Read Operation

Nine bytes must be written into the data register of the FDC in order to activate the command READ. The routine is shown in Fig. 8-21. The significance of each byte is indicated in the comments of the routine. Byte 8 is the programmable value of gap #3, which is the one between sectors. The 42 bytes of this gap, specified in the disk parameter table pointed to by INT 1EH, do not include the final 6 bytes of 0s.

The last byte, 0FFH, has no significance in this case, but it must be written. *After this operation, the command is executed automatically, and the program then waits for the signal from the FDC indicating completion of the read operation.* As in the case of both RECALIBRATE and SEEK, this signal is an interrupt request through level 6 of the 8259 interrupt controller. INT 0EH is activated, which sets the INT_FLAG of SEEK_STATUS.

Data transfer commences as soon as the ninth byte of the READ command has been received by the data register. The single-transfer mode of DMA is utilized. During byte transfers the CPU relinquishes control of the buses, but between trans-

```
86   ;
87   ; Execution of the READ Command
88          MOV   AH, 66H        ;READ command
89         CALL   FDC_OUT
90          MOV   AH, 1          ;drive B
91         CALL   FDC_OUT
92          MOV   AH, 9          ;track 9
93         CALL   FDC_OUT
94          MOV   AH, 0          ;head 0
95         CALL   FDC_OUT
96          MOV   AH, 3          ;start sector 3
97         CALL   FDC_OUT
98          MOV   AH, 2          ;code for 512 bytes per sector
99         CALL   FDC_OUT
100         MOV   AH, 9          ;last sector of track
101        CALL   FDC_OUT
102         MOV   AH, 2AH        ;gap #3 length = 42 bytes
103        CALL   FDC_OUT
104         MOV   AH, 0FFH       ;no significance
105        CALL   FDC_OUT
106        CALL   WAIT_FOR_INT
```

FIGURE 8-21. Routine for the READ command.

fers it has normal control and allows other units to implement DMA operations. The WAIT_FOR_INT procedure runs, except for brief intervals when buses float. *Periodic timer interrupts and memory refresh operations are implemented normally.*

In the program here the CPU does no other processing, but other programs can be used that provide for additional services. The WAIT_FOR_INT procedure can be redesigned, with desired processing implemented from within the loop. For example, between byte transfers it is possible to read data received from a communications line or to send data to a printer. The only requirement is that *the routine must poll INT_FLAG periodically,* at least once every 30 microseconds, while diskette reading is in progress.

After execution of the READ command, the program stays in the WAIT_FOR_INT procedure, and INT_FLAG is examined between each DMA transfer of a byte. *After the last byte is moved, the terminal count TC line from the DMA controller to the FDC goes high.* When this count of 0 is received, the FDC prepares 7 bytes of results for reading from the data register. It also loads its status register with appropriate data. Then interrupt request IRQ6 goes to the interrupt controller, INT_FLAG is set, and the READ routine terminates with the return from the WAIT_FOR_INT procedure of line 106.

Results

While the motor is still running, *7 bytes of results must be read from the data register of the FDC.* These data should be stored in the memory locations of FDC_STATUS. Prior to each reading of the data register, it is necessary to wait until it is ready, as indicated by a logical 1 state of bit 7 of the FDC status register. Therefore, the FDC status register is read first, and if bit 7 is set, the data register is read. This process is repeated seven times. The routine is given in Fig. 8-22.

The results must be read without delay, while the head is still reading ID information from the diskette. Otherwise reinitialization of the FDC is required prior to a subsequent read or write operation, which is done with execution of the SPECIFY and RECALIBRATE commands.

In order, the results read after program execution are bytes 1, 0, 0, 9, 0, 5, and 2. The first 3 are status bytes indicating drive B and no errors. Byte 9 denotes the

```
107 ;
108 ; Read the Results from the FDC Data Register
109          MOV  DI, OFFSET FDC_STATUS
110          MOV  BL, 7                    ;byte count
111          MOV  DX, 3F4H                 ;status register addr
112     A7:  IN   AL, DX                   ;get status
113          TEST AL, 80H                  ;data register ready?
114          JZ   A7                       ;if not, wait
115          INC  DX                       ;data register addr to DX
116          IN   AL, DX                   ;read data register
117          MOV  [DI], AL                 ;store byte in RAM
118          INC  DI                       ;increment RAM pointer
119          DEC  DX                       ;point to status port
120          DEC  BL                       ;decrement byte count
121          JNZ  A7                       ;loop seven times
```

FIGURE 8-22. Routine for reading the results.

present track location of the head, byte 0 is the current head address corresponding to side 0, byte 5 is the present sector location of the head, and the last one is code 2 for 512 bytes per sector. After the last byte is read, bit 6 of the status register changes from 1 to 0, indicating that the data register is now ready to receive a new command. Although not done here, it is appropriate to examine this bit before reading from or writing to the data register.

Motor Turn-Off and Terminate

The final routine is that of Fig. 8-23. Although the motor count was set for a maximum of 5 seconds, only about 1 second is actually required. Rather than allow the diskette to spin considerably longer than necessary, it is good practice to write a new count to byte location MOTOR_COUNT. The 2-second turn-off time of line 124 is chosen to keep the motor running awhile in case the read operation is followed by another diskette operation. Diskette reads and writes often occur in sequence, and the 2-second turn-off time *eliminates frequent starting and stopping*.

```
122 ;
123 ; Motor Turn-Off and Terminate
124          MOV   MOTOR_COUNT, 2*18    ;2 seconds
125          POP   DS
126          POP   DI
127          POP   DX
128          POP   CX
129          POP   BX
130          POP   AX
131          INT   20H                  ;terminate
132 CODE    ENDS
133 END     A1
```

FIGURE 8-23. Routine for motor turn-off and program terminate.

Diskette Write

With only minor modification, the program can be used for writing to the diskette (see Prob. 8-13). A 0.5-second delay must be inserted into the program immediately after the motor is turned on, at line 68. This delay can also be included in the READ program even though it is not needed.

There are two additional changes that must be made in the program. The DMA command 46H of line 47 must be replaced with 4AH, and the FDC READ command 66H of line 88 must be changed to the WRITE command 45H. With these changes, the new program will write 1024 bytes from memory at start location 0:8000 to sectors 3 and 4 of track 9 of side 0. Of course, the drive number, the track and sector numbers, and the memory address can be changed as desired.

INT 13H

The diskette I/O routine of INT 13H, which is stored in the BIOS ROM, is available for reset, status read, read, write, verify, and format operations. Although the READ routine is somewhat similar to the one presented here, it has checks at critical points with error indicators in event of failure, and there are timed exits from loops that avoid the possibility of endless looping in case of failure.

In addition, INT 13H maintains the status of memory variable MOTOR_STATUS, setting or resetting the appropriate bit when a drive is turned on or off. When a motor needs to be on, this variable is checked to see if it is already on, in which case the motor turn-on operation is bypassed. The operation to be implemented is specified by the byte of register AX when the routine is invoked, and the drive, head, track, and sector numbers are stored in CX and DX, with the number of sectors stored in AL. The memory address is that of ES:BX.

The first instruction of the program is STI, which turns the interrupts back on. As in the program examined here, this operation is essential, because the routine uses timer INT 8 through level 0 of the interrupt controller and also INT 0EH from the FDC through level 6. Both the timer interrupts and DMA memory refresh operations occur at periodic intervals between diskette byte transfers. When a memory refresh is implemented by timer 1 through DMA channel 0, signal $\overline{\text{DACK2}}$ is inactive, and therefore the terminal count TC of the refresh operation has no effect on the FDC.

8-6 DISK SPACE UTILIZATION

Unless otherwise stated, diskettes are assumed to be double-sided, with nine 512-byte sectors per track. Each side has 40 tracks numbered 0 through 39, and the sectors of a track go from 1 through 9. *Logical numbers* are used to identify sectors, with the first 27 logical numbers assigned as follows:

0 through 8 —sectors 1–9 of track 0 (side 0)
9 through 17—sectors 1–9 of track 0 (side 1)
18 through 26—sectors 1–9 of track 1 (side 0)

The sequencing continues, always going from a track of side 0 to the corresponding track of side 1. For single-sided diskettes, the numbers advance continuously through the tracks. This method also applies to tracks having only eight sectors.

Clusters

Diskette space is allocated by *clusters,* each of which is assigned a 12-bit number. All 12-bit numbers encountered in this section are represented by three hexadecimal digits without the H indicator. For one-sided diskettes, a cluster is exactly the same as a sector, and for two-sided ones, *it consists of two sectors with consecutive logical numbers*. Each file has a *starting cluster number* that is present in its listing in the diskette directory. The first two cluster numbers, 000 and 001, are reserved for special use to be described later, and the boot record, the file allocation table, and the directory, which precede the data area of a diskette, have no assigned numbers. Accordingly, *the first cluster of the data and program area has number 002*. If a diskette contains the system files, IBMBIO is the first file following the directory, and its starting cluster number is 002.

Clusters proceed in sequence through the sectors, following the logical numbers.

For identical track numbers, side 0 precedes side 1. For example, suppose cluster 005 contains sectors 7 and 8 of track 2 (side 1). Then cluster 006 has sector 9 of track 2 (side 1) and sector 1 of track 3 (side 0). Sectors 2 and 3 of track 3 (side 0) become cluster 007. Logical and cluster numbers are related as follows:

Logical number = (cluster number − 2) * (sectors/cluster)
+ logical number of first data sector

When data of memory are written to a new diskette file, clusters are selected starting with the lowest one available, *jumping over any clusters encountered that have already been assigned to files*. Normally, the clusters of a file are in sequence unless there has been considerable erasure and creation activity. For example, suppose data are being written to cluster 100 of a file, with cluster 009 having been vacated by an erasure. Then when cluster 100 is filled, the next cluster in the file sequence becomes 009, assuming no clusters with lower numbers are available.

DOS Diskette Areas

When DOS is booted, the bootstrap loader of the ROM BIOS writes the *boot record* from the DOS diskette into memory at 0:7C00 and then transfers control to it. The bootstrap loader was presented in Fig. 6-21. The boot record checks the diskette directory to determine that files IBMBIO and IBMDOS are present at the proper locations. These programs, and also COMMAND, are read into memory, and control is passed to the command processor. System programs are described in Section 11-7.

Listed in Fig. 8-24 are diskette programs and data of DOS, excluding the COM and EXE utilities classified as external commands. In the order listed, the boot record, the file allocation table (FAT), and the directory are present on every formatted diskette, using the lowest numbered sectors. Files IBMBIO, IBMDOS, and COMMAND must be present on the diskette used to boot the system. However, because they are loaded into memory by the boot record, they are not required on other diskettes. If IBMBIO and IBMDOS are present, *they must be in order immediately following the directory*. The locations of COMMAND and the external commands are unrestricted.

File Allocation Table

The *file allocation table (FAT)* has 12-bit entries corresponding to each of the possible cluster numbers of the diskette, from 000 through 163 hexadecimal. Thus,

1. Boot record —Always present on track 0, sector 1, side 0 of every formatted diskette.
2. FAT —File allocation table (two copies).
3. Directory —File names and other information.
4. IBMBIO.COM —Interface to ROM BIOS device routines.
5. IBMDOS.COM —High-level interface for user programs.
6. COMMAND.COM—A program called the *command processor*.

FIGURE 8-24. DOS diskette areas in order.

the table must contain 533 bytes, requiring two sectors. Because of the importance of the table, *two copies are maintained in case a defect occurs in one*. The four sectors of FAT have logical numbers 1-4. *FAT links the clusters of each file in the proper order and identifies those that are unused and available*. The set of binary numbers map the total space of the diskette.

The first two numbers, which correspond to clusters 000 and 001, are FDFFFF, assuming a double-sided diskette with nine sectors per track. For a single-sided diskette the code is FCFFFF, and other codes are used for other formats. *Thus, the first two clusters provide information about the format*. After these six digits come the three-digit words corresponding to the data clusters. As before, all 12-bit words are given in hexadecimal form.

Word 000 indicates that the corresponding cluster is not used and available, non-0 word XXX is the number of the next cluster of the file, and words FF8 to FFF denote the final cluster of a file. Reserved clusters are indicated by words FF0-FF7, but FF7 for a cluster not part of a chain is one that is defective. *The initial cluster of a file is provided by the directory*, and this value can be read by programs that need to locate the start of a file. Then the FAT entry corresponding to this cluster is examined, *word XXX gives the number of the next cluster, and so on*.

To illustrate, suppose it is desired to read into memory file ABC, stored sequentially in clusters 02A, 01A, and 03C. The command causes both the directory and FAT to be read into memory. First, the directory is searched, and from it the start cluster 02A is determined. The FAT entry corresponding to cluster 02A is examined, leading to the next cluster, 01A. Then the entry of this cluster is used to obtain cluster 03C. Finally, the entry corresponding to cluster 03C is examined, and value FFX indicates that cluster 03C is the final one of file ABC. *The loader uses this information to determine which clusters to read in order to load the complete file into memory*.

The preceding example shows how the FAT *provides the link that chains together the different sectors of a file*. In most cases the clusters are in order, but when there has been considerable erasure and creation activity, they can be quite scattered. Whenever a new sector is to be written to, the first one available as indicated by FAT is selected. FAT is not used when reading IBMBIO and IBMDOS, because these files have fixed locations.

Each next-cluster word XXX of a file reverts to 000 if the file is erased, and the first byte of the filename in the directory is replaced with byte E5H. However, data of the file *remain on the diskette until overwritten*. An erased file not overwritten can be restored provided the sequence of the clusters can be determined. In most cases, clusters are in sequence. Disk-look utility programs are available that aid in determining the cluster sequence of an erased file, and the DEBUG program can be used to load the sectors into memory, rename the file, and write it to the diskette.

Reading the FAT is complicated by the method of storage. The hexadecimal digits are arranged in 16-bit words, with the high byte at the higher address. The following algorithm gives a way to locate the clusters of a file.

1. Read the starting cluster from the directory.
2. Multiply it by 1.5, obtaining PRODUCT.
3. Read the 16-bit word from FAT at the offset equal to the whole part of PRODUCT.
4. If the start cluster is even, the low-order 12-bit word is the next cluster number. If the start cluster is odd, the high-order 12-bit word is used.

5. From this new cluster, similarly find the next one, repeating until FF8-FFF is found, denoting the end.

To illustrate, suppose the starting cluster is 02B, or 43D. Multiplication by 1.5 gives 64.5. Starting from offset 0 of FAT, the 16-bit word at offset 64 decimal is read. Suppose this word is 1A48H. Because the last cluster was odd, the next cluster is 1A4.

The Directory

Each *directory* entry consists of 32 bytes. Double-sided diskettes are allowed up to 112 entries, which are stored in seven sectors with logical numbers 5-11. The directory of a single-sided diskette has four sectors with 64 possible entries.

Contents of a 32-byte directory entry are formatted as shown in Fig. 8-25. At offsets 26-27 is the starting cluster number, with the least significant byte (LSB) first. The first word of the file size at offsets 28-31 is the least significant one, and each word has its LSB first. In order to be accessed, the directory must first be written into memory. Thus, the sector of the directory that contains the last file used is available in memory for reading.

If the first byte of a directory is 00H, the entry has never been used, and if it is E5H, the file has been erased. A value of 2EH signifies that the entry is the name of a subdirectory. A *subdirectory* is similar to a file, except that it contains names of subdirectory files. Any other character is the first one of a filename.

The file attribute is used to identify a file as either a normal one (00H), a read-only file (01H), a hidden file (02H), a system file (04H), or a subdirectory (10H). A value of 20H indicates that the file has been written to and then closed. A file may be marked *hidden* when it is created, and such files *are not listed in directory searches*. Both system files IBMBIO and IBMDOS are hidden, read-only, and written-to-and-closed. Their attribute is 27H, which is a combination of values. When the directory command DIR is issued, these files are not listed with the others.

The directory is properly called the *root directory,* or *base directory,* because subdirectories can be created. The directory system can be compared with a tree, with directories branching off from the root directory, others branching from these, and so on. All directories other than the root directory are in the form of files that can contain an almost unlimited number of subdirectories. The main purpose of this treelike directory system is to accommodate a *hard disk,* which can store 10

Offsets	Significance
0–7	Filename (E5H at offset 0 means erased entry)
8–10	Filename extension
11	File attribute (00 for normal files)
12–21	Reserved (all 00s)
22–23	Coded time of last update
24–25	Coded date of last update
26–27	Start cluster number (LSB first)
28–31	Byte size of file

FIGURE 8-25. Entry format of the DOS diskette directory.

MB or more of data. Thus, its requirement for directory names is far greater than that of a diskette. The hard disk is also called a *fixed disk* because it is fixed permanently in its drive.

The Line Printer

Before considering the DOS commands in the next chapter, it is appropriate to examine the characteristics of the printer. Several DOS commands are often used for activation and control of this output device. Because it is designed to store and handle the characters of a complete line as a basic unit, it is referred to as a *line printer*.

The adapter card that controls the monochrome monitor also provides a parallel interface for the printer. At the rear of the card is a 25-pin D-type connector that attaches at the back of the system unit to a shielded cable. The 6-foot cable carries signals between the adapter and the printer. Eight pins are for bidirectional data transfer, and bytes for the printer can be latched. Inputs from the printer are real-time signals (not latched). Nine pins provide control, and the other eight are grounded.

Commands from the processor are loaded into an 8-bit latch of the adapter card, and under program control the status of the printer is read, indicating when the next character can be written. Reading from and writing to the adapter are accomplished with processor IN and OUT instructions. During a power-on processor reset, the printer also receives a reset, providing initialization.

The IBM printer is an 80 character-per-second (CPS) wire matrix device that uses *bidirectional printing*. With serial impact, the easily replaceable nine-pin head prints characters in a 9 × 9 dot matrix, with most characters using only a 5 × 7 dot pattern. There are a maximum of 80 characters per line in the standard font; other fonts available to the programmer and controlled by the system unit provide 40, 66, and 132 characters per line. Margin requirements and paper size may reduce these numbers.

Paper width is from 4 to 10 inches. Especially suitable is continuous sprocket-feed paper 8.5 inches wide, plus 0.5-inch separable edges containing small sprocket holes 0.5-inch apart. Perforations every 11 inches lengthwise allow the hard copy to be divided easily into 8.5 × 11-inch pages, referred to as *forms*. Characters can be either condensed, normal size, or double width, and various types of print are allowed, including double-strike and bold.

Available characters are those of the standard ASCII 96 uppercase and lowercase sets, plus 64 block graphic forms used in graphics and 9 international symbols. However, a number of characters that can be printed on the CRT display simply appear as blank spaces on a hard copy. The lines per page are programmable from 1 to 66, and the lines per inch are either 5, 8, or 10. Used is a replaceable ribbon in a removable cartridge, with a ribbon life of about 10 hours of nonstop use. Although black ribbons are most common, colors are also available. When power is switched on, the printer runs a self-test.

In the circuits controlling the printer are two NMOS 8-bit microcomputers—the Intel 8041 and the 8049. Each of these 40-pin chips contains an 8-bit CPU plus ROM for program storage, RAM for data, two 8-bit I/O ports, and an 8-bit programmable timer/counter. The chips are designed for efficient interface of peripherals with microprocessor systems. Included on the control circuit board are 6 KB of EPROM and an 8155 chip with RAM and I/O ports.

REFERENCES

Component Data Catalog, Intel Corp, Santa Clara, Calif., 1980.
DOS Manual (2.10), IBM Corp., Boca Raton, Fla., 1983.
DOS Technical Reference, IBM Corp, Boca Raton, Fla., 1983.
Technical Reference, IBM Corp., Boca Raton, Fla., 1981.

PROBLEMS

SECTION 8-1

8-1. For the 8237A DMA controller, give the precise function of each pin labeled in Fig. 8-3. Also, for the PC, state the source of each input and the destination of each output.

8-2. (a) As used in Section 8-1, define the following: idle cycle, single transfer, block transfer, cycle stealing, verify transfer, dummy DMA operation, EOP, TC, and page register. (b) Give the major disadvantage of the DMA block transfer mode. (c) Compare DMA and non-DMA interrupt-controlled data transfer with respect to speed. (d) What happens in the PC if a memory refresh operation is requested during DMA transfer of a large block of data from diskette to memory? Why would the use of the block transfer mode for the diskette not work for the PC?

SECTION 8-2

8-3. List the assembly language instructions that initialize and start DMA memory refresh, and state precisely what each one accomplishes.

8-4. In sequence, bytes 46H, 46H, 0, 0, 80H, FFH, 3, and 2 are written to ports 0CH, 0BH, 81H, 4, 4, 5, 5, and 0AH of the 8237 and page register (port 81H). List the assembly language instructions and state precisely what each one accomplishes.

8-5. List assembly language instructions that initialize the DMA controller for a diskette write using channel 2, with $\overline{\text{DACK}}$ active low, DREQ active high, normal timing and write-pulse width, and fixed priority. Bytes from memory addresses 40:8000 through 40:9070 are to be written to the diskette using repeated single transfers. Use port 81H for the page register.

8-6. With reference to Fig. 8-7, explain what would happen if the flip-flop is removed, with timer 1 output connected directly to pin DREQ 0 of the 8237.

SECTION 8-3

8-7. Sketch to scale a circular diskette removed from its jacket. Show the innermost and outermost cylinders, the nine concentric sectors, and the index hole. With an arrow indicating the direction of rotation, identify the track and sector numbers.

8-8. For the diskette format of Fig. 8-10, determine the total number of bits and bytes of one complete track, including ID fields and gaps. With a rotational speed of 300 revolutions per minute, how many bits and bytes are read by a head in 1 second? During the time it takes the head to read 1 byte, how many system clocks occur, assuming a clock of 4.77 MHz?

8-9. With respect to the diskette system, define the following: heads 0 and 1, sides 0 and 1, spindle, head aperture, index hole, index access hole, write protect, track, sector, cylinder, step motor, soft-sectored, hard-sectored, single-density, double-density, FM, MFM, CRC, ID, gap field.

SECTION 8-4

8-10. Give a sequence of assembly language instructions that will turn on the motor of drive A and enable the DRQ2 and IRQ6 buffers. Also, identify all addresses of the diskette adapter.

8-11. Identify all input and output lines of the diskette drive adapter of Fig. 8-11, stating the source, destination, and function of each.

8-12. State the difference between a physical disk and a RAM disk, and give the main advantage of each.

SECTION 8-5

8-13. Convert DISK_READ of Section 8-5 into DISK_WRITE, so that the program writes the 1024-byte interrupt vector table from memory to sectors 3 and 4, track 9, side 0, of diskette B. Use a formatted blank diskette. Execute DISK_WRITE.COM, and then run DISK_READ using DEBUG to read the data from the diskette to memory. Examine the data read to memory.

8-14. When DISK_READ of Section 8-5 is run, identify the set of instructions that are executed the greatest number of times. What essential computer activities are being processed during the intervals of the WAIT loops? What does each bit of byte 2DH of line 66 of the program accomplish? With regard to the drive, what would you notice if the DISK_READ program was run in DEBUG with STI of line 36 replaced with NOP? Explain.

8-15. For a one-sided diskette, prepare an assembly language COM-format program that uses INT 25H to accomplish the same function as DISK_READ. Assemble, link, convert to COM, and run the program, using DEBUG to check the results. Also, give the DEBUG instruction that also does the task of DISK_READ.

SECTION 8-6

8-16. Identify the side, track, and sector numbers of cluster 0CAH for a single-sided and a double-sided diskette. Repeat for cluster 10FH. Assume that the logical number of the first data sector is 9 for a single-sided diskette, and assume 12 for a double-sided diskette.

8-17. A 32-byte entry into the directory of a one-sided DOS diskette is as follows: 44 45 42 55 47 20 20 20 43 4F 4D (thirteen 00 bytes) 04 03 37 00 A1 17 00 00. Determine the filename, its extension, the attribute, the initial track and sector, and the decimal number of bytes in the program. Repeat for entry 4C 49 4E 4B 20 20 20 20 45 58 45 (thirteen 00 bytes) 04 03 43 00 00 A9 00 00. Assume that the logical number of the first data sector is 9.

DOS Commands

DOS is a collection of programs designed to permit the operator to create and manage files, run programs, and use devices such as the diskette system and printer. Many of these programs have been examined, such as EDLIN, Disk BASIC, and DEBUG. The system files IBMBIO, IBMDOS, and COMMAND are described in Section 11-7. Here we shall restrict our study to the internal commands of COMMAND, the external commands that reside on the DOS diskette as COM or EXE files, and batch files. The chapter concludes with a BASIC program that handles large numbers of repetitive calculations.

Although the drive that serves as the default can be specified by entering "d:" after the prompt, with "d" denoting the drive letter, the default is assumed to be drive A. We begin with the internal commands, which are the utility programs stored in the command processor of the COMMAND program of RAM.

9-1

INTERNAL COMMANDS

In the DOS mode, when a word consisting of any sequence of characters is entered, the command processor searches a table of names contained within the COMMAND program for a match. If a match is found, the program branches to a routine that immediately executes the *internal command*. If no match is found, the

Internal Commands	Function
DATE and TIME	Read and change the date and time
DIR	List diskette directory entries
COPY	Copy a file
TYPE	Display a file
ERASE or DEL	Erase (delete) a file
RENAME or REN	Rename a file
CLS	Clear the screen
PROMPT	Change the prompt
CTTY	Change the console
BREAK	Specify checks for Ctrl-Break
VER	Display DOS version number
VERIFY	Verify correct recording of data
SET	Insert strings into environment

FIGURE 9-1. Selected internal commands of DOS.

directory of the diskette drive is loaded into memory, and filenames of COM, EXE, and BAT programs are searched for a match with the word. If a match is found here, the program is loaded from the diskette into memory and run. Otherwise the message "File not found" is displayed.

Figure 9-1 shows selected internal commands. Not included are seven batch commands presented in Section 9-3 and five that are described in the discussion of the directory command DIR. These five are the internal commands CD, MD, RD, VOL, and PATH. Those of the table are examined in the order listed. Many have various associated options, only some of which are described here. The DOS manual should be consulted for details.

Commands DATE and TIME. When DATE is entered, the current date known by the system is displayed, along with a request to enter new values if desired. Pressing ENTER keeps the displayed date. It automatically changes at midnight.

The command TIME is similar. The response is the current time as known by the system, accompanied by a request for the time update. It is maintained by the internal time-of-day clock. Both DATE and TIME are similar to the corresponding BASIC statements described in Section 4-5. Unless an external clock driven by a battery is incorporated into the system, the date and time should be entered after each system reset.

Directory Commands. Entering DIR displays the *directory of the default drive,* and DIR B: gives the *directory of drive B*. The display consists of *a list of the filenames and extensions in a column*. Alongside each is the filesize in bytes and the date of the last write to the file. System files IBMBIO and IBMDOS and other hidden files are not listed.

Often the list is too long to fit on the display. Scrolling can be stopped by using Ctrl-NumLock and resumed by pressing another key. There is a better way, however. If DIR/P or DIR B:/P is typed, the display pauses when the screen becomes full, and a message tells the operator to hit any key to continue the display. Another option is the /W parameter, which provides a listing with five filenames on one line, omitting sizes and dates. Other options allow the selection of particular types of files for listing.

Global characters can be used, as previously mentioned (Section 2-4). Thus, DIR *.ASM will list all diskette files with extension ASM. A second global character is ?, which is a replacement for any *single character*. Command DIR ABC is equivalent to DIR ABC.*, and DIR .XYZ is equivalent to DIR *.XYZ.

Subdirectories are allowed. The parameter *path* in DIR *path* denotes *a path of directory names and filenames,* with the names separated by backslash characters. For example, DIR \DIR1\DIR2 gets directory DIR1 from the root directory and then obtains DIR2 from DIR1 and displays its files. If the first backslash is omitted, the path starts from the *current directory,* which is either the root directory or the one defined by the CHDIR command.

The internal *change-directory* command CHDIR, or simply CD, is used *to specify the current directory*. For example, CD B: \DIR1 followed by DIR B: will list the files of DIR1. If CD is entered without parameters other than the drive, the current directory path is displayed. CD B: \ makes the root directory of drive B the current one.

To create a subdirectory, the internal *make-directory* command MKDIR, or MD, is used. For example, MD \LEVEL2 creates an entry in the root directory of the default drive for a new directory called LEVEL2. If this is followed by MD \LEVEL2\LEVEL3, then a subdirectory for LEVEL3 is entered in the LEVEL2 subdirectory. Only the number of available sectors limits the number of subdirectories that can be created this way.

To remove a subdirectory, the *remove-directory* command RMDIR, or RD, is available. The directory to be removed is the last one in the specified path, and it *cannot have any files*. An example is RD B: \DIR1\DIR2. *Neither the root directory nor the current one can be removed.*

When a diskette is formatted, it can be assigned a *volume label* for identification. The label is placed in the root directory and is given in the displays produced by the commands DIR, CHKDSK, and TREE. When one has many diskettes, the identification label is very helpful. The procedure is described in the discussion of the external FORMAT command. The internal command VOL, followed by the optional drive designation, displays the label of the diskette.

There is yet another internal command related to directories, which is the *search-directory* command PATH. The keyword PATH is followed by a list of drives and path names, with each combination separated from the next by a semicolon. After it is executed, at any time a command is entered and not found in the current directory of the specified drive, a search is made of the paths of the PATH command. For example, suppose PATH \LEVEL1\B:MYDIR is entered, with the program COMPARE.COM in the directory MYDIR. Then if COMPARE is entered, DOS searches through the path, finds COMPARE, and executes it. Note that subdirectory LEVEL1 is on the diskette of the default drive, whereas the specified file of this directory is one of drive B.

For the diskette, it is not usually advantageous to create subdirectories unless there are many small files, but in such cases, one or more subdirectories can be quite helpful. Their use increases the capacity above the limit of 112 applicable to the root directory. *For the fixed disk, they are essential.*

COPY a File. The COPY command is used *to copy data from one I/O device to another*. For example, data can be copied from the keyboard to the printer, from the keyboard to a diskette, and from a diskette to the printer. However, the most common application is for transfer of a file from one diskette to another. The command is executed by entering COPY *source-filespec destination-filespec*. A path can be specified for both the source and destination if there are subdirectories,

```
COPY ABC.ASM B:          ;copy file from drive A to B
COPY B:*.*               ;copy all files from B to A
COPY ABC.ASM AB.ASM      ;copy to same drive, change name
COPY A:XY    B:XY.EWF    ;copy and add extension
COPY B:CD.*  LPT1:       ;copy files to line printer
```

FIGURE 9-2. Several examples of the command COPY.

which is true for all of the commands. If the source file is not on the default drive and the destination filespec is omitted, the file is copied to the diskette of the default drive using the same name. Shown in Fig. 9-2 are several examples, with comments included. Either LPT1: or PRN: can be used as the filespec for the line printer.

The third command of Fig. 9-2 copies a file of the default drive to the same drive under a new name. This procedure is often useful. For example, suppose an assembly language program is needed that is nearly the same as one available. Perhaps the requirement is simply to generate a COM program from an EXE program without changing the EXE program. Using the COPY command and then EDLIN can save a lot of typing.

The keyboard is an input device and the display monitor is an output device. Both have filespec CON:, for *console,* but there is no conflict. A few examples are given in Fig. 9-3.

When copy is from the keyboard, the lines typed are stored in memory until function key F6 is pressed, followed by Enter. The lines are then written to the target device. To abort the command after entry, use Ctrl-Break. This key combination can be used to interrupt execution of any DOS command and return to the command level with the DOS prompt.

Sometimes it is desirable to combine several files into a single file. This operation is performed with the plus (+) sign, which gives *concatenation* as shown in the example that follows:

```
COPY A.TXT + B.LST + C.XYZ ONEFILE.TXT
```

There are various other options not described here. *The COPY command will not write over another file of a diskette.* If there is insufficient space available for storage of the file, the command is not executed, and an appropriate message is displayed. *In a system with only one drive, the physical drive is treated as two logical drives A and B.* A pause permits switching diskettes.

TYPE a File. The command TYPE *filespec* displays *the contents of the specified file on the standard output device. The screen is the default standard output device, but this can be changed by using the special character* >. To illustrate, TYPE *filespec* > ABC.XYZ creates the new file ABC.XYZ and writes the specified file to it, with no display. TYPE *filespec* is the same as COPY *filespec* CON:, and

```
COPY CON:   CON:     ;copy lines from keyboard to display
COPY CON:   B:XYZ    ;copy lines from keyboard to disk B
COPY *.ASM CON:      ;copy all .ASM files to display
```

FIGURE 9-3. COPY examples using keyboard/display.

the filespec can be that of any input device. For example, TYPE CON: corresponds to COPY CON: CON:. Text sent to the display by either COPY or TYPE can also be sent to the printer by pressing Ctrl-PrtSc prior to execution of the command. Press the keys again to inactivate the printer. Ctrl-Break will interrupt execution that is in progress. An alternate method is to use > LPT1 to redirect output to the printer, and LPT1 can be replaced by PRN, if desired.

ERASE or DEL a File. The format for the ERASE command is either ERASE *filespec* or DEL *filespec,* and of course, a path can be specified if the file is in a subdirectory. The file with the specified name is erased from the diskette. To erase all user files of a diskette, enter DEL *.*. *System files IBMBIO and IBMDOS cannot be erased.*

RENAME or REN a File. The command RENAME or the shorter form REN allows the operator *to change the name of a file*. The format is

REN *filespec filename*

A path can be specified, and an extension for the new filename is optional.

CLS. CLS is the same as the BASIC statement CLS, which *clears the screen.* A routine that cleared the screen was presented in the first part of the program in Fig. 5-11.

Change the System PROMPT. The internal command PROMPT is used *to set a new system prompt*. For example, if "PROMPT hello " is entered, the new prompt is "hello ", which replaces "A>". Special strings can be embedded in the new prompt by using $c, with the character *c* corresponding to a string contained in a table of the DOS manual. For example, "PROMPT hello$g " gives "hello> " and "PROMPT $t" displays the current time as the new prompt. The normal default value is restored by entering PROMPT with no additional characters.

Change the Console (CTTY). The internal *change console* command CTTY *changes the standard input device,* which is the keyboard, *and the standard output device,* which is the screen, *to the device specified*. To illustrate, entry of CTTY COM1 causes the communications port to serve as both the input and the output device. All commands that normally relate to the keyboard and screen henceforth relate to the communications port. Execution of CTTY CON reverts back to the default conditions. In Chapter 11, numerous function calls are mentioned that refer to standard I/O devices.

Specify Ctrl-Break Checks (BREAK). The command *control-break* BREAK is followed by either ON or OFF. Following execution of BREAK ON, whenever a program is run, *a check for key combination Ctrl-Break is made at each DOS function call*. These function calls are described in Chapter 11. Normally, checks are made only during standard input (keyboard), standard output (screen), output to a printer, and I/O to a communications port. Command BREAK OFF reverts to the normal condition. A check for Ctrl-Break that finds that the key combination has been activated usually breaks the running program and transfers control to the command processor.

Display DOS Version (VER). The *version* command VER does nothing but *display the version number of the DOS being used*. When the author acquired the PC in 1981, the version number was 1.0. Several new versions have been issued; the current one is 2.1, which will undoubtedly be updated at periodic intervals. Each new version adds additional commands and functions to the operating system.

VERIFY Data. When VERIFY ON is executed, *every time a diskette operation is performed, a check follows to determine that no errors occurred*. VERIFY OFF reverts to the normal condition, and VERIFY alone displays the on/off condition. Diskette operations are slower when VERIFY is on.

SET Strings into Environment. The command SET inserts character strings into the *environment* of the command processor. The format is SET *name* = *string*, and the entire string including the name *is put into a block of memory reserved for environment strings*. The name is converted into uppercase letters, but the rest of the string is inserted as entered. It is available for use by programs as desired. The starting address of the environment is placed in the program segment prefix of a program loaded into memory, which is described in Section 11-1. If SET is entered without a string, all strings of the environment are displayed.

9-2 EXTERNAL COMMANDS

External commands may be regarded as any COM, EXE, or BAT files of the DOS diskette, excluding COMMAND.COM, which has been loaded into memory. *No other extensions are recognized*. Simply by entering the name without the extension, the program *is loaded into memory and executed*.

Added to the DOS diskette by the author are numerous utilities that qualify as external commands, including DISK_LOOK, SUPERDRV, and ASTCLOCK (used for setting the battery-driven clock). These commands are not treated here. Also excluded from consideration are EDLIN, DEBUG, LINK, BASIC, and BASICA, which have already been described. Remaining are the external commands listed in Fig. 9-4 and special batch-processing commands, which are examined in the next section. Omitted are the fixed-disk commands BACKUP and RESTORE.

FORMAT a Diskette. Before a new diskette can be used, it must be formatted. *Formatting establishes the sectors* of the tracks, with gaps and ID information similar to those of Fig. 8-10. *It also writes the boot record into the first sector, sets up two copies of the FAT, initializes the diskette directory, and analyzes the diskette for defects*.

Four examples of the FORMAT command follow:

FORMAT   FORMAT B:/8   FORMAT /1/V   FORMAT B:/S/V

After entry of the command, a message is displayed telling the operator to insert the diskette to be formatted into the specified drive and press any key when ready. Formatting commences when this operation is done.

The first of the preceding commands formats the diskette of the default drive. If the drive has two heads, the format is that of a double-sided diskette. Otherwise the diskette is formatted for single-sided use. The second command specifies a

External Command	Function
FORMAT	Initialize the diskette
DISKCOPY	Copy an entire diskette to another
DISKCOMP	Compare the contents of two diskettes
COMP	Compare the contents of two files
CHKDSK	Display status report of a disk and RAM
MODE	Set mode for printer or color monitor
EXE2BIN	Convert EXE files to COM files
SYS	Transfer system files
ASSIGN	Assign a substitute diskette drive
FIND	Display specified strings of files
MORE	Send a screen full of data repeatedly
SORT	Read data, sort them, and then write them
TREE	Display all directory paths
RECOVER	Recover files from a defective diskette
PRINT	Print files concurrently
GRAPHICS	Print graphics screen

FIGURE 9-4. DOS external commands.

format of eight sectors per track for the diskette of drive B. The third one gives a format for single-sided use regardless of the drive type. In addition, the option /V prompts the operator for a *volume label* that may have from 1 to 11 characters. This disk ID is included in displays produced by the DIR, CHKDSK, and TREE commands, and the label is displayed by the command VOL.

The option /S of the last command causes the system files IBMBIO, IBMDOS, and COMMAND to be copied from the diskette of the default drive to the one of drive B being formatted. *This procedure allows the new diskette to be used for booting the system.* Replacing /S with /B does not copy the system files but reserves space for IBMBIO and IBMDOS, so that the files can be copied later if desired, using the SYS command. It is not necessary to reserve space for COMMAND, because it can be placed anywhere on a diskette. Option V is also included.

Formatting destroys any data previously put on the diskette. If a defect is detected, the track affected is marked as reserved, which inhibits its use. A formatted double-sided diskette cannot be used in a one-head drive, but use of a diskette formatted only on one side is unrestricted.

When the format of a diskette is done, the system displays a status report. Shown are the number of bytes of total disk space, the bytes used by the system files, the bytes of bad sectors, and the number available for storage of user programs. A report of bad sectors is very unusual.

Command DISKCOPY. The DISKCOPY command is used *to copy the entire contents of one diskette to another.* When the command is entered, a message appears instructing the operator to insert the source diskette in the source drive and the target diskette in the target drive. Two drives, although convenient, are not essential. If the specified source and target drives are the same, the operator simply complies with the displayed instructions, using the one drive as logical drives A and B. DISKCOPY is accomplished by first writing from the source diskette into memory and then transferring the data from memory to the target diskette. Three examples follow:

DISKCOPY   DISKCOPY A: B: DISKCOPY C: A:/1

The first of the listed commands uses the default drive for both the source and the target, and prompts request insertion of diskettes at the proper times. The next one specifies drive A as the source and drive B as the target. In the third command, drive C is the source and drive A is the target, with only side 0 copied regardless of the type of drive. *If the target diskette has not been formatted, DISKCOPY will do so while copying.*

Sometimes a diskette becomes very fragmented as a result of considerable file creation and deletion. For example, the clusters of a file may go in order from 12A to 011 to 130 to 02C, and so on. When the file is read, the head has to move excessively back and forth along the head slot and wait at times for the proper sector to be rotated to the read position, thereby introducing delays. In such cases, it is usually better to copy the files of one diskette to another with the command COPY *.* B:. This command writes the files from default drive A to the diskette of drive B, using available sectors in order. *Fragmentation is eliminated.* In addition, current files of the target diskette are not overwritten, as they are with DISKCOPY.

Command DISKCOMP. Following a DISKCOPY, it is appropriate *to compare the two diskettes* to ensure that no errors occurred. The DISKCOMP command is provided for this purpose. The format is

DISKCOMP A: B: /1/8 *(optional)*

Diskettes may be single- or double-sided, and in each case the parameters /1 and /8 are optional. The first of these specifies comparison only of side 0 even if the diskettes and drives are double-sided, and the second gives a comparison of only eight sectors per track even if there are nine. If no drives are specified, only the default drive is used, and the program prompts the operator when to change the diskette. All 40 tracks are compared, and any discrepancies are reported.

Compare-Files Command COMP. The *compare* command COMP *compares the contents of two files,* which may be on the same or different drives. When the command is entered, a message instructs the operator to insert the diskettes and press a key when ready. Only then does comparison commence. If filespecs are not entered, DOS prompts you for them. Two examples are

COMP   COMP A:XYZ.COM B:XYZ.COM

The specified files are compared byte by byte, and any discrepancies are reported. If as many as 10 discrepancies are found, the comparison is aborted. COMP is often used following a COPY command. Subdirectory paths may be included in the filespecs.

CHKDSK for Errors. The purpose of CHKDSK is to analyze the directories and the FAT of the designated drive, and *to give both a diskette and a memory status report.* The diskette status report states the number of bytes in total diskette space and the number that are free. It also reports hidden files, including one "hidden file" for the volume label, along with their bytes. In addition, the number of directories and user files and their bytes are given. The memory status report states the number of bytes in total memory and the number available for program use. Any errors found are reported. The format is simply the command followed by the drive specification, as in CHKDSK B:.

Sometimes clusters are lost and unavailable due to errors in a directory or in the FAT. They are often lost when improperly written programs are compiled or assembled. Lost clusters found by CHKDSK are reported, and if the /F option is used (CHKDSK/F), a message asks if the lost data should be recovered into files. For an affirmative response, the chains of lost clusters are put into named files of the root directory, where they can be examined and optionally deleted by the command DEL. A negative response simply restores the lost clusters without saving the data, which are usually of no value.

Specify the MODE. One option of MODE is used *to set the mode of operation on a printer*. Assuming only one printer is connected to the system, the format is as follows:

MODE LPT1: *n, m*

Parameter *n* must be either 80 or 132, denoting the number of characters per line, and *m* is either 6 or 8, indicating the lines per inch of vertical spacing. Default values are 80 characters per line and 6 lines per inch. The printer must be on when the command is executed. For printouts of listing programs generated by the assembler, the MODE command with *n,m* values of 132,8 is recommended.

A second option *sets the mode of operation on a display connected to the color/graphics adapter and provides for switching display adapters*. The format is as follows:

MODE *n, m, T*

Optional number *n* can be 40, 80, BW40, BW80, CO40, CO80, or MONO. The first of these numbers sets the display width to 40 characters per line, assuming the color/graphics adapter is used, and the next one gives 80 characters per line. The other values provide for switching control between display adapters when both monochrome and color display monitors are used.

BW40 and BW80 switch from the monochrome display to the color monitor in the black-and-white mode, with either 40 or 80 characters per line as specified. CO40 and CO80 are similar, except that color is enabled. Finally, MONO switches to the monochrome monitor, which always has 80 characters per line. When the author initially acquired the two displays, these commands were not available, and the programs MONO and COLOR described in Section 7-5 were used.

Optional *m* is either R or L, and option T gives a test pattern used for horizontal display alignment. There is no pattern unless *m* is specified. The test pattern consists of one line of decimal digits used for display alignment. Shifts are either right or left, as specified by the *m* parameter. Shifts are made when the response to a message asking if the alignment is satisfactory is N, and shifts are repeated as necessary. A response of Y ends the process.

There are two options related to the asynchronous communications adapter. One of them sets the baud rate, parity, data bits, and stop bits; the other takes output sent to the line printer and redirects it to the communications adapter. These options are examined in the next chapter, which treats communications systems.

EXE-to-Binary Command EXE2BIN. As we have learned, EXE2BIN *converts EXE files to COM files* provided certain conditions are satisfied. There must be no STACK segment, and the code and data of the EXE program loaded into memory must occupy less than 64 KB of space. In addition, *no instruction is allowed to contain code that depends on a segment address determined by the loader.* This restriction does not apply to an EXE file, because the header provides information that allows the loader to calculate these segment addresses and to insert them into the code at the proper places. The assembler statement ORG 100H must be placed ahead of the program, and the directive END must be followed by the label of the starting address. Filespecs of the command may contain paths to the files.

If the machine code of an instruction requires the address of a segment that is set by the loader, this load address must be determined in advance and put into the code before EXE2BIN can produce a COM file. This *segment fixup* was discussed in Section 7-6 at the end of the TICKS program.

Write-System-Files Command SYS. Application programs sold on diskettes without the version of DOS used by the system can be modified with the command SYS, provided the diskettes were formatted with either the /B or /S option. The command SYS allows the operator *to overlay the system files with the version of IBMBIO and IBMDOS of the DOS diskette.* The command can also be applied to formatted diskettes that are empty. With the DOS diskette in drive A and the properly formatted target diskette in drive B, the command SYS B: transfers the system files. COMMAND.COM is not transferred.

ASSIGN a New Drive. The command ASSIGN A = B instructs DOS *to route all requests for drive A to drive B,* and any valid drive letters may be used. With no drives specified, ASSIGN will undo the assignment.

Filter Command FIND. A *filter* is a command that *reads data, modifies it, and then writes it to a device.* The program is said to "filter" the data. The command FIND and the two that follow are classified as filters. FIND *sends to the standard output device all lines from specified filenames that contain a specified string.* For example, FIND "command processor" file1, file2.ewf will display all lines from file1 and file2.ewf that contain the string. The default standard output device is the display, but redirection with > is allowed. There are various options.

Filter Command MORE. Another filter command is MORE. When MORE is entered, *data are read from the standard input device, which is normally the keyboard, and the filtering sends* one screen full *of data to the standard output device, followed by the message "—More—".* Pressing any character key causes another screen full of data to be sent. The process continues until all input data are read or until Ctrl-Break terminates the process.

The *standard input device* can be specified by using the symbol <, just as the symbol > has been used to specify an output device. Three examples follow:

MORE   MORE <TEST.ASM >PRN   MORE <XYZ >FILE.ASM

The command at the left applies to the keyboard and the screen. The next one specifies file TEST.ASM as the standard input device, and the printer becomes the standard output device. At the right, file XYZ is used for input and FILE.ASM is the output. Execution transfers one screen full of data from one file to the other, even though the screen is not involved. LPT1 can replace PRN.

SORT Data of a File. The third filter command is SORT, which *reads data from the standard input device, sorts it using the ASCII collating sequence, and writes it to the standard output device*. Option /R reverses the SORT, putting Z ahead of A, and /+*n* starts the sort with column *n* instead of column 1. An example is SORT /R <FILE1.TXT >FILE2.TXT, which reads FILE1, does a reverse sort, and writes the output to FILE2.

Piping refers to *the chaining of programs with automatic redirection of standard input and output*. Chaining is done by separating programs of the command line *with the vertical bar (|) character*. DOS acts as a "pipeline" that directs the output of one program to the input of the next. DOS creates temporary files listed in the root directory of the default drive to hold the output data being piped. These temporary files have filespecs of the form %PIPEx.$$$. SORT is often piped with other commands. For example, DIR | SORT generates a sorted directory listing, and the command DIR | SORT>FILE sends the sorted directory to FILE.

Display the Directory TREE. *To display all directory paths on the specified drive,* the command TREE is entered, followed by a drive letter for one other than the default. Option /F causes the names of all files in each subdirectory to be displayed as well. To give a pause during output, press Ctrl-NumLock or pipe the output to the MORE filter. An example is TREE B: | MORE. To send the directory data to a file, follow TREE with >*filespec*.

RECOVER a Damaged File. The purpose of RECOVER is *to recover a file that has developed a bad sector,* excluding data of the bad sector. Also, it can recover all files of a diskette if the directory is damaged. Fortunately, these situations are rare. The format is simply RECOVER followed by an optional drive letter and an optional filespec, including a path.

PRINT a Sequence of Files. The command PRINT followed by a list of up to 10 filespecs *puts the files in a queue in memory and prints them while the system is used for other work* not involving the selected output print device. The files must be in the current directory, and global characters * and ? can be used in the filespecs. After entry of the command, a message prompts for the name of the output device. If Enter is pressed, PRN is the default. Specifying COM1 sends the files to the communications line. Files can be added to the queue after printing has started by entering another PRINT command.

The command terminates after all files have been printed or upon execution of PRINT /T, which stops printing and cancels the queue. Entry of PRINT with no parameters displays the names of files currently in the queue.

Print a GRAPHICS Display. When using the color/graphics adapter and the PC graphics printer, execution of the command GRAPHICS allows *for printing the contents of a graphics display*. Implementation is possible anytime after execution of the command by pressing key combination Shift-PrtSc. To invoke the screen print from a program, the sequence PUSH BP, INT 5, POP BP can be used. The interrupt was described in Section 6-5.

BATCH PROCESSING

As we have learned, only internal commands and machine language programs of the diskette with extension EXE or COM can be executed directly by the computer. However, it is possible to form a *batch* of such files and then instruct the computer to read the names and execute them one by one. The PC can perform this operation, provided the command programs assigned to a batch file are given a filename and the extension BAT.

Thus, a *batch file* is defined as *any file with extension BAT that contains one or more internal DOS commands or EXE, COM, or BAT programs of the diskette system*. A batch command is sometimes placed at the end of a batch file to transfer control to a second batch file. The commands of the file may include user-designed or purchased EXE and COM programs. Execution is accomplished by entering the drive and filename with the BAT extension omitted, and *commands of the file are processed as they are encountered*. Ctrl-Break can be used to terminate a batch file during its execution, if desired.

Here we investigate the creation of batch files, including the especially important *automatic-program-execution* file AUTOEXEC.BAT. If present, this file executes immediately following a system reset. In addition, we examine seven internal commands designed specifically for processing of batch files. These are often referred to as *subcommands,* and they are listed in Fig. 9-5.

Batch Processing Commands

The seven commands listed in Fig. 9-5 are used to control batch processing. All are internal, with their routines contained in the command processor. They are explained in the order listed, beginning with REM.

Display a Remark (REM). The command REM is used *to display remarks from within a batch file*. Following keyword REM can be any string up to 123 characters long, and the string is displayed when REM is executed. The command is often used without a string to create a line containing only the word REM, improving the readability of lines in succession.

Display a Message and PAUSE. PAUSE commands can be placed in a batch file *to display an optional message and then suspend processing until a key is pressed by the operator*. The optional message, up to 121 characters long, is placed immediately after PAUSE. When the command is executed, the optional message

Command	Function
REM	Display remarks
PAUSE	Provide a system wait
ECHO	Inhibit screen display
GOTO	Transfer control
FOR	Iterative execution of commands
IF	Conditional execution of commands
SHIFT	Provide replaceable parameters

FIGURE 9-5. Internal batch processing commands.

is followed by the message "Strike any key when ready...", and then the system waits. The delay can be used to change diskettes, to decide whether to end batch processing with Ctrl-Break, to provide time for reading a display, or for some other purpose.

Inhibit Display of Commands (ECHO). Commands of a batch file are normally displayed on the screen as they are executed. However, *the echo can be inhibited by placing* ECHO OFF *in the batch file*. Commands that follow are not echoed, including messages after REM and PAUSE, but messages generated by the commands are displayed. ECHO ON turns the echo on, and after its execution, commands are echoed normally. The default is ECHO ON.

Regardless of the current ON or OFF state, entry of an ECHO *message* displays the message on the standard output device. The current state of the echo is displayed in response to execution of ECHO with no parameters.

Transfer Control with GOTO. The statement GOTO *label* in a batch file transfers control to the command immediately following the specified label. *A label is any character sequence preceded by a colon.* No two labels are allowed to have the same first 8 bytes, which are the only ones checked. The label itself is not displayed during execution of the batch file. A label not used as the target of a GOTO statement can be used *as a batch file comment that is not displayed during execution.* Appropriate comments provide clarity when the batch file is typed to the display.

Iterative Execution with FOR. *For repeat of the same command with different parameters,* the FOR command is available. For example, suppose we want to copy from the diskette of drive A to that of drive B three files with names f1, f2, and f3 and assumed to have no extensions. This operation can be done with three separate COPY commands placed in sequence in a batch file. An alternate way is provided by the FOR command, written as follows:

```
FOR %%X IN (f1 f2 f3) DO copy %%X B:
```

The *single-character variable X* that follows %% is equated with the first member of the set in parentheses, and the COPY command is executed. Then the variable is equated with the next member of the set, and again COPY is executed. The process is repeated a third time. Note the keywords FOR, IN, and DO. The general format is

FOR %%*variable* IN (*set*) DO *command*

If a member of the set is a filespec with a global character, the variable is equated with each matching filespec. Only one FOR command can be specified on a command line, which implies no nesting, and path names cannot be used in the filespecs.

Conditional Execution with IF. *To implement conditional processing in a batch file,* the IF command can be used. Its format is

IF *condition command*

If the condition is true, the command is executed. Otherwise processing proceeds to the next command in sequence. Any DOS command can be used, but there are only three allowed conditions, two of which are described here. The third condition is described in Section 11-2.

One condition is EXIST *filespec*, and it is true if the filespec exists. Path names are not allowed. A second condition is *string1* = = *string2*, which is true if and only if the two strings are identical. Note the double equal signs of the string comparison. *The NOT operator can be used with either condition.* Some examples follow:

```
IF  EXIST  B:FILE1.TXT  TYPE  B:FILE1.TXT
if  not exist file2.xyz goto abc
if  %2 == Jodie echo Hello!
```

The first of these commands types the file if it exists, and the next one goes to the command following label *abc*, provided file2.xyz does not exist. In the third command, the *dummy parameter* %2 appears. Dummy parameters %0 through %9 can be used in a batch file, provided they are defined when the name of the batch file is entered for execution. If parameter %2 was defined as the string Jodie, the ECHO command is executed, sending the message to the display. In string comparison, uppercase and lowercase letters are not equal. Assignment of values to dummy parameters is described next.

Dummy Parameters. *Each dummy parameter used in a batch file is replaced with a character string supplied when the batch file is executed.* The parameter %0 is always replaced with the filespec, excluding the BAT extension, but the other dummy parameters are unrestricted. To illustrate, suppose the file B:ABC.BAT consists of the following three statements:

```
%1
TYPE  %0.BAT
BASIC %2
```

Assuming we want %1 to represent the command TIME and %2 the BASIC file B:XYZ with extension BAS, the batch file is executed with string values assigned by entering

```
A> B:ABC  TIME  B:XYZ
```

The prompt is shown in addition to the entry. Parameter %0 becomes B:ABC, parameter %1 becomes TIME, and parameter %2 becomes B:XYZ. Accordingly, the commands executed are

```
TIME
TYPE  B:ABC.BAT
BASIC B:XYZ
```

The current time is displayed, the batch file is typed on the display, and the BASIC program is loaded from the diskette of drive A, followed by the fetch and execution of the BASIC program B:XYZ.

Now if the batch file is executed a second time, different values can be assigned to the dummy parameters. Thus, *a batch file with dummy parameters can perform different functions with different data during each execution.* Values of up to 10 parameters can be assigned on the command line used for entry of the batch command. The first string that follows the filespec becomes %1, the next one becomes %2, and so on.

Additional Parameters with SHIFT. In case 10 replaceable parameters are insufficient, the command SHIFT can be used *to create more.* Each time SHIFT is executed, the parameters of the command line are reassigned by shifting values left one position. The old value of %1 becomes the new value of %0, and so on. The string that originally corresponded to the disallowed parameter %10 is assigned to %9. SHIFT can placed in the batch file as often as necessary.

Batch File Creation

For short batch files, perhaps the easiest way to create one is to copy it from the keyboard to the diskette. An example is given in Fig. 9-6. When key F6 is pressed, followed by Enter, the program is written to the diskette of drive B. An alternate way to develop the batch file ABC.BAT on the diskette of drive B is to enter EDLIN B:ABC.BAT. Then type the file using the insert mode of EDLIN, and store the new file on the diskette with EDLIN command E. Once created, the file can easily be replaced. It can be erased and a new one created. An alternate way is to use the COPY command, as in Fig. 9-6, to overwrite a batch file of the same name.

```
A> COPY CON: B:ABC.BAT
     REM   This is a batch file
     REM
     PAUSE Insert target diskette in drive B
     DATE
     TYPE  B:ABC.BAT
     BASIC B:XYZ
     (Press function key F6 and press Enter)
```

FIGURE 9-6. Creation of a batch file.

After file ABC.BAT is written to the diskette, it is executed by entering B:ABC. *In sequence, each command is displayed after the prompt A> and executed,* just as though it had been entered from the keyboard. No action is taken on execution of the REM commands. The last command of ABC loads and executes Disk BASIC and then immediately loads and executes B:XYZ.BAS. If a command is put in the batch file after BASIC, execution occurs following a SYSTEM command that returns control to DOS.

File AUTOEXEC.BAT

A batch file of special importance is AUTOEXEC.BAT. After every system reset, initiated by either the power switch or Ctrl-Alt-Del, *the command processor automatically searches the diskette directory of drive A for AUTOEXEC.BAT. If one is found, it is executed immediately,* and the normal prompts for the current date and time are omitted.

For automatically starting up programs, an AUTOEXEC.BAT program is quite

```
Astclock
PAUSE   Press Ctrl-Break only to avoid RAM disk.
Superdrv B:/1/i
PAUSE   Insert work diskette in right drive C.
Copy    C:*.ewf B:
EW
PAUSE   Ctrl-Break will fail to copy RAM disk drives.
Copy    B:*.* C:
PAUSE   Be sure all is OK prior to power off!!
```

FIGURE 9-7. An AUTOEXEC.BAT program.

useful. For example, the author's word processor EasyWriter program has the AUTOEXEC.BAT shown in Fig. 9-7.

When the system is booted with the EasyWriter diskette in drive A, the AUTOEXEC.BAT file is immediately executed. The first of the commands loads and executes ASTCLOCK.COM, which sets the date and time from a battery-driven clock on a multifunction board. Then the PAUSE provides the option of discontinuing the batch processing with an exit to the DOS mode. If the spacebar is hit, SUPERDRV.COM is executed, which inserts a RAM disk into memory, designates as it drive B, and converts the right-side physical disk to drive C. The next PAUSE tells the operator to insert the work diskette into drive C and strike a key when ready. All EasyWriter files, which have extension EWF, are then copied into the RAM disk. The EasyWriter program EW.COM is loaded into user memory and given control. The program assumes that the work diskette is that of drive B.

On completion of the editing session, which may edit files of the RAM disk and create new ones, an exit-to-DOS command is entered, and EasyWriter transfers control back to DOS. The batch program is still in control, and the PAUSE that follows EW is executed. The message gives the operator the option of terminating the batch program without copying the files from the RAM disk to the physical one. Assuming the spacebar is hit, all files of the RAM disk are copied to the diskette of drive C, overwriting those with the same names. The last PAUSE is a reminder to check that all files have been saved before turning off the power.

Figure 9-8 is an example of a batch file designed simply to illustrate some of the processing subcommands that have been described. The filespec is assumed to be B:ABC.BAT.

```
CLS
REM     This is a batch file
ECHO OFF
PAUSE   This message will not be displayed.
IF NOT %1 == file1.txt copy %3 B:
ECHO ON
FOR %%f IN(A: B:) DO VOL %%f
:This is a conditional end
IF %2 == xYz GOTO DONE
DATE
PAUSE   Press Ctrl-Break only to bypass TIME.
TIME
:DONE
```

FIGURE 9-8. Batch file B:ABC.BAT, illustrating parameter passing and use of six of the seven subcommands.

Suppose the following command is entered:

```
B:ABC FILE1.TXT xyz CLS.COM
```

The screen is cleared and the remark is displayed, but PAUSE, IF, and ECHO ON statements are not displayed. However, there is a pause with the message "Strike a key when ready...". After the key is struck, the IF statement is examined. String %1 is not equal to file1.txt because of the lowercase letters. Therefore, file CLS.COM is copied from the diskette of drive A to that of drive B. Next, the FOR statement executes the VOL command twice, first for drive A and then for drive B. The comment after the colon is not displayed. The GOTO statement is ignored because the two strings are not identical.

9-4

INTEGRATION USING BASIC

We depart from the subject of DOS commands to examine BASIC further. An example is presented here that illustrates *the ability of a digital computer to manipulate numbers rapidly*. In addition, several new features of BASIC are introduced, including the PRINT USING statement and the TAB function for formatting output.

Physical systems can often be modeled by differential equations, which can sometimes be solved only by approximate integration techniques using high-speed computers. The equation to be solved here is that of (9-1) for time $t > 0$, with initial conditions given. Because the highest derivative is the second derivative of x with respect to time t, the equation is said to be second order. Also, the coefficients are constants, independent of time, and the equation is linear.

$$d^2x/dt^2 + 2 * dx/dt + 64 * x = 64 \qquad (9\text{-}1)$$

with $x(0) = 0$ and $dx/dt(0) = 0$

The values of x and dx/dt at time 0 are 0. These are the *initial* conditions. This elementary equation can be solved by analytical methods, and the exact solution is given without proof as Eq. (9-2).

$$x(t) = 1 - \sqrt{64/63} * e^{-t} * \sin(\sqrt{63} * t + \arccos 0.125) \qquad (9\text{-}2)$$

The method of approximate integration to be used here can be applied not only to this elementary example but also to differential equations of higher order n that are nonlinear and perhaps have time-varying coefficients. For readers unfamiliar with differential equations, the subsection that follows can be omitted.

The State Model

In order to solve the differential equation by approximate integration techniques, it is first necessary to reduce the second-order equation to two first-order ones. This is done by letting $y = x$ and $z = dx/dt$. It follows that the second derivative of x with respect to t equals dz/dt. From these relations and (9-1), we easily obtain

$$\begin{aligned} dy/dt &= z \\ dz/dt &= -2 * z + 64 * (1 - y) \end{aligned} \qquad (9\text{-}3)$$

This set of first-order differential equations contains the same information as the second-order equation of (9-1). The set is called the *state model,* and the new variables y and z are *state variables*. The method used here to reduce the given equation to a set of first-order equations can be applied to most differential equations of higher order. To illustrate, for a third-order equation, let the state variables be y, z, and w, with $y = x$, $z = dx/dt$, and $w = dz/dt$ (see Prob. 9-16). With these selections, the equation can be reduced to three first-order ones following the method used to obtain (9-3).

To solve (9-3), the initial values of y and z must be found. Clearly, from (9-1), these initial conditions are

$$y(0) = 0 \text{ and } z(0) = 0 \tag{9-4}$$

Approximate Integration

Figure 9-9 shows a plot of a continuous variable $w(t)$ versus time t. If H denotes a very small increment of time, then from time t to $t + H$ the very small segment of the curve approximates a straight line and its slope equals $[w(t+H) - w(t)] / H$. This is reasonable only for a very small value of H.

In (9-3) the derivatives represent the slopes of the curves of $y(t)$ and $z(t)$ at time t. In terms of the small time increment H, the equations of (9-3) become

$$\begin{aligned} [y(t+H) - y(t)]/H &= z(t) \\ [z(t+H) - z(t)]/H &= -2*z(t) + 64*[1 - y(t)] \end{aligned} \tag{9-5}$$

The values of y and z in (9-5) are taken at the start of the incremental segment. Equations (9-5) are approximate, but for a sufficiently small time increment H they are reasonably accurate. Rearranging terms gives:

$$\begin{aligned} y(t+H) &= y(t) + H*z(t) \\ z(t+H) &= (1 - 2*H)*z(t) + 64*H*[1 - y(t)] \end{aligned} \tag{9-6}$$

Only discrete intervals of time are used in the calculations, with the increments being H. Thus $t = K * H$, with K denoting an integer. For $K = 0$ the preceding equations become

$$\begin{aligned} y(H) &= y(0) + H*z(0) \\ z(H) &= (1 - 2*H)*z(0) + 64*H*[1 - y(0)] \end{aligned} \tag{9-7}$$

From the known initial values of y(0) *and* z(0), *the values of* y(H) *and* z(H) *can be found from (9-7).*

For $K = 1$, or $t + H = 2H$, the equations become

$$\begin{aligned} y(2H) &= y(H) + H*z(H) \\ z(2H) &= (1 - 2*H)*z(H) + 64*H*[1 - y(H)] \end{aligned} \tag{9-8}$$

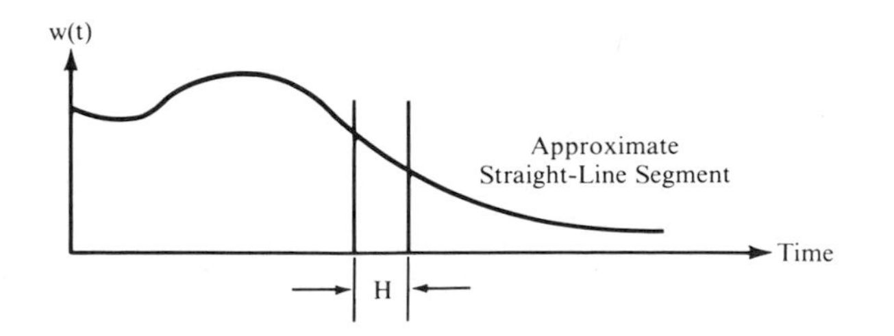

FIGURE 9-9. A plot of variable *w(t)* versus *t*.

From these and the previously calculated values of y(H) *and* z(H), *values of* y(2H) *and* z(2H) *are found. The process can be continued for as many time intervals as desired.* For a total time interval of 5 seconds with H chosen to be 0.001, the equations would have to be evaluated 5000 times. This is easily done on a digital computer.

Let us review the general procedure. The differential equation of order n is subdivided into a set of n first-order differential equations. From these equations and the initial conditions, the slopes of the curves of the state variables are found at time 0. Multiplying the slopes by the small time increment H and adding the results to the initial values give the values of the variables at the end of the first increment. Next, the slopes are calculated here and multiplied by increment H; as before, the results are added to the values found at the start of the increment to give the values at the end of the second increment. *The process is repeated again and again for as long as desired.*

This approximate integration method is called *Euler's method*. There are other approximate integration procedures, most of which utilize the computer more efficiently. However, Euler's method is simple and a high degree of accuracy is obtained, provided H is sufficiently small.

The BASIC Program

From (9-7) and (9-8) we deduce that the general form of the equations to be programmed can be expressed as

$$Y(J) = Y(J - 1) + H*Z(J - 1) \qquad (9\text{-}9)$$
$$Z(J) = (1 - 2*H)*Z(J - 1) + 64*H*[1 - Y(J - 1)]$$

Integer J has values 1, 2, 3, and so on (corresponding to $K + 1$). Variable $Y(J)$ equals $y(J * H)$, which is the value of $y(t)$ at $t = J * H$, and product $J * H$ denotes *the number of time increments*. Variable $Z(J)$ is related to $z(t)$ similarly. Equations (9-9) are easily programmed in BASIC.

In Fig. 9-10 is the program. In the remark of line 100, XDD and XD denote the second and first derivatives of X, respectively. The next remark states that the values of CI, N, and FT to be entered from the keyboard are to be 0.2 second, 400, and 3 seconds, respectively. The first of these is the *communication interval CI,* which is *the increment of time used for sending data to the printer.* Within each communication interval are N calculations, and hence the *calculation interval H* is CI/N, or 0.0005 second. The *final time FT* is to be 3 seconds. For these keyboard entries, in the 3-second interval there are 15 sets of data sent to the printer and 6000 sets of calculations from the programmed equations corresponding to (9-9).

Declaring I, J, and N as integers in line 130 is not essential, but the program runs appreciably faster when this is done. The three input statements of lines 140-160 for the entry of values for CI, N, and FT allow the operator to run the program using different parameters. Next, the array variables Y(N) and Z(N) are dimensioned so that each array can store the values calculated during a communication interval. At the end of each communication interval, values are printed and *the variables that have been calculated in the loop no longer need to be stored.* Initial values of the state variables are given in line 180, and in 190 the calculation interval H is determined, with H = CI/N.

```
100  REM Solve XDD + 2*XD + 64*X = 64; X(0) = XD(0) = 0
110  REM Input CI = 0.2, steps N = 400, final time FT = 3
120  '
130  DEFINT  I, J, N
140  INPUT  "Enter the communication interval CI: ", CI
150  INPUT  "Enter the number of steps N: ", N
160  INPUT  "Enter the final time FT: ", FT
170  DIM  Y(N), Z(N)
180  Y(0) = 0: Z(0) = 0
190  H = CI/N
200  '
210  LPRINT
220  LPRINT Tab(8); "t(s)"; Tab(21); "x";
                     Tab(34);  "dx/dt";  Tab(47);  "xx"
230  LPRINT
240  LPRINT  USING  "########.####"; 0, 0, 0, 0
250  '
260    FOR  I = 1  TO FT/CI
270  '
280          FOR  J = 1  TO  N
290            Y(J) = Y(J - 1)  +  H*Z(J - 1)
300            Z(J) = (1 - 2*H)*Z(J - 1)  +  64*H*(1 - Y(J - 1))
310          NEXT J
320  '
330      XX = 1 - SQR(64/63) * EXP(-H*N*I)
                     * SIN (SQR(63)*H*N*I + ATN(SQR(63)))
340  '
350      LPRINT  USING  "########.####";  H*N*I, Y(N), Z(N), XX
360      Y(0) = Y(N): Z(0) = Z(N)
370  '
380    NEXT  I
390  '
400  END
```

FIGURE 9-10. **Program to solve $d^2x/dt^2 + 2dx/dt + 64x = 64$ using approximate integration.**

Let us now consider the four LPRINT statements of lines 210-240. These and the LPRINT statement of line 350 should initially be PRINT statements so that output will go to the display. Then when the results are found to be satisfactory, conversion to LPRINT statements will send output to the printer. Statements of lines 210 and 230 provide blank lines, and the one of line 220 gives headings for the columns of output data. The TAB(n) function *fixes the position on the printed line of the variable that follows it.* An alternate way to space the string variables is simply to include the proper number of blank spaces within the quotation marks. The PRINT USING statement, which has not been used before, is described next.

PRINT USING. The statement PRINT USING is used for printing strings or numbers using a specified format, and LPRINT USING is similar except that output goes to the line printer. In the program of Fig. 9-10, only numeric fields are formatted with this command. With reference to lines 240 and 350, *each number sign (#) represents a digit position.* If the number to be printed has fewer digits than the positions specified, it is right justified.

A decimal point may be placed anywhere within the field. With the format of lines 240 and 350, there are eight digit positions allocated to the left of the decimal point and four digit positions to the right. Each of the four 0s that follow the semicolon of line 240 is printed as 0.0000, with seven leading blank spaces. The horizontal spacing of the numeric columns shown in the program results is that specified by the statements of lines 240 and 350.

For formatting strings, the number signs are replaced with two back slashes separated by blank spaces. An example using four blanks between the slashes is

```
PRINT USING "\    \"; "Two", "cubed", "is eight"
```

A field of character spaces equal to the number of blanks plus two is assigned to each string variable that follows the semicolon, and *the strings are left justified.* Thus, execution of the preceding statement gives the display

```
Two   cubed is eig
```

Other options of PRINT USING are available.

Calculations. With N = 400, there are 400 calculation intervals within a communication interval, which are performed within the nested FOR loop of lines 280-310. An exit from the loop occurs at the end of the communication interval. Then the true value, designated XX, is calculated from the exact equation (9-2), *which allows a check on the accuracy of the approximate method.* Sent to the display or printer by the statement of line 350 are four columns of results. The first of these is II * N * I, which is the *current value of time* t. Next are the current values of the state variables Y(N) and Z(N), corresponding to x and dx/dt, and the last one is the true value of x.

In line 360 the initial values Y(0) and Z(0) are assigned their current values, and the outer FOR loop is repeated. When I becomes FT/CI, the corresponding time t, or (H * N * I), equals the final time FT. Thus, the program ends. For the specified input values, the program passes through the outer loop 15 times, and it goes through the inner loop 6000 times. The final statement END is not required.

Results. When the program is run with results sent to the screen, the display is that of Fig. 9-11. Note the very close agreement between the exact values xx and the values of x obtained by the approximate method. Of course, the accuracy depends on the calculation interval H. If it is too large, accuracy suffers, but if it is too small the computer may take an excessive time to run the program.

```
Enter the communication interval CI: .2
Enter the number of steps N: 400
Enter the final time FT: 3
     t(s)          x              dx/dt          xx

     0.0000        0.0000          0.0000        0.0000
     0.2000        0.9109          6.6211        0.9105
     0.4000        1.6770         -0.1897        1.6728
     0.6000        1.0407         -4.4603        1.0416
     0.8000        0.5422          0.2557        0.5479
     1.0000        0.9856          3.0011        0.9844
     1,2000        1.3092         -0.2584        1.3035
     1.4000        1.0008         -2.0167        1.0022
     1.6000        0.7914          0.2321        0.7965
     1.8000        1.0054          1.3536        1.0041
     2.0000        1.1405         -0.1953        1.1363
     2.2000        0.9923         -0.9073        0.9935
     2.4000        0.9054          0.1578        0.9088
     2.6000        1.0079          0.6075        1.0068
     2.8000        1.0635         -0.1238        1.0609
     3.0000        0.9929         -0.4061        0.9938
```

FIGURE 9-11. Displayed results.

For the numbers used here, program execution following the last keyboard entry took 155 seconds. The equations of lines 290 and 300 were calculated 6000 times, and the one of line 330 was calculated 15 times. With LPRINT replacing PRINT, the printed headings and data have the same format as Fig. 9-11, and execution time increases by about 12 seconds.

The column *dx/dt* of the results of Fig. 9-11 is the slope of the curve of $x(t)$ versus t; it changes sign at maximum and minimum points of the curve. Within the given interval, the sign changes seven times, corresponding to four maxima and three minima. In order to plot accurately the curve of $x(t)$, more data are needed (see Prob. 9-13).

To indicate further the power of this method, let us note that introducing other forcing functions and perhaps even nonlinearities does not necessarily complicate the solution. For example, suppose the specified equation of (9-1) is changed to the following nonlinear one:

$$d^2x/dt^2 + 2x\ dx/dt + 64x^2 = 64 \sin t \tag{9-10}$$

On new line 285 of the program of Fig. 9-10, assign time T the value (N*I − N + J − 1) * H. Then in line 300 replace (1 − 2*H) with (1 − 2*H*Y(J − 1)) and also replace (1 − Y(J − 1)) with (sin(T) − Y(J − 1)^2). Delete line 330. These are the only required modifications (see Prob. 9-14).

The new program, with the same input parameters, takes a little longer to run. The method can be applied to a set of differential equations with more than one dependent variable. The equations can be nonlinear with time-varying coefficients (see Prob. 9-15), and there is no restriction on the number of input forcing functions. However, a high-speed digital computer is a necessity.

REFERENCES

BASIC Manual, IBM Corp., Boca Raton, Fla., 1981.
DOS Manual (2.10), IBM Corp., Boca Raton, Fla., 1983.
DOS Technical Reference, IBM Corp, Boca Raton, Fla., 1983.
Technical Reference, IBM Corp., Boca Raton, Fla., 1981.

PROBLEMS

SECTION 9-1

9-1. List the five internal directory commands described in Section 9-1 and state the function of each.

9-2. Suppose the following sequence is executed:

```
MD \DIR1
CD \DIR1
COPY B:ABC A:
```

Give a sequence of commands that will now eliminate subdirectory DIR1, with file ABC placed in the root directory.

9-3. List COPY commands that copy from the diskette of drive A, without changing any filenames, (a) all files to drive B; (b) all files having names starting with CH to drive B; (c) all files with extensions starting with T to drive B; (d) all files to the line printer; and (e) files having names starting with XY to the display.

9-4. Give the command that appends file DEF.TXT to the end of file ABC.TXT, with the new file being XYZ.TXT. All are on drive B.

9-5. Give a command that deletes all files of the root directory of drive A with extension EXE. Also give one that changes the prompt to "ABC> ".

SECTION 9-2

9-6. Explain how to place the volume identification label DATA BASE A on a diskette and state the names of all commands that will display the label.

9-7. Identify two different ways, using different commands, for copying all files from the diskette of drive A to an unformatted diskette of drive B, and state the relative advantages of each.

9-8. In a system having both IBM monochrome and color monitors, explain how to switch from one to the other in the DOS mode.

9-9. What does execution of MORE > LPT1 < B:ABC accomplish, and how does it differ from COPY B:ABC PRN: ?

9-10. Define piping of commands. Explain what command TYPE ABC.TXT | MORE > PRN: accomplishes.

SECTION 9-3

9-11. Assume the existence of default drive A with file AA.BAT on the diskette of drive B. Suppose B:AA Jo_Pat is entered in the DOS mode. For the nine commands of AA.BAT that follow, state exactly what each one does.

```
REM Hello %1
Pause
Type %0.BAT
Pause
B:
Dir/p
Pause
A:BASIC SORT
A:
```

9-12. With TREE.COM on default drive A and ABC.BAT on drive B, command B:ABC z z b is executed. The batch program follows:

```
REM  This batch program passes parameters and
REM       uses several subcommands.
REM
PAUSE  If you are ready to proceed,
if exist tree.com tree b:
if not %2 == a copy %0.bat
for %%Y in (a: b:) do dir/p %%Y
pause
if %3 == b goto A2
time
:A1
REM  A forever loop - use Ctrl-Break for exit
:a2
goto a1
```

State precisely what each instruction does and identify all commands that are executed.

SECTION 9-4

9-13. Execute the program of Fig. 9-10 with a communication interval of 0.1 for time 0-3 seconds, selecting an appropriate calculation interval. Sketch and dimension both $x(t)$ and dx/dt.

9-14. Prepare and execute a BASIC program that gives a solution to the nonlinear differential equation (9-10). Print data for x and dx/dt at intervals of 0.1 second for time 0-2 seconds. Select a calculation interval H so that the first two significant figures of the calculated values of x, when rounded, do not change when H is doubled.

9-15. Solve (9-1), with the constant coefficient 2 replaced with 2 * EXP($-t$).

9-16. By the numerical method of Section 9-4, solve
$$d^3x/dt^3 + 2d^2x/dt^2 + 3dx/dt + 4x = 5 + 6 \sin t$$

with $x = 1$ and $dx/dt = d^2x/dt^2 = -1$ at $t = 0$. Use CI = 0.5 second, N = 100, and FT = 20 seconds. To obtain the three first-order differential equations, let $y = x$, z = XD, and w = XDD, with XD and XDD denoting first- and second-order derivatives. Give the program and a nicely formatted printout of values of t, x, XD, and XDD.

Communications

10-1

SERIAL COMMUNICATIONS FUNDAMENTALS

Generally, a computer should be able to communicate with another nearby computer, either directly or through a modem that transmits over long distances via telephone lines. Communication at a distance is called *telecommunication*. Such data transfer is normally done with the bits of the bytes transmitted serially over a pair of conductors. Personal computers have this capability, provided suitable communications circuits are connected to the I/O channel through an expansion slot of the system board. In the author's system, the asynchronous communications adapter is part of a multifunction board that supplies additional memory and other features, using only a single slot. Numerous such boards are available from various vendors. In addition, connected to the system is a Hayes Smartmodem 1200.

Asynchronous Communications

In *asynchronous* data transfer, the required synchronization between the source transmission and the reception at the destination is accomplished by means of *start* and *stop* bits. These bits locate both the beginning and the end of a character. Each transmitter and receiver has its own clock, and the clocks are not synchronized.

One serial train representing a character may be immediately followed by another, or there may be a gap between two transmitted characters. In any event, each character is preceded by a start bit and is terminated by one or more stop bits. Asynchronous transfer is especially well suited for transmission directly from a keyboard because of the time intervals between the key closures.

In communications, bits are said to be either in the *space* state (0) or the *mark* state (1). A start bit has the space (0) state and a stop bit has the mark (1) state. During intervals of no transmission, communications lines have the mark state. When the receiver of a modem detects a change in the line from the mark to the space state, the start bit of an incoming character is indicated. After the specified number of character bits has been received, the line goes to the mark state for at least one stop bit. If there is no character immediately following the one received, the line remains in the mark state until the next start bit arrives.

Figure 10-1 shows one possible format of a serial character. There are seven character bits representing the letter *F* (code 46H), plus one stop bit and a parity bit, with even parity assumed. The parity bit is generated so that the 8 character and parity bits have an even number of 1 states. The bits move to the right as time increases. Note that the LSB of the character is sent first and received first, following the start bit, and the MSB precedes the stop bit. The start and stop bits *frame* the character. The levels of the figure are simply the representation of mark (1) and space (0) states. Because negative logic is used on communications lines, the levels do not denote voltages.

In the sample format of Fig. 10-1, a second character is indicated as following the 10-bit stream of the letter *F*. Otherwise, at the extreme left, the space state of the next start bit would be replaced with the mark state of an idle line. Seven character bits, even parity, and 1 stop bit per character are very common. Both ends of the line must, of course, use exactly the same format. As shown in the figure, there are 10 bits in the serial train, including the start and stop bits. When a file is transmitted, there are no intervening idle states between the characters. In this case, with a transmission rate of 300 bits per second (bps), 30 characters are sent per second, or 1800 per minute. *The bit rate in bits per second nearly equals the number of words per minute.*

Because communication between computers and terminals is usually asynchronous, we will devote most of our attention to this type. Two communicating devices will be referred to as *local* and *remote* units, even if they are close together.

Synchronous Communications

In *synchronous* systems there are no start and stop bits, thus giving a more rapid transfer of data. The bits of one character immediately follow those of the one preceding it. Data are usually sent in blocks at high bit rates. They are normally buffered prior to transmission, and the received data go to a buffer for storage until they can be processed. To implement the required synchronization, the transmitter

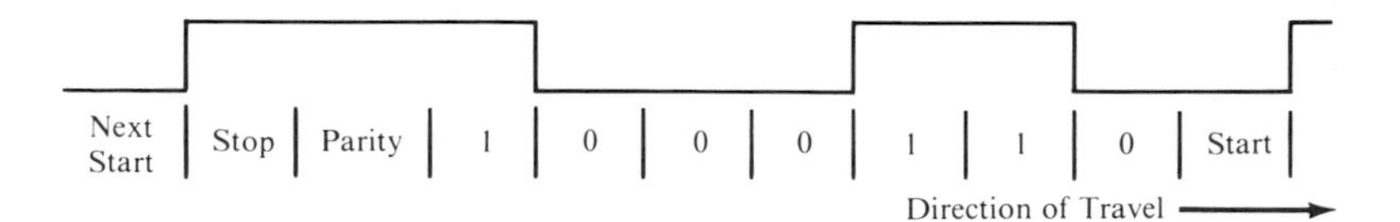

FIGURE 10-1. Sample format of a serial data word.

and the receiver must use clocks that are precisely synchronized with one another. *Synchronization is accomplished by means of signals sent with the data.* Because the required equipment is more expensive, synchronous communication is practical only for systems that must transfer large amounts of data at high speed. There are many such systems that communicate synchronously.

Communications Lines

Lines that connect computers and terminals are classified by the direction and manner in which they move data. The three types are simplex, half-duplex, and full-duplex.

A *simplex* system transmits data in one direction only. Familiar examples are TV and radio. In contrast, a duplex line gives two-way communications. Simplex lines are seldom used even for one-way data transmission because control signals usually must be returned.

Half-duplex (HDX) provides communications in two directions, but not simultaneously. Only one channel is used, and the direction of data transfer is controlled by handshake signals. When one end of the line is transmitting, the receiving end must wait for transmission to terminate before it can send data. It is not possible to transmit and receive at the same time. An example of half-duplex communications is CB radio. Two conductors are adequate.

Most widely used is *full-duplex (FDX),* which employs two channels. Each channel carries data one way, but the directions are opposite, allowing simultaneous communication. An example is a telephone conversation. Basically, a full-duplex line consists of two simplex channels. Although separate pairs of wires can be and are sometimes used, most full-duplex systems put the channels on a single pair of conductors. In fact, on the standard telephone line connected to the local telephone exchange, only two wires are usually available. Normally, the cable that connects a telephone to the wall jack has four conductors, but one of these is simply a ground not used in communication, and another is unconnected. *All data transfers, including handshake signals and rings, travel over a single pair of wires.*

Most full-duplex systems provide *echoplex.* This system causes a character sent from one end to be echoed by the receiving end back to the source for display. The operator views the returned characters, and any characters garbled or lost in transmission are readily revealed. With echoplex, it is not necessary to send parity bits or to employ other ways of detecting errors.

The common telephone circuit is called a *switched* line, because the local loop is normally switched to another line at the central office. There are leased, or private, lines that provide nonswitched permanent connections between stations. These are installed by the phone company, require no dialing, and give faster and higher-quality communications. Then there are direct-connect in-house lines, which often connect multiple computers or terminals over relatively short distances to a mainframe. In addition to the two-wire line of the local telephone loop, data are frequently moved over three-wire and four-wire lines. Four-wire leased-line data communication links are fairly common.

Modems

A *modem,* or *data set,* provides the interface between the circuits of the communications adapter and the telephone or other serial line. The word *modem* is a contraction of *modulator-demodulator*. *Modulation* converts the digital signals of the computer into analog signals, or tones, suitable for transmission over telephone lines. This conversion is done by changing some property of a sinusoidal carrier signal. In contrast to a digital signal that has only two allowed states, *analog signals are continuous,* with an infinite number of states. Because telephone lines and their switching circuits would distort digital signals beyond recognition, modulation is essential. *Demodulation* is the reverse process, which changes the analog signals back to their original digital form. *Modems allow data transfer between dissimilar computers.*

Baud

Because the digital signals that are converted into analog signals before transmission are restricted to two discrete states, the modulation changes abruptly at the instant between bits or groups of bits. *The number of times that the modulation is fixed per second is called the baud.* If each transmitted bit fixes the corresponding modulation, the bit rate in bits per second (bps) and the baud rate are equal.

In many cases, however, the modulation is fixed only for groups of bits, in which case the bits per second and the baud differ. For example, when operating at 1200 bps, the Smartmodem has a modulation rate that is fixed for pairs of bits called *dibits,* and the baud rate is 600. In fact, on telephone lines at and above 1200 bps, most data sets have a baud rate that is less than the bit rate. Although technically inaccurate, many engineers refer to the bits per second as the baud. Modems and modulation methods are examined in Section 10-3.

ASCII Control Characters

Given in Fig. 1-22 were codes for a number of ASCII characters. However, most of the characters used for controlling communications, as well as printers and other peripherals, were omitted. The 33 *ASCII control codes* are shown in Fig. 10-2. Each code can be produced by holding down the Ctrl key while pressing another, which is a second reason for calling them *control characters.*

The table gives the mnemonic and identifying word names, the activating key combinations, and the codes. The actions of many of these codes can be observed by pressing the key combinations in the DOS mode (see Prob. 10-1).

To illustrate, code 06H is generated by Ctrl-F of the PC to indicate an acknowledgment. Perhaps the computer at the far end sends a handshake signal to the PC requesting the use of a half-duplex line for transmission, and the response from the PC is 06H. This response may be recognized by the computer at the far end as an *affirmative acknowledgment.* Code 15H for NAK is interpreted as a *negative acknowledgment,* which serves as an indicator that a received request or message contained errors and should be repeated.

Ctrl-L produces code 0CH for *form feed,* which can be sent to a printer to direct it to advance printing to the top of the next page. A transmitted backspace character with code 08H may cause the cursor of the display unit at the receiving end to move one space backward, with the prior character deleted by replacing it with a

NUL	Null	Ctrl-2	00H
SOH	Start of heading	Ctrl-A	01H
STX	Start of text	Ctrl-B	02H
ETX	End of text	Ctrl-C	03H
EOT	End of transmit	Ctrl-D	04H
ENQ	Enquiry	Ctrl-E	05H
ACK	Acknowledge	Ctrl-F	06H
BEL	Bell	Ctrl-G	07H
BS	Back space	Ctrl-H	08H
HT	Horizontal tab	Ctrl-I	09H
LF	Line feed	Ctrl-J	0AH
VT	Vertical tab	Ctrl-K	0BH
FF	Form feed	Ctrl-L	0CH
CR	Carriage return	Ctrl-M	0DH
SO	Shift out	Ctrl-N	0EH
SI	Shift in	Ctrl-O	0FH
DLE	Data line escape	Ctrl-P	10H
DC1	Device control 1	Ctrl-Q	11H
DC2	Device control 2	Ctrl-R	12H
DC3	Device control 3	Ctrl-S	13H
DC4	Device control 4	Ctrl-T	14H
NAK	Negative acknowledge	Ctrl-U	15H
SYN	Synchronous idle	Ctrl-V	16H
ETB	End-of-transmit block	Ctrl-W	17H
CAN	Cancel	Ctrl-X	18H
EM	End of medium	Ctrl-Y	19H
SUB	Substitute	Ctrl-Z	1AH
ESC	Escape	Ctrl-[	1BH
FS	File separator	Ctrl-\	1CH
GS	Group separator	Ctrl-]	1DH
RS	Record separator	Ctrl-6	1EH
US	Unit separator	Ctrl -	1FH
DLE	Data link escape	Ctrl-space	20H

FIGURE 10-2. ASCII control characters.

blank (20H) at the new location of the cursor. Several NUL characters with code 00H are sometimes inserted into a stream of data simply to give a time delay, following a control such as a CR to a mechanical printer that requires time for execution.

The ENQ control character (05H) serves as a request for a response from a remote station, possibly seeking status information or identification. Control BEL (07H) usually activates an audible attention device. Horizontal tab HT with code 09H may advance the cursor of the display at the far end or the position of the head of a printer to the next predetermined character position of the same line. To advance the active position to the same location of the next line, vertical tab VT, with code 0BH, is transmitted.

Control CAN, with code 18H, indicates that the data with which it is sent are erroneous and should be disregarded. The precise meaning of this character, as well as that of the other characters, must be defined for the particular application. Escape (ESC) is usually used as a prefix affecting the interpretation of the character that follows it. Most of the control characters were developed for teletypewriters and carried over to computer usage.

Adapter Circuitry

Figure 10-3 is a block diagram of the circuits of the *asynchronous communications adapter* of the PC. Except for power lines that supply 5, 12 and -12 V, connections at the I/O slot are shown at the left side. In addition to the 10 address and 8 data lines, there are 5 controls. These controls are $\overline{\text{IOR}}$, $\overline{\text{IOW}}$, AEN, RESET, and IRQ4.

Addresses. As indicated in the figure, chip-select $\overline{\text{CS}}$ is enabled when address bits A9–A3 are set, provided address-enable AEN is low. AEN is always maintained low by the processor, except when the hold-acknowledge HLDA line is high during a DMA operation. Bits A2, A1, and A0 feed directly into the controller for addressing the internal registers. Thus, the addresses of the communications controller are 3F8H to 3FFH, but 3FFH is not used. With the chip enabled, the data bus transceiver is also enabled by the low input to terminal $\overline{\text{G}}$. When the chip is disabled, the 8-bit bidirectional data bus and the control lines float in the three-state mode.

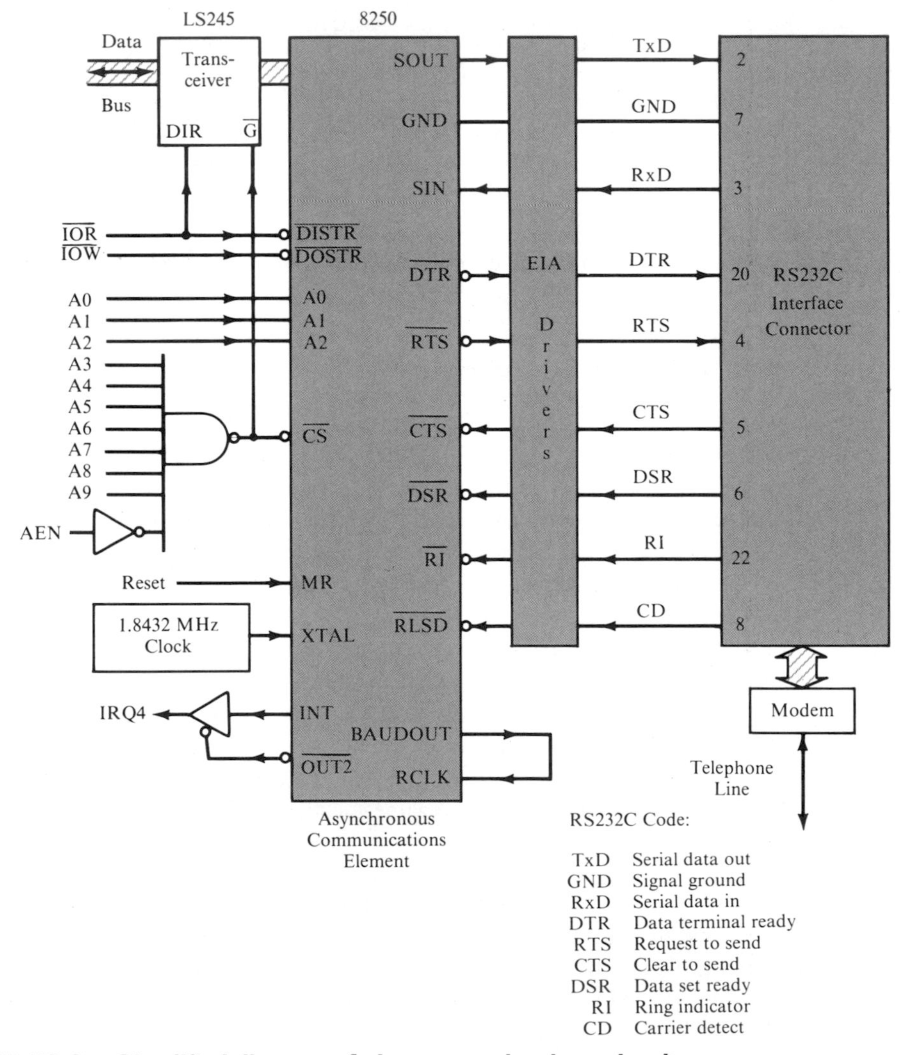

FIGURE 10-3. Simplified diagram of the communications circuits.

Read and Write. For a read operation, the I/O read signal $\overline{\text{IOR}}$ from the CPU must activate the data-input-strobe pin $\overline{\text{DISTR}}$, with the write signal $\overline{\text{IOW}}$ inactive. Bytes are transferred from the controller to the processor. The low signal at the direction pin DIR of the transceiver sets this chip for right-to-left flow of data, as required for reading.

On the other hand, with $\overline{\text{IOR}}$ inactive and $\overline{\text{IOW}}$ strobing the data-output-strobe pin $\overline{\text{DOSTR}}$, the transceiver direction is reversed, and the write operation moves data from the CPU to the controller. The 74LS245 transceiver with three-state outputs provides both directional control and buffering.

Reset and IRQ4. An active reset from the processor provides a signal to the master-reset MR pin of the controller. This signal inactivates all control signals and clears the line control register, the modem control register, and the interrupt enable register, along with appropriate bits of status registers.

Interrupt request IRQ4 from the communications controller to level 4 of the interrupt controller, corresponding to INT 0CH, passes through a three-state buffer enabled by output signal $\overline{\text{OUT2}}$. The program designed to use the interrupt loads the vector address into four locations in RAM starting at 00030H; this vector address must point to a service routine.

The RS-232-C Interface

Transmission between the communications adapter and the peripheral, assumed here to be a modem, is via the Electronics Industry Association (EIA) standard *RS-232-C* communications interface. Unlike the I/O interface chips, this hardware interface does not provide control and buffering. Rather, it simply joins two units with a connector that is attached to a cable, *with specified voltage levels and signal parameters*.

Although the standard interface has 25 pins, which are numbered and assigned functions and signal levels, the application determines which of them are connected. Here only 13 are used, with 9 of the numbered pins indicated in Fig. 10-3. The acronyms for these pins are defined by the RS-232-C code at the bottom of the figure. The other four are connections from the system to spare lines of the interface, and they are available for current loop operation. Current loops are described in Section 10-5. The communications circuits of the adapter in Fig. 10-3 constitute the data terminal that connects the CPU with the modem. This system is shown in Fig. 10-4.

Standards such as the widely accepted RS-232-C are essential for the interconnection of equipment made by different manufacturers. Figure 10-3 shows *EIA drivers*. These drivers convert the TTL voltages to EIA standards, providing both buffering and inverting. In the negative logic system that is specified, the space (0) state must have a positive voltage between 3 and 15 V, and the mark state is represented by a negative voltage ranging from -3 to -15 V, with values between

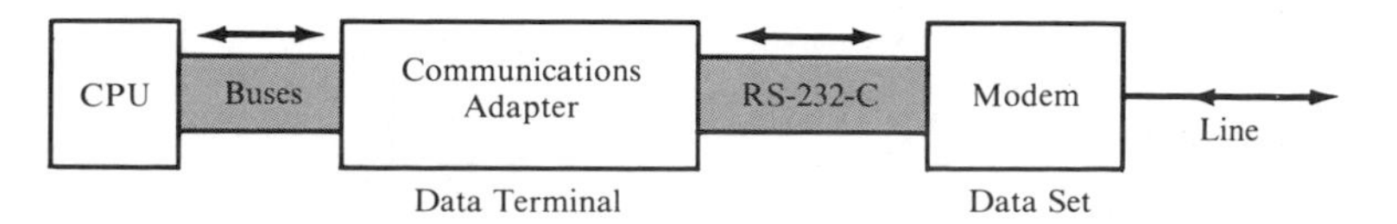

FIGURE 10-4. The data terminal and data set.

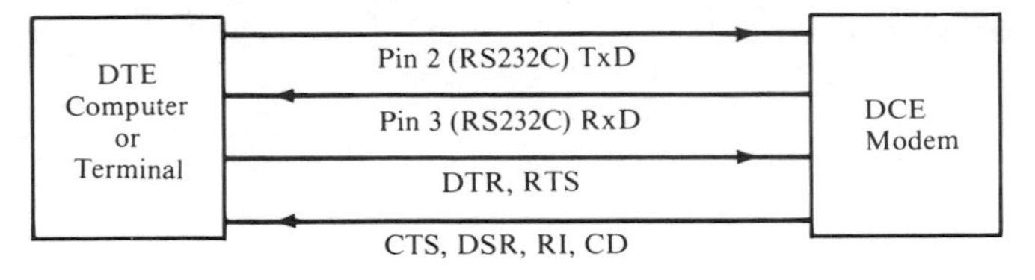

FIGURE 10-5. DTE and DCE.

−3 and +3 V undefined. These values were selected in preference to the TTL levels, because errors in transmission are reduced. The EIA drivers of the PC supply +12 and −12 V for states 0 and 1, respectively.

Data Terminal and Data Communications Equipment. Equipment that sends out data on pin 2 of the RS-232-C interface and receives on pin 3 is referred to as *data terminal equipment (DTE)*. Such equipment also transmits the data-terminal-ready DTR signal and/or the request-to-send RTS signal. Usually received are one or more of the signals clear-to-send CTS, data-set-ready DSR, ring indicator RI, and carrier-detect CD. Most computers and terminals are connected as DTE, as indicated by the RS-232-C connections of Fig. 10-5.

On the other hand, equipment that transmits on pin 3 and receives on pin 2 is called *data communications equipment (DCE)*. Also, the DSR signal and/or the CTS signal are transmitted. In addition, DCE may send other signals, such as RI and CD. Signals from the DTE are received by the DCE, and those from the DCE are received by the DTE. In most cases, the DCE is a modem, as in Fig. 10-5.

Figure 10-6 shows a diagram of a telecommunications system consisting of two PCs, two data sets, and a telephone line. The lines between the data sets and their local telephone switching stations consist of two wires, and transmission can be either half-duplex or full-duplex.

Signals TxD and RxD. The *transmitted-data (TxD)* line from the computer to its modem is held in the mark condition during intervals between characters and at all times when no data are being transmitted. In order to transmit data, it is essential that signals DTR, DSR, RTS, and CTS be active, if these lines are connected.

The *received-data (RxD)* line supplies the serial input data from the modem to the computer, but signals DTR and DSR must both be active. The line is held in the mark (1) condition at all times when CD is inactive.

Signals DTR, RTS, and CTS. The *data-terminal-ready DTR* signal always goes from the DTE, assumed here to be a computer, to the DCE, or assumed modem. The active (ON) condition prepares the modem for connection to the communication channel, and it maintains the connection once established by a call or an answer. When the DTR goes OFF, the modem is normally removed from

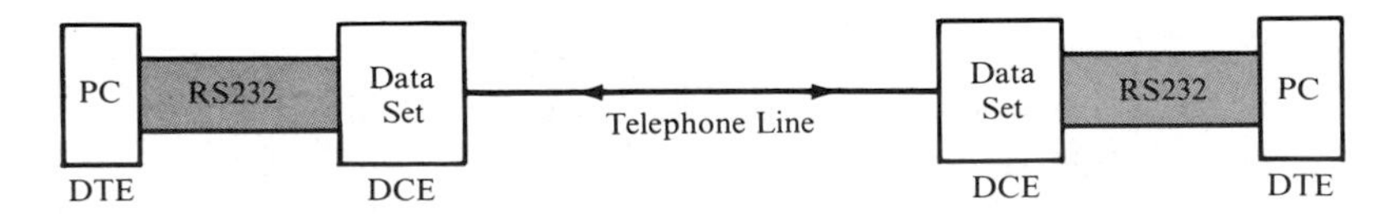

FIGURE 10-6. A communications system.

the communication channel following completion of an operation in progress. RI is not affected by DTR. Usually, DTR is ON at all times after the communications circuits have been initialized.

Signal *request-to-send RTS* from the computer (DTE) to its modem (DCE) is used to condition the local modem for data transmission. With full-duplex lines, RTS is held active at all times, *which enables the transmitter of the modem.* On simplex lines it is held either ON or OFF, depending on the direction of communication. When it is OFF, the transmitter of the data set is inhibited.

On a half-duplex channel, the ON condition of RTS notifies the data set that its local computer is ready to transmit, and the OFF condition places the modem in the receive mode. When RTS goes OFF to ON, the modem takes action as necessary, and if it is not receiving data, it sends a *clear-to-send (CTS)* signal to the computer, with the transmitter of the modem enabled and its receiver disabled. An ON-to-OFF transition instructs the modem to complete any transmission in progress and then to inhibit the transmitter. The data set responds by turning off CTS to the computer.

The CTS signal from the data set indicates to its local computer that the transmitter is ready to receive data for transmission to the far end. The computer must not send serial data to the modem when CTS is OFF. If RTS is not connected through the RS232C interface, it is assumed by the modem to be ON at all times, and the CTS response is based on this assumption.

Signals DSR, RI, and CD. The *data-set-ready (DSR)* signal from the data set to its computer indicates the status of the local modem. When ON, the indicator usually signifies that the data set is connected to a communications channel, with any prior call operation having been completed. With switched service as in the case of telephone lines, the modem is said to be ''off hook'' when DSR is active, whereas it is ''on hook'' when DSR is inactive. These modes are illustrated in Fig. 10-7.

In the on-hook mode, the data set disregards all signals except RI. If DSR goes from active to inactive during a call, the computer should interpret this action as an indicator of a lost or aborted connection, equivalent to ''hanging up'' the telephone, and the computer should then terminate the call.

When ON, the *ring indicator (RI)* from the modem to its local computer indicates that a ringing signal is being received on the communication channel. RI is active during the actual rings. Between ring cycles and at other times when no ringing is in progress, RI is OFF. The ring indicator is not affected by the status of the DTR signal.

The RS-232-C standard refers to *carrier detect (CD)* by the name *received line signal detect (RLSD).* It is sent from the modem to its local computer. An active CD signal indicates that the data set is receiving a satisfactory carrier from the remote end, whereas an inactive state denotes either no carrier or one that is unsatisfactory. Whenever CD is OFF, the serial input data channel is clamped to the mark (1) condition. On half-duplex channels, signal CD is held OFF by the modem whenever the RTS signal is active.

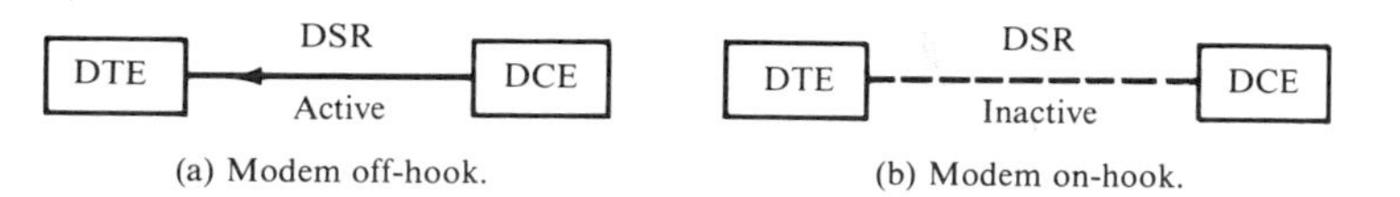

FIGURE 10-7. Off-hook and on-hook modes.

8250 ORGANIZATION

Here we shall examine carefully the 8250 *asynchronous communications element,* which is the controller chip of the adapter of Fig. 10-3. The IC of the 40-pin DIP is fabricated with NMOS technology. With associated circuits, it provides the interface between the parallel data bus of the computer and the serial data lines of the modem. Parallel-to-serial conversion of data is implemented by a shift register for transmission to the modem. A second shift register provides serial-to-parallel conversion when receiving from the modem.

The 8250 is commonly referred to as a *universal asynchronous receiver/transmitter (UART).* Some communications chips are designed for control of either synchronous or asynchronous transfers and these are called *universal synchronous/asynchronous receiver/transmitter (USART).*

Registers

The 10 addressable byte registers of the 8250 require 9 port numbers for input (READ) and 8 for output (WRITE). Because the three pins dedicated to internal addressing provide only eight combinations, the MSB of the line control register is used to distinguish between ports with the same numbers. This bit is called the *divisor latch access bit (DLAB)* because it must be set in order to read or write the two divisor latches with port numbers 3F8H and 3F9H. When DLAB is 0, addresses 3F8H and 3F9H relate to other registers. Whenever an address that depends on the state of DLAB is given, it will be followed by the state in parentheses. Figure 10-8 lists the 10 registers and their hexadecimal addresses.

Block Diagram

Figure 10-9 shows a block diagram of the 8250 element, with most of the internal control lines omitted. Immediately above each addressable register is the address, including in parentheses, where appropriate, the state of the DLAB. With DLAB low, address 3F8H applies to either the receiver buffer or the transmitter buffer register, depending on whether the operation is READ or WRITE, and with DLAB

Address	Register	Read/Write
3F8 (0)	Tx buffer	Write only
3F8 (0)	Rx buffer	Read only
3F8 (1)	Divisor latch LSB	Read/write
3F9 (1)	Divisor latch MSB	Read/write
3F9 (0)	Interrupt enable	Read/write
3FA	Interrupt ID	Read only
3FB	Line control	Read/write
3FC	Modem control	Read/write
3FD	Line status	Read/write
3FE	Modem status	Read/write

FIGURE 10-8. 8250 registers and addresses.

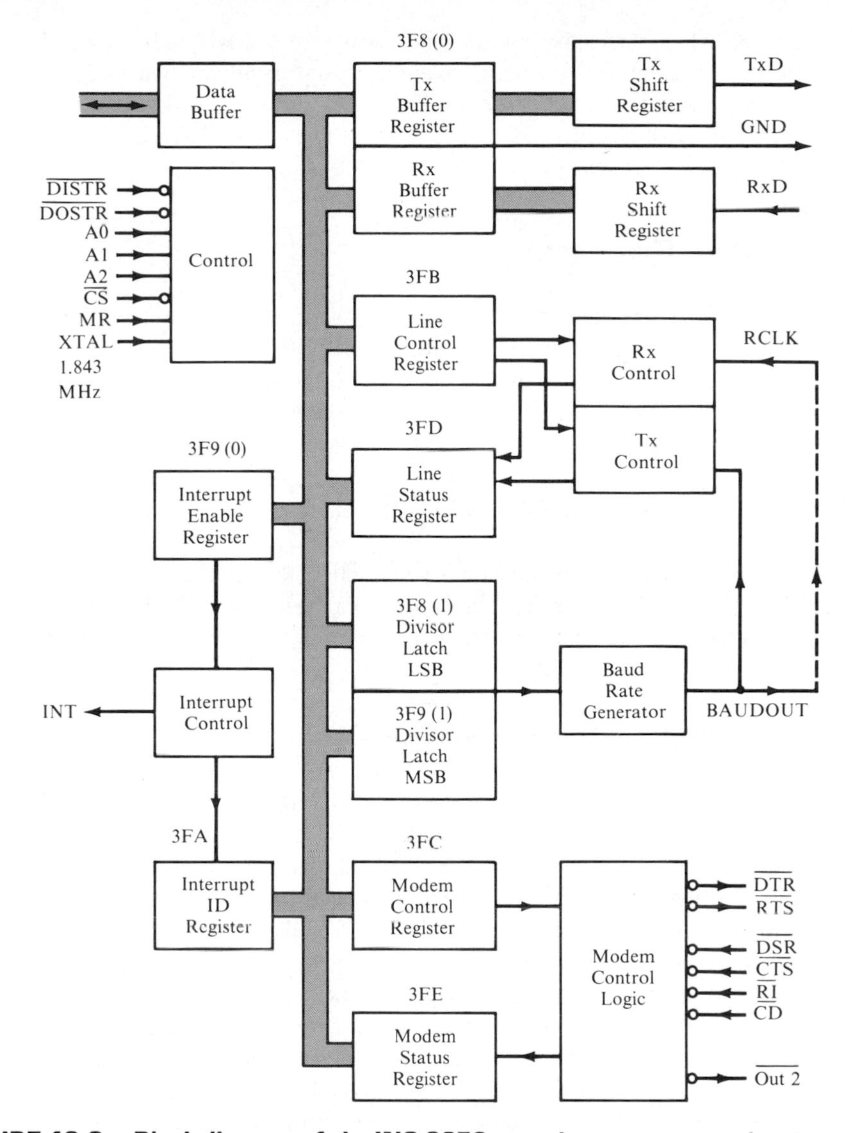

FIGURE 10-9. Block diagram of the INS 8250 asynchronous communications element.

high, the address accesses the low byte of the divisor latch. Address 3F9H is used for the interrupt enable register when DLAB is low, but when the bit is high, the most significant byte of the divisor latch is addressed.

In the block diagram 29 pins are indicated, including the 8 pins of the data bus. In addition, two pins provide the 5-V supply, and the other nine are connected directly to either a logical 0 or a logical 1, or simply not connected.

Connected internally to the transmitter control block is the BAUDOUT line from the *baud rate generator,* which provides a pulse train to the transmitter clock. The vertical dashed line at the right is an external connection of BAUDOUT to the receiver clock (RCLK). The connection makes the bit rate for reception the same as that for transmission, which is the usual condition. The actual bit rate equals the frequency of the pulse train of BAUDOUT divided by 16.

The XTAL input to the control block is a 1.8432-MHz pulse train. This pulse train is generated from a circuit having a crystal oscillator that has 10 times this frequency, with the output of the oscillator passed through a divide-by-10 network. The frequency of BAUDOUT is that of the XTAL input divided by the number of the two divisor latches.

Baud Rate Generator

Figure 10-9 shows a baud rate generator with an input from the two divisor latches. In addition, the generator receives the 1.8432-MHz pulse train from the XTAL input. Division of this frequency by the 16-bit number of the divisor latches gives the frequency of the pulse train of BAUDOUT. Because the frequency of BAUDOUT also equals 16*bps, it follows that the bit rate is

Bit rate = 1,843,200 / (16 * divisor) bps

For example, if the divisor number is 0060H (96 decimal), the bit rate is 1200 bps. The divisor must be written into the latches during the initialization of the chip prior to communications. The maximum allowed bit rate is 9600 bps.

TxD and RxD Buffer Registers

When a character is written to the *transmitter buffer register,* it is held until the *transmitter shift register* has shifted out serially the last character sent to it. Then the character of the buffer is moved to the shift register, and the buffer becomes empty. Writing to it when it is not empty replaces the prior character, which should be avoided, of course. The LSB is data bit 0, and it is the first one shifted to the communications line. Because the buffer holds a character until it can be transmitted, it is often called the *transmitter holding register*.

The *receiver buffer register* contains the last character moved to it from the *receiver shift register*. Its LSB is the first one that serially entered the shift register from the communications line. Unless the character is read before the next one is received, it is destroyed, which is the *overrun* error. Both the transmit and receive buffer registers have address 3F8H, with DLAB zero. There is no conflict, because one is an input port and the other is an output port. Because the transmitter and receiver have buffer registers in addition to their shift registers, they are said to be *double buffered*.

Line Control Register

The format of the asynchronous data word is specified by writing to the *line control register*. Also, the state of DLAB is defined by bit 7. The contents and encoding are shown in Fig. 10-10. The port number is 3FBH, and both writing and reading are allowed.

The *set break* bit 6 of the line control register is set when it is desired to alert a distant computer or terminal for attention. It simply changes the output line from the normal mark state to the spacing state. Set break is disabled by writing a 0 to bit 6.

7	6	5	4	3	2	1 0
DLAB	Set Break	Stick Parity	Parity Select	Parity EN	Stop Bits	Word Length

Encoding:

Bit 6:	When 1, the TxD output line is forced to the space state and remains there as long as bit 6 is 1.
Bit 5:	When 1 with bit 3 also 1, the parity bit is 0 if bit 4 is 1, and it is 1 if bit 4 is 0.
Bit 4:	Even parity = 1; odd parity = 0.
Bit 3:	Parity enable = 1; no parity bit = 0.
Bit 2:	Two stop bits = 1 (for 6- to 8-bit words); one stop bit = 0.
Bits 1, 0:	5 bits = 00; 6 bits = 01; 7 bits = 10; 8 bits = 11.

FIGURE 10-10. Line control register.

Three of the bits relate to parity. If bit 3 is 0, there is no parity bit, and bits 4 and 5 are don't-cares. However, if bit 3 is 1, there is a parity bit, with choices of even, odd, 1, or 0. The selection depends on the states of bits 4 and 5. If bit 5 is 0, the parity is either even or odd, as specified by bit 4. Even parity is most often used.

However, with parity enabled and bit 5 set, the parity bit is always 0 if bit 4 is 1, but it is always 1 if bit 4 is 0. Some systems with echoplex do not allow parity error checking and may require a mark parity or perhaps a space parity.

Bit 2 is used to specify either 1 or 2 stop bits for words having lengths of 6 to 8 bits. However, 5-bit words can have only 1 or 1.5 stop bits, and in this special case, a value of 1 for bit 2 gives 1.5 stop bits. Most common is 1 stop bit.

Each transmitted or received serial character can have a specified length of either 5, 6, 7, or 8 bits. Usually, text characters have 7 bits and binary codes have 8 bits. Encoding of bits 1 and 0 specifies the word length.

The proper data must be written into the line control register before communication is possible. Once the word format has been set, it is not necessary to write to the register again, or to read it, except to set or clear DLAB. The instruction sequences that follow will set or clear the bit without changing the other bits of the line control register:

```
MOV DX, 3FBH        MOV DX, 3FBH
IN  AL, DX          IN  AL, DX
OR  AL, 80H         AND AL, 7FH
OUT DX, AL          OUT DX, AL
```

Modem Control Register

Port 3FCH accesses the read/write *modem control register,* which controls the interface with the data set. Indicated in Fig. 10-11 are its contents. Bits 3 and 2 control the outputs of pins $\overline{OUT2}$ and $\overline{OUT1}$, which are available for control as desired. In the PC, an active $\overline{OUT2}$ enables the output buffer of the IRQ4 interrupt, and $\overline{OUT1}$ is not used.

In the full-duplex mode, bits 0 and 1 are written with 1s, thereby turning ON the data-terminal-ready and request-to-send signals of output pins DTR and RTS. These remain ON at all times during communications. When half-duplex is used, the RTS bit is set when the computer wants to send data, but once the transmission has ended, it is reset to return the modem to the receive mode. This operation is called *line turnaround*.

7	6	5	4	3	2	1	0
0	0	0	Loop	OUT2	OUT1	RTS	DTR

Encoding:

Bits, 7, 6, 5: Always 0.
Bit 4: When 1, the output of the transmitter shift register is looped back into the receiver shift register.
Bit 3: The complement of the bit goes to pin $\overline{\text{OUT2}}$.
Bit 2: The complement of the bit goes to pin $\overline{\text{OUT1}}$.
Bit 1: The complement of the bit goes to pin $\overline{\text{RTS}}$.
Bit 0: The complement of the bit goes to pin $\overline{\text{DTR}}$.

FIGURE 10-11. Modem control register.

To enable interrupt IRQ4 from the 8250, bit 3 must be set. Resetting the bit disables the interrupt line out of the buffer.

Bit 4 provides a loopback feature for diagnostic testing of the 8250. For normal communications, the bit is reset. When it is set, however, the output of the transmitter shift register is fed internally into the receiver shift register, so that a word written to the transmitter buffer register returns immediately via the receiver buffer register. Transmitting a character and then comparing it with the one received provides a way of verifying the transmit and receive data paths of the chip.

When the loopback feature is activated, the serial output line TxD is set to the mark (1) state, and the serial input line RxD and inputs $\overline{\text{DSR}}$, $\overline{\text{CTS}}$, $\overline{\text{RI}}$, and $\overline{\text{CD}}$ from the modem are disconnected. In addition, outputs $\overline{\text{DTR}}$, $\overline{\text{RTS}}$, $\overline{\text{OUT1}}$, and $\overline{\text{OUT2}}$ are internally connected to respective inputs $\overline{\text{DSR}}$, $\overline{\text{CTS}}$, $\overline{\text{RI}}$, and $\overline{\text{CD}}$. In the diagnostic mode, all interrupts are operational and can be tested.

The Divisor Latches

Port 3F8H is that of the least significant byte (LSB) of the *divisor latch,* and port 3F9H is that of the MSB. The DLAB bit must be set prior to addressing, and both reading and writing are allowed. The latches are programmed prior to communications to give the desired bit rate, which is 1,843,200 divided by the product of 16 and the divisor.

Line Status Register

The read/write *line status register* with address 3FDH provides status information concerning data transfer. Its contents are indicated in Fig. 10-12. Bit 7 is always 0, and bit 6 is read-only, not affected by write operations.

Whenever a character is moved from the transmitter buffer register to the transmitter shift register, bit 6 is reset. It is set only after the character has been serially shifted out. Thus, state 1 signifies an empty shift register.

State 1 in the bit 5 position indicates that the transmitter buffer (holding) register is empty and ready to accept a new word for transmission. *This bit should always be examined prior to writing to the Tx buffer register in order to avoid writing over a character that is waiting to be transmitted.* In addition, when the bit is set, the output interrupt pin is active if bit 1 of the interrupt enable register is set.

The *break-interrupt (BI)* bit 4 is set when the RxD input line is held in the space state for longer than a full word transmission time. When set, it generates an

7	6	5	4	3	2	1	0
0	TSRE	THRE	BI	FE	PE	OE	DR

Encoding:

Bit 6: Tx shift register empty (TSRE) indicated by 1.
Bit 5: Tx holding register empty (THRE) indicated by 1.
Bit 4: Break interrupt (BI) indicator.
Bit 3: Framing error (FE) indicator (no stop bit).
Bit 2: Parity error (PE) indicator
Bit 1: Overrun error (OE) indicator
Bit 0: Receiver data ready (DR) indicator.

FIGURE 10-12. Line status register.

interrupt, provided bit 2 of the interrupt-enable register is set. Recall that bit 6 of the line control register can be set to force the serial output line to the space state to alert another computer. The BI bit of the line status register is the detector of such a signal on the serial input line. BI is reset when the serial input line returns to the normal mark condition.

Bits 3, 2, and 1 are error indicators. A *framing error (FE)* occurs when a received character is without a stop bit, in which case the character is improperly framed. A *parity error (PE)* is detected when the parity is not the one specified, and an *overrun error (OE)* results when a character is transferred into the receiver buffer register before the last one was read. Each of these errors causes the corresponding bit of the line status register to be set, and each generates an interrupt if bit 2 of the interrupt-enable register is set. Reading the line status register clears the error indicators.

The *data-ready (DR)* bit 0 is set whenever a complete incoming character has been received in the receiver buffer register. It is reset when the character is read by the CPU, and it can also be reset by writing to the line status register. Bit DR should always be examined before reading the receiver buffer register in order to determine if a valid character is present. When bit 0 is set, an interrupt is generated provided bit 0 of the interrupt-enable register is set.

Modem Status Register

Port 3FEH is that of the read/write *modem status register*. The current state of each of the control lines CTS, DSR, RI, and CD from the modem to the CPU can be read from a bit of the high nibble of this register.

In addition, each bit of the low nibble of the register reveals whether or not the corresponding input signal *has changed state since the last reading of the modem status register*. Whenever a control input changes state, its associated low-nibble bit is set, and whenever the modem status register is read, it is reset. Also, a bit that is set generates an interrupt if bit 3 of the interrupt enable register is set. The contents of the modem status register are shown in Fig. 10-13.

7	6	5	4	3	2	1	0
CD	RI	DSR	CTS	Delta CD	Delta RI	Delta DSR	Delta CTS

FIGURE 10-13. Modem status register.

Status bits 7-4 of the figure are the complements of the bits received at the input pins. The word *Delta* denotes a change. For the bit 2 ring indicator, the Delta RI bit is set only when the RI input to the chip goes from ON to OFF, and it is reset when the register is read.

Prior to a transmission, the modem status register should be read to ascertain that the data set is ready and that a character can be sent. Bits 5 and 4 are the indicators to be examined. Also, bit 5 of the line status register should be checked to determine that the transmitter buffer register is empty. *Before reading a character from the receiver buffer register, the modem status register should be read for inspection of the DSR bit.* Then the data-ready bit 0 of the line status register should be checked.

Interrupt Enable Register

Port 3F9H with DLAB zero is the address of the read/write *interrupt-enable register*. The bits of the high nibble are always 0, but those of the low nibble provide for enabling individually the four allowed types of interrupts, with a bit set to 1 for enable. The contents of the interrupt enable register are indicated at the top of Fig. 10-14, and the interrupt priorities are shown in the table of the figure. Note that the bit numbers of the table are out of order.

All interrupts are disabled by writing 00H to the register; when this is done, line IRQ4 is inactive. Interrupts can also be inhibited by a reset of the OUT2 bit (3) of the modem control register, which disables the external buffer that feeds line IRQ4. The different interrupt types can be selectively enabled, of course, by setting the OUT2 bit and writing to the interrupt enable register.

Interrupt ID Register

The read-only *interrupt identification (ID) register* with port number 3FAH has 5 high-order bits that are always 0, as shown in Fig. 10-15. Thus, only the 3 lowest bits have significance. Bit 0 is 0 whenever an interrupt is pending, which implies that pin INT is active. When it is 1, there is no pending interrupt.

The output of the INT pin of the PC is used to generate a hardware interrupt through level 4 of the 8259 controller, implementing INT 0CH in the PC. An alternate method is to poll bit 0 periodically with software to determine if an interrupt is pending. The hardware method provides a more efficient use of CPU time.

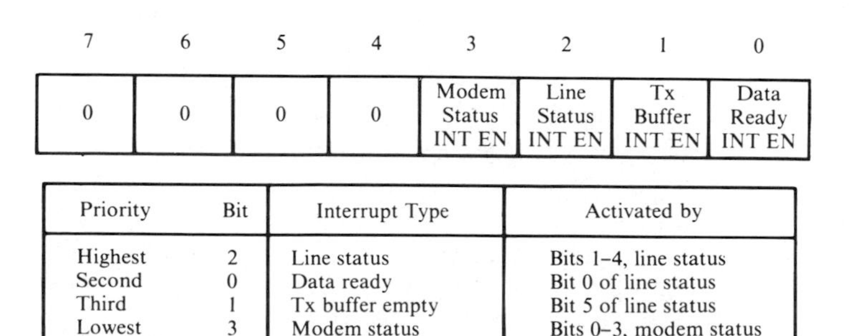

7	6	5	4	3	2	1	0
0	0	0	0	Modem Status INT EN	Line Status INT EN	Tx Buffer INT EN	Data Ready INT EN

Priority	Bit	Interrupt Type	Activated by
Highest	2	Line status	Bits 1–4, line status
Second	0	Data ready	Bit 0 of line status
Third	1	Tx buffer empty	Bit 5 of line status
Lowest	3	Modem status	Bits 0–3, modem status

FIGURE 10-14. Interrupt-enable register and interrupt priorities.

7	6	5	4	3	2 1	0
0	0	0	0	0	INT ID bits	Pending

Encoding:

Bits 2, 1: ID of pending interrupt of highest priority.
Bit 0: 0 if an interrupt is pending; otherwise 1.

FIGURE 10-15. Interrupt ID register.

Bits 2 and 1 simply identify which pending interrupt has the highest priority. Bit values 11, 10, 01 and 00 correspond to priorities 1, 2, 3, and 4, respectively, with the priorities defined in Fig. 10-14. A reading of the interrupt ID register allows a program to determine whether or not an interrupt is pending, and if so, to ascertain the highest priority.

10-3

MODEMS

Modems can be classified by speed, housing design, type of coupling, features and intelligence, and protocol. Usually modems with bit rates below 1200 bps are considered *low speed,* from 1200 to 9600 bps is *medium speed,* and over 9600 is *high speed.* Both *stand-alone* and *expansion board* modems are available. The former is a separate unit that has LED status indicators and can be used with any system, whereas the latter requires an expansion slot and is designed for a particular type of computer.

The expansion board modem is less expensive, can be located on the board with the communications adapter circuits, is more portable, and requires neither an RS-232-C cable nor an external power connection. However, it increases the internal heat, has no visible indicators, and cannot be transferred to a dissimilar computer. Still, it is quite popular for use with microcomputers. Various classifications are shown in Fig. 10-16.

Coupling is either direct or acoustic. A *direct-coupled* modem connects directly through a cable to the wall jack of the telephone line. An *acoustic coupler* is coupled by means of sound waves. After a number is dialed, a data switch of the acoustic coupler is turned on, and the telephone handset is placed firmly into two cylindrical rubber grommets of a cradle containing a small speaker and a microphone. Data transmission is from the speaker to the telephone mouthpiece via audio tones, and reception is from the telephone receiver through sound waves to the microphone of the cradle. In general, acoustic couplers are less expensive, but they

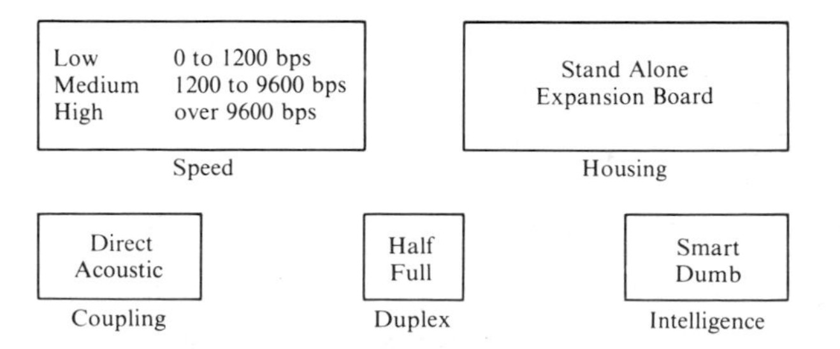

FIGURE 10-16. Various classifications of data sets.

are more susceptible to noise, are less convenient to use, and are limited by bandwidth to low-speed applications.

Data sets are classified by features and capabilities in many different ways. One such classification is *automatic-dial* or *manual-dial*. Some sets provide automatic dialing and an answering service, with a directory of telephone numbers stored in a modem memory. Those controlled by a microprocessor and containing an operating system stored in an internal ROM are called *smartmodems,* or *intelligent modems*. These have numerous special features, including automatic-dial, file transfer, and local echo. In contrast, a *dumb terminal* is a system connected to a modem not having the intelligence provided by a microprocessor and a stored program. Normally, a dumb terminal transmits only from the keyboard, and all received data are sent to the display, with manual dialing and answering.

Protocol

Protocol consists of a set of rules designed to ensure reliable communication that is both efficient and nearly error free. Both equipment manufacturers and various organizations prepare protocols, most of which are general to allow flexibility. Computers that communicate with each other must use the same protocol and standards.

Typically, a protocol may be required to do the following:

1. Identify the code of the character set.
2. Fix operation as simplex, half-duplex, or full-duplex.
3. Define handshake signals.
4. Specify error-checking and error-correcting procedures.
5. Establish rules for ringing a destination.
6. Set rules for answering a call.
7. Provide a way for one end to interrupt the other end while it is transmitting.

With the defined handshake signals, a way is provided to verify that a channel has been established and is clear for data transfer. Also, a handshake protocol defines how one end will notify the other end that it is ready, or perhaps momentarily busy. Usually, the modems handle the protocols, with the operations being transparent to the computers.

Two protocols are described briefly here. These are known as the Bell 103J and the Bell 212A.

Bell 103J Protocol

The most common low-speed data sets used with PCs are those compatible with the 103J protocol. Full-duplex asynchronous operation on two channels of a two-wire switched network is provided, with RS-232-C interfacing. A 117-V, 60-Hz power supply is required, with a dissipation of 10 W. Typical dimensions are 2 × 6 × 10 inches, and the weight is about 4 pounds.

Baud and bit rates are usually 300, giving a transfer rate of about 300 words per minute. Although relatively slow, *this rate is advantageous when communication is by keyboard with a system that charges for the connection time*. Costs per unit time are generally somewhat greater for modems with higher bit rates;

furthermore, the cost of modems increases with increased bit rate capability. However, when large files are sent in blocks, 1200 bps is more economical. A 103J modem can operate only from 0 to 300 bps.

Originate and Answer Modes. Within the telephone bandwidth from about 300 to 3300 Hz is adequate space for two separate full-duplex channels using a modulation technique called *frequency shift keying (FSK)*. Each channel requires a bandwidth that is 1.5 times the baud rate. For a baud of 300, the two channels will use a total of 900 Hz. In the 103J, channel 1 extends from about 950 to 1400 Hz, and channel 2 uses the band from 1950 to 2400 Hz.

The system that originates a call always transmits on one channel, designated here as channel 1, and receives on the other, designated here as channel 2. It is said to be in the *originate mode*. On the other hand, the end receiving the call transmits on channel 2 and receives on channel 1. It is in the *answer mode*. During the total time interval of the communication, the designations remain unchanged.

There are *originate-only* and *answer-only* data sets. Two originate-only devices cannot communicate with one another; nor can two answer-only devices. Fortunately, most are *originate-answer* devices, with the ability both to transmit and to receive on both channels. However, when two originate-answer data sets are communicating with each other, one is always in the originate mode and the other in the answer mode. Some originate-answer sets can initiate communications with an originate-only set by sending a signal indicating to the destination that a connection is desired; it then reverts to the answer mode to await the call.

The 103J is an originate-answer data set. It has seven front panel LEDs that provide status information. The automatic answer option enables the data set to answer a telephone call unattended, provided the data-terminal-ready DTR line is ON. Another option causes the data set to terminate a call whenever the received carrier is lost for more than 275 milliseconds. Several other options are also available.

RS-232-C Signals. For the data set to enter the data mode with a connection to a channel, the DTR signal must be active. After a call is made and the data set is switched to the data mode, the modem-ready (MR) lamp turns on. Unless the received carrier is detected within about 15 seconds, the data set changes from off-hook to on-hook, and the call is aborted. An active carrier-detect CD signal always indicates that the data carrier has been received for at least 100 milliseconds. Anytime the received carrier is lost for more than 10 milliseconds, the CD signal goes OFF. Following a connection, if DTR becomes inactive for more than 5 milliseconds, the data set disconnects irreversibly.

When the data set is in the data mode and is connected to the communication channel, the data-set-ready DSR line is active. An active clear-to-send CTS signal indicates that the data set is ready to transmit any character sent to it on the serial output line. CTS may become inactive at the completion of an appropriate handshaking sequence, and it always becomes inactive when the data set disconnects. Ring indicator RI is ON whenever a ringing signal is received with the modem on-hook. The ON condition of RI is approximately coincident with the rings of the ringing cycle. RI turns OFF when ringing stops or when the line goes off-hook.

Frequency Shift Keying (FSK) Modulation. Let us suppose a connection is established at 300 bps between 103J compatible data sets. When no data are being transferred, the modem in the originate mode transmits a pure sinusoidal signal at

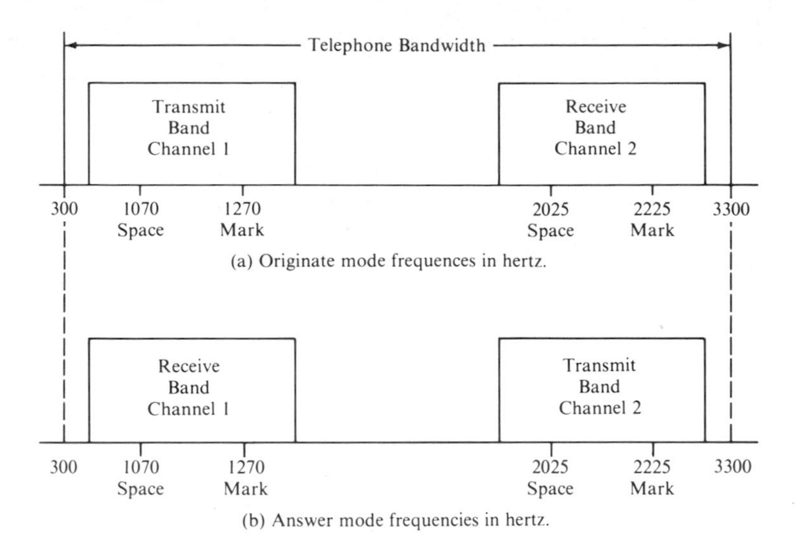

FIGURE 10-17. Low-speed modem frequencies.

1270 Hz on channel 1, and the answer-mode device transmits a similar signal on channel 2 at 2225 Hz. Each channel has the mark state. Both the transmitted and the received carriers use the same two wires of the telephone circuit.

If the operator listens to the sound in the telephone receiver, the pure tone at 2225 Hz is heard. When data are being received, however, the sound changes, with a warbling superimposed on the steady tone. This sound is a consequence of the space states that appear between some of the mark states. The space state of the received channel 2 has a frequency of 2025 Hz. For channel 1 it is 1070 Hz.

Figure 10-17 shows the frequencies used by the two modems for the space and mark states. Frequency-shift-keying (FSK) modulation sends one frequency for a mark and a different frequency for a space, with the lower frequency of each band denoting the space state. Whenever a bit changes state, the frequency of the channel shifts to the other value. This is a form of frequency modulation (FM). Each band requires a width of 450 Hz for transmission of the space and mark frequencies without undue distortion.

Figure 10-18 indicates the way the frequency is shifted when a bit change occurs. The levels of the figure denote space and mark states, not voltages, and the character is 4AH, or the letter J, traveling to the right. Seven character bits, even parity, and 1 stop bit are assumed, and the character is not immediately followed by another one. It should be noted that the modulation is fixed for each bit, with one frequency for a space and another for a mark. Thus, *the baud and the bit rate are identical*.

Originating a Call. With manual calling, the usual telephone dialing procedure is followed. If the remote end answers in the talk mode, the caller verbally requests a transfer to the data mode *in which a carrier is generated*. Stations having automatic answering will answer in the data mode. In any event, after reception by the caller of a 2225-Hz tone, which is the answer-mode mark state, the CD and CTS signals are turned ON. Then the data key is depressed, and the originate-mode modem transmits its mark-state tone at 1270 Hz, completing the operation. The exchange of tones between the originating and answering data sets is called *handshake*.

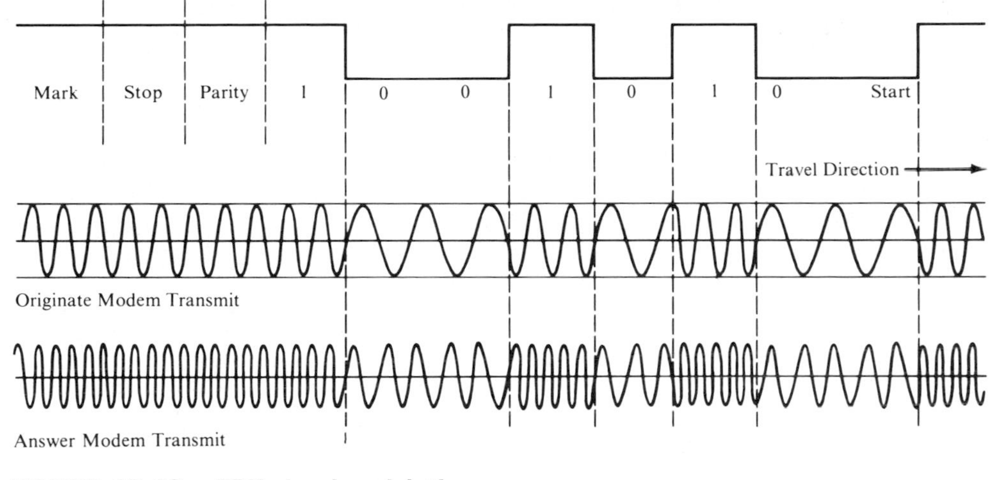

FIGURE 10-18. FSK signal modulation.

The automatic call option allows the telephone number to be written from the keyboard into memory, with a stored program used to dial it. If the continuous tone of the mark state at 2225 Hz is received within 15 seconds, the modem sends its mark state at 1270 Hz, and data transfer can commence. An abort occurs if the 2225-Hz carrier is not detected by the originate-mode modem within the time limit.

Answering a Call. When a ringing voltage is received by an on-hook modem, the RI signal is turned ON during ring intervals and sent to the communications controller. First, let us assume manual answering. In response to the rings, the line key of the telephone set is depressed and the handset is lifted. Momentarily depressing the data key connects the modem to the line, and the DSR line is driven active. Then within 2 seconds, the 2225-Hz mark tone is transmitted. After a short delay, the originating station responds with its 1270-Hz mark tone, which must be received within 15 seconds to avoid an abort. Following reception of the mark tone, the CD and CTS signals are turned ON, and data transfer can begin.

With automatic answering, if the DTR signal is active when a ring is received, the data set is automatically connected to the line at the end of the ring. Then the DSR signal is turned ON, and after 2 seconds the 2225-Hz mark tone is transmitted. The remainder of the procedure is the same as that for manual answering.

Full-Duplex Operation. Once a call is established in the full-duplex mode, the RTS and CTS signals are maintained active at all times, as are DTR, DSR, and CD. With echoplex, characters sent are echoed and appear on the screen. Any local echo provided by the local modem or the program should be inhibited; otherwise the characters typed appear on the display in duplicate. In fact, with echoplex and local copy by both the modem and the program, characters appear in triplicate. The two methods for local copy are illustrated in Fig. 10-19. Without echoplex, one of them should be used.

In some cases, reflected characters may differ from the original ones. For example, some systems return only uppercase letters even though lowercase letters were received. The remote station may elect not to return all characters, such as confidential passwords and account numbers. If the remote station is transmitting when characters are sent to it, the echoed characters may be appended to the end

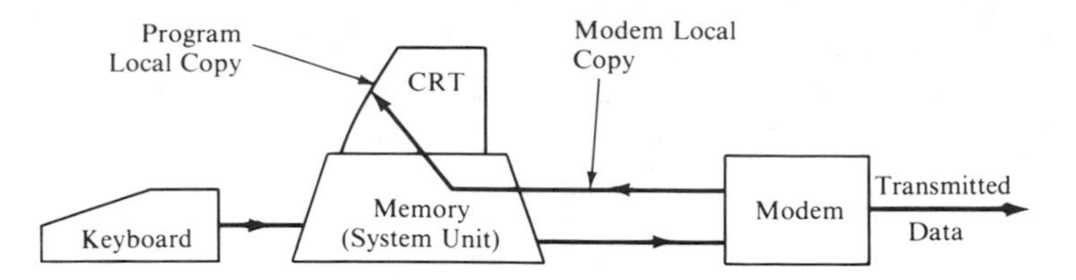

FIGURE 10-19. Program and modem local copy.

of the block transmission; alternatively, they may be interspersed with the characters of the transmission in progress. The sender has no control over what is returned.

Half-Duplex Operation. Although the 103J is a full-duplex modem with different transmitting and receiving channels, it can operate in a half-duplex manner. In this mode, a block of data is sent out, followed by reception of a block of data. The process is repeated again and again. Thus, *a half-duplex protocol is observed on a full-duplex link.*

To illustrate, suppose communication is between a PC with a 103J data set and an IBM VM/370 system in the half-duplex mode. Special data characters called *turnaround characters* are used to switch the direction of transmission. Such a character can be appended by the data set to each block of transmitted data. When the receiving end detects the turnaround character, it knows that the block has ended and that the sending end will now wait for a response. The line is not turned around, as it would be if only one channel were present. Both channels are used, one for sending data in one direction and the other for sending in the opposite direction, but *the two channels do not transmit data at the same time.*

In the preceding example, *turnaround was done with software.* For modems not capable of full-duplex operation, turnaround is a hardware operation. *With only one channel available, turning a line around requires switching the transmitter off and the receiver on at one end and switching the transmitter on and the receiver off at the other end.* Thus, the direction of the carrier is switched. Proper timing must be used, which complicates the operation. In highly interactive environments the efficiency of the system is substantially reduced. Because most modems are capable of full-duplex operation, hardware turnarounds are not often necessary.

Suppose the PC is receiving a very large block of data in the half-duplex mode of a full-duplex link. To interrupt the process, the PC can transmit a series of space states to the remote end as a break signal. This type of break is referred to as a *hardware break,* in contrast to the *software Ctrl-Break* key combination. The break is not possible on a half-duplex link. Some modems designed to transmit and receive on the same frequency band must operate in the half-duplex mode on a two-wire link but are able to operate full-duplex on a four-wire line, using one pair for transmission and one for reception.

Bell 212A Protocol

A data set with the 212A protocol actually consists of two data sets. One of them is a low-speed modem compatible with the 103J protocol, and the other is a medium-speed modem that can function at 1200 bps. Although the latter can be used for either synchronous or asynchronous communications, only asynchronous operation is examined here.

The FSK modulation used for the 103J data set requires a bandwidth equal to 1.5 times the baud rate, and the baud rate is the bit rate. Thus, for a bit rate of 1200 bps, the bandwidth must be 1800 Hz, or 3600 Hz for the two bands needed for a full-duplex line. Clearly, the telephone band is too narrow for FSK at 1200 bps.

Phase-Shift-Keying (PSK) Modulation. In contrast to frequency-shift-keying (FSK) modulation, which encodes 1 data bit per baud, *phase-shift-keying (PSK)* modulation, with four levels of modulation, encodes 2 bits per baud. The four levels refer to phase shifts of 0, 90, 180, and 270 degrees, *with each phase shift corresponding to a baud.* When a sinusoidal carrier is abruptly shifted in phase, its maximum and minimum values and all values in between are displaced in time, with no change in either frequency or amplitude. To illustrate, Fig. 10-20 shows a carrier with abrupt phase shifts of 90 and 180 degrees.

The 0 and 1 bits are grouped into pairs called dibits, and each of the four possible phase shifts corresponds to one of these dibits. *The dibits are 00, 01, 10, and 11, and the corresponding phase shifts are 90, 0, 180, and 270 degrees,* respectively. Each phase shift is said to be *differential,* because it is relative to the phase immediately prior to the shift. Clearly, a fix of the modulation represents 2 bits. Accordingly, a baud of 600 gives 1200 bps. *Most 1200-bps data sets used on telephone lines operate at 600 baud.*

The word length plus the parity bit must equal 8, and these are divided into 4 dibits. For binary code the parity bit is omitted. If 7-bit characters are used without parity, an eighth bit must be included, selected as 0 or 1. *The modem always groups the 8 bits that follow the start bit into 4 dibits for PSK modulation and for demodulation.* The originate-mode data set transmits on a carrier of 1200 Hz and receives at 2400 Hz, whereas the answer-mode set sends data on a carrier of 2400 Hz and receives at 1200 Hz.

In comparison with low-speed modems, the increased bit rate requires transmitter and receiver circuits of greater complexity. Also, there is more sensitivity to noise and distortion introduced on telephone lines. A 600-baud channel requires a bandwidth of about 900 Hz, and there is ample space for two such channels within the bandwidth of the telephone line. Except for the use of PSK modulation in place of FSK, operation of the medium-speed modem is similar to that of the low-speed one. The audio sound of the 1200-bps modulated carrier signal resembles noise, in contrast to the tones of the 300-bps signals.

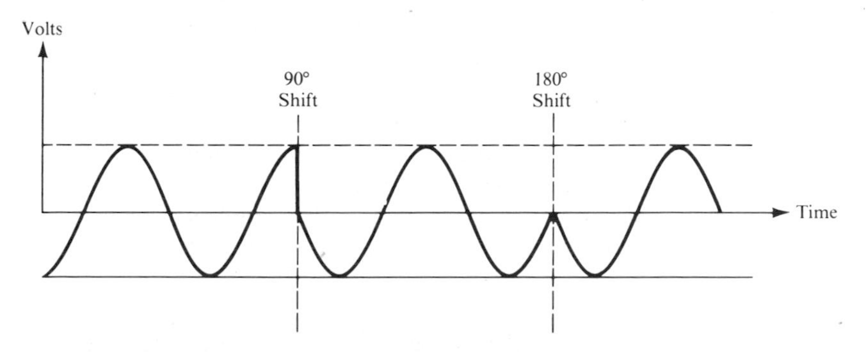

FIGURE 10-20. Phase shifts of 90 and 180 degrees.

The Hayes Smartmodem 1200

A brief description is given here of the Hayes Smartmodem 1200, which is quite similar to many other 1200-bps modems. It operates either at 1200 bps with the asynchronous 212A protocol or at 300 bps with a protocol compatible with the 103J. Thus, *it is two modems in one case.*

On the back side is a power switch and a connection to an 18-W power pack that gives 13.5 V AC from a 120-V, 60-Hz supply. An internal regulated supply has DC outputs of 5 and 10 V. Also on the back panel is a 25-pin RS-232-C female connector attached to an 8-foot RS-232-C cable, which goes to the communications adapter. Then there is a telephone socket connected to the telephone wall jack by means of a modular cable. A volume control knob for the speaker is provided. Federal Communications Commission (FCC) registration permits connection to the nationwide telephone system.

The direct-coupled modem has an internal Z8 microprocessor, 4 KB of ROM that contains a control program, and a set of 17 addressable read/write 8-bit registers. The contents of the registers control certain parameters, such as the duration and spacing of Touch-Tones and the carrier-detect response time. Among the many intelligent functions are automatic answering of a ring and dialing of a number using either pulses or tones. Many telecommunications software programs are available that integrate with the modem to enhance its functions and to simplify its use. There is a 2-inch speaker in the modem that monitors the progress of a call. Although you can listen to the distant rings and the answer to a call, *you cannot talk back through the modem.*

RS-232-C Signals. The modem is connected with 10 pins of the RS-232-C connector to the communications adapter. One is a signal ground and one is a protective ground, and both are connected to the metal frame of the system unit. A high-speed indicator at pin 12 has no connection to the PC. Of course, there are also the serial data lines TxD and RxD.

The remaining five lines consist of the DTR line from the computer and the DSR, CTS, CD, and RI signals to the computer. With the normal position of one of the eight modem switches, the DSR and CD lines are held active whenever the modem power is ON. *An active CD signal allows the computer to send commands to the modem when it is in a local command state, even if there is no carrier from the remote station.* The CD signal causes the computer to function as though the modem is on-line.

LED Indicator Lights. On the front panel of the modem are eight lights that permit a visual check of the modem's status. These LED indicators are illustrated in Fig. 10-21. The receive-data RD light is turned ON when data received from the remote station are sent from the modem to the local computer. Similar is the send-data SD light, which is ON when data are being sent from the local computer to the modem, either for transmission or for interpretation as a command. These two lights give a visual indication of data transfer.

An illuminated TR light indicates that the data terminal is ready, provided switch 1 of the eight internal configuration switches is in the UP position. Other switches have the factory setting.

The On-Line State. The data set has two functional states, the *local command state* and the *on-line state*. In the on-line state, data are transmitted and received

HS	AA	CD	OH	RD	SD	TR	MR

Encoding:

HS: High-speed (1200 bps indicator).
AA: Auto-answer mode.
CD: Carrier detect (carrier from remote station).
OH: Off-hook indicator (modem using telephone line).
RD: Receive-data indicator.
SD: Send-data indicator.
TR: Terminal ready (DTR active with switch 1 up).
MR: Modem ready (modem power switch on).

FIGURE 10-21. Modem indicator lights.

over the communications line, and the modem is in either the originate mode or the answer mode. Exit from the on-line mode to the command mode occurs whenever the received carrier is lost or when a proper escape code is entered. The default escape code is + + +, entered after at least a 1-second delay following the last transmitted character. *This escape code switches the modem to the local command mode without breaking the connection with the remote station.*

The Local Command State. Whenever power is turned ON, the modem assumes the local command state. All bytes sent to the modem over the serial data line, either from the keyboard or from a program in memory, are interpreted as command data. Commands are in ASCII, *starting with* AT *in uppercase*. The prompt is OK. Entering AT alone gives the prompt OK, which is a clear indication that the modem is in the local command state. The buffer that receives the command can store up to 40 characters.

A few of the many commands are given here to illustrate some of the capabilities of a smartmodem. Each must follow AT, and spaces are optional. Multiple commands can be lumped together in one statement. The command D followed by a number causes the number to be dialed, using pulse or tone dialing, as last specified. If desired, P or T can be included after D for specification of pulse or tone dialing. The command R for *reverse* is placed at the end of the telephone number when calling an originate-only modem.

For example, the command ATDP 703 555 1212 followed by the Enter key will cause the modem to pulse dial the 10-digit number. The spaces can be omitted. After the number is dialed, the modem waits for a carrier tone from the remote station. *If a carrier is received within 30 seconds, a CONNECT result code is sent to the display, and the modem goes on-line.* Otherwise, a NO CARRIER message is displayed with the line released.

Command O returns the modem to the on-line state following an earlier escape. F0 specifies half-duplex, and when on-line in this mode, the smartmodem automatically echoes transmitted characters back to the display. In half-duplex with a remote station having full-duplex echoplex, double characters appear on the display. For full-duplex, the command is F1, and there is no on-line echo from the modem. In the full-duplex mode, when communicating with a remote station not using echoplex, characters entered from the keyboard do not appear on the display. If either of the last two situations occur, the escape code + + + can be used to exit the on-line state, followed by AT F*n* O, with *n* either 0 or 1. This code gives a return to the on-line state with the undesirable situation corrected.

The E0 command turns off the smartmodem echo in the command state, and E1 turns the modem echo on. E1 is the default value. Normally, *E0 is used only if the program provides local echo.* With program echo and E1 active, in the local command state the display shows double characters. The E command has no effect in the on-line state.

To read one of the 17 registers of the modem, use command S*r*?, with *r* denoting the decimal number of the register from 0 to 16. Writing is done with S*r* = *n*, with *n* denoting the decimal value of the byte written, from 0 to 255. For example, ATS5 = 8 puts the number 8 into register S5, which stores the character that is echoed when the backspace key is pressed in the local command state. Processing of a backspace consists of echoing the code back to the terminal, which moves the cursor back over the last character entered. An ASCII space (20H) is sent to the display, which erases the character but advances the cursor. Then another backspace character is sent to back up the cursor again. In addition, the last character entered into the command buffer is deleted.

Some of the registers hold timing parameters. UART status data can be read from register 13. Register 16 is used to initiate a self-test mode in which the modem tunes the transmitter to the same frequency as the receiver, allowing transmitted data to be received by the smartmodem.

There are many additional commands. They provide for repeating the last command, setting the transmitted carrier ON or OFF, inserting pauses into dialed number sequences, controlling the on-hook and off-hook modes, controlling the speaker, eliminating result codes from data sent to a printer, disabling the automatic-answer mode, and activating various other functions.

Short-Haul Modems

Many modems are not designed for interfacing with the switched network of the telephone system, but instead are intended for *direct-connect point-to-point systems.* Lines usually range from a few feet up to 15 miles long. The data transfer rate is inversely proportional to the length of the conducting cable. The modems are often located in university environments and industrial plants. Frequently, they connect computers together in a local network.

Various modulation methods are utilized. In fact, some modems can transfer the digital signals without conversion into analog form. Two-wire, three-wire, and four-wire lines are common. *Usually the baud and the bit rate are the same,* with the modulation fixed for each serial bit. High speeds are easily obtained over short lines having wide bandwidths, using relatively simple modulation methods. A common one is differential PSK modulation using phase shifts of 0 and 180 degrees for the 0 and 1 bits. Bit rates up to 10 million bps or more are possible.

10-4 PROGRAMMING

Most communications are controlled by sophisticated programs that are user friendly, *designed so that the local modem is almost transparent.* Although assembly language is common, many programs are written in BASIC or some other high-level language. BASIC programs at 300 bps can be run on the interpreter, or for increased speed, they can be compiled into machine code and then run. To illustrate some of the principles involved in programming, we shall examine in detail a short and rather simple assembly language routine for initializing and controlling commu-

```
10  OUT &H3FB, &H80     ' set DLAB
20  OUT &H3F9, 1        ' divisor high
30  OUT &H3F8, &H80     ' divisor low
40  OUT &H3FB, &H1A     ' line control register
50  OUT &H3F9, 0        ' interrupt-enable register
60  OUT &H3FC, 3        ' modem control register
```

FIGURE 10-22. Initialization routine.

nications between a PC and an IBM VM370 mainframe, which is assumed here to dictate *the half-duplex mode of a full-duplex link*. BASIC programs are presented in the next section.

Initialization

Before communications can be started, the 8250 must be initialized, and there are four major steps. First, the bit rate is established by writing into the divisor latches. The second step is a write operation to the line control register to establish the word format, specifying parity, number of stop bits, and character length. Third, all communications interrupts are masked by writing to the interrupt-enable register. The control routine following the initialization does not utilize these interrupts. Finally, the $\overline{DTR}$ and $\overline{RTS}$ lines from the computer to the RS-232-C interface are turned ON by writing to the modem control register. Because the modem is assumed to be compatible with the 212A protocol, there are two channels. Therefore, the request-to-send $\overline{RTS}$ line can be maintained active at all times, even when operating in a half-duplex mode.

Figure 10-22 is a BASIC routine that initializes the COM1 port, including all four of the steps that have been mentioned. Following each OUT is the port number and next is the byte. A similar assembly language program can be used (see Prob. 10-17).

Dumb Terminal Program

For communication with the virtual machine (VM) of the IBM 370 in the *conversional-monitor-system (CMS)* mode, the half-duplex mode of the full-duplex link is used. Because this mode eliminates echoplex, *local echo is essential*. Echo is supplied by the program; even though it is not needed for the 212A modem, which can supply local echo. It follows, of course, that the echo from the modem must be inhibited. The routine in assembly language is given in Fig. 10-23, with line numbers added for reference. After conversion into machine code, the program is named COMM1.COM. The format assumes 300 bps, even parity, 7-bit characters, and 1 stop bit.

Once the program has been assembled, converted to type COM, loaded on a diskette, and run, the VM370 machine can be called by dialing manually the correct number, using the telephone in the normal manner. With the Hayes data set, this must be done in the local command state, but after the first ring, the modem should be placed off-hook (on-line) with the command ATO, and the telephone is returned on-hook without delay. Although other modems may have somewhat different procedures, the program can still be used for manual dialing. As soon as the connection is made, the proper *logon* messages are sent from the keyboard to the remote station, and the responses are received. Communication then proceeds normally.

Data can be sent only from the keyboard, with the typed characters echoed by the program to the display, and the data received from the remote station are sent

```
01  CODE SEGMENT
02    ASSUME CS:CODE, DS:CODE
03    ORG    100H
04    START: STI
05  ;
06  ; Initialization of the UART
07           MOV DX, 0     ; 300 bps, even parity,
08           MOV AH, 0     ;     1 stop bit,
09           MOV AL, 5AH   ;     7-bit characters
10           INT 14H
11  ;
12  ; Polling Loop Routine (Check Rx Buffer and KB_BUFFER)
13       A1: MOV DX, 3FDH ; get the line status
14           IN AL, DX
15           TEST AL, 1    ; is Rx data ready?
16           JNZ A2        ; if so, jump to service
17           MOV DL, 0FFH
18           MOV AH, 6     ; check KB_BUFFER for byte
19           INT 21H       ;    if available, move it to AL
20           JZ A1         ;    otherwise repeat
21  ;
22  ; Send Character from KB_BUFFER to COM Port and Echo
23           MOV DX, 0
24           MOV AH, 1     ; send character to COM port
25           INT 14H
26           CALL CRT
27           CMP AL, 3     ; Ctrl-Break?
28           JZ  A3        ; if so, jump to end
29           JMP A1        ; return to polling loop
30  ;
31  ; Get the Received Character and Send It to the Display
32       A2: MOV DX, 0
33           MOV AH, 2     ; get character from COM
34           INT 14H
35           CALL CRT
36           JMP A1        ; go back to polling loop
37       A3: INT 20H       ; terminate program
38  ;
39     CRT  PROC NEAR      ; send byte of AL to display
40           MOV AH, 0EH
41           INT 10H       ; BIOS-ROM video I/O routine
42           CMP AL, 8     ; backspace character?
43           JNZ A4        ; jump to return if not
44           MOV AX, 0E20H; character for blank space
45           INT 10H       ; transmit erasure byte
46           MOV AX, 0E08H
47           INT 10H       ; send another backspace
48       A4: RET
49     CRT  ENDP
50  CODE  ENDS
51  END   START
```

FIGURE 10-23. Communications program COMM1.ASM.

only to the display. Thus, *the local computer is programmed simply as a dumb terminal.*

With the smartmodem there is no need for manual dialing. Instead, the program is run and the call is initiated by dialing from the keyboard. The command is ATD *number,* which replaces manual dialing, and the modem goes on-line when the connection is made (see Prob. 10-19). Normally, a dumb terminal program is not often used, because inexpensive programs are available that allow files to be transmitted and received, data to be stored in memory, received data to be printed, and other things of importance. However, with the elementary program presented here, the modem can be put into either the on-line state or the local command state. Thus, *all features of the smartmodem are available.*

Let us assume that both the receiver buffer register of the UART and the main keyboard buffer KB_BUFFER (described in Section 1-5) are initially empty. Following initialization, the program loops while polling the two buffers, looking for a signal that a byte is available. *It goes from one buffer to the other, over and over, until a valid character is detected.*

Suppose a character is found in the receiver buffer. It is sent to the display, and polling is then resumed. When one is found in KB_BUFFER, it is checked for the DOS indicator 03H of Ctrl-Break, and if found, the program ends. Otherwise the character is sent to the modem, and again polling resumes. *Most of the time the program is in the wait loop looking for a character.*

Function 6 of INT 21H

Section 6-6 gives a brief description of function 6 of INT 21H. For byte FFH in DL on entry, the function checks KB_BUFFER. If a character is available, it is transferred to AL; otherwise AL is cleared and flag ZF is set. This operation is used in lines 17-19 of the program of Fig. 10-23. The jump of line 20 is not taken if a valid character is found.

For values other than FFH in DL on entry, the function sends the character represented by the code of DL to the display. Key combination Ctrl-Break has no effect.

INT 14H

In Section 6-5 the communications interrupt type 14H was briefly mentioned. It is referred to as the *RS232_IO* routine, and its routine in the BIOS ROM can be used to initialize the communications port, to transmit and receive characters over the communications line, and to return the status of the communications port. Assuming only one port is present, register DX must contain 0000H on entry to the routine. Register AH must have either 0, 1, 2, or 3, depending on the function to be implemented.

Initialize the Port. On entry to function 0, which gives the initialization, AL must contain parameters, as indicated in Fig. 10-24. The byte placed in AL in the program is 5AH. For 1200-bps communications, this value should be changed to 9AH. Mentioned earlier were four initialization steps; the fourth one consists of turning on the signals DTR and RTS. Although these signals are not activated by function 0, they are turned on by the other functions when needed.

7 6 5	4 3	2	1 0
Bit Rate	Parity	Stop Bit	Word Length

Encoding:

Bits 765: 000–110, 001–150, 010–300, 011–600, 100–1200, 101–2400, 110–4800, 111–9600.
Bits 43: X0 = no parity, 01 = odd, 11 = even
Bit 2: 0 = 1 stop bit, 1 = 2 stop bits.
Bits 10: 10 = 7 bits, 11 = 8 bits.

FIGURE 10-24. Initialization parameters of AL.

The initialization routine first sets DLAB to 1, gets the baud rate divisor from a look-up table in the ROM, and writes the divisor into the divisor latches. Then it writes the parameters of the word format, as specified by the contents of AL, into the line control register, with the DLAB and the set-break bits cleared. Next, it masks off all communications interrupts. Finally, it reads the contents of the line status register into AH and those of the modem status register into AL. Function 0 of INT 14H is implemented in the program of Fig. 10-23 in lines 7-10.

Transmit a Character. Function 1 provides for sending a character from AL to the transmitter buffer register. The byte of AL is preserved, and the line status is returned in AH. The MSB of the line status register is always 0 and gives no useful information. Thus, only the 7 low bits of the line status register are put in AH. The MSB is used as an indicator, being set if the routine was unable to transmit the data.

To transmit the character, bits 0 and 1 of the modem control register are set, thereby activating DTR and RTS. The modem status register is read, and the DSR and CTS bits are tested. If they are not active, the program waits until they are, but for no more than 0.5 second. Next, the line status register is read. The byte is checked by testing bit 5 to determine if the transmitter buffer register is empty. If it is not, the routine waits for up to 0.5 second. The Tx buffer normally will be ready within this time interval, and the byte in AL is written to this output data port. Function 1 is implemented in lines 23-25 of the program of Fig. 10-23.

Receive a Character. The routine of function 2 is used in lines 32-34 of the program to transfer a character from the receiver buffer register of the UART to AL. On exit, bits 7 and 4-1 of register AH contain status information, with bit 7 set if the operation failed and the other 4 bits being the break-interrupt and error bits of the line status register.

First, the routine sets the RTS bit 1 of the modem control register and then reads the modem status register to find out if the DSR bit 5 is active. The program will wait for up to 0.5 second, if necessary. Assuming the bit is active within this time interval, the line status register is read and the data-ready bit 1 is tested. The program will now *wait indefinitely* for this bit to be set, indicating that the receiver buffer has a character to be read. A Ctrl-Break can exit the wait loop and terminate the routine. In the program of Fig. 10-23 there is no way to get stuck in the loop, because the data-ready bit must be active prior to the jump to the interrupt. The test is made in line 15. The character of the receiver buffer is read to AL.

Return the Port Status. Function 3 simply reads the line status and modem status registers to AH and AL, respectively. This function was not used in the program of Fig. 10-23. Although the functions of INT 14H simplified the program of Fig. 10-23 and reduced its length, the savings were not great, because the routines are elementary and straightforward (see Prob. 10-18).

Procedure CRT

Subroutine CRT transmits the character of AL to the display and handles the backspace key properly. Function 0EH of INT 10H is utilized. This is the video I/O routine of the BIOS ROM that sends the character of AL to the display and returns with all registers saved. It was used earlier in the program of Fig. 6-11 and described in Section 6-5. An alternate way is provided by function 2 of INT 21H, which sends the byte of DL to the screen. This function was described in Section 6-6.

DOS recognizes code 08H, which is that of the backspace key, to be a command to move the cursor to the left. The program sends 3 bytes to the display. The first one moves the cursor one space to the left, using code 08H, and the second one writes a blank (20H) to the display, eliminating the character at the new cursor location. Because this operation advances the cursor, it is necessary to transmit another backspace code 08H. After this is done, *the cursor is one space to the left of its original position, under a blank.*

Comments

The simplicity of the communications program should be noted. Although limited in scope, it is very fast and easy to use, and it operates satisfactorily at either 300 or 1200 bps. Because echo is included in the program, when using the Hayes or a similar data set, the modem echo of data sent to it in the local command state should be eliminated by executing the command ATE0. Recall that entry from the on-line state is by means of the escape code + + +. The command ATE0 has no effect on the modem echo when on-line. Because the data set automatically echoes all transmitted characters when placed in the on-line half-duplex mode, the full-duplex mode should be specified with the command ATF1. The two commands ATE0 and ATF1 serve to eliminate all echoes from the data set.

Even though the modem is placed in the full-duplex mode, *operation is still half-duplex on a full-duplex link,* with handshake signals governing the direction of transmission. The CMS mode of the VM machine is assumed here to dictate half-duplex operation. The full-duplex command turns OFF the modem on-line echo without changing the operating mode. Because there is no echoplex in the assumed half-duplex protocol, *all echoes to the display come from the program,* in both the local command and on-line states.

A more flexible program can be designed. It should include a default value of the word format in memory, and it should display messages that instruct the operator to enter any desired changes, such as odd parity or 8-bit characters. This operation is left as an exercise (see Prob. 10-16). Then both receiver and keyboard buffering is desirable so that files may be transferred. The program of this section is as simple as possible.

DOS Commands

The DOS commands can, of course, be used in communications. Especially important are the external commands MODE and PRINT and the internal commands CTTY, COPY, and TYPE.

Command MODE. The use of MODE for specifying the mode of operation of the printer and the color monitor adapter was described in Section 9-2. Another option provides for *initializing the asynchronous communications adapter*. The format is

MODE COM1: *baud, parity, data bits, stop bits*

An alternate name for COM1 is AUX. The protocol parameters that follow COM1: are selected as desired, and execution of the command initializes the port. The baud must be included, with choices being 110, 150, 300, 600, 1200, 2400, 4800, and 9600 bps. Only the first two digits are required. Parity can be specified as N, O, or E, corresponding to *none, odd,* and *even,* with even parity being the default. There may be either 7 or 8 data bits, with a default value of 7, and either 1 or 2 stop bits are allowed. For all baud except 110, the default is 1 stop bit, and for 110 it is 2.

For example, entry of the command MODE COM1: 1200 initializes the port for 1200 bps, even parity, 7 data bits, and 1 stop bit. Command MODE COM1: 12,,8 is the same, except for specification of 8 data bits.

Another option of MODE *redirects output to the printer to the communications port*. The format is MODE LPT1: = COM1, and after execution all output directed to the printer will go instead to the communications port. Initialization of the port is required prior to the redirection. The command is cancelled by repeating it with = COM1 omitted.

Other DOS Commands. As described in Section 9-1, execution of CTTY COM1 causes DOS to use the communications port as the standard input and standard output device. After execution, any DOS function or command that refers to the "standard device" for either an input or an output operation will use the communications port for the operation. For example, TYPE FILE.TXT will send the file to the port rather than to the display. Execution of CTTY CON reverts to the normal condition. In the next chapter, several interrupt function calls that refer to a standard I/O device are described.

The external PRINT command can be used to queue a set of files and transmit them to the communications port while the computer does other things. The first time the PRINT command is executed, a message requests the name of the device to be used, and COM1 or its alternate name AUX can be specified.

The command TYPE always sends the specified file to the standard output device. For sending the file to the communications port, it is not necessary to use CTTY. Redirection can be used, as in the command TYPE FILE1.TXT > AUX. Reserved device names can be used with the internal COPY command. Thus, COPY FILE1.TXT AUX is the same as the preceding TYPE command. Another example is COPY AUX CON, which copies data received by the communications port to the display.

COMMUNICATIONS WITH BASIC

The first BASIC program in this section parallels rather closely the instructions of the assembly language of the previous section. This program is followed by one that includes a communications buffer, treating both the COM1 and SCRN: ports as numbered files. We also examine communication with a teletypewriter via current loops and the use of interrupts.

The First Program

In the BASIC program of Fig. 10-25, the six OUT statements at the beginning are those of the initialization. These statements are followed by the polling loop, which looks for a character from either the receiver buffer register or the keyboard.

The Polling Loop. The four statements in lines 220-250 are executed repeatedly in sequence while the routine waits for a character. The first statement reads the line status from port 3FDH. Used is the BASIC function INP(*n*), with *n* denoting a port number from 0 to 65535, and INP performs the same function as the assembly language instruction IN. The IF statement of line 230 tests the data-ready bit of the status to determine whether or not the receiver buffer has a character to be read. If so, a branch occurs from the wait loop to the service routine starting at line 360.

In lines 240-250 the routine checks the keyboard buffer and returns to the start of the loop if nothing is found. When the keyboard buffer has a character, it is assigned to the string variable Z$. In this case, there is no loop, and the routine goes to the service routine that follows line 250.

Function INKEY$ is used to check KB_BUFFER and to read a character if one is available. The value assigned to string variable Z$ in line 240 is a character string of length 0, 1, or 2. A null string of length 0 indicates that no character is pending, and a two-character string indicates a special extended code, in which case the first byte is the null code 00H. The 2-byte extended codes, described in Section 1-5, apply to certain ALT key and Ctrl key combinations, function keys, cursor movement keys, and others. The extended characters are neither transmitted nor echoed in the communications program of Fig. 10-25. Only the first byte 00H is accepted. *Note the difference between a null string, which has no characters, and a null code, which is byte 00H.*

Similar to INKEY$ is the function INPUT$. The statement that follows assigns to variable x$ a string of *n* characters read from file number *m*, with file number 0 denoting the keyboard:

x$ = INPUT$(*n*,*m*)

If *m* is omitted, 0 is assumed. For example, INPUT$(1) returns one character from the keyboard, with 2-byte extended codes included. Unlike INKEY$, if no valid characters are available, INPUT$ waits for the specified number of characters to be entered, whereas INKEY$ simply returns a null string.

KB_BUFFER Processing. Lines 290-320 process the character read from KB_BUFFER and assigned to Z$. Initially, integer B is assigned the ASCII code

```
100  REM  Initialization-300 bps, even parity, 1 stop, 7 bits
110  '
120   OUT &H3FB, &H80                       ' set DLAB
130   OUT &H3F9, 1                          ' divisor high
140   OUT &H3F8, &H80                       ' divisor low
150   OUT &H3FB, &H1A                       ' line control register
160   OUT &H3F9, 0                          ' interrupt EN register
170   OUT &H3FC, 3                          ' modem control register
180   DEFINT A - Z
190  '
200  REM  The polling loop that looks for a character.
210  '
220   A = INP(&H3FD)                        ' get line status
230   IF (A AND 1) <> 0 THEN 360            ' if RxD ready, go 360
240   Z$ = INKEY$                           ' get KB byte if there
250   IF   Z$ = "" THEN 220                 ' go back if no character
260  '
270  REM  Send KB byte to line, echo it, and go back to polling
280  '
290   B = ASC(Z$)
300   OUT &H3F8, B                          ' transmit ASCII byte
310   GOSUB 430                             ' echo to display
320   GOTO 220
330  '
340  REM  Get byte from RxD buffer, display it, go to polling
350  '
360   C = INP(&H3F8): Z$ = CHR$(C)
370   IF Z$ = CHR$(10) THEN 220
380   GOSUB 430                             ' echo to display
390   GOTO 220
400  '
410  REM Subroutine DISPLAY for display of character
420  '
430      IF Z$ = CHR$(8) THEN LOCATE ,POS(0) - 1,1: PRINT " ";:
            LOCATE ,POS(0) - 1,1 ELSE PRINT USING "!"; Z$;
440   RETURN
```

FIGURE 10-25. Communications program in BASIC.

of the first byte of Z$ by ASC(Z$). For the special extended codes, this is the null character 00H. The code is transmitted to the communications data port in line 300 and echoed to the screen by subroutine DISPLAY called in line 310. Then the program branches back to the polling loop.

Received Data Processing. The first of the four lines from 360 to 390 gets the character from the receiver buffer and assigns it to integer variable C. Then the CHR$(C) function converts the numeric value of C into the string character Z$, which is sent to the display by the subroutine call of line 380, provided it is not a line feed. In line 390 is a branch back to the polling loop.

Normally, each carriage return character from the receiver buffer is followed by a line feed. However, the PRINT and PRINT USING statements automatically attach a line feed to each carriage return. To avoid double spacing on the display screen, *the line feeds from the receiver should be deleted*. This operation is accomplished by the statement of line 370.

Subroutine DISPLAY. The subroutine for echoing a keyboard character and for transferring a received character to the display first checks the string variable Z$ for the backspace code 08H. When the PRINT USING statement displays a string, the display for 08H is a special symbol, not a backspace as in DOS. Thus, the program must shift the cursor one space to the left *with the LOCATE statement,* print the space 20H, and then shift the cursor once again to the left. These operations are accomplished in line 430 of the subroutine. The POS function of line 430 returns the current column position of the cursor. The argument (0) is a dummy but it must be included. Thus, execution of x = POS(0) assigns to x the current column number. Occurring twice in line 430 is the statement

```
LOCATE , POS(0) - 1, 1
```

Because the row entry is blank, it is not affected. The column position is POS(0) − 1, which moves the cursor one space to the left. The cursor value is 1 for ON. The LOCATE statement is described further in Section 6-7.

The "!" that follows PRINT USING specifies that *only the first character of the given string is to be printed.* The semicolons following the string variables to be printed cause the characters *to be displayed on the same line* until a carriage return is encountered.

The Second Program

The program here treats COM1: and SCRN: as files, with the OPEN COM1: statement allocating buffers for receiving characters from the communications line and for transmitting characters to the line. The transmit buffer normally holds 128 bytes. The receive buffer is often referred to as the *communications buffer,* or simply *buffer.* Figure 10-26 gives the buffered communications program. In many

```
100  DEFINT A - Z
110  OPEN "COM1: 300, E, 7" AS 1
120  OPEN "SCRN:" FOR OUTPUT AS 2
130 '
140 REM Get characters from KB_BUFFER and transmit
        them as long as the communications buffer is
        empty.
150 '
160  Y$ = INKEY$
170  IF  Y$ <> ""   THEN PRINT #1, Y$;
180  IF  EOF(1)  THEN 160
190 '
200 REM Input all characters of the communications
        buffer and display them, continuing until
        buffer is empty
210 '
220  Z$ = INPUT$(LOC(1), #1)
230  D = 0
240  D = INSTR (D + 1, Z$, CHR$(10))
250  IF  D > 0  THEN  MID$(Z$, D) = " ": GOTO 240
260  IF  Z$ = CHR$(8)  THEN LOCATE , POS(0) - 1, 1:
     PRINT #2, " ";: LOCATE , POS(0) - 1, 1
     ELSE PRINT #2, Z$;
270  IF LOC(1) > 0 THEN 220
280 GOTO 160
```

FIGURE 10-26. Program with a communications buffer.

ways it is similar to the previous one, but there are important differences. In BASIC, device names COM1:, LPT1:, KYBD:, and SCRN: are recognized, and the colon is required. Not accepted are DOS names AUX, PRN, and CON.

When entering the BASIC command, the size of the buffer can be specified up to 32,767 bytes by placing /C:size after the command word. The default value is 256 bytes, but 1024 bytes are recommended for high-speed lines. A major problem in high-speed communications is the processing of incoming data files before the buffer overflows. With execution by the interpreter, the bit rate is restricted to low speeds, but higher speeds are possible with compiled BASIC.

The OPEN Statements. In BASIC, the OPEN statement must precede any I/O to a file. It allocates a buffer to the specified file and assigns the file a number. After variables are defined as integers in order to increase program speed, device COM1: is opened in line 110 as file #1, with the bit rate initialized at 300 bps. Parity is specified as even, and the character length is 7 bits. By default there is 1 stop bit. Also, the DTR and RTS lines of the RS-232-C interface are activated. There are various other options that can be included in the OPEN statement. Any BASIC statements for random access and sequential files, which are described in the next chapter, can be used for read/write of the COM1: file.

For example, RTS can be turned OFF, timeout intervals for input signals DSR, CTS, and CD can be specified, and the processor can be instructed to send out an LF character following each transmitted CR. The OPEN statement not only establishes a memory buffer for communications but also provides the required initialization of the port. Values "300, E, 7" in line 110 can be omitted because these are the defaults. For the COM1: file, statements GET1,*n* and PUT1,*n* can be used to read and write *n* bytes, with *n* normally restricted to 128 bytes. Both GET and PUT are described in the next chapter.

The device SCRN: is opened in line 120 as file #2 for output. This operation establishes a memory buffer and allows the use of the PRINT# statement for output to the display. The purpose of the PRINT# statement is *to write data to a file or a device opened as a file*. The number sign # is *always* followed by the file number and a comma, as shown in line 260. Statement PRINT# writes to the opened file, which is the display for the case here, precisely as PRINT normally writes to the display.

The Wait Loop. The three lines from 160 to 180 constitute the polling loop. The function INKEY$ in line 160 checks for a character. If none is available, a check of the communications buffer is made next, using EOF(1) in line 180. In general, EOF(*filenum*) returns -1 (true) if the end of the file has been reached, and *for a communications file a true value indicates that the buffer is empty*. In this case, a branch back to line 160 is made, and the keyboard buffer is checked. The polling procedure continues until a character is found.

Keyboard Input and Transmission. Suppose INKEY$ of line 160 returns a character to string variable Y$. This information is sent to the transmit buffer by PRINT#1 of line 170, which writes data sequentially. Any characters in the transmit buffer are, of course, transmitted as soon as they are permitted by the protocol.

Receiver Input and Display. Whenever EOF(1) of line 180 is false, indicating that the communications buffer has one or more characters waiting to be read, the program continues to line 220. Function LOC(1) appears here. For communications files only, *it returns with the number of bytes in the buffer*. Accordingly, the INPUT$ function of line 220 assigns the entire string of characters of the buffer to Z$, but not in excess of 255, which is the maximum number allowed for a string variable. For other types of files, the LOC function is described in Section 11-3.

The next three lines replace all line feeds of the string with spaces having ASCII code 20H. Backspace characters are handled in line 260 as before, and the modified string is written sequentially to file SCRN: by PRINT #2 of line 260. A test of the communications buffer is then made in line 270, and if new characters are present, the routine branches back to line 220 to process them. Otherwise a jump is made to the keyboard buffer check of the wait loop. *Note the higher priority given to the communications buffer*. Characters must not be allowed to fill it, or input data will be lost. We now examine the deletion of the line feeds of string Z$.

Function INSTR and the MID$ Statement. The BASIC function INSTR(*n*, X$, Y$) searches for the first occurrence of string Y$ in X$ and returns the position at which the match is found. The search starts at the optional offset *n*, and if *n* is omitted, the search starts at the first character. In case Y$ is not found, INSTR returns 0. In line 240 is the statement

```
D = INSTR (D + 1, Z$, CHR$(10))
```

Thus, integer variable D is assigned a value of 0 if Z$ has no line feed, but if one is found, D is assigned its offset value. In line 250 is the MID$ statement

```
MID$ (Z$, D) = " "
```

This statement replaces the characters of Z$ starting at offset D, with the characters of the string at the right side. In this case, there is only one character at the right, which is a space with ASCII code 20H. The line feed detected in line 240 is replaced with this space. The GOTO statement at the end of line 250 returns the routine to line 240, with D now equal to the offset of the space replacing the line feed, and a search is made for the next line feed. *The process continues until all line feeds are removed from the string*.

The two programs in this section give results that are quite similar for 300-bps dumb terminal communications. The second one provides buffers, which is an advantage, especially when files are received at higher speeds. No program echo was included, however. Thus, modem echo is essential. With the Hayes Smartmodem, execution of commands ATE1 and ATF0 in the local command state provides echoes in both the local command and on-line states.

The Teletypewriter

A popular *teletypewriter TTY*, or simply *teletype*, that utilizes serial communications is the Model ASR33 Teletype, which has a keyboard, a paper printer, a paper reader, and a paper punch. The keyboard is similar to that of a typewriter, except for the inclusion of special controls and functions. The LINE switch connects the teletype for use as a computer I/O device, and characters received from the CPU are printed.

When code is sent to the computer, the character of a pressed key is printed only if the computer program returns the received code, using an output instruction.

This is the method used for echo. For use simply as a typewriter, the Local switch is activated. The keyboard can also be used to punch paper tape.

In order to type special characters or symbols shown on the upper portion of selected keys, the Shift key must be held depressed. Also on the upper portion of some keys are certain special function keys such as Space, Return, Line Feed, and Rubout. These keys respectively insert a space in a line, return the carriage to the start of a line without advancing, move the printer carriage one line forward, and delete a character. They are implemented by holding down the Ctrl key when depressing the function key.

Bits are transmitted and received serially over a pair of wires referred to as a *20-mA current loop,* and the transfer is asynchronous. A transmitted word has 11 bits, including a start bit, 7 character bits, a parity bit, and 2 stop bits, with characters encoded in ASCII. The maximum bit rate is 110 bps, corresponding to transmission of about 110 words per minute.

In the half-duplex mode, a single current loop transmits characters in either direction. With full-duplex there are two current loops. Each is a simplex line that carries information in one direction only. Current loops tolerate considerable noise and can transmit digital signals over distances up to about a mile. Logic states of the teletype wires are represented by 0 and 20 mA.

Current Loop Interface

Provided by the communications adapter card of Fig. 10-3 is a 20-mA current interface that has the circuit shown in Fig. 10-27. It is used for attachment of certain types of serial printers. The light-emitting diode (LED) of the *opto-isolator* emits with a forward current of 20 mA and a diode voltage drop of 2 V, and the light activates the *phototransistor*. The transmit and receive circuits of the remote peripheral are identical to those in the figure.

Let us consider transmission of a logical 1. A pulse of 5 V is transmitted and inverted, giving a voltage of 0 at the output of the inverter. The 5-V source drives a current of 20 mA clockwise around the loop, with 2 V across the LED of the opto-isolator of the I/O device. The phototransistor of this opto-isolator is activated, and the output of the transistor is inverted to provide a 5-V signal to the peripheral.

On the other hand, if the input to the transmit circuit is low, the output of the

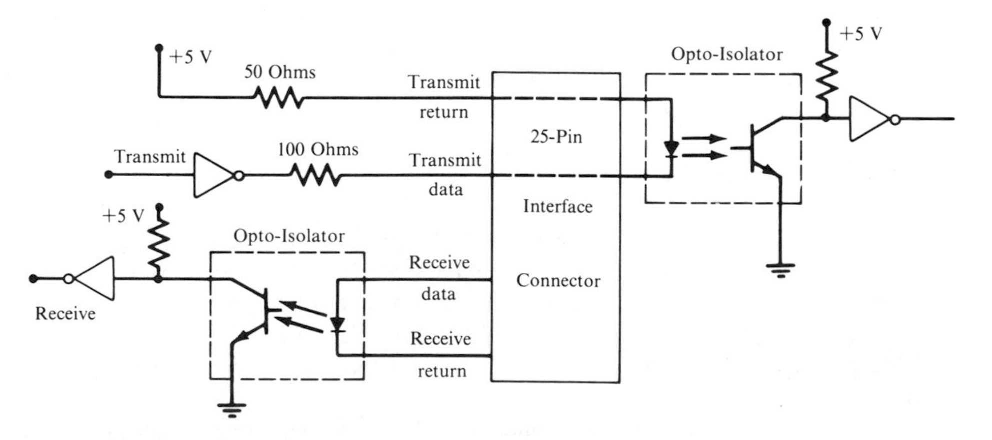

FIGURE 10-27 A 20-mA current loop interface.

inverter of the figure is 5 V, and there is no current in the loop. Accordingly, the phototransister of the opto-isolator of the output device is not activated, which provides a low voltage to the peripheral. Transmission in the reverse direction is similar. Because of *optical coupling,* the communications adapter and the I/O device are not electrically connected. On the adapter card is a jumper block that enables the operator to select manually either the EIA RS232C voltage interface or the 20-mA current loop interface.

REFERENCES

Bell System Technical Reference, Pub. 41106, *103J Data Set,* American Telephone and Telegraph Co., 1977.

Bell System Technical Reference, Pub. 41214, *212A Data Set,* American Telephone and Telegraph Co., 1977.

E. G. BROONER AND P WELLS, *Computer Communication Techniques,* H. W. Sams & Co., Inc., Indianapolis, Ind., 1983.

Component Data Catalog, Intel Corp., Santa Clara, Calif., 1980.

Data Sheets, *INS8250 Asynchronous Communications Element,* National Semiconductor Corp., Santa Clara, Calif., 1978.

Electronic Industries Association, *EIA Standard RS-232-C,* Washington, D.C., 1969.

J. MARTIN, *Telecommunications and the Computer,* Prentice-Hall, Inc., Englewood Cliffs, N.J., 1976.

J. E. MCNAMARA, *Technical Aspects of Data Communications,* Digital Press, Bedford, Mass., 1978.

E. A. NICHOLS, J. C. NICHOLS, AND K. R. MUSSON, *Data Communications for Microcomputers,* McGraw-Hill Book Co., New York, 1982.

W. D. SCHWADERER, *Digital Communications Programming on the IBM PC,* John Wiley & Sons, Inc, New York, NY, 1984.

Technical Reference, IBM Corp., Boca Raton, Fla., 1981.

PROBLEMS

SECTION 10-1

10-1. In the DOS mode, press all key combinations listed in Fig. 10-2 and tabulate the results of each operation.

10-2. Using the format of Fig. 10-1, give the bit stream for character 'A' and also for 'p', using 7 character bits, even parity, and 1 stop bit. Repeat using 8 character bits, odd parity, and 2 stop bits. For the latter format, how many characters can be sent per minute at 300 bps?

10-3. Explain the difference, if any, between baud and bps, modem and data set, start and stop bits, simplex and half-duplex, HDX and FDX, synchronous and asynchronous communications, digital and analog, modulation and demodulation, echoplex and parity checking, and EIA drivers and TTL buffers.

10-4. As defined in Section 10-1, explain the difference, if any, between space and mark, DTE and DCE, DTR and DSR, SIN and SOUT, RTS and CTS, CD and RLSD, DISTR and DOSTR, characters 00H and 20H, CD and LF, on-hook and off-hook, and control and noncontrol characters.

SECTION 10-2

10-5. List a sequence of assembly language instructions that will establish a bit rate of 110 bps for the 8250. The initial state of DLAB is unknown. Also, for nominal bit rates of 110, 300, and 2000 bps, determine the divisor and the percentage of error of the difference between the desired and actual bit rates.

10-6. Suppose a PC program reads in order the hexadecimal bytes 8F, 03, 00, 06, 69, and A9 from respective ports 3FB, 3FC, 3F8, 3F9, 3FD, and 3FE. Then it writes 0BH to port 3FBH, followed by an input of 0FH from port 3F9. List each of the 7 bytes read, and alongside each state its significance. Also, determine the byte that would be read from port 3FAH, and identify all pending interrupt types of the 8250.

10-7. With DLAB previously cleared and without affecting other conditions, list a sequence of assembly language instructions that will

(a) change the current state of pin RTS.
(b) enable pin OUT2.
(c) wait until the transmitter buffer is empty and then write a CR character to it.
(d) wait until the receiver buffer has a valid character and then read it into the accumulator.
(e) clear all error bits of the line status register.
(f) determine if the modem is in a transmit mode.
(g) determine if signal CD has changed since the last reading of the modem status register.
(h) determine if a line status interrupt is pending, and if so, decide whether the source is a parity error.

10-8. If port 3F9(0) is written with 0AH and the bytes present in the registers of ports 3FD and 3FE are 4FH and BAH, respectively, what interrupt types are pending and what is their highest priority?

10-9. Suppose hexadecimal bytes 5B, AA, 00, and 20 are moved to absolute RAM locations 30H through 33H, respectively. Identify the precise significance of the bytes.

SECTION 10-3

10-10. With respect to the modem, explain the difference between (a) low speed, medium speed, and high speed; (b) stand-alone and expansion board; (c) direct and acoustic coupling; (d) smart and dumb; (e) automatic-dial and manual-dial; (f) echoplex, modem echo, and program echo; (g) 103J and 212A; (h) FSK and PSK.

10-11. In a form similar to the FSK modulation in Fig. 10-18, sketch the transmitted character A. However, for the state-level diagram, show the plus and minus 12 V states corresponding to the transmitted bits. Assume 7-bit characters, even parity, 1 stop bit, and preceding and following idle states. Include approximate modulation waveforms that clearly show modulation changes.

10-12. In PSK modulation, sketch the bands and show the carrier frequencies used by the Hayes Smartmodem 1200 in a form similar to that of Fig. 10-17. Identify band extremities in hertz.

10-13. With operation in the half-duplex mode, explain the major differences between the protocols of a half-duplex link and a full-duplex link. As analogies, relate single-lane and dual-lane highways to the links, with traffic flow control corresponding to byte transfer control.

10-14. In a form similar to that of Fig. 10-18, but with PSK replacing FSK and plus and minus 12 V levels used, sketch the transmitted character K with even parity, 1 stop bit, and leading and following idle states. Identify each dibit and the corresponding differential phase shift. Also, illustrate the phase shift modulation for the originate-mode modem.

10-15. For the Hayes Smartmodem 1200 in the local command state, give commands that will

(a) Touch-Tone dial 555-5555.
(b) pulse dial 805-555-5555 to an originate-only modem.
(c) verify that the modem is in the local command state.
(d) return to the on-line state with a duplex change to half-duplex.
(e) turn off the command-state modem echo.
(f) read register 13.
(g) write byte 123 to register 2.

SECTION 10-4

10-16. Revise the initialization program of Fig. 10-22 to allow the operator to modify any of a set of displayed default format parameters. Allow bit rates of 110, 300, and 1200 bps, 7- or 8-bit characters, and 1 or 2 stop bits. Parity choices are even, odd, 1, 0, and none. The program should be user friendly.

10-17. Give an assembly language routine that parallels as closely as possible the BASIC initialization of the 8250 UART shown in Fig. 10-22. Use 1200 bps, 8-bit words, no parity, and 1 stop bit.

10-18. Rewrite COMM1.ASM without using either INT 14H or INT 10H of the ROM BIOS. Function 2 of INT 21H is available.

10-19. Suppose that CALL DIAL is placed in line 11 of the COMM1.ASM program of Fig. 10-23. Assuming the use of the Hayes Smartmodem 1200, design procedure DIAL for insertion between lines 49 and 50. DIAL should prompt the user to enter from the keyboard a telephone number consisting of up to 25 digits and characters. Following the carriage return, the modem should dial the number automatically. Save the registers and include messages in the subroutine. The CRT subroutine of COMM1 may be used.

10-20. With the Hayes Smartmodem 1200, assume that CALL DIRECTORY is placed in line 11 of the COMM1.ASM program. Design DIRECTORY to display a column containing digits 1 to 5, with space at the right for entry of five telephone numbers. The operator is given the options either entering a number up to 25 digits or dialing a number previously entered. Once a number is entered, it becomes part of the directory until it is changed or the power is turned off (unless the new directory is written to a diskette). For dialing, the operator enters the appropriate single digit. The CRT subroutine of COMM1 may be used.

SECTION 10-5

10-21. Add keyboard echo to the program of Fig. 10-26. Assume the revised program is run and a connection is made to a mainframe in the half-duplex mode, with double characters displayed. Give all commands required to delete the double characters in both modem operating states, with a return to the on-line condition.

10-22. Write an assembly language program that parallels as closely as possible the BASIC program of Fig. 10-26.

10-23. Write a BASIC program that clears the screen and displays the cursor at the four corners of a rectangle in the middle of the screen. The vertical sides should span 10 rows and the horizontal sides should span 30 columns.

10-24. For the routine that follows, give the display after execution. Repeat if " " of line 30 is replaced with CHR$(0), and repeat again for "".

```
10  Z$ = "AABBXAZA": D = 3
20  D  = INSTR(D + 1, Z$, CHR$(65))
30  MID$(Z$, D) = " ": PRINT Z$
```

CHAPTER 11

File Management

The main objective of this chapter is to show how one uses the DOS software interrupts to manage diskette files. The approach consists of introducing and describing of a number of short COM programs containing the interrupts and their functions. Some are followed by an equivalent BASIC program. It is not possible to include all details and functions in one chapter. However, sufficient information is presented so that the reader can continue the investigation at a more advanced level.

In the 2.1 version of DOS, there are 88 different function calls in the software interrupt type 21H, many of which are designed for file control. In the file control group are two subsets that duplicate and overlap one another. The functions of the older subset, referred to as the *traditional* ones, establish a *file control block* for each file of a program. Data put in the block by the traditional open call, and by a running program, provide control of read and write operations. In the newer set, which has greater power and flexibility, the file control block is not used. Only these preferred functions are described in this chapter and *all file control statements refer to them exclusively*. Consequently, a detailed study of the file control block is omitted.

If execution of a file control function is successful, the carry flag CF is cleared. A failure not only sets the flag but also puts an error code in register AX. Although CF should be checked after each operation, for simplicity this is not done in the programs here. Fortunately, errors are quite infrequent. We begin with an examination of the program segment prefix, which has an important role in file control.

PROGRAM SEGMENT PREFIX

To load a program, the command processor first sets up a *program segment prefix (PSP)* at the lowest available address in memory, and then the program is loaded at offset 100H relative to the start of the program segment prefix. Only COM programs are considered in this chapter. In these programs, each of the four segments starts at the beginning of the program segment prefix, with IP pointing to code commencing at offset 100H and SP set to the end of the program's segment. For EXE programs, registers DS and ES also point to the program segment prefix, but values for CS, SS, IP, and SP are set by the linker. To use the DOS interrupts, an understanding of the program segment prefix structure is essential.

To illustrate, the significant portions of the program segment prefix established by DOS when one enters the command COMP abc.DEF b:Wxyz.k are shown in Fig. 11-1. Most bytes not included are either 0s or of no significance. The program segment prefix consists of 256 bytes from offset 0 through FFH, with code beginning at 100H. The first 2 bytes are code CD20 for INT 20H. Bytes numbered 2 and 3 give the memory size in *paragraph* form (multiples of 16). The actual size is 40000H, or 256 KB. Byte number 4 is reserved.

The word of offsets 6 and 7 is FFF0H, which denotes the number of bytes available in the segment less 16. The 5-byte combination from 5 through 9 is code 9AF0FF0DF0 for a long CALL instruction to location F00D:FFF0. *The address is precisely the same as 0000:00C0.* At this location in the region of the vector-

0	1	2	3	4	5	6	7	8	9
CD	20	00	40	00	9A	F0	FF	0D	F0
INT 20H		RAM Size		Long Call to DOS					

A	B	C	D	E	F	10	11	12	13	14	15
8C	02	3A	05	99	02	3A	05	E2	04	3A	05
INT 22H Address				INT 23H Address				INT 24H Address			

5C	5D	5E	5F	60	61	62	63	64	65	66	67
00	41	42	43	20	20	20	20	20	44	45	46
Drive	A	B	C						D	E	F

6C	6D	6E	6F	70	71	72	73	74	75	76	77
02	57	58	59	5A	20	20	20	20	4B	20	20
Drive	W	X	Y	Z					K		

80	81	82	83	84	85	86	87	88	89	8A	8B	8C	8D	8E	8F	90	91	92
11	20	61	62	63	2E	44	45	46	20	62	3A	57	78	79	7A	2E	6B	0D
Bytes		a	b	c	.	D	E	F		b	:	W	x	y	z	.	k	CR

FIGURE 11-1. Portions of a program segment prefix.

interrupt table is code EA140BE300 for a FAR jump to the DOS dispatcher at location 00E3:0B14 within IBMDOS.

Byte locations at hexadecimal offsets A, B, C, and D contain the INT 22H vector address (53A:28C) that was in effect when the program segment prefix was created. This is the address in COMMAND to which control returns when either INT 20H or INT 27H is executed. If the program loads and runs a second one, it may be desirable first to change the terminate address of INT 20H so that control returns to the main program when the second one ends. However, *termination of the original program transfers control to DOS at the address stored within the program segment prefix, and this address is automatically returned to the interrupt table*. The stored exit addresses for INT 23H (Ctrl-Break) and INT 24H (critical error) are handled similarly. An example illustrating the use of an interrupt exit address is given later.

As indicated in Fig. 11-1, there are two parameter areas that are formatted as standard unopened file control blocks. One of these areas consists of the 12 bytes starting at offset 5CH, and the other starts at 6CH. If the command does not include filespecs, these parameter areas are filled with hexadecimal 20s.

The initial byte of the unopened file control block at offset 5CH is 00H, denoting the default drive. If drive A had been included in the command, followed by a colon, this byte is 01H, and the corresponding code for drive B is 02H. Characters below the blocks correspond to the stored code. Note that the filename is left justified, with 20H replacing each of the allowed eight characters that were omitted. If the extension has fewer than three characters, it is also left justified within its field, with spaces supplied as fillers.

The 128 bytes from 80H to 100H constitute the default *disk transfer area (DTA)*. It receives those characters of a command line entered from the keyboard that follow the actual command. Byte 11H at offset 80H equals the number of bytes between this one and the first carriage return 0DH. *The characters are those of the entered filespecs and paths, if any, including all spaces and punctuation starting from the last character of the command word* (see Prob. 11-1). Parameters (<, >, and |) and associated filenames are excluded. The region can be used for other purposes by a program, such as for a stack with SP initialized at 100H.

ASCIIZ String

Several of the function calls require an *ASCIIZ string* as an input. This is a character string containing an optional drive specifier, a directory path, usually a filename, and a final byte of 00H. No more than 64 characters, including the null byte, are allowed, and leading spaces are prohibited. When a command such as COMP File1.ABC B:File2.XYY is entered, before the two files whose names follow the command COMP can be opened and read into memory, *the filespecs must first be converted into ASCIIZ strings*. Paths are included if subdirectories are involved. For development of the required ASCIIZ strings, a short procedure is presented here. It will be called when needed by programs presented in this chapter.

On entry to procedure ASCIIZ, *register DX must contain the offset address in segment DS pointing to a reserved memory location initialized with 64 bytes of 00H*, and if the program requires two strings, *a second identical memory space must immediately follow the first*. Continuing with this procedure, up to five strings are allowed (see Prob. 11-2), which permits a program to open up to five files for reading or writing. In addition, the strings as entered by the command must still be present on the command line of the DTA area of the program segment prefix; that is, they must not have been overwritten.

On exit, *properly constructed ASCIIZ strings will be present in the designated memory spaces corresponding to the filespecs and paths of the command line*. Although DOS allows various characters to serve as filespec separators, the procedure here permits only spaces for this purpose. The contents of flags and all registers are preserved. The routine is given in Fig. 11-2.

If the procedure of Fig. 11-2 is executed with the program segment prefix being that of Fig. 11-1, the two ASCIIZ strings are abc.DEF and b:Wxyz.k, and each terminates with the initialized value of 00H as required. The offset of the second string is 64 bytes above the start of the first one.

```
ASCIIZ  PROC   NEAR
      PUSHF
      PUSH DX
      PUSH CX
      PUSH DI
      PUSH SI
      CLD
      MOV CL, 5                  ;no more than five strings
      MOV SI, 81H                ;start at offset 81H of PSP
  P1: MOV DI, DX
  P2: CMP BYTE PTR [SI], 20H  ;eliminate leading spaces
      JNZ P3
      INC SI
      JMP P2
  P3: CMP BYTE PTR [SI], 13   ;end if carriage return
      JZ  P4
      MOVSB                      ;move byte, increment SI,DI
      CMP BYTE PTR [SI], 20H
      JNZ P3                     ;if space, form next string
      ADD DX, 64
      DEC CL
      JNZ P1
  P4: POP SI
      POP DI
      POP CX
      POP DX
      POPF
      RET
ASCIIZ   ENDP
```

FIGURE 11-2. Procedure for forming ASCIIZ strings.

Environment Strings

A word of special importance not shown in Fig. 11-1 is the segment address of the *environment*. The 2 bytes are present at offset 2CH. In this figure, the word is 0603H and the environment starts at 0603:0000. Beginning here are stored strings placed in the environment by DOS and by the DOS internal command SET. A program may read the segment address at offset 2CH of the program segment prefix, and look up and use named strings that were previously placed in the environment. All string names are in uppercase letters, and each string ends with the null character 00H. The last one is followed by 2 null bytes. When DOS starts up, it puts strings into the environment with the names PATH and COMSPEC, and they give the path that DOS uses to reload the command processor when necessary. Space in the environment is limited to 32 KB.

Another word not shown in Fig. 11-1 is code CD21H for INT 21H located at offset 50H, which can be called from a program. Other bytes of the gaps preceding offset 5CH are reserved for the DOS system and should not be used by a program. However, locations 5CH to 100H are available for any desired purpose. The program segment prefix of a program is easily examined by loading the program with DEBUG, followed by execution of the DEBUG command D 0. Vector addresses of interrupts 22H and 23H will point to locations in DEBUG.

11-2

COMPARISON OF FILES

To introduce several file control functions, the program B:COMPARE is presented and examined. The assembly language is given in a form for conversion into a COM file, and six functions of INT 21H are included. These are *print-string* 09H, *open-file* 3DH, *read-file* 3FH, *get-attribute* 43H, *terminate-program* 4CH, and *get-file-information* 4EH. The main purpose of the routine is to illustrate these interrupt functions. An equivalent BASIC program is investigated in the next section. The COM program is run by entering the command

```
B:COMPARE  filespec1  filespec2
```

Paths can be included with the filespecs if subdirectories are present. The program compares the two files and prints a message giving the result. Thus, it is basically the same as the DOS command COMP. However, with 362 bytes, it is much shorter and simpler than the 2534-byte COMP program, but the penalty is reduced flexibility. The pseudo code is given in Fig. 11-3.

Assembly Language (COMPARE)

The complete program is presented in Figs. 11-4 and 11-5. At the beginning, location FILE_SIZE is reserved for receiving in order the number of bytes of each

```
PROGRAM  Compare        {program to compare two files}
begin main
  construct two ASCIIZ strings             {CALL ASCIIZ}
  get file1 attribute and size
  get file2 attribute and size
  if sizes are unequal
    then display error message and terminate program
  end if
  open and read each file
  compare the files
  print appropriate message
end main
begin ASCIIZ
  construct the two ASCIIZ strings
end ASCIIZ
```

FIGURE 11-3. Pseudo code for the program COMPARE.

```
TITLE  Compare
;
CODE  SEGMENT
    ASSUME CS:CODE,DS:CODE,ES:CODE,SS:CODE
 ORG 9AH
    FILE_SIZE DW ?
 ORG  100H
;
  START:  JMP A1
;
    PROG1   EQU   300H
    PROG2   EQU  8000H
    ASCIIZ1 DB   64 DUP(0)
    ASCIIZ2 DB   64 DUP(0)
    SIZE1   DW   ?
    MSG1    DB   'Files compare OK$'
    MSG2    DB   'An error has occurred-'
            DB   'Files may not be identical.',7,24H
;
  ASCIIZ  PROC  NEAR
    ....                    ;for code of the procedure
    ....                    ;   see Fig. 11-2
  ASCIIZ  ENDP
```

FIGURE 11-4. Part 1 of the program COMPARE.ASM.

of the two files to be compared. When the size of file 1 is received, it is saved in location SIZE1. (Word SIZE is not allowed.) The equates establish the buffers for the two programs. The first program is written to the starting location at offset 300H, which is substantially beyond the end of COMPARE.COM, and the second one is loaded at 8000H. The programs must fit into the assigned spaces.

Two 64-byte locations are initialized to 00H and reserved for storage of the ASCIIZ strings of the two files to be compared. The procedure that constructs the strings is given in Fig. 11-2. At the end of the error message, byte 07H produces a speaker beep, calling attention to the error. The initial JMP instruction with the label START branches the program around the data, messages, and procedure, and code of the main program starts at label A1 of Fig. 11-5.

A suggested way to test the program is to copy CHKDSK.COM from the DOS diskette of drive A to the UTILITY subdirectory of drive B. To create the subdirectory, to copy the program, and to test COMPARE.COM identified in the root directory of drive B, the following commands should be executed:

```
MD  B:\UTILITY
COPY  CHKDSK.COM  B:\UTILITY
B:COMPARE  CHKDSK.COM  B:\UTILITY\CHKDSK.COM
```

The first command creates the subdirectory. Execution of the COM file COMPARE should report identical files. Comparison of unlike files will generate the error message of the program (see Prob. 11-3).

Program Analysis

Analysis of the program consists of an examination of the included functions of INT 21H and several related ones. Seven functions are described here, and many others will be considered in later sections.

```
    A1: STI
        MOV DX, OFFSET ASCIIZ1
        CALL ASCIIZ
        MOV AX, 4300H
        INT 21H                      ;get attribute byte
        JC  A2
        MOV AH, 4EH
        INT 21H                      ;get file1 data
        MOV AX, FILE_SIZE
        MOV SIZE1, AX                ;store byte size of file1
;
        MOV DX, OFFSET ASCIIZ2
        MOV AX, 4300H
        INT 21H                      ;get attribute byte
        JC  A2
        MOV AH, 4EH
        INT 21H                      ;get file2 data
        MOV AX, FILE_SIZE
        CMP AX, SIZE1
        JNZ A2                       ;if sizes differ, end
;
        MOV DX, OFFSET ASCIIZ1
        MOV AX, 3D00H
        INT 21H                      ;open file1
        MOV BX, AX
        MOV DX, PROG1
        MOV CX, SIZE1
        MOV AH, 3FH
        INT 21H                      ;read file1 to PROG1
;
        MOV DX, OFFSET ASCIIZ2
        MOV AX, 3D00H
        INT 21H                      ;open file2
        MOV BX, AX
        MOV DX, PROG2
        MOV CX, SIZE1
        MOV AH, 3FH
        INT 21H                      ;read file2 to PROG2
;
        MOV SI, PROG1
        MOV DI, PROG2
        REPE CMPSB                   ;compare the files
        JNZ A2
        MOV DX, OFFSET MSG1
        JMP SHORT A3
    A2: MOV DX, OFFSET MSG2
    A3: MOV AH, 9                    ;print the message
        INT 21H
        MOV AX, 4C05H
        INT 21H                      ;terminate the program
CODE  ENDS
END   START
```

FIGURE 11-5. Final part of the program COMPARE.ASM.

Function 43H (Get/Change File Attribute). It is necessary to obtain the sizes of the two files to be compared, which is done with function 4EH of INT 21H. However, on entry, the attribute byte of the file should be present in register CX. Because this operation is accomplished with function 43H, it is executed first.

On entry to function 43H, DS:DX must point to the ASCIIZ string of the file, and register AL should contain either 00H or 01H. If AL has 00H, the file attribute is read from the directory of the diskette and placed in CX. If AL is 01H, the reverse occurs, with the attribute set into the directory to the value in CX. *This procedure allows the operator to change the attribute of a file.* File attributes are defined in the discussion of the directory in Section 8-6.

Function 4EH (Get File Data from Directory). When 4EH is called, DS:DX must point to the ASCIIZ string. Also, CX should contain the file attribute. On execution, the specified directory is searched, and if the file is found, the current DTA is filled with 43 bytes. The first 21 bytes are reserved for DOS usage on a subsequent call for a search for another file identified by the string, which may have global characters. The next byte is the attribute found from the directory of the disk.

Then 4 bytes give in order the time and date of the last write to the file. These data are encoded. To illustrate, for program CHKDSK.COM the date is given as 0754H, with the high-order byte in the higher location. The 7 high-order bits have a numeric value from 0 to 119, corresponding to the years 1980–2099. The next 4 bits give a number from 1 to 12 denoting the month, and the 5 low-order bits represent days 1–31 of the month. Thus, CHKDSK is dated Oct. 20, 1983. For the time, which precedes the date, the 5 high bits give the hours from 0 to 23, the next 6 bits give the minutes, and the low 5 bits give 2-second increments.

The 4 bytes that follow are the size of the file, with the low word at the lower address and the low byte of each word at the lower address. For files less than 64 KB, as assumed here, only the word at offset 1AH of the DTA is significant. With the default DTA at offset 80H, the file size is placed in the word at offset 9AH from DS. This is location FILE_SIZE in the program. The last 13 bytes consist of the name and extension of the file, with blanks removed, followed by a byte of 0s.

After the size of each file has been obtained, the sizes are compared. If they are unequal, the error message is displayed and the program ends.

Function 3DH (Open File). *A file must be opened before a read or write operation can be implemented,* and this is done with function 3DH. On entry, DS:DX must point to the ASCIIZ string, and AL contains an *access code*. If AL contains 0, the file is opened only for reading. A value of 1 opens the file for writing, and value 2 provides both reading and writing. In the program COMPARE, AX was loaded with 3D00H, because only reading was desired. *It is not possible to write to a file with an access code of 0, nor can a file with an access code of 1 be read.*

On return, a 16-bit file *handle* is contained in AX. *This handle becomes a reference to the file,* and it is needed in order to implement various file operations. When a file is opened, *a read/write pointer is set at the first byte of the file, and the record size is 1 byte*. Read/write operations transfer bytes in records, and a record size of 1 means that single bytes can be read or written. Transfers to and from a diskette, however, consist of whole sectors.

Handles with values from 0 to 4 are *predefined* by DOS, and files with these handles *do not need to be opened prior to read/write operations*. If one of these files is opened, a handle is returned, but not the normal value. Handles not predefined start with 0005H, and values of the predefined handles are from 0 to 4 as noted in Fig. 11-6.

0 Standard input device (keyboard unless redirected)
1 Standard output device (screen unless redirected)
2 Standard error output device (screen)
3 Standard auxiliary device (AUX, or COM1)
4 Standard printer device (PRN, or LPT1)

FIGURE 11-6. Predefined handles.

Function 42H (Move the Read/Write Pointer). The read/write pointer can be changed with function 42H. On entry, the file handle must be in BX, and the desired signed offset is in CX and DX, with CX containing the most significant part. If AL has 0 on entry, the offset is from the start of the file. For a value of 1, it is from the current location, and for 2, the pointer is moved to the end of the file plus the signed offset. On return, the new location of the pointer from the start of the file is in DX and AX, with DX having the most significant part. If 2 is placed in AL with the offset specified as 0, the return in DX and AX is *the size of the file*. This is an alternate way to obtain this information. The file must first be opened, however.

Function 3FH (Read File). A file or device can be read with function 3FH. On entry, BX contains the file handle, which implies that files without predefined handles must first be opened. Register CX contains the number of bytes to be read, and DS:DX has the *buffer address*. Reading starts at the location of the read/write pointer, which can be set with function 42H. On return, AX contains the actual number of bytes read. This value is 0 if the pointer is at or beyond the end of the file. For a device, the specified number of bytes may not always be read. For example, reading from the keyboard is restricted to one line of text at most.

In COMPARE, file1 is opened and read, with the buffer at PROG1, or 300H in DS. The read pointer is at the start of the file after opening, and the entire file is read. The assigned handle is 0005H. Then file2 is opened and read to PROG2, or 8000H. Its assigned handle is 0006H. The two files are compared byte by byte using instruction REPE CMPSB, with CX initially containing the number of bytes in each file. The prefix REPE can be replaced with its equivalent REP, and the SHORT directive of the JMP instruction near the end of the program is optional, saving only 1 byte of code. If this directive is used with the JMP instruction with the label START, an assembler error results. Do not use SHORT if there is any doubt about the length of the jump.

Transfer of data between a diskette and memory is done by complete sectors. Many data base and BASIC programs specify a record size, which is typically 128 bytes, and reading and writing of records occur in a random fashion. To read one record, a 512-byte sector must be read to a special diskette buffer. Other records can be read from this buffer if they are present. The initial buffering is bypassed when the read operation encompasses one or more complete sectors. Whenever a full sector is read, it is transferred directly to the designated file buffer, whereas a partial sector is always first moved to the diskette buffer. Write operations are processed similarly.

Consider a file that occupies sectors 1 and 2 of track 3, and assume that only one diskette buffer is present, which can hold one sector of bytes. Suppose a program reads a record from sector 1 and writes one into it, reads from sector 2 and writes into it, and writes into sector 1 in the order stated. First, sector 1 is read into memory, and the read/write takes place from the diskette buffer. Because the new data must be saved, before reading from sector 2 the 512 bytes of the buffer are returned to sector 1 of the diskette. Sector 2 is then moved to the buffer. After the read/write operation, it is returned to the diskette, and sector 1 is moved to the buffer again.

The frequent transfer of sectors back and forth between a diskette and a buffer is called *thrashing,* which is time-comsuming. Random block read and write routines using blocks that are as large as possible minimize thrashing. Another way to improve speed is to use more than one diskette buffer. DOS provides two by default, and others can be specified in a configuration file, to be described shortly.

Function 4CH (Terminate Program). Call 4CH is a preferred replacement for INT 20H, previously used for terminating a program. Control is transferred to the invoking process. There are several advantages over INT 20H. *All files opened by call 3DH are automatically closed, and it is not necessary for the CS register to hold the segment address of its program segment prefix. Also, an optional return code can be placed in AL on entry.* This code can be interrogated by the function call 4DH and by the batch subcommand IF/ERRORLEVEL, which are described next. In some applications, the feature is quite useful. The return code of COMPARE was arbitrarily chosen to be 05H.

Function 4DH (Get Exit Code). On execution, the *exit code* of another process is returned in AX. *The low byte of the exit code is the return code of call 4CH used to end a program.* A similar code is supplied by the terminate-and-stay-resident function 31H, described in Section 11-6. The high byte is 00H for normal termination, 01H for termination by Ctrl-Break, 02H for a critical device error termination, or 03H for termination by function 31H. It returns the exit code only once.

Batch Subcommand IF/ERRORLEVEL. In Section 9-3 the IF-*condition* subcommand for two conditions was described. There is also a third condition related to the return code of calls 4CH and 31H. The format is

IF ERRORLEVEL *number command*

The condition ERRORLEVEL *number* is true if the return code of the last program executed is equal to or greater than the specified number. The logical operator NOT can be put ahead of the condition if desired.

To illustrate, suppose the program COMPARE is run with the specified exit code of 05H. After execution, entry of the statement IF ERRORLEVEL 6 DIR does nothing, but if number 6 is changed to 5 or less, the directory of the default drive is displayed.

Configuration File

During each reset, DOS automatically searches the root directory of the default drive for the *configuration* file CONFIG.SYS. If it is found, DOS reads and executes the special commands of the file. Allowed commands with comments are listed in Fig. 11-7.

In Fig. 11-7, the maximum value of each *xx* is 99. Whenever DOS must read a record, it checks the diskette buffers, and if the record is present, it is transferred

BREAK = ON/OFF	(described in Section 9-1)
BUFFERS = *xx*	(default number is 2)
DEVICE = filespec	(name of file with a device driver)
FILES = *xx*	(default number is 8)
SHELL = filespec	(user-designed command processor)

FIGURE 11-7. Allowed commands of file CONFIG.SYS.

to the proper place in memory without reading the diskette. To write to a record, if the data of the sector is in a buffer, the write operation transfers the bytes from memory to the buffer. In each case, if the data of the specified sector are not in a buffer, the sector must first be read. Increasing the number of diskette buffers above the default value of 2 increases the speed, provided there is plenty of memory. The buffers reduce the memory available for programs.

The FILES command is used to specify the number of files that can be open at the same time. When a file is opened, DOS constructs a control block in an area of memory set aside for this purpose. The size of this area depends on the value specified in the FILES command.

The DEVICE command, although very useful, will not be considered here. It allows special character sequences to be put in a program for *controlling the screen cursor,* and it provides for *reassignment of the meaning of any key of the keyboard.* It also provides for automatic loading at startup time of files designed to service devices. The SHELL command is used to replace the command processor with one designed by the operator. All configuration commands are described in detail in the *DOS Manual* and the *DOS Technical Reference Manual.*

11-3

COMPARISON IN BASIC

The files considered here are called *random access* because the open process *allows individual records to be accessed in any order* for reading and writing. In contrast, *sequential access always starts with the first byte of the file, and bytes are read or written in sequence.* For example, to read a byte near the end of the file requires the file to be read from its beginning.

Here we examine a comparison program in BASIC that is nearly identical to the one of the preceding section. The program is given in Fig. 11-8. Included are three new statements and two new functions. The new statements are OPEN, FIELD, and GET. The new functions are LOF(n) and LOC(n), with n denoting the file number.

After the two filespecs are entered in lines 120 and 130, the first file, designated as file1, is opened for random access with an allocated DTA. The unspecified record size has the default value of 128 bytes. In line 160, variable Y is equated to the byte length of the file, and in lines 170 and 180, variable X is set to the number of records, including the partial record at the end of file1 if the length is not a multiple of 128 bytes. String variables C$(M) and D$(N) are dimensioned so that each subscripted variable can be equated to a record having the corresponding number.

The FIELD statement of 220 allocates the entire space of the 128-byte random-access file buffer to variable C$(M), and the GET statement that follows reads record number M from file 1 into the DTA buffer and assigns the data to string

```
100  REM  Program COMPARE in BASIC
110  M = 1:   N = 1
120  INPUT  "Enter the first filespec:  ", A$
130  INPUT  "Enter the second filespec:  ", B$
140  '
150  OPEN  "R", 1, A$
160  Y = LOF(1)                              'length of file
170  X = Y / 128
180  IF X > INT(X)  THEN X = INT(X) + 1
                    ELSE X = INT(X)
190  DIM  C$(X), D$(X)
200  '
210  WHILE  M * 128 <= Y                     'read file1
220    FIELD 1, 128 AS C$(M)
230    GET 1, M
240    M = M + 1
250  WEND
260  '
270  OPEN  "R", 2, B$
280  Z = LOF(2)
290  IF Z <> Y  THEN PRINT "Files are different sizes.": END
300  '
310  WHILE N * 128 <= Z                      'read file2
320    FIELD 2, 128 AS D$(N)
330    GET 2, N
340    N = N + 1
350  WEND
360                                          'compare files
370  FOR  J = 1  TO  N - 1
380    IF C$(J) <> D$(J)  THEN  PRINT "Files not the same.": END
390  NEXT
400  '
410  PRINT:  PRINT Y; "bytes and"; LOC(1); "records read."
420  PRINT  "Files compare OK."
```

FIGURE 11-8. The COMPARE program in BASIC.

variable C$(M). Subscript M is incremented and the WHILE-WEND loop is repeated. This process continues until all records have been read. The same procedure is followed for file2 in program lines 310–350.

In line 290 the sizes of the two files are compared; if they are unequal, the program ends with an error message. The statement in line 410 prints the size in bytes of the identical files. Also printed is function LOC(1), *which gives the number of the last record read from file1*. In BASIC, record numbers start with 1. The comparison takes place in the FOR-NEXT loop. Corresponding records are compared in order, and the message that is printed depends on the outcome. We now examine the new BASIC statements and functions.

The OPEN Statement. In Section 10-5 it was mentioned that the OPEN statement must precede any I/O to a file, *with a buffer allocated to the specified file and a number assigned*. To open a file for random I/O, the following form of the statement can be used:

OPEN "R", *n*, filespec, *m*

If the filespec is not found in the diskette directory, a new file is created. This action requires a write operation to the directory, and if the diskette is write-protected, the program ends with a message.

The letter *R* enclosed in quotation marks specifies a random file, which allows direct access to any specified record. In contrast, a sequential file permits read/write of records only in sequence; for example, to read record 9, it is first necessary to read records 1–8. Thus, sequential files are not as flexible as random files. The form of the OPEN statement for sequential files is given when sequential files are examined in Section 11-5.

In Disk and Advanced BASIC, numeric *n* is an assigned file number from 1 to 3, and the optional parameter *m* denotes the record length. If *m* is omitted, the preceding comma should also be omitted. The assigned record length cannot exceed the buffer size for random files. If BASIC is entered without specifying this buffer size, it is set at 128 bytes, which is the maximum value allowed for the record size *m*. However, by including the option */S:number* after BASIC or BASICA, with *number* having a value from 1 to 32,767, the buffer size is set to the specified value. Thus, with this option, the record length can be specified up to 32,766.

The number of files that can be open at one time can also be changed by an option of the BASIC statement. The format is */F:files,* with *files* denoting a number from 1 to 15. Each file requires 188 bytes for the BASIC file control block plus the number of bytes reserved for the buffer. For example, the statement BASIC /S:512 /F:10 sets up 10 buffers with 512 bytes each for up to 10 files that can be simultaneously open. The file control blocks and buffers reserve 7000 bytes of memory.

An alternate form of OPEN is

OPEN filespec AS *n* LEN = *m*

In both forms, the number sign (#) can precede the file number *n* if desired. The only device that can be opened for random access is the printer, with filespec "LPT1:" and used only for output. Up to three printers are allowed, with numbers 1, 2, and 3. The keyboard and the display screen can be opened only as sequential files, and a communications file is a special case that was examined in the preceding chapter.

The FIELD Statement. *Space in a random file buffer can be allocated to string variables by means of the FIELD statement.* To illustrate, consider the following statement:

```
FIELD 2, 20 AS COST$, 5 AS ID$, 50 AS NAM$
```

This statement allocates the first 20 bytes of a record of file2 to COST$, the next 5 bytes to ID$, and the next 50 bytes to NAM$. *The statement does not specify a specific record, and there is no transfer of data.* The actual record is the one that is referred to in a subsequent statement. For example, if the FIELD statement is followed by GET 2, 8, record 8 is read to the buffer and the specified bytes are then transferred to the variables. The total number of bytes allocated must not exceed the record length. No more than 255 bytes can be assigned to a string variable.

Any number of FIELD statements may be executed for the same file number. A new FIELD statement redefines the buffer from the first character position, but preceding statements remain in effect. Thus, the same bytes of a record can be allocated to more than one variable.

The GET Statement. *To read a record from a random file, the statement GET is used.* For example, GET 3, 25 will read record 25 from the random file assigned number 3 (see Prob. 11-8). In BASIC the first record is number 1. If the record number is omitted, the record following the last GET statement is read. Thus, if GET 3 follows the preceding GET statement, record 26 is read into the buffer. GET can be used with communications files, in which case the "record number" of the statement becomes the number of bytes to be read from the communications buffer.

Function LOF(*n*). In lines 160 and 280 of the program, function LOF(*n*) *returns the number of bytes allocated to the file with number n*. Files with extension BAS are always assigned a length that is a multiple of 128 bytes.

Function LOC(*n*). *With random files, LOC(n) returns the record number of the last record read or written*. It is present in line 410 of the COMPARE program. For comparison of CHKDSK.COM with B:CHKDSK.COM, the displayed value of 50 indicates that the last record read was number 50. CHKDSK has 6400 bytes and requires a reading of fifty 128-byte records.

With sequential files, the function returns the total number of records read from or written to the file after the file is opened. When a file is opened for sequential input, the first sector of the file is automatically read, and LOC(*n*) will return 1. The use of LOC(*n*) with communications files was described in Section 10-5.

11-4

FILE CREATION

Emphasis here is on the assembly language program NEW_FILE, designed for creating a file. *The record size and all data are entered from the keyboard, and the created file is put on a diskette with the filespec B:DATA-n*. Following this program is a similar BASIC program. Also included are short routines that erase a file.

The NEW_FILE program uses seven functions of INT 21H. Only two are new: 3CH for creating a file and 40H for writing to a file. Call 4EH searches the diskette directory to verify that the file to be created does not exist, and 4CH terminates the process. In addition, function 9 for display of a message appears four times, 0AH for buffered keyboard input twice, and 01H for single-character keyboard input once. The assembly language program that deletes a diskette file calls 41H. Although not used here, the close-file function 3EH is described.

NEW_FILE in Assembly Language

Assume that NEW_FILE will be stored on the drive B diskette and that file DATA-*n*, to be created, is on the same drive. After NEW_FILE has been assembled and converted into a COM type, it is activated by entering B:NEW_FILE B:DATA-*n*. The NEW_FILE program is given in Figs. 11-9 through 11-12. At the front of the program are reserved areas for the ASCIIZ string, the handle of file DATA-*n* to be created, the record size to be entered from the keyboard, a file buffer, two keyboard buffers, and five messages. These data are followed by the ASCIIZ procedure defined in Section 11-1.

Keyboard Buffers. Keyboard buffer KB1_BUF has five byte locations. The first byte is initialized to 3, thereby allowing a maximum of 3 characters, including CR, to be placed in the buffer. It is used to receive from the keyboard a one- or two-digit decimal number from 1 to 99 that specifies the number of characters that can be typed into a record, excluding CR and LF. The second byte of the buffer is set automatically to the number of characters actually received, excluding CR. Thus, it will be either 0, 1, or 2.

The second keyboard buffer has 101 byte locations. The first byte, which is set by the program, cannot exceed 99. The selected sizes of the two buffers allow the entered record size to be specified with no more than two decimal digits, thereby simplifying the program. *The actual record size will be the entered size plus two, which allows for the addition of LF and CR characters.* If the data entered into a record are less than the specified number of characters, *blanks are used as fillers*. Two lines are required for the data of messages 2 and 3, because the assembler does not allow bytes of a line to precede a string.

Program Discussion

The program is divided into 14 sections numbered at the left. In section 1, a test is made to determine if a filename is present in the ASCIIZ1 string, including an optional path to a subdirectory. None will be present if B:NEW_FILE is entered without a second filespec. If no filespec is found, the program jumps to section

```
TITLE New_File
;
CODE SEGMENT
   ASSUME  CS:CODE, DS:CODE, ES:CODE, SS:CODE
   ORG  100H
;
START: JMP A1
;
   ASCIIZ1  DB  64 DUP(0)
   HANDLE   DW   ?
   REC_SIZE DW   ?
   BUFFER   DB  101 DUP(?)
;
   KB1_BUF  DB  3,?,?,?,?
   KB2_BUF  DB  101 DUP(?)
;
   MSG1 DB  ' Enter record size (1-99 characters): $'
   MSG2 DB    10, 13
        DB  ' Enter data of the record.', 10,10,13,24H
   MSG3 DB    10, 10, 13
        DB  ' Another record? (Y/N): $'
   MSG4 DB  'New filespec was not specified.$'
   MSG5 DB  'File to be created already exists.$'
;
ASCIIZ    PROC    NEAR
   ....                (Code given in Fig. 11-2)
   ....
   ....
ASCIIZ    ENDP
```

FIGURE 11-9. Part 1 of the NEW_FILE program.

```
1  ;Construct ASCIIZ1 String and Test for Filespec
     A1: STI
         MOV  DX, OFFSET ASCIIZ1
         CALL ASCIIZ
         CMP  ASCIIZ1, 0          ;is a string present?
         JNZ  A2
         JMP  B4
   ;
2  ;Verify That File To Be Created Does Not Already Exist
   ;
     A2: MOV CX, 0
         MOV AH, 4EH              ;search diskette directory
         INT 21H                  ;    for matching filename
         JC  A3
         JMP B5
   ;
3  ;Create the File Named DATA-n
   ;
     A3: MOV AH, 3CH              ;function create-file
         INT 21H
         MOV HANDLE, AX           ;save file handle
   ;
4  ;Print MSG1 "Enter desired record size..."
   ;
         MOV DX, OFFSET MSG1
         MOV AH, 9
         INT 21H
   ;
5  ;Provide for Keyboard Input of One or Two Digits
   ;
         MOV DX, OFFSET KB1_BUF
         MOV AH, 0AH
         INT 21H
   ;
6  ;Determine the Record Size
   ;
         MOV AL, KB1_BUF + 3
         CMP AL, 13
         JNZ A4
         MOV AH, 30H
         MOV AL, KB1_BUF + 2
         JMP SHORT A5
     A4: MOV AH, KB1_BUF + 2
     A5: SUB AX, 3030H            ;record size minus 2
         AAD
         INC AX
         MOV WORD PTR KB2_BUF, AX  ;record size minus 1
         INC AX
         MOV REC_SIZE, AX         ;save record size
```

FIGURE 11-10. Part 2 of the NEW_FILE program.

```
7  ;Print MSG2 "Enter data of the record."
   ;
          MOV DX, OFFSET MSG2
          JMP SHORT A7
      A6: MOV DX, OFFSET MSG2 + 2
      A7: MOV AH, 9
          INT 21H
   ;
8  ;Provide for Keyboard Entry of Data for Record
   ;
          MOV DX, OFFSET KB2_BUF
          MOV AH, 0AH
          INT 21H
   ;
9  ;Move Data of Record to Reserved Buffer
   ;
          MOV CX, REC_SIZE
          SUB CX, 2
          MOV SI, OFFSET KB2_BUF + 2
          MOV DI, OFFSET BUFFER
      A8: MOV AL, [SI]                ;move data until CR
          CMP AL, 13                  ;   is found or count
          JZ  A9                      ;   becomes 0
          MOV [DI], AL
          DEC CX
          INC SI
          INC DI
          JMP A8
      A9: OR  CX, CX
          JZ  B2
      B1: MOV BYTE PTR [DI], 20H      ;fill buffer with blanks
          INC DI
          DEC CX
          JNZ B1
      B2: MOV WORD PTR [DI], 0A0DH
   ;
10 ;Write the Record to File DATA-n
   ;
          MOV BX, HANDLE
          MOV CX, REC_SIZE
          MOV DX, OFFSET BUFFER
          MOV AH, 40H                 ;write function
          INT 21H
   ;
11 ;Print MSG3 "Another record? (Y/N)"
   ;
      B3: MOV DX, OFFSET MSG3
          MOV AH, 9
          INT 21H
```

FIGURE 11-11. Part 3 of the NEW_FILE program.

```
12 ;Provide Keyboard Input for Y/N Response
   ;
         MOV AH, 1
         INT 21H
         CMP AL, 79H                  ;lowercase y?
         JZ  A6
         CMP AL, 59H                  ;uppercase Y?
         JZ  A6
         CMP AL, 6EH                  ;lowercase n?
         JZ  B7
         CMP AL, 4EH                  ;uppercase N?
         JZ  B7
         JMP B3
   ;
13 ;Print Error Message If Appropriate
   ;
     B4: MOV DX, OFFSET MSG4
         JMP SHORT B6
     B5: MOV DX, OFFSET MSG5
     B6: MOV AH, 9
         INT 21H
   ;
14 ;Terminate the Program, Closing File DATA-n
   ;
     B7: MOV AX, 4C00H                ;terminate function
         INT 21H
   ;
   CODE  ENDS
   END   START
```

FIGURE 11-12. Part 4 of the NEW_FILE program.

13, prints message MSG4 using function 9, and ends. Because the jump is long, *it must be unconditional*. Assuming filespec B:DATA-*n* was included in the command, a check is made in section 2 to verify that the file does not already exist. *Otherwise the new file would overwrite the current one*. Execution of function 4EH gives a set carry flag CF if the file is not found. Next, the new file DATA-*n* is created.

Function 3CH (Create File). On entry, DS:DX must point to the ASCIIZ string containing the filespec and its optional path, and CX must have the attribute of the file to be created. In this case, CX is loaded with 00H for a normal file. The diskette directory is read and searched for a matching entry. If it is found and if the access code allows writing, the directory is initialized to a file of 0 length. This action effectively eliminates the old file. If it is not found, a new entry is set up for a file of 0 length. In either case, the file is opened and *assigned an access code for reading and writing*. The assigned handle is returned in AX.

If the file cannot be created, flag CF is set, and an error code is put in AX. The error code table, which applies to all file management functions, is given in Chapter 5 of the *DOS Technical Reference Manual*. Possible reasons for failure include a full directory or an existing file marked for read-only.

Function 3CH is identical to the open function 3DH, except that the size of the file is 0 and the date is the current date. Function 3CH *creates a new file and opens it*. It is not necessary either to create or to open a file for a device with a predefined handle, as explained in Section 11-2.

Record Size. Sections 4 and 5 of the program provide for entry of the record size. First, function 9 is used to print a message requesting the operator to enter the desired size from 1 to 99. Then function 0AH sets up a keyboard buffer for receiving entries of one- or two-decimal digits. Note that it is not possible to enter a number greater than 99. The true record field is the entered value plus two, allowing for addition of CR and LF at the end of each record. Thus, the field may be from 2 to 101, corresponding to 0 to 99 characters entered by the operator.

In section 6 the keyboard entry of one- or two-decimal digits is converted into a binary number. Then the proper size of KB2_BUF is established, and *the true record size is placed in location* REC_SIZE. Thus, the keyboard buffer will not accept more characters than can be put into a record, and the record size provides room for LF and CR characters.

Data Entry to BUFFER. The reserved location BUFFER has adequate space for storage of a record. In section 7, function 9 is again used to print a message, which instructs the operator to enter the data of a record; function 0AH of section 8 sets up the keyboard buffer to receive the data. The record is moved to BUFFER by instructions of section 9, with the LF and CR characters added. If the keyboard entries are less than the specified record size, the remaining spaces are filled with blanks (20H). For example, *if the stated size is 90 characters and only 40 are entered, the 92-byte record consists of the 40 characters typed from the keyboard, 50 blanks, and LF and CR*. This program is designed so that each record has the same size, regardless of the number of keyboard entries.

Function 40H (Write to a File). From BUFFER, the data of the record are transferred to the diskette by the write function 40H. This action is accomplished in section 10 of the program. On entry, BX must have the file handle, and CX contains the number of bytes to write. Also, *DS:DX must point to the buffer* that contains the data to be written. Register AX returns the actual number of bytes written, which will be less than that specified if there is insufficient space on the diskette. The write function can be used to write to a device, if desired, and redirection from one standard device to another is allowed.

When transfer from the buffer to the diskette consists of data with fewer than 512 bytes, the transfer is indirect, with the data fed first to one or more 512-byte diskette buffers in memory. Transfer from the buffers to the diskette occurs when they become full. If the last record entered leaves data in the buffers, these data are flushed to the diskette when the file is closed, and the diskette directory is updated. Accordingly, *whenever writing to a file occurs, it is essential to close the file;* otherwise data remaining in the buffers are lost. It is not necessary to close a file that has been opened unless writing has occurred. A file can be closed with function 3EH, but function 4CH, used to terminate a program, also performs this action. Thus, the close-file function 3EH is not present in NEW_FILE. It is described next.

Function 3EH (Close File). On entry, BX must contain the handle of the file or device to be closed. *All internal buffers are flushed to the file or device, and the directory is updated.*

Additional Records. Program lines of sections 11 and 12 provide for writing of additional records. As many records as desired can be placed into the file. Following the message asking if another record is to be entered, function 1 is called to provide a wait for for keyboard entry of a single character, which goes to AL and is echoed. This procedure was described in Section 6-6. The Y/N response is checked and appropriate action taken. After the last record has been entered, the program jumps to section 14. The call of function 4CH terminates the program and closes file DATA-*n*.

A more complete program contains instructions that allow the operator to enter the number of a record to be displayed, edited, and rewritten. As many files as desired can be created simply by changing the value of *n*. To examine the created file, enter TYPE B:DATA-*n*. Each record is displayed on a new line because of inserted CR and LF characters.

NEW_FILE in BASIC

Figure 11-13 is the BASIC program B:NEW_FILE.BAS, which gives the same result as B:NEW_FILE.COM. The response to the INPUT statement of line 110 determines the record length M in characters, excluding the CR and LF to be added. The new file B:DATA is opened as #1 for random access by the statement of line 120, with a record length equal to M + 2. In the FIELD statement, M characters are assigned to A$ and two (CR and LF) are assigned to B$. The symbol N denotes the record number (see Prob. 11-12).

```
100  'Program B:NEW_FILE for creating B:DATA
110   INPUT  "Enter record size (1-99 characters): ", M
120   OPEN   "R", 1, "B:DATA", M + 2
130   FIELD   1,  M AS A$,  2 AS B$
140   N  = 1
150   R$ = "Y"
160   WHILE  R$ = "Y" OR R$ = "y"
170     PRINT
180     INPUT  "Enter data of record: ", X$
190     PRINT
200     LSET  A$ = X$
210     RSET  B$ = CHR$(10) + CHR$(13)
220     PUT   1, N
230     N  =  N + 1
240     R$ =  ""
250     WHILE  R$<>"Y"  AND  R$<>"y"  AND R$<>"N"  AND  R$<>"n"
260       INPUT  "Another record? (Y/N): ", R$
270     WEND
280   WEND
290   LSET  A$ = CHR$(&H1A)
300   PUT   1, N
310   CLOSE
```

FIGURE 11-13. NEW_FILE in BASIC.

LSET and RSET. The purpose of the BASIC statements LSET and RSET is *to move data from memory to a random-access file buffer*. LSET left justifies the string in the field, and RSET right justifies the string. *Extra positions are padded with spaces*. If X$ in line 200 is longer than the M spaces reserved for A$, characters are dropped from the right. Only string variables are allowed.

LSET and RSET are not restricted to string variables of FIELD statements. For example, suppose G$ = SPACE$(25), which allocates 25 byte spaces to G$, and suppose H$ has 10 characters. Then the statement RSET G$ = H$ fills the 10 left-hand spaces of G$ with the characters of H$.

In line 210, string variable B$ is equated with the two characters LF and CR, *with the plus sign concatenating the two strings*. The result is right justified in the random file buffer. Thus, the characters of the record start at the left, with spaces added if the number of characters is less than M. The last two characters of the record are LF and CR.

Statement PUT. *Transfer of a record from the random-access buffer to a file is done with the PUT statement.* The first number after PUT is the file number, which may be preceded by the number sign if desired. The second number is the record number; if this number is omitted, the next one after the last PUT statement is assumed. Record numbers from 1 to 32,767 are allowed. The transfer is buffered until enough data to fill a sector are available.

The End-of-File (EOF) Mark. After the last record has been entered, the program inserts the *end-of-file (EOF) mark* 1AH at the left side of the final record. This operation occurs in lines 290–300. Recall that BASIC files are always assigned a length that is a multiple of 128 bytes. Without the EOF mark, when the command TYPE B:DATA is entered from the DOS mode, all 128 bytes of the last block are displayed, which most likely include data that are not part of the file. *The EOF mark prevents this result.*

Function MKI$(*x*) and Related Ones. It is often necessary to convert a numeric variable to a string variable. For example, a numeric value that is to be placed in a random-access file buffer with LSET or RSET must first be converted to a string. Function MKI$ *converts an integer into a 2-byte string.* For example, suppose numeric variable AMT is an integer that is to be placed in field A$ of a random-access file buffer using LSET. This action is done with the statement LSET A$ = MKI$(AMT). If AMT is 2321H, ASCII codes 23H and 21H are assigned to A$, corresponding to characters # and !, respectively.

Execution of the statement PRINT MKI$(&H2321) gives a display of !#, because characters of a string are shown in order of increasing addresses and the low-order byte 21H has the lower address. In contrast, PRINT (&H2321) displays 8993, which is the decimal value of the number.

The function MKS$*(x) converts the numeric value of the single precision number* x *to a 4-byte string,* and MKD$*(x) converts the double precision number* x *to an 8-byte string.*

The corresponding inverse functions are CVI, CVS, and CVD. Function CVI *converts a 2-byte string into an integer,* and CVS and CVD are similar except that *they operate on 4-byte and 8-byte strings,* respectively. Thus, the statement that follows displays 8993:

```
PRINT CVI (CHR$(&H21) + CHR$(&H23))
```

Statement CLOSE. The BASIC statement CLOSE is used to conclude I/O to a file or device. If it is followed by numbers separated by commas, only the specified files are closed. In the absence of file numbers, all opened files are closed. A CLOSE of a file or device *opened for sequential output causes the file buffers to be written to the file or device,* and for a file, the diskette directory is updated. However, *for files opened for random access, the PUT statement must be executed for this purpose.*

When any of the commands END, NEW, RESET, or SYSTEM are executed, all files and devices are automatically closed. RESET is the same as CLOSE without file numbers, except that RESET closes only diskette files. It does not affect device files.

Program DELETE

We have examined assembly language and BASIC programs that create a new file. An elementary COM program for deleting a file is presented here. The program DELETE consists of the instructions in Fig. 11-14, along with the ASCIIZ procedure in Fig. 11-2. The 64-byte location ASCIIZ1 must be reserved for receiving the string from the procedure. Details are left as an exercise (see Prob. 11-13).

```
MOV  DX, ASCIIZ1      ;offset for ASCIIZ string
CALL ASCIIZ
MOV  AH, 41H          ;delete-a-file function
INT  21H
MOV  AX, 4C00H        ;terminate function
INT  21H
```

FIGURE 11-14. A portion of the program DELETE.

Function 41H (Delete File). On entry, DS:DX points to an ASCIIZ string, which must not contain global characters. The directory entry is removed. *Read-only files cannot be deleted without first changing the attribute of the file to 0,* using call 43H described in Section 11-2. In the DOS mode, a file can be deleted by means of command DEL or ERASE, and in BASIC the command is KILL.

11-5 SEQUENTIAL FILES

Data written to a sequential file are stored and read sequentially. To read a record, data are normally read from the start of the file until the desired record is found; writing is also done from the beginning. The program CON2FILE will be used for illustration. It is designed to create the DATA file by writing sequentially to the diskette from the keyboard, and in many ways it is similar to the random access NEW_FILE of the preceding section.

In assembly language there is no real distinction between a sequential and a random-access file, because *reading and writing of any read/write file can be either sequential or random as desired.* This is not the case with BASIC files, however. *In BASIC, a file is either sequential or random access,* depending on the open statement used.

Program CON2FILE

The assembly language program is designed to be converted into a COM file after assembly. The keyboard buffer is set up to hold 128 characters plus CR, which is an arbitrary choice. Any size will do. Regardless of the specified size, *characters entered from the keyboard are written sequentially to the data file without any carriage returns, line feeds, or spaces.* Such a file occupies less diskette space than one that fills extra spaces of records with blanks.

With the files on drive B, sequential writing to the data file is initiated by entering

```
b:con2file b:data
```

If DATA is already present on the diskette of drive B, it is overwritten. Otherwise a new file is created. The program is given in Fig. 11-15.

```
Title   CON2FILE
;
CODE    SEGMENT
   ASSUME  CS:CODE, DS:CODE, ES:CODE, SS:CODE
   ORG     100H
;
   START: JMP A1
;
   ASCIIZ1    DB     64        DUP(0)
   KB_BUF     DB    129, 130   DUP(?)
;
 ASCIIZ    PROC    NEAR
    .....
    .....              (code is given in Fig. 11-2)
 ASCIIZ    ENDP
;
  A1: STI
      MOV   DX, OFFSET ASCIIZ1
      CALL  ASCIIZ
      MOV   CX, 0
      MOV   AH, 3CH                   ;create file
      INT   21H
      JC    A3
      MOV   BX, AX                    ;file handle to BX
;
  A2: MOV   DX, OFFSET KB_BUF
      MOV   AH, 0AH
      INT   21H                       ;wait for keyboard input
;
      MOV   CL, KB_BUF + 1
      ADD   DX, 2                     ;DX points to first byte
      MOV   AH, 40H
      INT   21H                       ;write to data file
;
      MOV   DL, 0AH                   ;LF character
      MOV   AH, 2
      INT   21H                       ;send LF to screen
;
      MOV   DI, OFFSET KB_BUF + 2
      MOV   CX, 128
      MOV   AL, 1AH                   ;EOF mark is 1AH
      REPNZ  SCASB                    ;scan buffer for EOF mark
      JNZ   A2                        ;end program on EOF mark
;
  A3: MOV   AX, 4C00H
      INT   21H                       ;close file and end
;
CODE   ENDS
END    START
```

FIGURE 11-15. Program CON2FILE with sequential writing from the keyboard to a data file.

Function 3CH near the beginning of the program creates the data file, which may have a path. In the event of failure, which will occur if the data filespec is omitted from the command, the carry flag is set and the program ends. The handle of the created file is moved to BX, and function 0AH is called to provide keyboard entry. Up to 128 characters can be entered. *Following the return, the entire contents of the buffer are scanned for the end-of-file (EOF) character 1AH. If it is found, the program closes the data file and ends.* Otherwise, function 0AH is again called for an additional keyboard entry of up to 128 characters. *The process continues indefinitely until the EOF indicator is found.*

Between each call of the keyboard buffer function, a line feed is sent to the screen with function 2, which was described in Section 6-6. If this were not done, each display of characters entered from a buffer call would overwrite the preceding entry, *because the buffer returns only a carriage return at the end of the character string.*

The EOF mark 1AH is placed in the buffer by key F6 followed by CR. It is ASCII for Ctrl-Z, which can also be used to give the EOF mark. When a file is read to the screen with the TYPE command of DOS, *characters that follow the EOF mark are not displayed.* Similarly, the DOS command COMP will not compare beyond an EOF mark if one is encountered.

CON2FILE with Devices. If we enter B:CON2FILE CON, the program creates the file CON for output, which is the screen (see Prob. 11-14). The result is similar to the command COPY CON CON. However, each record of the CON2FILE program is written to the screen just as it would be to a diskette, whereas in the COPY command the data is buffered until the EOF mark is entered on a separate line.

Although LPT1 can be used as the output file, this choice is not satisfactory. Because no carriage returns and line feeds are included in the records, the printer writes one record on top of another. (The display inserts CR and LF at the end of each line.) In addition, the buffer of the printer supplies 80 characters at a time, which are printed only when the buffer becomes full or when a CR or LF is encountered. Hence, the last partially filled buffer of the printer is not printed. The program was not designed for writing from the keyboard to this device.

CON2FILE in BASIC

Figure 11-16 shows a BASIC program quite similar to the one just considered. In line 110, file B:DATA is opened as number 1 for sequential output. "O" denotes the output mode. Each record is subdivided into three fields, designated A$, B$,

```
100  REM     Program B:CON2FILE - create sequential file DATA
110  OPEN    "O", 1, "B:DATA"
120  WHILE  A$ <> "$"
130    INPUT  "Enter data of field #1: ", A$
140    INPUT  "Enter data of field #2: ", B$
150    INPUT  "Enter data of field #3: ", C$
160    PRINT#1, A$; B$; C$;
170  WEND
180  CLOSE
```

FIGURE 11-16. CON2FILE.BAS for creating a sequential file.

and C$. The INPUT statements allow for assignment to the string variables, up to 255 characters each. Line 180 ends the program when the operator responds to the previous INPUT statement with $ for A$. No spaces, line feeds, or carriage returns are written to the DATA file (see Prob. 11-18).

The OPEN Statement. In the OPEN statement "O" denotes *sequential output* and "I" denotes *sequential input*. A file opened for output must be closed before it can be opened for input and read. Specification of the record length is invalid for sequential files. Although opening a file for sequential output causes all data of the file to be lost, it is possible to APPEND data to a sequential file. The statement for this operation is

```
OPEN "B:DATA" FOR APPEND AS #1
```

The record number is set to the end of the file, so that the data written extend the length of the file without replacing prior data. APPEND is valid only for diskette files.

In BASIC, the devices "KYBD:" and "SCRN:" can be opened only for sequential operations, but "LPT1:" may be opened in either the sequential or random-access mode. *BASIC inserts a line feed after each carriage return sent to the printer.* The program of Fig. 11-16 can replace "B:DATA" in the OPEN statement with "LPT1:", provided the semicolon at the end of line 160 is omitted and the total number of characters in each three-field record does not exceed the printer line width of 80 characters. Omission of the semicolon causes a carriage return to be placed at the end of each record.

The PRINT# and INPUT# Statements. As mentioned briefly in Section 10-5, the purpose of PRINT# is *to write data sequentially to a sequential or random-access file or device*. Conversely, INPUT# *reads items from a sequential or random-access file or device and assigns them to program variables*. For each, a file number and a comma must follow the number sign #. PRINT#1, writes to opened file 1 precisely as PRINT writes to the screen. INPUT#1, reads items from opened file 1 and assigns them to variables similar to the way INPUT reads items from the keyboard and assigns them to variables.

To illustrate, consider the program of Fig. 11-17. Numeric variables A, B, and C are assigned in line 100, and string variables A$, B$, and C$ are assigned in line 110. File "B:DATA" is opened as #1 for sequential output, and three PRINT# statements follow. These statements write sequentially to the file.

```
100  A = -5: B = -8: C = 6
110  A$ = "BASIC is ": B$ = "fun": C$ = CHR$(33)
120  OPEN  "O", 1, "B:DATA"
130  PRINT#1, A; B; C
140  PRINT#1,
150  PRINT#1, A$; ","; B$; C$
160  CLOSE
170  OPEN  "I", 1, "B:DATA"
180  INPUT#1, D, E, F, G, X$, Y$
190  PRINT    D; E; F; G; X$; Y$
200  END
```

FIGURE 11-17. Program to illustrate PRINT# and INPUT#.

Then file 1 is closed and reopened for sequential input. The INPUT# statement reads four numeric variables and two string variables from file 1 and assigns them to D, E, F, G, X$, and Y$. The PRINT statement displays the values of the variables, and END closes the file and ends the program.

The PRINT# statement of line 130 writes 11 bytes to the file. In hexadecimal, these bytes are 2D 35 20 2D 38 20 20 36 20 0D 0A, corresponding to −5, −8, and 6. A space follows each number and a space precedes each positive number, as indicated by the code. CR and LF bytes always end a PRINT# statement. If the two semicolons are replaced with commas, blanks are inserted between the numbers so as to place each in a 14-character print zone. This procedure wastes diskette space.

In line 140 the PRINT# statement does nothing but write bytes 0D 0A to the file, which create a blank line when the file is typed to the screen in the DOS mode. *The comma is required.* In BASIC the item is treated as a number with a value of 0.

PRINT# in line 150 writes 16 bytes to the file, which are 42 41 53 49 43 20 69 73 20 2C 66 75 6E 21 0D 0A. These bytes correspond to "BASIC is ,fun!" plus CR and LF. One more byte is added when the file is closed, which is the EOF mark 1AH. Thus, the file has a total length of 30 bytes, although the size listed in the directory is the minimum BASIC value of 128 bytes. When the file is displayed using the DOS command TYPE B:DATA, the display is as follows:

```
-5 -8  6

BASIC is ,fun!
```

Although the comma is improper here, it is useful when items are assigned with the INPUT# statement.

In line 180 the INPUT# statement reads items from the file and assigns them to variables D, E, F, G, X$, and Y$. There are four numeric values and two strings. It is essential that the type of each variable match the corresponding variable type of the file. Variable D becomes −5, variable E is −8, variable F is 6, and G is 0, corresponding to the BASIC "entry" of line 140. If G is omitted in line 190, the program stops with an error message. Whenever an assignment by INPUT# leaves an empty buffer, the next record is automatically read into it.

String variable X$ is "BASIC is ", and Y$ is "fun!". Without the comma in line 150, variable X$ would be "BASIC is fun!", and Y$ would not exist. Thus, the comma serves to delimit the strings, allowing them to retain their separate identity. Because no comma was inserted between strings B$ and C$ in line 150, these were merged into a single string. Execution of the PRINT statement of line 190 gives the display

```
-5 -8  6  0 BASIC is fun!
```

The comma is not printed; it serves only as a string separator.

In line 110, if the space following "is" is deleted and a space is inserted ahead of "fun", bytes 2C for the comma and 20 for the space are simply interchanged in the file. However, execution of the PRINT statement of 190 now gives "−5 −8 6 0 BASIC isfun!". When the INPUT# statement assigns values to variables, leading spaces, carriage returns, and line feeds are ignored. Accordingly, *spaces placed in strings to separate words should be placed at the end of a string, not at the beginning.* In the DOS mode, delimiters are not needed, and spaces at the beginning of a string are valid.

A major difference between sequential and random-access files is now rather obvious. To read the *n*th item of a sequential file requires reading all *n* items from the start, using the INPUT# statement. On the other hand, the item can be read directly from a random-access file with GET. Any file can be used in either a sequential or random-access manner.

Statement PRINT# USING. In Section 9-4 it was shown that the USING option can be used with PRINT to control the format of a display. It can be used similarly with PRINT# to control the format of a diskette file. An example is

```
PRINT#1, USING "###.##"; A, B, C
```

Figure 9-10 is a BASIC program that solved a differential equation by approximate integration, and the results were transmitted directly to the line printer. *A better method is to send the results to a diskette file*. This action is performed by inserting the statement OPEN "O",1,"B:DATA" in line 200 and changing all occurrences of LPRINT to PRINT#1,. The result gives an example with the PRINT# USING statement (see Prob. 11-16). If a hard copy is desired, tap key PrtSc with key Ctrl depressed, and then enter TYPE B:DATA. The file is printed simultaneously on both the screen and the line printer. On completion, activate Ctrl-PrtSc again to toggle the print mode off. Turning off the power to the printer is inadequate, because the computer continues to try to send data moved to the screen to the display.

Statement WRITE#. WRITE# is very similar to PRINT#. The difference is that WRITE# *includes quotation marks around each string, inserts commas between items, and does not put a blank space in front of a positive number*. To illustrate, suppose that each PRINT# of the program in Fig. 11-17 is replaced with WRITE#. In addition, delete the comma delimiter of line 150, and add Z$ to the end of lines 180 and 190. Three string items are placed in the file. Note that explicit string delimiters need not be included in the list that follows WRITE#. When the modified program is run, the resulting display is exactly the same as before.

However, there are significant changes in the bytes of the file. These changes are shown by the DOS command TYPE B:DATA, which yields the following display:

```
-5,-8,6
"BASIC is ","fun","!"
```

Note the commas between the items and the quotation marks around each string. There are 35 bytes in the file, consisting of the 28 that are explicitly indicated by the display, 3 carriage returns, 3 line feeds, and the EOF mark 1AH.

11-6

PROGRAM TRANSFERS

Three main objectives in this section are *to demonstrate a way to change the Ctrl-Break function* so as to provide a user application, *to show how program A can load and execute program B*, and *to illustrate the transfer of control from program*

B back to program A. The second program can read from and write to files opened by the first one, manipulating these files in some desired manner in the process. This is a very useful operation.

Program DISPLAY

To illustrate the methods, two programs are given here. Program B is assigned the filespec B:DISPLAY.COM. When loaded into memory either by the normal DOS loader or by program A, it reads up to 256 bytes from the *standard input device* and then writes the bytes to the *standard output device*. The number was selected arbitrarily and can be changed, if desired. Without redirection, the program reads up to 127 bytes from the keyboard, which is the size of the keyboard buffer, and writes them to the screen. Although this operation may appear trivial, important applications of the method are described. First, let us briefly examine the routine, which is given in Fig. 11-18 in a form suitable for conversion to DISPLAY.COM.

Prior to writing to the standard output device, the data read from the standard input device can be manipulated. For example, the data read can be divided into records, sorted, and then written in a specified order. *The standard devices can be specified by the program that calls DISPLAY*. With DISPLAY.COM on the diskette of drive B, entering B:DISPLAY gives a wait for data entry from the keyboard. Up to 127 characters can be entered. They are echoed as the keys are pressed, and then when Enter is pressed, the entire data are displayed at one time, and the program ends. We next examine a program that can be used to call DISPLAY, with redirection of its input.

```
TITLE  Display
CODE  SEGMENT
   ASSUME  CS:CODE
   ORG  100H
;
   B1:  STI
        MOV BX, 0          ;handle for standard input device
        MOV CX, 100H       ;maximum number of bytes to be read
        MOV DX, 0          ;offset of buffer
        MOV AX, 0C00H
        MOV DS, AX         ;segment address of buffer
        MOV AH, 3FH
        INT 21H            ;read up to 256 bytes to 0C00:0000
;
        MOV CX, AX         ;put number of bytes read in CX
        MOV BX, 1          ;handle for standard output device
        MOV AH, 40H
        INT 21H            ;write bytes to standard output
        MOV AX, 4C00H
        INT 21H            ;terminate and return to caller
;
CODE  ENDS
END   B1
```

FIGURE 11-18. The program DISPLAY.

Program GET

The program GET is designed to serve as the parent one that calls DISPLAY. When put on the diskette of drive B as a COM file, execution is by entry of

```
b:get b:data.txt b:display.com
```

Pseudo code for GET is given in Fig. 11-19.

```
PROGRAM  Get
begin
  construct ASCIIZ strings for files DATA and DISPLAY
  open text (DATA) file for reading
  instruct operator to press either Ctrl-Break or other key
  change Ctrl-Break exit address to point to load routine
  wait for response from operator

  if response is not Ctrl-Break
    then
      terminate program
    else
      free memory for the load of DISPLAY
      designate text file as the standard input device
      load and run DISPLAY
         {data from text file are read to the display}
      restore registers
  end if
end
```

FIGURE 11-19. Pseudo code for the program GET.

Reserved Data Area

At the beginning of part 1 of the GET program shown in Fig. 11-20 is a defined word label at offset 2CH. This label identifies the location of the segment address of the environment. Strings of the environment are passed to the called program by placing the segment address in its program segment prefix at offset 2CH. The two ASCIIZ areas provide for the strings of the text and DISPLAY files. The text file is unrestricted. However, it will be opened, and its first 256 bytes will be displayed. Therefore, a file of printable ASCII characters is appropriate. One possible choice is B:GET.ASM, and in fact, any ASM file will do. A text file is easily created using the DOS command COPY CON B:ABC.

The BLOCK of the data area is used by the function that loads and executes DISPLAY. It is discussed in conjunction with the description of this function. The word location HANDLE is for storage of the handle of the opened text file that is to be read. There is one message, which is displayed on two lines of the screen when the program is run. Finally, the procedure ASCIIZ, with the code given in Fig. 11-2, provides for construction of the two ASCIIZ strings from data of the command line of the DTA area.

```
TITLE   Get
;
CODE    SEGMENT
  ASSUME  CS:CODE, DS:CODE, ES:CODE, SS:CODE
;
  ORG 2CH
    SEG_ADDR   LABEL   WORD
;
  ORG 100H
    START:      JMP A1
;
    ASCIIZ1      DB   64 DUP(0)
    ASCIIZ2      DB   64 DUP(0)
    BLOCK        DW    ?, 0080H, ?, 005CH, ?, 006CH, ?
    HANDLE       DW    ?
;
    MSG DB     'To load and run DISPLAY, press Ctrl-Break.', 10, 13
        DB     'Otherwise, press any key.', 10, 13, '$'
;
    ASCIIZ      PROC NEAR
        ......        (Code given in Fig. 11-2)
    ASCIIZ      ENDP
```

FIGURE 11-20. Part 1 of the program GET.

The Main Routine

The jump preceding the data is to label A1 at the beginning of the main routine, shown in Fig. 11-21. After the interrupt system is turned on, the two ASCIIZ strings are established, the text file is opened for reading only, and the assigned handle is saved. The following message then appears on the screen:

```
To load and run DISPLAY, press Ctrl-Break.
Otherwise, press any key.
```

Function 25H (Set Interrupt Vector). Functions 25H and 35H, which respectively get and set the interrupt vector, were described in Section 6-6. On entry to call 25H of INT 21H, the interrupt type is in AL and the 4-byte vector address to be placed in the interrupt table is that of DS:DX. In the program, the Ctrl-Break exit address CS:IP of INT 23H is replaced with DS:OFFSET A3. *Then a Ctrl-Break keyboard interrupt transfers the program to the instruction at label A3*. DOS saves the previous vector address and restores it when the program terminates.

After displaying the message and changing the Ctrl-Break exit address, the program waits for a response from the operator. This response is provided by function 1, which was described in Section 6-6. If it is desired to load and run DISPLAY, the operator presses Ctrl-Break; if not, any other key is pressed, which terminates this illustrative program. For a key other than Ctrl-Break, function 1 places the character in AL and echoes it to the display. There is no wait for a carriage return. The program *shows how the exit address of INT 23H can be used to provide a different operation for the Ctrl-Break key combination*. Let us suppose that the combination is pressed, advancing the program to the instruction at label A3.

Function 4AH (SETBLOCK). The instructions at label A3 set up and call the SETBLOCK function 4AH. When DOS loads a program such as B:GET.COM into memory, usually *all of the available memory space is assigned to the loaded program*. This space cannot be used by another program unless a portion of it is

```
 A1: STI
     MOV   SP, 400H
     MOV   DX, OFFSET ASCIIZ1
     CALL  ASCIIZ                ;string of a text file
;
     MOV   AX, 3D00H             ;open text file for reading
     INT   21H
     MOV   HANDLE, AX            ;save handle
;
     MOV   DX, OFFSET MSG
     MOV   AH, 9
     INT   21H                   ;display message
;
     MOV   DX, OFFSET A3
     MOV   AX, 2523H             ;change Ctrl-Break exit address
     INT   21H
;
     MOV   AH, 1
     INT   21H                   ;wait for 1-byte KB input
;
A2:  MOV   AX, 4C00H
     INT   21H                   ;terminate program
;
A3:  MOV   BX, 40H               ;400H of RAM allocated for GET
     MOV   AH, 4AH
     INT   21H                   ;SETBLOCK of memory
;
     MOV   BX, HANDLE
     MOV   CX, 0
     MOV   AH, 46H               ;text file is now the
     INT   21H                   ;   standard input device
;
     MOV   AX, SEG_ADDR          ;set up the BLOCK
     MOV   BLOCK, AX
     MOV   BLOCK +  4, CS
     MOV   BLOCK +  8, CS
     MOV   BLOCK + 12, CS
;
     MOV   DX, OFFSET ASCIIZ2
     MOV   BX, OFFSET BLOCK
     MOV   AX, 4B00H
     INT   21H                   ;load and run B:DISPLAY.COM
;
     MOV   AX, CS                ;restore registers as desired
     MOV   DS, AX
     MOV   ES, AX
     MOV   SS, AX
     MOV   SP, 400H
     JMP   A2                    ;go to the terminate function
;
CODE  ENDS
END   START
```

FIGURE 11-21. Main routine of the program GET.

freed. There are several ways to accomplish this task, but the program uses function 4AH.

On entry, ES must contain the segment address of the memory block allocated to the loaded program. For both COM and EXE files, ES has the address at the start of the program segment prefix unless it is changed by the program. Also, *BX must contain the requested block size in paragraphs*. The program uses a value of 40H, providing 400H bytes of space, including the stack. Note that the stack pointer of GET was initialized at 400H. When the function is executed, DOS tries to either increase or decrease the specified block. If there is insufficient memory to increase a block to the specified size, the block becomes as large as possible, and the size is returned in BX.

When SETBLOCK is executed in the GET program, the block allocated to GET is reduced from all of memory above ES to 400H bytes above ES, and space higher than ES:0400 *becomes available for other programs*.

Function 46H (Assign a Handle). Following SETBLOCK, function 46H is called. *The purpose of the call in this program is to designate the open text file as the standard input device* so that DISPLAY will read from the text file rather than from the keyboard. This operation is performed by assigning the text file a duplicate handle with a value of 0000H, which is the predefined handle for the standard input device.

On entry, BX must contain the handle of the opened file, and CX has the duplicate handle that is to be assigned. On return, *both file handles refer to the file*. If the CX handle is that of another opened file, the file is automatically closed by the call, thus releasing the handle for the reassignment.

The function allows a program *to open files for standard input or standard output*. The handle for the standard output device is 0001H.

The EXEC Function (4BH)

The loading and execution of DISPLAY are done by the same function used by DOS to load and execute COM and EXE files. It is called the *execute (EXEC)* function, and on entry, DS:DX must point to the ASCIIZ string of the file to be loaded. Also, ES:BX must point to a *parameter block* with the proper data present, and AL must contain a value that selects one of two options. The function uses the loader portion of COMMAND.COM to perform the loading. The two options are examined in order.

The First Option. With AL containing 0 on entry, the EXEC function *loads and executes a program*. In doing so, a program segment prefix is established, and *its terminate exit address is set to point to the instruction that follows the INT 21H instruction initiating the call*. For the program GET, the new program segment prefix created for DISPLAY contains a terminate exit address that returns to the main program, similar to the return from a FAR CALL. The Ctrl-Break exit address of the new program segment prefix is the one last placed in the vector table, which is the pointer to the instruction of Fig. 11-21 with label A3. *When a program terminates, the exit address from the program segment prefix is first put in the interrupt vector table, and this address is the one used to transfer control*.

The block pointed to by ES:BX has 14 bytes. The first 2 bytes must be the segment address of the environment string, located at offset 2CH in the program segment prefix of the parent program. With reference to the program of Fig. 11-21, the first word of BLOCK in the reserved data area becomes this segment address. This action occurs when the program executes the two instructions that move the word at label SEG_ADDR to AX and then to BLOCK.

The next 4 bytes constitute the doubleword pointer CS:IP *to the command line that is to be moved into the default DTA area at offset 80H of the new program segment prefix*. The segment address has the higher location. The pointer is established by the instruction that moves CS to BLOCK + 4, along with the initialized value 0080H.

Two more doubleword pointers follow. The first one points to the default file control block at CS:005C, and the second one points to default FCB at offset 6CH. Although the newer functions do not use these file control blocks, the pointers are included in GET.

After the function has been executed, the new program segment prefix created for the loaded program DISPLAY contains the segment address of the environment, the command line, and two FCBs, which are replicas of those in the program segment prefix of GET. *The program DISPLAY is loaded into memory and run, and control then returns to GET*. When control returns to the parent program, CS:IP points to the instruction following the function call, but *all other registers are changed in a rather arbitrary way,* including those of the stack. Registers SS and SP should be restored, and other registers may need to be restored, depending on the needs of the program. Although GET simply ends after the return, several of the registers are restored by instructions at the bottom of Fig. 11-21, to emphasize the possible need to do so in other programs.

The Second Option. With 3 in AL when the EXEC function is called, the second program is loaded but not run. *No program segment prefix is created for the loaded file*. The option can be used for overlaying a portion of a program of memory with the loaded one. For this option, the block pointed to by ES:BX consists of only two words. The first word is the segment address of the environment, as for the first option. The second word is a relocation factor specifying the offset at which the overlay is to be placed. The overlay should be within the memory block assigned to the parent program. After it is loaded, *it becomes part of the parent program,* with control transferred by program instructions, not by DOS.

Discussion. Although the programs used here to illustrate certain file management techniques were kept short and quite simple, the methods are of considerable practical importance. Programs often change the Ctrl-Break exit address in order to provide a user-designed service routine that may serve a special need. More important, however, is the ability of a program to call another program and execute it, with control returned to the calling program.

For example, using the principles described in GET, a program might open a read-only data file as the standard input device and create a new file as the standard output device. Then control might be transferred to a sort program designed to read records from the standard input device, sort them into alphabetical order, and write the records into the standard output device. Execution of the program would transfer the records from the data file to the newly created one, stored in the desired order (see Prob. 11-21).

We note that *open files are duplicated in the one created by the EXEC call and strings of the environment are passed*. The second program is able to call a third one if the need arises. *The parent program can control the meanings of standard input and standard output for devices used in the called program,* which is a powerful feature.

Memory Management

Memory is allocated by DOS 2.1 in blocks, and three function calls are available for controlling the blocks. Normally, when a program is loaded by the EXEC function, which is the way programs are loaded by DOS, *the allocated memory block starts at the lowest available location and includes all memory above this area*. The program segment prefix starts at offset 0 in this block. The loaded program has this entire region for its exclusive use, and the region includes the transient command processor with its loader. The command processor occupies the highest region of memory, and its space can be used by the loaded program if needed.

It is good practice to include in a program the SETBLOCK function 4AH *to shrink the allocated memory block down to the size needed, thereby freeing a portion of memory for use by other programs*. As was done in GET, the stack should be established before SETBLOCK in order to keep the stack within the allocated block. When a running program exits in the normal manner (function 4CH or INT 20H), the block of memory initially allocated to it is freed for use by other programs. In addition to SETBLOCK, functions 48H and 49H control memory allocation. These functions are described next.

Function 48H (Allocate Memory). A program *can request an additional allocation of memory by means of function 48H*. On entry, BX contains the number of paragraphs requested, and on return, AX:0 points to the allocated block. There must be enough memory available.

Function 49H (Free Allocated Memory). On entry, ES contains the segment of the block that is to be returned. *The returned memory block is free for other use*. An application of 49H is given following a description of the terminate and stay resident function 31H.

Function 31H (Terminate and Stay Resident). Call 31H is a preferred replacement for INT 27H, which was used earlier in the program TICKS. The exit code is placed in AL, and DX *must store the memory size value in paragraphs of 16 bytes that are to be protected from overwrite*. Specifically, DX should contain the size of the program that is terminated, given in paragraphs, from the start of its program segment prefix. The advantages over INT 27H are the same as those stated for call 4CH. Care must be taken that the stack used by a program protected from overwrite is not overwritten. This action can be performed by including the stack within the protected region.

To illustrate the use of the terminate-and-stay-resident function, let us refer to the program TICKS.EXE in Fig. 7-24. In TICKS, interrupt type 27H was used. It can be replaced by either of the sequences in Fig. 11-22. The one on the left uses

```
A4: MOV AH, 49H        A4: MOV BX, 30H
    INT 21H                MOV AH, 4AH
    MOV DX, 30H            INT 21H
    MOV AX, 3101H          MOV DX, 30H
    INT 21H                MOV AX, 3101H
                           INT 21H
```

FIGURE 11-22. Possible replacements for INT 27H in TICKS.

AH	Function
0EH	Specify the default drive, return number of drives
19H	Return the current default disk
2EH	Set/reset the verify switch for diskette writes
30H	Get the DOS version number
33H	Get/set the Ctrl-Break state (as for command BREAK)
36H	Get the number of available diskette clusters
38H	Get international information
39H	Create a subdirectory
3AH	Remove a subdirectory
3BH	Change the current directory
44H	I/O control for devices, including byte transfers
45H	Get a second file handle
47H	Get the path and name of the current directory
4FH	Find the next matching file (similar to 46H)
54H	Get the state of the verify switch (refer to 2EH)
56H	Rename a file (see Prob. 11-22)
57H	Get/set the date and time of a file

FIGURE 11-23. Additional functions of INT 21H.

function 49H to free memory not used by TICKS, and the one on the right uses the SETBLOCK function to reduce the allocated memory block to 300H bytes. Both call 31H with an exit code arbitrarily chosen to be 1. In this case, INT 27H provides a simpler program.

Other Functions

In this chapter, 17 functions of INT 21H have been defined (31, 3C–43, 46, 48–4E), in addition to 19 that were presented in Section 6-6 (0–C, 25, 2A–2D, 35). Excluding 13 reserved exclusively for DOS and 22 older functions not normally used in programs written for DOS 2.1, there are 17 additional ones. These are listed in the table of Fig. 11-23.

11-7 DISK OPERATING SYSTEM

We have examined the major elements of the operating system of DOS 2.1. Here we take a look at the entire system, emphasizing interactions among the various components.

Because DOS uses the low-level interrupts of the BIOS ROM, this firmware operating system may be considered as a part of DOS. In fact, from a general viewpoint, DOS consists of the components listed in Fig. 11-24. Those not previously described are examined in this section.

1. The firmware operating system in the BIOS ROM
2. The boot record
3. The file allocation table (FAT)
4. The diskette directory
5. Program IBMBIO.COM
6. Program IBMDOS.COM
7. The command processor (COMMAND.COM)
8. All COM and EXE utility programs of the DOS diskette

FIGURE 11-24. Components of DOS.

The COM programs IBMBIO, IBMDOS, and COMMAND are not present on all diskettes, but *they must be on the one that boots the system* and located in the designated sectors. COMMAND.COM can be located in any available space of the diskette.

Firmware Operating System

In the 8-KB section of ROM, from address FE000H through FFFFFH, is the Basic I/O System (BIOS), also called the *firmware operating system*. In addition to a self-test and a bootstrap loader that loads the boot record of the diskette, it contains many routines that aid the assembly language programmer by providing system services and software interfacing to I/O devices. These routines were examined in Section 6-5. The program is designed to adjust to certain changes in the hardware, such as the addition of standard options connected to the I/O slots of the system board.

Development of the BIOS program was done in assembly language, rather than in a high-level language, in order to conserve ROM space and to provide the fastest possible execution. The functions can be divided into five major groups, which are listed in Fig. 11-25. Only the first two are examined here, the others having been described.

1. Self-test
2. Bootstrap loader
3. I/O support
4. System services
5. Graphics character generator

FIGURE 11-25. Major divisions of the BIOS ROM.

Self-Test

When power is turned on, or when a system reset is initiated with Ctrl-Alt-Del, a positive pulse is applied to the RESET pin of the 8088 microprocessor. As explained in Section 5-1, the CPU is designed to clear all status and control flags, to empty the instruction queue, and to load CS with FFFFH, while clearing the other segment registers and IP. Accordingly, *the instruction at absolute location FFFF0 is initially fetched and executed.* At this address in ROM the 5-byte instruction EA5BE000F0 gives a jump to F000:E05B, which is the start address of the *self-test* routine. The routine and its procedures occupy 1777 byte locations from FE000 through FE6F1, terminating with a jump to the BIOS bootstrap loader if no errors are detected.

Phase 1 (Registers, ROM, RAM, Peripherals, Refresh). There are four phases in the self-test. In phase 1, the processor flags and registers are checked by writing and reading bit patterns. Because the bits of all read/write storage locations at power-on are random 0s and 1s, the parity of each location may be even or odd. To avoid parity error messages prior to writing all locations, the nonmaskable interrupts (NMI) are disabled by writing 0 to the NMI mask register with port address 0A0H. A checksum is performed on the ROM monitor, which consists of adding together the 8K bytes and examining the low byte of the result for the correct value of 0. Then proper operation of timer 1 used for memory refresh is verified. After the registers of the 8237 DMA controller have been tested, the chip is initialized, and memory refresh commences.

With the read/write memory now in operation, procedure STGTST (storage test) is called. It loads and then reads each location of the low 16 KB of RAM with hexadecimal bit patterns FF, 55, AA, 01, and 00. The test verifies storage addressability and proper functioning of this portion of RAM. Phase 1 ends with initialization of the 8259 interrupt controller and storage of the NMI vector address F000:E2C3 at address 0:8, corresponding to INT 2.

Phase 2 (Interrupts, Counter 0). Phase 2 begins by loading the vector address F000:FF54 of INT 5 into RAM at address 0:14H. INT 5 is the Print-Screen service routine of ROM. Next, a test is performed on the 8-bit interrupt mask register by writing and reading 1s and 0s. Then the register is loaded with 1s, thereby masking off all external interrupts, and the interrupt system is enabled with instruction STI. At the end of a 0.5-second delay, a check is made to determine that no hot (unexpected) interrupts have occurred. Interrupt level 0, which is supplied by the time-of-day general-purpose timer (counter 0) is masked on, and a routine verifies that the system timer is neither too fast nor too slow.

Phase 3 (BASIC, Switches, Video, RAM). Included in the tests of phase 3 is a checksum of each of the four 8-KB ROMs that store the BASIC interpreter. Then the switch settings of DIP switch SW1 are read from input port A of the PPI, with the result stored in the low byte of EQUIP_FLAG at offset 10H in the data segment DS (40H). The encoded hardware information is given in the subsection on INPUT PORT A in Section 5-5.

Next, the 6845 CRT controller is initialized and started, and data patterns are written into and read from the video display buffer, located at absolute address B0000H on the monochrome display adapter card. The video is enabled, the mode is set using INT 10H (VIDEO_IO) of the ROM, and the CRT interface lines are tested. Again using INT 10H, the blinking cursor is placed on the screen in the home position. The time is now about 4 seconds after the start.

By reading port C of the PPI, the I/O RAM size is found. The result is stored at DS:15H as variable IO_RAM_SIZE. It is also added to the on-board RAM size, determined from EQUIP_FLAG, and stored as MEMORY_SIZE at offset 13H in DS (40H). Finally, procedure STGTST is used to write and read data patterns for each 16-KB increment of RAM, both on-board and off-board, except for the one previously checked. The memory check is the most time-consuming part of the self-test, taking about 35 seconds for the 256-KB memory of the author's system. *It is bypassed for a reset initiated by Ctrl-Alt-Del.*

Phase 4 (Keyboard, Vector Addresses, Disk, Printer). Phase 4 begins with a keyboard test. A software reset is sent to the keyboard, and the proper response to the CPU should be scan code 0AAH, which is checked. Stored in ROM at 96 locations from FFEF3 through FFF52 are 24 doublewords that represent vector addresses for interrupt types 08 - 1FH. During phase 4, these data are moved into the interrupt area. For interrupt types A, B, C, D, F, and 1F, each vector address of the table is 00000000H, which implies that the addresses of these interrupts must be initialized by a system or user program, not BIOS.

The audio cassette, if present, is tested; also tested is the diskette system. The floppy disk controller is reset, and its status is checked. After the motor of drive 0 is turned on, the SEEK procedure of the diskette I/O routine of ROM is called, with track 1 specified. Because the drive has not previously been accessed during the current session, the routine automatically recalibrates the FDC. Then with track 34 specified, the SEEK procedure is again called. This action is followed by motor turn-off. Various reserved locations in segment DS are initialized, such as BUFFER_HEAD and BUFFER_TAIL for KB_BUFFER. If a printer is attached, it is set up for proper operation. Next, the NMI interrupts for parity errors are enabled; detection of such an error activates INT 2, which prints a message and halts computer operation by executing instructions CLI and HLT. The self-test ends with a short beep of the speaker and a jump to the bootstrap loader.

The entire self-test operation from power-on to the beep requires 5 to 60 seconds, depending on the amount of installed memory. A Ctrl-Alt-Del reset takes only about 9 seconds, because all memory storage tests are omitted. If an error is detected at any point in the self-test, it is indicated by either additional beeps of the speaker, a printed error message, a computer halt, or combinations of these actions. Along with the computer comes a diagnostics program on a diskette. It contains a series of tests that help the user to locate a problem and to determine if service is needed.

Segments of the Self-Test Routine. Segment ABSO, with 31 KB of reserved space, is established at address 0 for storage of interrupt pointers. For the self-test, a temporary stack of 256 bytes is needed. It is set up at address 30H, which is stored in SS and called the STACK segment. A DATA segment is used, with register DS initialized with 0040H. This address defines a BIOS data area immediately above the 1-KB memory region reserved for the 256 possible interrupt vector addresses. An extra segment XXDATA is assigned address 0050H, and the CODE segment has F000H in CS. The initial value of IP is E05BH. In addition, VIDEO_RAM is specified at segment address B800H, with space reserved for 16 KB. This segment resides in the video memory area.

Boot Record

Upon completion of the reset self-test, the bootstrap loader of ROM, which has the code given in Fig. 6-21, loads the boot record from sector 1 of track 0 of side 0 into RAM at BOOT_LOCN. In segment ABSO (0000H), BOOT_LOCN is defined at offset 7C00H as a FAR label, identifying it for use as the target of a FAR jump. Then instruction JMP BOOT_LOCN, with code EA007C0000, implements a direct intersegment jump from the loader of ROM to BOOT_LOCN. The boot record, which contains numerous error messages in case of booting problems, is executed.

The diskette I/O interrupt type 13H is called a number of times. The first call resets the diskette. Then the first sector of the directory is read and examined for the names IBMBIO.COM and IBMDOS.COM. Assuming these names are found in the proper places, the next several calls read IBMBIO and IBMDOS into memory. The FAT is not used, because the sectors of the files are known to DOS. The system files must be stored in the designated sectors of the diskette of the default drive whenever a reset takes place; otherwise an error message is sent to the display. File IBMBIO has 4736 bytes, and IBMDOS has 17,024 bytes. All diskettes must contain the boot record, but for those without IBMBIO and IBMDOS, the only purpose of the boot record is *to provide an error message in case the diskette is in the default drive when a reset is implemented.*

Next, a portion of COMMAND.COM is read to the region of memory that follows that of IBMDOS. After the transfers are made, the program executes an indirect intersegment jump to the first instruction of IBMBIO. After the jump, program boot record in memory has no further use, and the storage region is available for other purposes.

Program IBMBIO.COM

The 4736-byte IBMBIO routine at 0070:0000 contains code that *provides a software interface between the diskette system and the BIOS device routines.* Also, there is code *for initializing attached devices,* such as the printer and the RS232 adapter circuits, and *code for setting the vector addresses of several low-numbered interrupts.* Some of these vector addresses point to code within IBMBIO. An equipment check is made, using INT 11H.

The first instruction of IBMBIO is a direct jump to code that implements the initialization. Following the initialization, its code is no longer needed. To save memory space, the IBMBIO routine relocates IBMDOS downward, overlaying part of IBMBIO. The program then calls the first instruction of IBMDOS, which gives a direct intersegment jump to its initialization code. After IBMDOS has completed its initialization, using the absolute memory space from 500H to 700H as a DOS communication area, control is returned to IBMBIO. *The command processor is now loaded* as directed by the IBMDOS initialization. The program then jumps to the first instruction of COMMAND.COM.

Program IBMDOS.COM

When IBMDOS at E3:0 is called by IBMBIO, it initializes its internal working tables and interrupt vectors for INT 20H through INT 27H. Before the return to IBMBIO, the IBMDOS routine determines suitable addresses for locating various data buffers, and it builds a program segment prefix for the command processor at the lowest available segment address. When no longer needed, the initialization code of IBMDOS is overlaid by a data area and the command processor.

Processed within IBMDOS are numerous interrupts, including type 0 (divide-by-0) and hexadecimal types 20, 21, 25, 26, 27, and others reserved for exclusive use by DOS. *Routines for all functions of INT 21H are in the IBMDOS program.* The functions of INT 21H frequently call, through the IBMBIO interface, the low-level interrupts of the BIOS ROM, which control operation of the various devices attached to the system. For example, the buffered keyboard input function 0AH calls the keyboard I/O routine of INT 16H for each entry. As stated earlier, when the computer is waiting for input of a character from the keyboard, which is most of the time, the program is in the WAIT loop of INT 16H, waiting for an interrupt.

The Command Processor

The command processor, with the name COMMAND.COM, consists of 4959 bytes of code. It has three major parts. Described first is the one referred to as the resident portion.

Resident Portion. The program segment prefix of the *resident portion* is located in memory above a data area. The initial address is 053A:0000H, corresponding to absolute decimal address 21408. Code starts at offset 100H. Included are routines for interrupt types 22H, 23H, and 24H, described in Section 6-6, along with many error messages. *All standard DOS error handling is done in the resident part,* including the display of error messages and action required following the operator's response to inquiries.

Another important function of the resident portion of the command processor is related to the transient portion that is located at the highest end of memory. The transient part can be overlaid by another program if the memory space is needed. However, if it is overlaid, it must be replaced before the next command is entered. Accordingly, *whenever a program ends, a checksum procedure is automatically implemented by the resident portion to determine whether or not the transient portion has been overlaid. If so, it is immediately reloaded from the diskette.*

The Initialization Portion. Following the resident code of COMMAND, the *initialization portion* is stored. It receives control during the reset, and it determines *the lowest segment address available for storage of programs.* Typically, this address is 606:0, or decimal 24,672. Thus, system programs occupy a significant part of total memory. The initialization portion contains the setup routine for the AUTOEXEC file processor. Also included is a routine that prompts for entry of the current date at startup when there is no AUTOEXEC file on the diskette. After initialization is done, this part of COMMAND has no use. Hence, *it is overlaid by the first program loaded by the command processor.*

The Transient Portion. At the highest end of memory are about 13 KB of code and messages that constitute the *transient portion* of COMMAND. In the author's system, with 256 KB of memory, the program segment prefix for the transient portion starts at 3CBF:0. The transient portion may be regarded as the actual command processor because it processes all commands entered by the operator.

The transient portion has four major functions. First, *it interprets commands* entered by the user. Second, *it executes internal commands.* Third, *it processes batch files* and executes any internal commands included. Fourth, *it loads and executes external commands* entered from either the keyboard or presented from a batch file.

Routines for all internal commands are located in the command processor, along with a loader for loading programs of external COM and EXE commands stored as files on the DOS diskette. The processor generates the system prompt.

Once a command is entered, *a search is made from a table in the command processor that contains the names of all internal commands of DOS.* If a match is found, a service routine within the command processor is called. Otherwise the command is assumed to be external, and *the diskette directory is read and searched.* If a match is found here, a program segment prefix is set up following the resident portion. Then the program is loaded into the new segment using the EXEC function of INT 21H, and control is transferred to offset 100H. If no match is found, an error message is displayed.

The program COMMAND can be modified or replaced by the user, if desired. Also, as we have seen, new external commands can be created and added, with the machine code written onto the DOS diskette. DOS makes no distinction between an external command supplied with the system and any other EXE or COM file.

REFERENCES

D. L. BRADLEY, *Assembly Language Programming for the IBM Personal Computer,* Prentice-Hall, Inc., Englewood Cliffs, N.J., 1984.

Disk Operating System Manual, version 2.10, IBM Corp., Boca Raton, Fla., 1983.

DOS Technical Reference Manual, version 2.10, IBM Corp., Boca Raton, Fla., 1983.

P. NORTON, *Inside the IBM Personal Computer,* (including software and documentation, especially disklook program), Robert J. Brady Co., Bowie, Md., 1983.

Personal Computer Technical Manual, IBM Corp., Boca Raton, Fla., 1981.

PROBLEMS

SECTION 11-1

11-1. Suppose the DOS command COMP DATA1.ABC b:data2.z is entered. Give the hexadecimal bytes of portions of the program segment prefix in the form of Fig. 11-1, extending through the carriage return that enters the command. Assume a 384-KB RAM. Identify the significance of each byte.

11-2. Suppose a B:ABC.COM program is needed that will open and write to six data files of drive B, named DATA-0 through DATA-5. Give the assembly language and the command that activates the corresponding COM program, that forms the ASCIIZ strings for each of the six files. Do not use subroutines, save all registers and flags, and include all directives and data areas. This program should only *construct the strings and then end* with INT 20H. Assemble and run the program, using DEBUG to examine the constructed strings.

SECTION 11-2

11-3. With EDLIN, create the file COMPARE.ASM on a diskette. Then assemble and convert it to type COM, and run it. Also, trace it in DEBUG, using command G-breakpoint to run through interrupt routines. Select two identical files for comparison, with at least one in a subdirectory. Also, compare two unlike files.

11-4. After execution of function 4EH, suppose the 18 hexadecimal bytes starting at offset 95H of the program segment prefix are 20 00 60 A7 04 80 A3 00 00 4C 49 4E 4B 2E 45 58 45 and 00. Identify the related file, give its attribute and decimal size, and find the time and date of creation.

11-5. Repeat Prob. 11-4 for the 21 bytes 20 00 60 54 07 80 45 00 00 43 4F 4D 4D 41 4E 44 2E 43 4F 4D and 00.

11-6. (a) In DOS, suppose COMPARE is entered with filespec2 omitted. Explain exactly what happens. At what point does a jump occur, and to what target? What are the observed results? (b) Repeat if the command is B:COMPARE CHKDSK.COM MODE.COM, assuming the files to be compared are on drive A. (c) What is the effect of careless omission of $ at the end of MSG1?

SECTION 11-3

11-7. Enter, run, and test the BASIC program of Fig. 11-8. Compare both identical and dissimilar files.

11-8. Write and execute a BASIC program that reads record #5 of 128 bytes of CHKDSK.COM. Assign the first 5 bytes to A$, the next 60 to B$, and the remaining bytes to C$, and print the strings (some characters may not be printable). Execute PRINT ASC(A$), which gives ASCII for the first byte of A$.

SECTION 11-4

11-9. With EDLIN, create the file NEW_FILE.ASM. Then assemble and convert it to a COM file, and run it. Also, trace the execution in DEBUG, using G to run through the interrupt routines. Use the TYPE command to examine the created DATA file.

11-10. Give pseudo code and sketch a detailed flowchart for NEW_FILE.

11-11. State the objectives of the first nine lines of section 6 and all lines of section 9 of NEW_FILE, including the precise purpose of each instruction.

11-12. Write the BASIC program TEST for creating a DATA file with a record size of 20 characters plus CR and LF. Assign A$, B$, C$, and D$ to 5, 12, 3, and 2 bytes in order. With occasional blanks at the start of a field, input data for A$, B$, and C$ for each of five records, and use left justify. Insert CF and LF at the end of each record and an EOF mark at the end of the file. Display the file with the DOS TYPE command.

11-13. Prepare an assembly language program for deleting a read/write file, based on the routine of Fig. 11-14. Convert it to COM and run the program, deleting a file. Then modify the program so that it will delete a read-only file in addition to read/write files.

SECTION 11-5

11-14. With EDLIN, create the file CON2FILE.ASM. Then assemble and execute B:CON2FILE COM, noting that whole records are written to the display as soon as they are entered.

11-15. Modify CON2FILE of Fig. 11-15, inserting a CR LF sequence into the created data file at the end of each entry from the keyboard buffer. Assemble and convert it to a COM program. Run the program and examine the DATA file with the TYPE command.

11-16. Modify the BASIC program of Fig. 9-10 so that the results go to a diskette file rather than to a printer. Run the program. Then in DOS, type the file to the display and obtain a hard copy.

11-17. With six INPUT statements, write a BASIC program to create a sequential file. Run it and input the characters for storage of six lines of Fig. 11-15, starting at label A1. Use function CHR$(*n*) for inserting CRs and LFs. Type the created file to the display in DOS. The format of the displayed lines should exactly match that of the figure.

11-18. In a form similar to that of Fig. 11-17, write a program that creates a sequential file that stores the first paragraph of Section 11-5. After creation, the program should read the file and print it to the display. Then type the created file to the display in DOS. The format should match that of the text.

SECTION 11-6

11-19. With EDLIN, create both DISPLAY and GET, and convert them to COM files on drive B. Enter B:GET B:GET.ASM B:DISPLAY.COM and note the results. Then trace execution in DEBUG, noting that DEBUG replaces the Ctrl-Break exit address with its own when it stops at a breakpoint. Implement the Ctrl-Break jump *by changing IP*. Also, run B:DISPLAY and enter data from the keyboard.

11-20. Using COPY CON B:DATA1, create file DATA1 with 50 alphabetic characters chosen at random. Design an assembly language program that opens DATA1 for reading as the standard input device and creates DATA2 for writing as the standard output device. Then the program should call DISPLAY to transfer the data from DATA1 to DATA2. Assemble and run the program, using DEBUG for debugging.

11-21. Repeat the preceding problem, but use a substitute for DISPLAY that divides the 50 characters into five records of 10 characters each, sorts the strings alphabetically, adds LF and CR characters to each string, and then writes them to the standard output device DATA2. When DATA2 is displayed using the DOS command TYPE, *the strings should appear in a column in alphabetic order*. Refer to the bubble sort of Section 4-5.

11-22. Prepare an assembly language program NEW_NAME that changes the name of the file specified in *filespec1* to that of *filespec2* when the command B:NEW_NAME *filespec1 filename2* is entered. Assemble and verify that execution yields the specified results. Use function 56H, which has on entry DS:DX pointing to the ASCIIZ string of the file and ES:DI pointing to that of the new filespec.

11-23. Run TICKS (Fig. 7-24) using each of the replacements of Fig. 11-22 for INT 27H.

11-24. Prepare MSG.ASM, assemble it, and convert it to MSG.COM. The program should contain a WAIT for a one-character keyboard input, preceded by a message to hit either Ctrl-Break or any other key. Each choice should display an appropriate message, which ends the program. Run the COM program.

SECTION 11-7

11-25. Explain the difference between the bootstrap loader and the boot record. Where is each program located in memory while executing? Also, identify the segment names and addresses of segments used by the self-test. What does a failure of the BASIC test indicate?

11-26. Enumerate the functions of the following programs: boot record, IBMBIO, IBMDOS, and COMMAND with subdivisions resident, initialization, and transient.

APPENDIX 

8086/8088 Machine Instruction Encoding

Encoding Key

Identifier	Explanation
word	16 bits (2 bytes).
doubleword	32 bits (4 bytes).
/	Slash denotes "or".
mod	2-bit mode field.
reg	3-bit register field.
r/m	3-bit register or memory field.
EA	Effective address (16 bits).
CS	New code segment (16 bits).
IP	New IP word.
IP-inc8	Signed (+/−) byte increment to IP.
IP-inc16	Signed (+/−) word increment to IP. Contents of IP, after execution of the present instruction, are incremented.
disp	Displacement, for finding EA. It is either not present or 8 or 16 bits, as determined from Table A-2. Disp8 is sign extended through 16 bits.
data	Byte or word (data8 or data16)
SR	Seg reg code: ES = 00, CS = 01, SS = 10, DS = 11.
d	If d = 0, reg field is source. If d = 1, reg field is destination.
w	If w = 0, operand is a byte. If w = 1, operand is a word.
sw	If sw = X0 (00 or 10), data are a byte. If sw = 01, data are a word. If sw = 11, data are a low-order byte, but the CPU inserts a high-order byte, with all bits equal to the MSB of the low byte.
v	If v = 0, the shift/rotate count is 1. If v = 1, the count is specified in CL.

TABLE A-1. Field Encoding, reg and r/m, with mod = 11

reg or r/m	w = 0	w = 1
000	AL	AX
001	CL	CX
010	DL	DX
011	BL	BX
100	AH	SP
101	CH	BP
110	DH	SI
111	BH	DI

TABLE A-2 Field Encoding, r/m with Memory Addressing

	Memory EA Calculation		
r/m	mod = 00	mod = 01	mod = 10
000	(BX) + (SI)	(BX) + (SI) + disp8	(BX) + (SI) + disp16
001	(BX) + (DI)	(BX) + (DI) + disp8	(BX) + (DI) + disp16
010	(BP) + (SI)	(BP) + (SI) + disp8	(BP) + (SI) + disp16
011	(BP) + (DI)	(BP) + (DI) + disp8	(BP) + (DI) + disp16
100	(SI)	(SI) + disp8	(SI) + disp16
101	(DI)	(DI) + disp8	(DI) + disp16
110	disp16	(BP) + disp8	(BP) + disp16
111	(BX)	(BX) + disp8	(BX) + disp16

Instruction Set Notes

1. Code is given in either binary or hexadecimal.
2. Even if "disp" is included in the instruction set, it is omitted in the instruction if no displacement is shown in Table A-2. Disp8 is sign extended by the processor.
3. Direct memory addressing uses mod 00 with r/m = 110.
4. Parentheses around a register designation in Table A-2 signify that the contents of the register are used as a pointer.
5. Sixteen-bit displacements, increments, addresses, and data are given in an instruction, with the low-order byte preceding the high-order byte. Thus, the low byte is at the lower address.

8086/8088 Instruction Set (Alphabetical Order)

Instruction		First Byte	Other Bytes [Comments]
AAA	ascii add adjust	37	
AAD	ascii div adjust	D5	0A
AAM	ascii mul adjust	D4	0A
AAS	ascii sub adjust	3F	
ADC	add with carry		
reg/mem with register		000100dw	mod-reg-r/m disp
immediate to reg/mem		100000sw	mod-010-r/m disp data
immediate to AL/AX		0001010w	data
ADD			
reg/mem with register		000000dw	mod-reg-r/m disp
immediate to reg/mem		100000sw	mod-000-r/m disp data
immediate to AL/AX		0000010w	data
AND			
reg/mem with register		001000dw	mod-reg-r/m disp
immediate to reg/mem		1000000w	mod-100-r/m disp data
immediate to AL/AX		0010010w	data
CALL	subroutine		
direct in segment		E8	IP-INC16
indirect in segment		FF	mod-010-r/m disp
direct intersegment		9A	IP CS
indirect intersegment		FF	mod-011-r/m disp
CBW	byte-word convert	98	
CLC	clear carry flag	F8	
CLD	clear direction	FC	
CLI	clear interrupt	FA	
CMC	complement carry	F5	
CMP	compare		[subtract (flags only)]
reg/mem and register		001110dw	mod-reg-r/m disp
immediate with reg/mem		100000sw	mod-111-r/m disp data
immediate with AL/AX		0011110w	data
CMPS	compare strings	1010011w	[subtract (flags only)] [CMPSB, CMPSW allowed]
CWD	word to doubleword	99	
DAA	decimal adjust, add	27	
DAS	decimal adjust, sub	2F	
DEC	decrement		
register or memory		1111111w	mod-001-r/m disp
16-bit register		01001reg	
DIV	unsigned divide	1111011w	mod-110-r/m disp
ESC	external escape	11011xxx	mod-yyy-r/m disp [code xxxyyy for transfer]
HLT		F4	
IDIV	integer divide	1111011w	mod-111-r/m disp
IMUL	integer multiply	1111011w	mod-101-r/m disp

Instruction		First Byte	Other Bytes [Comments]
IN	input to AL/AX		
immediate port		1110010w	byte port address
port address in DX		1110110w	
INC	increment		
register/memory		1111111w	mod-000-r/m disp
16-bit register		01000reg	
INT	interrupt		
number is specified		CD	data8
Type 3		CC	
INTO	overflow interrupt	CE	
IRET	interrupt return	CF	
JA/JNBE	jump if above	77	IP-inc8 [(CF or ZF) = 0]
JAE/JNB/JNC	not below	73	IP-inc8 [CF = 0]
JB/JNAE/JC	below	72	IP-inc8 [CF = 1]
JBE/JNA	below/equal	76	IP-inc8 [(CF or ZF) = 1]
JCXZ	jump on CX zero	E3	IP-inc8
JG/JNLE	greater than	7F	IP-inc8 [(SF xor OF) or ZF = 0]
JGE/JNL	greater/equal	7D	IP-inc8 [(SF xor OF) = 0]
JL/JNGE	less than	7C	IP-inc8 [(SF xor OF) = 1]
JLE/JNG	less/equal	7E	IP-inc8 [(SP xor OF) or ZF = 1]
JNS	jump on not sign	79	IP-inc8 [SF = 0]
JNZ/JNE	not 0	75	IP-inc8 [ZF = 0]
JO	jump on overflow	70	IP-inc8 [OF = 1]
JPE/JP	even parity	7A	IP-inc8 [PF = 1]
JPO/JNP	odd parity	7B	IP-inc8 [PF = 0]
JS	jump on sign	78	IP-inc8 [SF = 1]
JZ/JE	jump on 0	74	IP-inc8 [ZF = 1]
JMP	unconditional		
direct in segment		EB	IP-inc8
direct in segment		E9	IP-inc16
indirect in segment		FF	mod-100-r/m disp
direct intersegment		EA	IP CS
indirect intersegment		FF	mod-101-r/m disp
LAHF	flags to AH	9F	[for 8085 compatibility]
LDS	pointer to register	C5	mod-reg-r/m disp
LEA	load EA to register	8D	mod-reg-r/m disp
LES	pointer to register	C4	mod-reg-r/m disp
LOCK	bus lock prefix	F0	
LODS	load to AL/AX	1010110w	[pointer is SI, updated] [LODSB, LODSW allowed]
LOOP	loop CX times	E2	IP-inc8
LOOPZ/LOOPE	while 0	E1	IP-inc8
LOOPNZ/LOOPNE	not 0	E0	IP-inc8

Instruction		First Byte	Other Bytes [Comments]
MOV			
reg/men to/from reg		100010dw	mod-reg-r/m disp
immediate to reg/mem		1100011w	mod-000-r/m disp data
immediate to register		1011wreg	data
memory to AL/AX		1010000w	EA [16-bit address]
AL/AX to memory		1010001w	EA [16-bit address]
register/memory to SR		8E	mod-0SR-r/m disp [MOV to CS not allowed]
SR to register/memory		8C	mod-0SR-r/m disp
MOVS	string move	1010010w	[pointers:
MOVSB	byte-string move	A4	for source string, SI
MOVSW	word-string move	A5	for destn string, DI]
MUL	unsigned mul	1111011w	mod-100-r/m disp
NEG	2s complement	1111011w	mod-011-r/m disp
NOP	no operation	90	
NOT	complement	1111011w	mod-010-r/m disp
OR			
reg/mem and register		000010dw	mod-reg-r/m disp
immediate to reg/mem		1000000w	mod-001-r/m disp data
immediate to AL/AX		0000110w	data
OUT	out from AL/AX		
immediate port address		1110011w	byte port address
port address in DX		1110111w	
POP	pop from stack		
register/memory		8F	mod-000-r/m disp
16-bit register		01011reg	
segment register		000SR111	[SR = seg reg code] [POP to CS not allowed]
POPF	pop flags	9D	
PUSH	push to stack		
register/memory		FF	mod-110-r/m disp
16-bit register		01010reg	
segment register		000SR110	[SR = seg reg code]
PUSHF	push flags	9C	
RCL	rotate carry left	110100vw	mod-010-r/m disp
RCR	rotate carry right	110100vw	mod-011-r/m disp
REP	[MOVS,STOS]	F3	[repeat while CX not 0]
REPE/REPZ	[CMPS,SCAS]	F3	[repeat while ZF = 1]
REPNE/REPNZ	[CMPS,SCAS]	F2	[repeat while ZF = 0] [and CX not 0]
RET return from call			
within segment		C3	
in segment, add to SP		C2	data16
intersegment		CB	
intersegment, add to SP		CA	data16
ROL	rotate left	110100vw	mod-000-r/m disp

Instruction	First Byte	Other Bytes [Comments]
ROR rotate right	110100vw	mod-001-r/m disp
SAHF store AH in flags	9E	[For 8085 compatibility]
SAR arith shift right	110100vw	mod-111-r/m disp
SBB subtract with borrow		
reg/mem and register	000110dw	mod-reg-r/m disp
immediate from reg/mem	100000sw	mod-011-r/m disp data
immediate from AL/AX	0001110w	data
SCAS scan string	1010111w	[pointer is DI, updated]
		[SCASB, SCASW allowed]
Segment Override Prefix		[not an instruction]
ES:	26	[prefix precedes instruction]
CS:	2E	
SS:	36	
DS:	3E	
SHL/SAL shift left	110100vw	mod-100-r/m disp
SHR shift right	110100vw	mod-101-r/m disp
STC set carry	F9	
STD set direction	FD	
STI set interrupt	FB	
STOS store from AL/AX	1010101w	[pointer is DI, updated]
		[STOSB, STOSW allowed]
SUB subtract		
reg/mem and register	001010dw	mod-reg-r/m disp
immediate from reg/mem	100000sw	mod-101-r/m disp data
immediate from AL/AX	0010110w	data
TEST operation AND		[flags updated]
reg/mem and register	1000010w	mod-reg-r/m disp
immediate and reg/mem	1111011w	mod-000-r/m disp data
immediate and AL/AX	1010100w	data
WAIT wait state	9B	
XCHG exchange		[source,dest operands]
reg/mem with register	1000011w	mod-reg-r/m disp
16-bit register with AX	10010reg	
XLAT translate table	D7	[BX,AL point to byte]
XOR exclusive-or		
reg/men and register	001100dw	mod-reg-r/m disp
immediate to reg/mem	1000000w	mod-110-r/m disp data
immediate to AL/AX	0011010w	data

APPENDIX 

8086/8088 Summary of Instructions

This appendix contains a brief summary of the instruction set. All instructions are included, along with a few comments and examples. Affected status flags are shown in the tables at the far right, using letters O, S, Z, A, P, and C. Encoding into machine language is described in Appendix A, and all instructions used in the text are listed in the index.

TABLE B-1 Instructions for I/O, Data Transfer, Address Transfer, and Flag Operations

Instruction	Comments/Examples
	Input/Output
IN accumulator, port OUT accumulator, port	IN AL, 0A4H IN AX, DX OUT 89, AX OUT DX, AL
	Data Transfer
MOV dest,source XCHG dest,source XLAT source (table) PUSH source POP destination	MOV ES:MASK [BX][SI], CS XCHG AX, CX XLAT SOURCE_TABLE PUSH ES PUSH ALPHA POP CX POP BETA[SI]
	Address Transfer
LEA reg16,source(mem) LDS reg16,source(mem) LES reg16,source(mem)	Load EA of memory operand Load pointer to register and DS Load pointer to register and ES
	Flag Operations
PUSHF POPF STC CLC CMC STD CLD STI CLI	Push flag word to stack Pop stack word to flags Set CF Clear CF Complement CF Set DF Clear DF Set IF Clear IF

TABLE B-2 Arithmetic Instructions

Instruction	Comments/Examples	
	Addition	
ADD dest,source	ADD AL, 5 ADD BETA, CL	OSZAPC
ADC dest,source	Add with carry; ADC BX, 8	OSZAPC
INC destination	Increment by 1; INC CX	OSZAP
AAA	ASCII adjust for addition	AC
DAA	Decimal adjust	OSZAPC
	Subtraction	
SUB dest,source	SUB ES:[BP − 9], CL	OSZAPC
SBB dest,source	Subtract with borrow	OSZAPC
DEC destination	Decrement by 1; DEC ARRAY[SI]	OSZAP
NEG destination	Form 2s complement	OSZAPC
CMP dest,source	Compare destination to source	OSZAPC
AAS	ASCII adjust for subtraction	AC
DAS	Decimal adjust for subtraction	SZAPC
	Multiplication	
MUL source	Unsigned multiplication	OC
IMUL source	Integer multiplication, signed	OC
AAM	ASCII adjust for multiplication	SZP
	Division	
DIV source	Unsigned division; DIV BX	
IDIV source	Integer division, signed	
AAD	ASCII adjust for division	SZP
CBW	Convert byte to word	
CWD	Convert word to doubleword	

TABLE B-3 String Instructions and Prefixes

Instruction	Comments/Examples	
	String Instructions	
MOVS dest-string,source-string	MOVS A2, B5	
CMPS dest-string,source-string	Compare	OSZAPC
SCAS dest-string	Scan	OSZAPC
LODS source-string	Load (AL or AX)	
STOS dest-string	Store (AL or AX)	
	Prefixes	
REP/REPE/REPZ	Repeat while CX not 0 (and ZF = 1)	
REPNE/REPNZ	Repeat while CX not 0 (and ZF = 0)	
(ZF status is tested only by CMPS and SCAS)		

TABLE B-4 Logical Instructions

Instruction	Comments/Examples	
	Logicals	
NOT destination	Logical NOT; NOT WORD_A	
AND dest,source	AND AX, 101000B	OSZPC
OR dest,source	OR CMD_WORD[BX], AX	OSZPC
XOR dest,source	XOR OLD_CODE, 0D0ABH	OSZPC
TEST dest,source	(AND with no operand change)	OSZPC
	Shifts	
SHL/SAL dest,count	Shift left; SHL AL,1	OC
SHR dest,count	Shift logical right	OC
SAR dest,count	Shift arithmetic right	OSZPC
	Rotates	
ROL dest,count	Rotate left; ROL SI, CL	OC
ROR dest,count	Rotate right; ROR BETA, 1	OC
RCL dest,count	Rotate left through carry	OC
RCR dest,count	Rotate right through carry	OC

TABLE B-5 Conditional Jump Instructions

Mnemonic	Jump if Condition is True
JC/JB/JNAE	CF = 1
JNC/JAE/JNB	CF = 0
JZ/JE	ZF = 1
JNZ/JNE	ZF = 0
JP/JPE	PF = 1
JNP/JPO	PF = 0
JS	SF = 1
JNS	SF = 0
JO	OF = 1
JNO	OF = 0
JBE/JNA	(CF or ZF) = 1
JNBE/JA	(CF or ZF) = 0
JL/JNGE	(SF xor OF) = 1
JNL/JGE	(SF xor OF) = 0
JLE/JNG	(SF xor OF) or ZF = 1
JNLE/JG	(SF xor OF) or ZF = 0

TABLE B-6 Unconditional Transfer Instructions

Instruction	Comments/Examples
Unconditional Transfers	
CALL target	Call procedure; CALL S2
RET pop-value (if any)	Return from procedure
JMP target	Jump; JMP BETA [BX]

TABLE B-7 Instructions for Iterations, Interrupts, Multiprocessor Control, Halt, and No Operation

Instruction	Comments
Iteration Controls	
LOOP short-label	Loop until CX = 0
LOOPZ/LOOPE short-label	Loop if ZF = 1, CX not 0
LOOPNZ/LOOPNE short-label	Loop if ZF = 0, CX not 0
JCXZ short-label	Jump if CX is 0
Interrupts	
INT *n*	Activate interrupt, type *n*
INTO	Interrupt if OF = 1
IRET	Interrupt return
Multiprocessor Control	
WAIT	Wait while TEST = 1
ESC external-opcode,source	Escape
LOCK (instruction prefix)	Locks bus
Halt and No Operation	
HLT	Halt until interrupt
NOP	No operation

Introduction to Logic Design

The subject of logic design is broad, and only a brief introduction is given here. However, foundations are laid for the design of both combinational and sequential circuits. Study of the material in this appendix can directly follow Chapter 4, if desired. We begin with examination of logic expressions and two-level logic systems.

C-1 TWO-LEVEL CIRCUITS

Two basic forms of logic expressions are examined here. One of them gives an expression as a *sum of products,* with the products consisting of Boolean products of variables. The other type of an expression is a *product of sums.* In Section 4-3, two identities known as DeMorgan's laws were described. Because of their special importance in logic design, they are repeated in (1) and (2) for three variables.

$$(A + B + C)' = A'B'C' \quad (1)$$
$$(ABC)' = A' + B' + C' \quad (2)$$

We examine first the sum-of-products form of a logic expression.

Sum-of-Products Form

A logic equation can always be expressed in the form of a sum of products, with the function Y written as the sum of terms, each of which consists of a product of logic variables. An example is the relation

$$Y = AB' + AC + A'BC' \tag{3}$$

The right side of (3) has the sum-of-products form. Assuming that both the true variables A, B, and C and their complements are available as inputs, the logic can be implemented by either circuit of Fig. C-1.

In Fig. C-1a, the respective outputs of the AND gates from top to bottom are AB', AC, and $A'BC'$, and the three-input OR gate gives the desired function Y. The corresponding outputs of the NAND gates of Fig. C-1b are $(AB')'$, $(AC)'$, and $(A'BC')'$. It follows that Y is

$$Y = [(AB')'(AC)'(A'BC')']' = AB' + AC + A'BC' \tag{4}$$

The right side of (4) is deduced from DeMorgan's law (2), with variable A in (2) replaced with $(AB')'$, variable B with $(AC)'$, and variable C with $(A'BC')'$.

The configurations of Fig. C-1 are referred to as *two-level* logic systems, because each input passes through two logic operations before reaching the output. When a gate switches state, the switch is not instantaneous. There is a small *gate propagation delay,* perhaps only a few nanoseconds, but these time delays limit the speed of the system, and hence the maximum clock frequency in synchronous circuits. When both the true variables and their complements are available as inputs, the sum-of-products form of a logic expression can always be implemented by a two-level system, following the method used to implement (3). If additional product terms are added to the right side of (3), the diagrams corresponding to those of Fig. C-1 have additional input gates, *but the logic remains two-level.*

The equivalence of the logic configurations of Fig. C-1 should be carefully noted. In general, any two-level AND-OR circuit such as that of the figure can be replaced with a two-level NAND network, and the output Y *is the sum of the products of the inputs of the first-level gates.*

All inputs to the positive logic gates of Fig. C-1 are active high. Input A is high when A is 1, and input A′ is high when A′ is 1, corresponding to $A = 0$. The functions are converted to the mixed logic system simply by replacing the negation indicators with polarity indicators; the logic with respect to voltage levels is unchanged. As before, input A′ is high when A′is 1, corresponding to the low-level value of A.

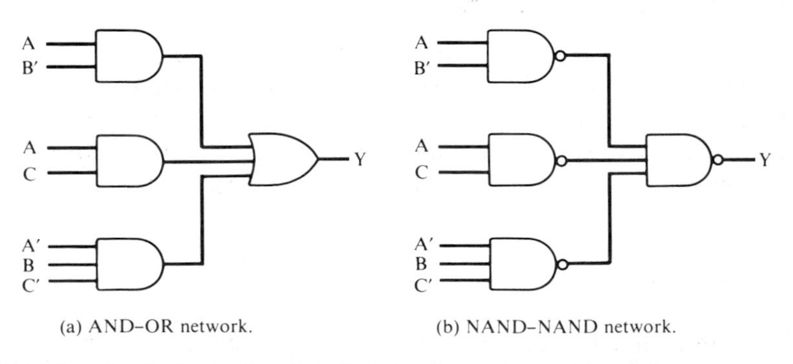

(a) AND–OR network. (b) NAND–NAND network.

FIGURE C-1. Equivalent circuits that implement equation (2).

Conversion of a logic expression into a sum-of-products form can be accomplished by means of Boolean identities, including DeMorgan's laws. The technique is illustrated in the following example.

EXAMPLE 1

For $Y = [(AB + C)' + A(B' + C')]'$, find a sum-of-products form.

SOLUTION

It is usually best to remove the outermost prime first, by using DeMorgan's law. This procedure gives $Y = (AB + C)[A(B' + C')]'$. The outermost prime of this expression is eliminated by means of DeMorgan's law (2), giving $(AB + C)[A' + (B' + C')']$. Next, we apply (1), obtaining $(AB + C)(A' + BC)$. The product can be multiplied term by term, giving $AA'B + ABBC + A'C + BCC$. This is a sum-of products form, but it can be simplified with the aid of identities to

$$Y = A'C + BC \tag{5}$$

■

A sum-of-products form of a logic expression is easily found from a truth table. Suppose we wish to find the logic expression represented by the truth table of Fig. C-2. Each combination of variables in a truth table is called a *minterm*. The Boolean function in the output column of the table is 0 for some minterms and 1 for others. Included at the right side of the table are the minterm products. Observe that a variable having the 0 state is complemented in the product. *The sum of the minterm products that give logical 1s for Y equals Y.*

In the second data row of the truth table of Fig. C-2, word ABC is 001, and minterm $A'B'C$ is 1 for these values. For all other combinations of inputs, the minterm $A'B'C$ gives 0. From this result, it follows that *ORing the logical 1 product expressions of the table gives the correct relation.* The sum-of-products form is shown in the figure, but it is not in its simplest form. With the aid of Boolean identities it can be reduced to $A'C + BC$, which is that of the preceding example.

The procedure is general. For each row of a truth table having an output of 1, we form a minterm product consisting of the variables, *using the complement of a variable that is 0*. The sum of the products is the output. If there is more than one output, the method is applied to each. We now have a direct method for implementing any logic expression with a two-level circuit. We simply develop the truth table and use it to find a sum-of-products form. If possible, the expression should be simplified before implementation with either an AND-OR or a NAND-NAND circuit.

A	B	C	Y	Minterms
0	0	0	0	A′B′C′
0	0	1	1	A′B′C
0	1	0	0	A′B C′
0	1	1	1	A′B C
1	0	0	0	A B′C′
1	0	1	0	A B′C
1	1	0	0	A B C′
1	1	1	1	A B C

$$Y = A'B'C + A'BC + ABC$$

FIGURE C-2. A truth table and its logic expression.

Product-of-Sums Form

It is also possible to implement a logic expression with two-level OR-AND or NOR-NOR circuits. The procedure is to express the function Y as a product of several terms, each of which consists of a sum of one or more input variables. To illustrate, we shall implement with NOR gates the equation

$$Y = AB + AC + B'C \qquad (6)$$

The first step is to find the complement of Y, or Y′, as a sum of products. This is accomplished by using a truth table with output Y′. An alternate method is first to get Y in the sum-of-products form, complement Y, and then apply DeMorgan's law. Using this method, from (6) we find that

$$Y' = (AB + AC + B'C)' = (AB)'(AC)'(B'C)'$$

or

$$Y' = (A' + B')(A' + C')(B + C') = A'B + A'C' + B'C'$$

The sum-of-products form for Y′ has now been found. It can be simplified with the aid of Boolean identities to give $Y' = A'B + B'C'$; although desirable, this operation is not essential. Next, we complement Y′ to get Y and then apply DeMorgan's laws, first (1) and then (2). This procedure leads to

$$Y = (A'B + B'C')' = (A'B)'(B'C')' = (A+B')(B+C) \qquad (7)$$

The expression at the right consists of a product of two terms, each of which is a sum of Boolean variables. This is the product-of-sums form. Both sum-of-products and product-of-sums Boolean expressions are said to be in *canonical form,* provided each input variable is present in each product of the former or each sum of the latter.

With positive logic gates, both circuits of Fig. C-3 implement the logic of (7). The system is converted to the mixed logic convention by replacing the negation indicators with polarity indicators. This action has no effect on the logic with respect to voltage levels. Outputs of the OR gates are $A + B'$ and $B + C$, and the AND gate that follows provides the Boolean multiplication. The NOR-NOR arrangement is an alternate choice, being equivalent to the OR-AND circuit.

The procedure is general. First, *the complement Y′ of Y is expressed as a sum of products.* A truth table relating Y′ to the inputs can be used for this purpose. Next, *DeMorgan's laws are applied to Y′ to obtain Y in the form of a product of sums,* such as that of (7). Implementation is done with *either* OR-AND or NOR-NOR circuits.

The methods presented in this section can be used to design two-level combinational networks that will implement a desired logic function. However, the techniques do not always give the best solution. There are usually a number of equiv-

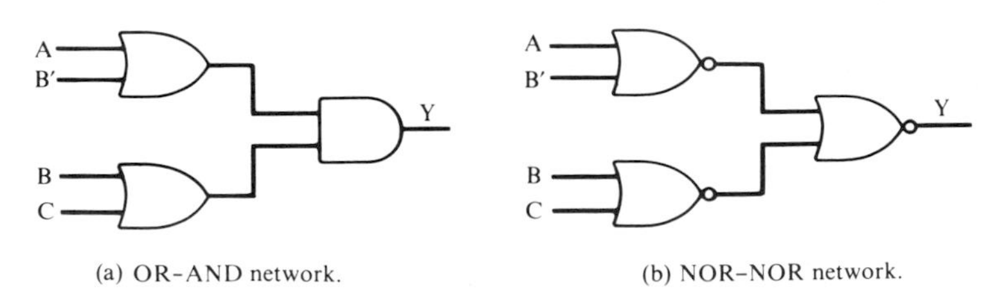

(a) OR–AND network. (b) NOR–NOR network.

FIGURE C-3. Equivalent circuits that implement equation (6).

alent logic expressions. Each occurrence of a variable or complemented variable in a logic expression is referred to as a *literal*. For example, the expression of (6) for Y has six literals. The simplest circuit is obtained from the *minimal expression,* defined as the one *with the smallest number of terms containing the fewest number of literals*.

When AND-OR or NAND-NAND is used, the sum-of-products expression for Y is minimized, and when OR-AND or NOR-NOR logic is selected, the product-of-sums expression is minimized. Often the minimal equation can be found by application of Boolean identities to a specified logic relation. This procedure is not always easy and may require considerable intuition. Fortunately, there is a direct method known as *Karnaugh mapping*. We shall study this procedure later.

EXAMPLE 2

Function Y = A′B′C′ + A′BC + AB′C′ + AB′C + ABC′. With negation indicators as needed, find a NOR-NOR network that implements the logic, assuming complements of the variables are not available as inputs.

SOLUTION

Figure C-4 presents the truth table for both Y and Y′, obtained from the given expression for Y. It shows Y′ in the sum-of-products form to be A′B′C + A′BC′ + ABC. Hence, Y is (A′B′C + A′BC′ + ABC)′. Two applications of DeMorgan's laws give the product-of-sums form of Fig. C-4, and the circuit implementing this result is also shown.

For the input NOR gate at the bottom of Fig. C-4 the output is 0 if at least one of the inputs is 0, and the output is 1 if and only if all inputs are 1s. This is the AND function. It follows that an OR gate with negation indicators at all terminals is equivalent to an AND gate. Similarly, an AND gate with negation at all terminals implements OR. ■

Exercise 1. Y = ABC + A′B′C + AB′C + ABC′. (a) Implement the logic using three two-input NAND gates. (b) From a truth table find Y′ in the sum-of-products form, simplifying the result to the sum of two product terms. Implement the logic with three two-input NOR gates. Use inverters to obtain necessary complements of input variables A, B, and C.

Exercise 2. Y = ABC′ + A′BC + AC. Implement the logic with a two-level NOR network, using only three two-input NOR gates at the first level and one three-input gate at the output.

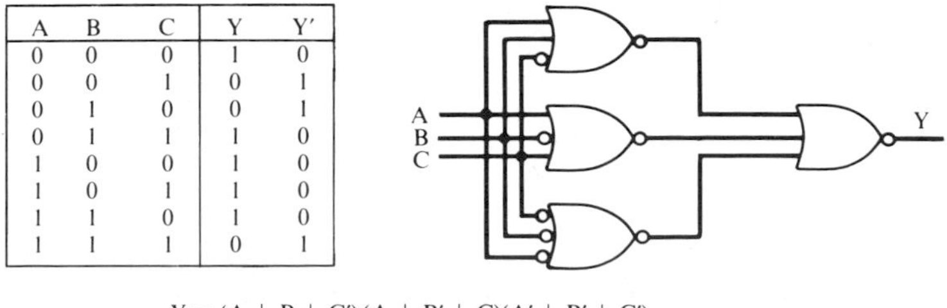

A	B	C	Y	Y′
0	0	0	1	0
0	0	1	0	1
0	1	0	0	1
0	1	1	1	0
1	0	0	1	0
1	0	1	1	0
1	1	0	1	0
1	1	1	0	1

Y = (A + B + C′)(A + B′ + C)(A′ + B′ + C′)

FIGURE C-4. Truth table and results of Example 2.

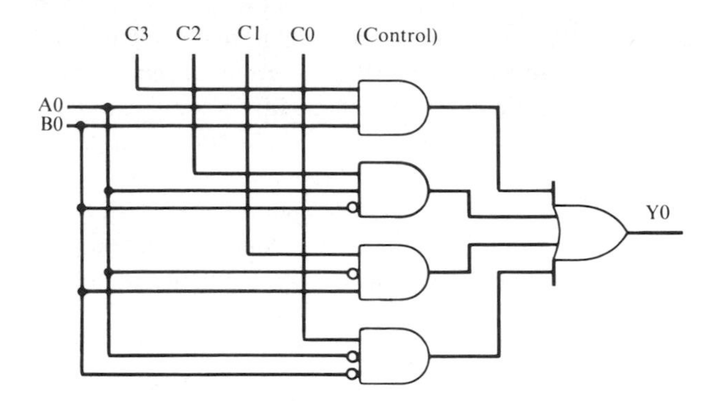

FIGURE C-5. Hardware logic operations for bits A0 and B0.

Logic Operations

Most of the common logic operations on binary words have been examined. They are used extensively in computer operations to manipulate the bits of words. Designing a computer with the capability of performing the basic logic functions on pairs of corresponding bits of words A and B allows the implementation of more complicated functions, not provided by hardware, by means of a program. In fact, any function that can be implemented with logic gates can also be implemented with software control of the proper sequence of logic operations.

Figure C-5 shows a circuit that will implement the following operations *on bits A0 and B0 of words A and B:* AND, OR, NAND, NOR, XOR, XNOR, A0, B0, NOT A0, NOT B0, CLEAR, SET, A0 IMP B0, B0 IMP A0, and complements of the two IMP functions. Operation CLEAR gives an output of 0, and SET does the opposite. The particular function is determined by the control word C.

Exercise 3. For the logic circuit of Fig. C-5, find the output Y in terms of input words A and B for each of the 16 possible control words. Simplify the expressions.

C-2

FUNCTIONAL CIRCUITS

In medium-scale integrated (MSI) packages, there are numerous combinational and sequential circuits that are designed to accomplish logic functions in a wide variety of applications. They are often referred to as *functional logic circuits*. Examples are adders, data selectors, code converters, flip-flops, and registers. Emphasis here is on the data selector and the decoder/demultiplexer. We begin with the data selector.

The Data Selector

A *data selector*, or *multiplexer*, is a combinational circuit that selects one of several input words and supplies the selected word to an output. With n selection lines, the number of possible selections is 2 raised to power n. Data selectors are extensively used in digital systems for connecting at different times two or more sources to the same destination. Another useful application is implementation of combinational logic. A multiplexer with n selection lines can provide any desired combinational function of $n + 1$ variables.

Figure C-6 shows the standard symbol and the function table for a four-to-one data selector with two select inputs A and B. The decimal numbers at these inputs are their *weights*, and those at the data inputs correspond to the sum of the weights of the active select inputs. For example, if word AB is binary 10, the sum of the weights of the active select inputs is 2, being that of A, and data line D2 is selected. In this case, output Y equals D2 and the optional output Z is its complement. The decimal numbers of the figure are sometimes replaced with binary numbers. Flow arrowheads are optional.

When $\overline{DIS}$ is 0, output Y is 0 and Z is 1. In this condition, the values of the select inputs have no effect and are don't-care states. An equivalent label is EN for enable, provided both the inhibiting input and negation indicators are removed. The symbols and function table of the positive logic data selector of Fig. C-6 are converted to mixed logic by replacing the negation indicators with polarity indicators and by using states H and L in place of 1 and 0, respectively, in the function table.

Data selectors are often used to organize a bus system. Figure C-7 shows a circuit that connects the 4-bit words of four registers through data selectors to a four-wire bus. Connections to the registers are indicated. When the select word XY is 10, word C is transferred to the four-wire bus, and the word stored in C is unchanged. If the bits of register C change with time, those of the bus also change. The process of selecting one input word from several available words and effectively connecting it to the output is called *multiplexing*. If the data selectors of the figure are inhibited by appropriate inputs to enable terminals, all bus lines are low level.

Many data selectors are made with three-state outputs. For outputs such as those of Fig. C-7, an active inhibit input floats the four outputs, effectively disconnecting the multiplexers from the bus. *This procedure allows the designer to connect several multiplexed systems to the same bus*. Those not selected for communication are floated. In Fig. C-8 is the symbol for a data selector having three 4-bit input words and both true and complemented three-state outputs. It can be made from four 4-to-1 cascaded elements, with one input of each unused.

There are various types of data selectors in MSI packages. A 16-to-1 element has 16 input data lines and four select lines with weights of 1, 2, 4, and 8. A dual four-to-one chip has two separate four-to-one circuits, although each is controlled by the same select inputs A and B and by the same enable. The symbol is similar to that of Fig. C-8, but there are only two blocks below the control block and each

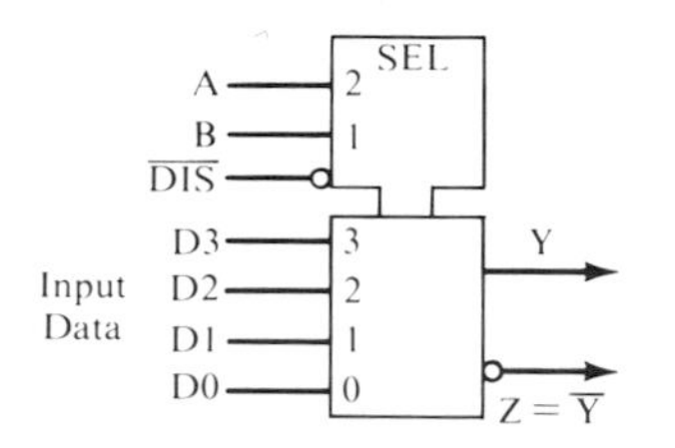

Disable $\overline{DIS}$	Select A	B	Output Y
0	X	X	0
1	0	0	D0
1	0	1	D1
1	1	0	D2
1	1	1	D3

Function Table
(X = don't care)

FIGURE C-6. A four-to-one data selector.

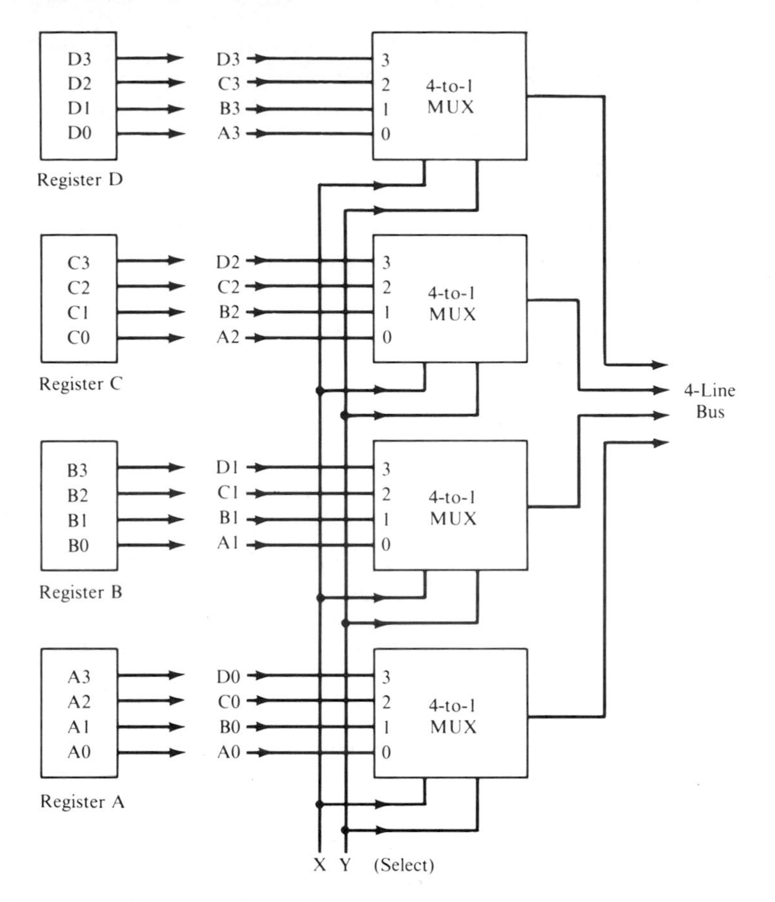

FIGURE C-7. A four-register bus system.

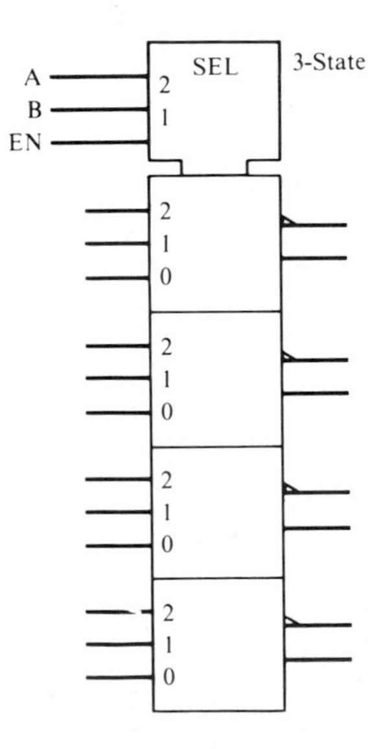

FIGURE C-8. Quad three-to-one data selector.

function block has four inputs. The 2-bit output can be connected to any one of the four input words.

Exercise 1. (a) Sketch a two-level AND-OR network, with two inverters, that implements the logic of the function table of Fig. C-6. Use the minimum number of four-input gates. (b) Sketch the standard symbol for a quad four-to-one data selector with select inputs A and B, four 2-bit input words C, D, E, and F, and output word Y. Include an enable input EN and give the function table.

Logic Operations

From the function table of Fig. C-6 with the circuit enabled, the logic expression for Y is found to be

$$Y = ABD_3 + AB'D_2 + A'BD_1 + A'B'D_0 \quad (8)$$

This sum-of-products form can be used to design a suitable two-level AND-OR or NAND-NAND circuit that implements the data selector of Fig. C-6. Such a circuit was examined in the previous section and shown in Fig. C-5. Although the network of Fig. C-5 was presented as one for implementing logical operations on a pair of corresponding bits of two binary words, it is in fact a data selector that gives logic equivalent to that of (8). Multiplexers are often used to implement logic operations, and multiplexers for this purpose are common components of VLSI data processing chips.

The data selector of Fig. C-6 gives an output expressed in (8) as a sum of products of all possible minterms of a truth table having two input variables A and B. Each minterm is multiplied by D_n, which can be made 1 or 0. It follows that any logic expression with two input variables can be implemented with a four to one data selector. Similarly, an eight-to-one data selector with three select inputs can implement the logic of any three-input truth table.

It is possible to do better by using one variable and its complement as data inputs, with the remaining variables being the select inputs. With this procedure, *any function of* $n + 1$ *variables can be generated with a* 2^n*-to-1 data selector*. To illustrate, suppose an eight-to-one data selector is to be used to implement the function Y of four variables:

$$Y = A'B'C' + A'C'D + ABCD + AB'CD' \quad (9)$$

Choosing A, B, and C as select inputs, we need to determine the data selector inputs that give the desired logic operation.

A straightforward procedure consists of arranging the minterms of a truth table in the form shown in Fig. C-9. The states of data input variable D are placed at the left side, and those of the select input variables are placed along the top of the table, with the minterm values entered in the squares. These values are found from (9). If both minterms of a column are 0s, a 0 is placed just below the table, and if both are 1s, the entry is 1. If the top minterm of a column is 0 and the bottom one is 1, the entry below the table is D′, and if the values are reversed, the entry is D′. The data below the table are the proper data selector inputs that correspond to select word ABC.

Exercise 2. On two standard symbols of an eight-to-one data selector with select inputs ABC having respective weights 4, 2, and 1, label the data inputs 0

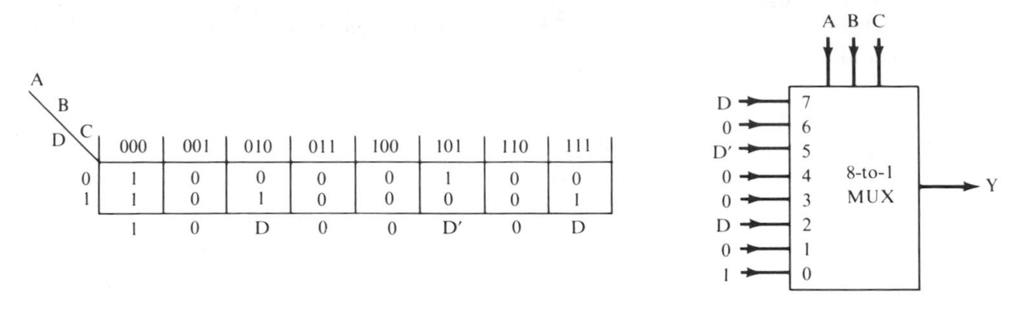

FIGURE C-9. Data selector that implements the logic of equation (9).

through 7. On the diagrams show the 0, 1, D, and D′ inputs required to give an output Y that is (a) ABCD + A′BC′D + A′B′C′D′ and (b) A + B′C′D′.

Decoders/Demultiplexers

In general, a digital *coder* is a combinational logic function with multiple inputs and outputs in which the relationship between the inputs and outputs is described by a truth table or some other equivalent technique. Frequently, the relationship is specified by means of a line-weighting procedure using decimal or binary numbers. Coders are also called *code converters* or *translators*.

A type of coder of special importance has one and only one output line active at any given instant, with the particular active output determined by the state of the *select word* at the input. This functional circuit is the *decoder*. Another type not considered here is the display decoder, which may have several outputs simultaneously active. Display decoders drive light-emitting diode (LED) displays.

The Decoder

Figure C-10 shows a mixed logic circuit for a *two-to-four decoder*. At the right is the symbol. Select inputs A and B are buffered by the double inverters, and enable $\overline{\text{EN}}$ is active low. The letter *G* in the symbol denotes *gate*. When word AB has state LL with the circuit enabled, output Y_0 is active at the low level. Other outputs are, of course, inactive. When AB is LH, line Y_1 is low, and so on.

The decimal numbers of the symbol in the figure denote weights. *The active output is identified by a weight equal to the sum of the weights of the high inputs.* When $\overline{\text{EN}}$ is inactive, all outputs are inactive at the high level. The circuit is often called a *one-of-four decoder,* indicating that only one of four outputs is selected. Implementation requires four three-input positive logic NAND gates and five inverters.

Available in the TTL, ECL, and CMOS families are MSI packages with dual 2-to-4 decoders and single 3-to-8 and 4-to-16 decoders, in addition to various other types. Some are designed to decode BCD data, such as the TTL 7442A 4-to-10 decoder. It activates with a low voltage 1 of 10 outputs corresponding to the BCD input word. For example, input word 0111 selects the output line with decimal weight 7. Invalid inputs from 9 to 15 make all outputs high. The circuit consists of 8 inverters and 10 four-input NAND gates in a 16-pin package.

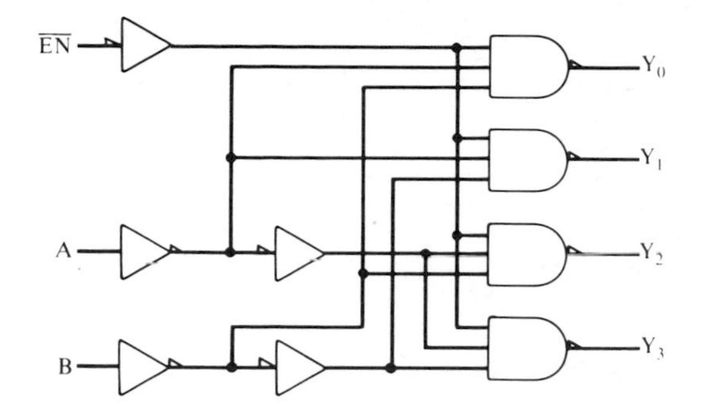

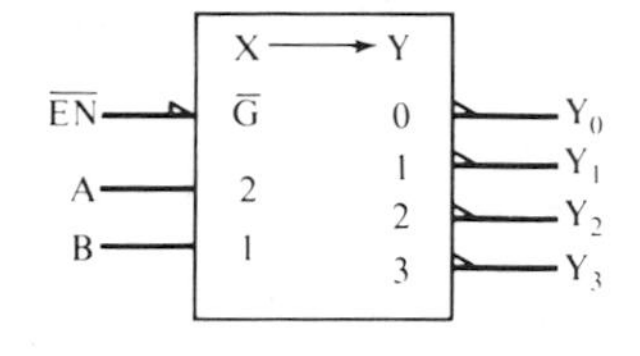

FIGURE C-10. Two-to-four decoder and symbol.

The enable inputs of decoders allow expansion of two or more packages into a decoder of larger capacity. Two 4-to-16 decoders can be used to implement a 5-to-32 decoder. The four select inputs of each package are connected together, providing four of the five inputs. The fifth input is the enable terminal of one package that is connected through an inverter to the enable terminal of the second IC. Any of the 32 outputs can be activated. Enable inputs of IC packages are often used in this manner.

Decoders are common components of MSI and LSI chips, serving to reduce the number of package pins. To illustrate, let us recall that the 8-bit shift register of Fig. 1-13 has control inputs for shift in, shift out, hold (store), and load. These controls are mutually exclusive in that one and only one is active at any given time, assuming they are not overridden by other controls such as clear or disable. By placing a two-to-four decoder on the chip, the four terminals can be reduced to two, with inputs of 00, 01, 10, and 11 activating the respective functions. This technique can always be used to reduce the number of terminals for mutually exclusive controls.

Exercise 3. Show connections for implementing a positive logic 5-to-32 decoder using two 4-to-16 decoders with enable inputs. Label select inputs EDCBA, with E assigned weight 16, and label outputs Y0 through Y31. All signals are active high.

Address Decoding

One of the most important uses of decoders is address decoding. To illustrate, address decoding for some of the peripherals of the PC is examined here. Figure C-11 shows a 74LS138 decoder used for this purpose.

Address bits A19 through A10, as well as bit A4, are not used in the addresses of the peripherals considered here, and thus become don't-cares. For convenience, we will treat them as 0s. As indicated by the figure, bits A9 and A8 are 0s in the addresses. Accordingly, the three high-order hexadecimal digits of the addresses are 000, and only the low byte of each address needs to be considered. Although address enable input $\overline{AEN}$ is defined as active low, it is driven high by the CPU whenever an operation requires use of the data bus by the processor.

For the DMA chip, address bits A3 through A0 connect directly to address pins of the chip, giving addresses 00H through 0FH. For the interrupt controller, only address bit A0 connects to the chip. Treating bits A4 through A1 as logical 0 don't-

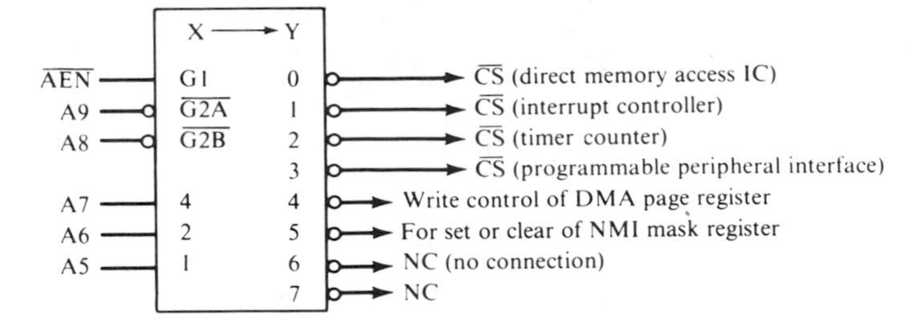

FIGURE C-11. Address decoder for peripherals.

cares, the addresses of the controller are 20H and 21H. The timer addresses are 40H - 43H, those of the PPI are 60H - 63H, and addresses of the DMA page register are 80H - 83H, with don't-care bits A4 - A2 taken to be 0 in these cases. Address 101XXXXX, or A0H, is used for the NMI mask register.

Combinational Logic

The two-to-four AND gate decoder in Fig. C-12 can implement any desired logic function of two variables by connecting selected outputs to the inputs of a second-level OR gate. The AND gates generate the four minterms of inputs A and B, and the OR gates provide the Boolean sum as required.

For example, to implement the equivalence function AB + A′B′, outputs Y_0 and Y_3 are connected to a two-input OR gate, with EN = 1. This is AND-OR logic.

An alternative to AND-OR logic is NAND-NAND logic, and the two are equivalent. The NAND gate decoder generates the *complements* of the minterms of the inputs.

Four basic ways to implement combinational logic have been presented. The circuits are as follows:

1. Two-level AND-OR gates (or NAND-NAND)
2. Two-level OR-AND gates (or NOR-NOR)
3. Multiplexers
4. Decoders and gates

EXAMPLE

Design a full adder with an AND gate decoder and OR gates.

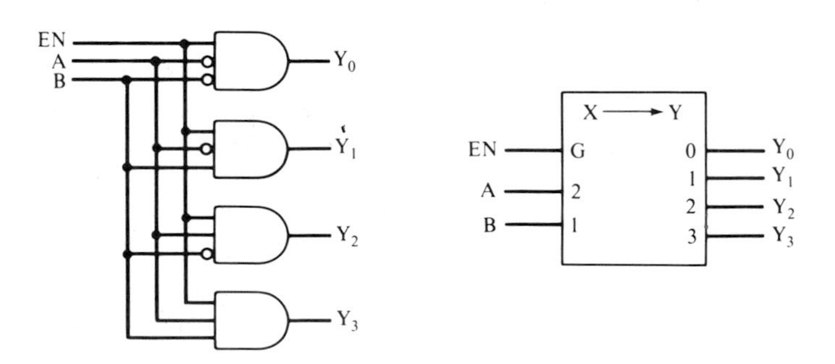

FIGURE C-12. AND gate decoder and symbol.

SOLUTION

Because the full adder has the three inputs A, B, and carry C, a three-to-eight decoder is required to generate the eight minterms. Output line 7 gives minterm ABC, line 6 gives ABC′, and so on. From the truth table of Fig. C-13, it follows that the sum S is the Boolean sum of minterms ABC, AB′C′, A′BC′, and A′B′C, which correspond to output lines 7, 4, 2, and 1. Also, the output carry K is the Boolean sum of minterms ABC, ABC′, AB′C, and A′BC, corresponding to output lines 7, 6, 5, and 3. The circuit is also given Fig. C-13. ■

The decoder gives all minterms of the input variables or perhaps the complements of the minterms. Consequently, the decoder method of logic design is best suited *for circuits having many outputs, each of which has only a few minterms* to minimize the number of OR gate inputs.

An alternate procedure for using an AND gate decoder to generate combinational logic consists of using the minterms of the complement of the function. The corresponding output lines are fed into a NOR gate. This method sometimes gives a simplified circuit.

Demultiplexers

The function of a *demultiplexer* is the reverse of that of a multiplexer in that data from a single input line are transferred to one of several output lines. The choice is determined by select inputs. *A decoder becomes a demultiplexer when an enable input is used for data.* For example, let us consider again the 2-to-4 decoder in Fig. C-10. If enable is low, the selected output is low, and if it is high, the selected output is also high. Outputs not selected are always high. Because the element can be used as either a decoder or a demultiplexer, it is often called a *decoder/demultiplexer*. Demultiplexers are also known as *data distributers* or *data routers*.

The 4-to-16-line TTL 74154 decoder/demultiplexer has a 24-pin DIP with 16 outputs, 4 select inputs, 2 enable inputs, and 2 pins for the 5-V supply. The two enables allow cascading of packages into a demultiplexer of larger capacity. Two such packages can be interconnected to give 32 outputs. An example of a similar circuit is illustrated in Fig. C-14 in which two 2-to-4 decoders are connected to provide a demultiplexer with eight output lines. Implementation is possible with the TTL 74155 chip, which consists of a pair of 2-to-4 decoder/demultiplexers with enable inputs as shown in the figure.

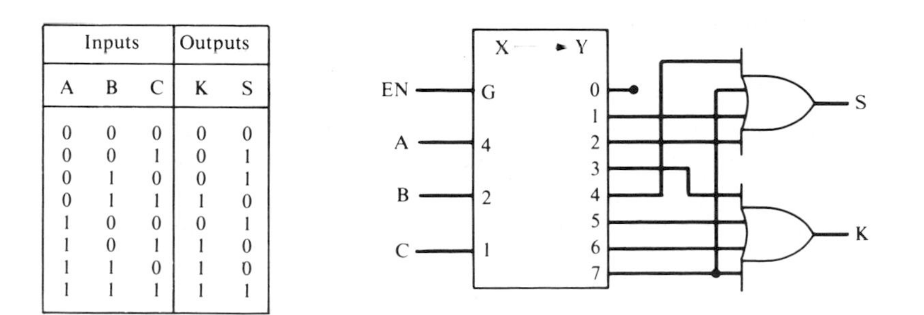

Inputs			Outputs	
A	B	C	K	S
0	0	0	0	0
0	0	1	0	1
0	1	0	0	1
0	1	1	1	0
1	0	0	0	1
1	0	1	1	0
1	1	0	1	0
1	1	1	1	1

FIGURE C-13. Full adder truth table and circuit.

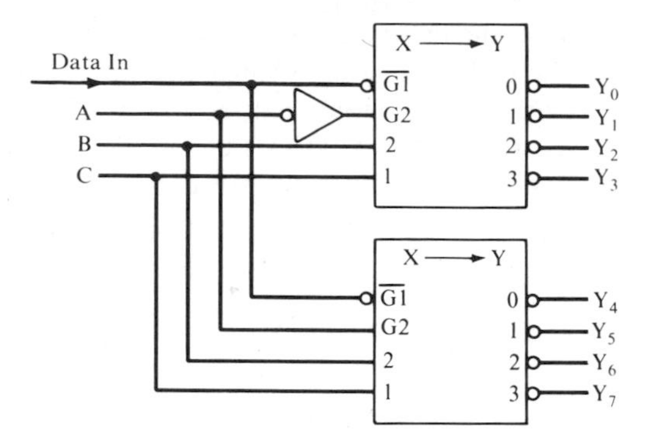

FIGURE C-14. Three-to-eight demultiplexer from two 2-to-4 functions.

Suppose ABC is 000. The bottom circuit is disabled, and output line Y_0 of the top circuit is selected. The bit of Y_0 is the same as that on the input data line, and all other outputs are high. For select word ABC of 100, the top circuit is disabled, and line Y_4 is selected.

Frequently, words of *n* bits must be demultiplexed. Eight 3-to-8 decoder/demultiplexers can rout an 8-bit input word to one of eight output buses, each with eight lines. Computer diagrams often show a block labeled MUX that serves as an interface between a bus and a register array. Transfer of data from the bus to a selected register is done with demultiplexer circuits within the MUX block, whereas transfer from a selected register to the bus lines is done with data selector circuits.

C-3 KARNAUGH MAPS

Karnaugh mapping provides a direct way of reducing a Boolean logic function to a minimal expression. From this expression, which has the smallest number of terms and the fewest number of literals, can be deduced the simplest circuit that will implement the logic. A Karnaugh map is essentially a truth table *arranged in a form suitable for determining by inspection the minimal sum-of-products expression of a logic function*. Information contained in the map is identical to that of the truth table.

One-Variable Maps

Suppose Y is a function of the single variable A. The truth table is normally arranged with a column for A at the left side, containing the two possible values 0 and 1. At the right side is the column for Y, giving its values corresponding to those of A. Almost identical are the one-variable Karnaugh maps of Fig. C-15.

As shown in the figures, the variable A and its possible 0 and 1 values are placed at the left of the box, or map, and the corresponding values of Y are placed in the squares, or cells. A blank entry is understood to be 0. *The expression for Y is found from the logical 1 entries,* with the 0s ignored. Single-variable expressions are so elementary that one-variable maps are seldom used, but their presentation here illustrates concepts applicable to maps having additional variables.

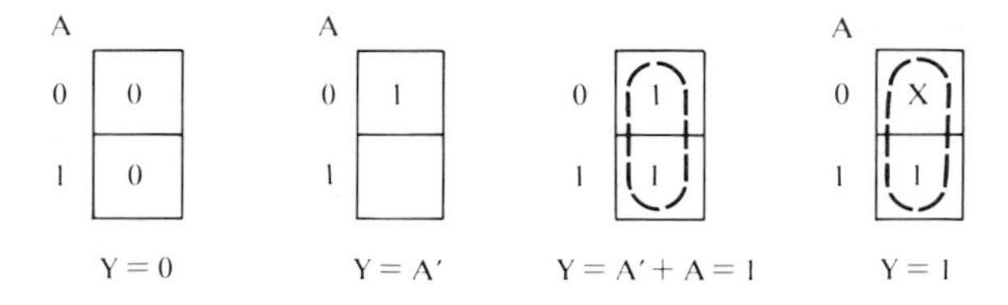

FIGURE C-15. Four one-variable Karnaugh maps.

In the map at the left, there are no 1 entries, which implies that Y is 0. In the next map, Y is 1 in the square corresponding to a logical 0 for A, and hence $Y = A'$. Note the blank used in place of 0. For the map with two 1's, the logical 1 corresponding to 0 for A gives A′ for Y, and the next 1 gives A. One condition or the other applies. Thus, Y is A′ OR A, written $Y = A' + A$, which is 1. This equation illustrates an important relation. *Two 1s in adjacent squares can be grouped into a unit that gives 1 for Y for either value of A*. It is customary to circle *groups*, as shown. The don't-care condition X of the last map is selected to be 1. This decision allows the two 1s to be grouped together as before, and Y is 1.

Two-Variable Maps

Suppose we need to find the minimal expression for

$$Y = A' + AB \tag{10}$$

A truth table can be set up with logical 1 minterms determined to be A′B′, A′B, and AB. The table can be arranged in the form of the two-variable Karnaugh map shown in Fig. C-16. As before, the possible 0 and 1 values of A are at the left side of the map, but the values of B are at the top. *Adjacent squares* are defined as *those having a common side*. Note that *only one variable changes its value from any cell to an adjacent one*.

In the map of Fig. C-16, the 1s are placed in the squares in accordance with the logic of (10). The first equation at the right side of the map gives Y as the sum of the logical 1 minterms, and the next one groups the minterms corresponding to the circled groups of the map. Note that A′B is used twice; this decision is justified by the identity $A + A = A$. Clearly, a logical 1 in a cell adjacent to two other logical-1 cells can be included in two groups. Thus, *groups can overlap*.

The bottom expression of the figure is the minimal one. It can be deduced directly from the groups of the map. The row group gives $Y = 1$ for $A = 0$, regardless of B, and Y is A′. The column group gives $Y = 1$ for $B = 1$, regardless of A, and Y is B. Because either condition makes $Y = 1$, we write $Y = A' + B$.

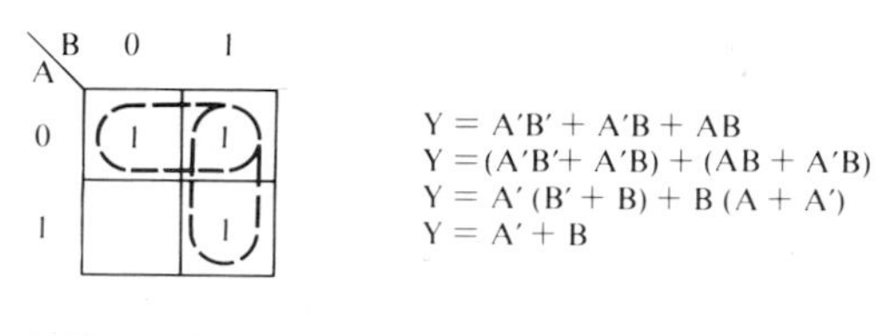

FIGURE C-16. Two-variable map and expressions.

Three-Variable Maps

To illustrate a three-variable map, suppose output Y of a system equals A′B′C OR ABC, with the system designed so that input combination A′BC is never allowed to occur. Then we have

$$Y = A'B'C + ABC \tag{11}$$

with A′BC = X (don't-care). The three-variable map corresponding to (11) is shown in Fig. C-17a. The squares are identified by number in Fig. C-17b, *with each decimal number denoting the value of minterm ABC*. This numbering system provides a convenient way to refer to a particular square.

The expression for Y has three variables. The first, variable A, has its 0 and 1 states along the side boundary. Variables B and C are placed at the top of the map. Their respective values, from left to right, are 00, 01, 11, and 10, as shown along the top boundary. The arrangement of the minterms of A, B, and C is such that *exactly one input variable changes when passing from any square to an adjacent one*.

Corresponding squares at the extreme left and right are considered adjacent, because there is a change in only one variable. For example, the ABC values of cell 0 are 000, and those of cell 2 are 010; thus, only B changes. The values of ABC along the top and side boundaries must be exactly as shown. Entries in the squares of (a) are the values of Y corresponding to the minterms, as specified by (11).

Each cell in Fig. C-17 is equivalent to one row of a truth table. We have learned how to express Y as a sum of products from inspection of a truth table. This can also be done from a Karnaugh map. The logical 1 in location 1 corresponds to ABC values of 001, leading to the minterm A′B′C. The X in the adjacent cell, if chosen to be 1, leads to A′BC, and the remaining 1 gives ABC. The Boolean sum of these values is Y. Each cell with a 1 contributes a product term of three variables.

The group formed by squares 1 and 3 give the sum of A′B′C and A′BC. This sum equals A′C(B + B′), or simply A′C. In the expression for Y, the product term A′C can be substituted for the sum of the two product terms. This deduction is evident from the map. Because the two squares contain 1s with B equal to both 0 and 1, it is clear that the terms represented by these adjacent squares are independent of B. Because A is 0 and C is 1, the two adjacent squares represent the expression A′C. This observation can be generalized as follows. *Any two adjacent squares with 1s represent a single product term of two variables, with the deleted variable being the one that changes from one square to the next.*

According to the rule, the 1s of the group containing squares 3 and 7, with X taken to be 1, give the product BC. This is the sum of A′BC and ABC. Accord-

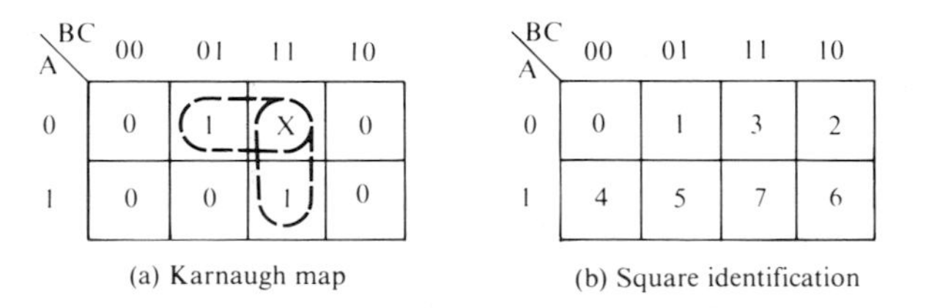

(a) Karnaugh map (b) Square identification

FIGURE C-17. Map for the logic of equation (11) and identification numbers of squares.

ingly, the minimal expression for Y in the sum-of-products form is

$$Y = A'C + BC \tag{12}$$

This equation is the simplified form of (11), having only two product terms and four literals. The X was taken as 1 and used twice. Logical expressions developed from Karnaugh maps can utilize overlapping groups of squares. Furthermore, *don't-care states should be selected so that each 1 is present in a group with the maximum number of squares.*

We have found that a logical 1 in a square of a three-variable map represents a product term having three variables. *If a variable is 0, its complement appears in the product term, and if its value is 1, the true variable is present.* Two adjacent squares with 1's correspond to a product term with two variables, and squares at the left edge are considered to be adjacent to those of the right edge.

Figure C-18 shows Karnaugh maps with logical 1s in groups of four. Blanks represent 0s. In (a) the 1s of cells 4 and 5 give AB′, and the X and the 1 of cells 7 and 6 give AB, using 1 for the don't-care. The sum is A(B′ + B), or A. We deduce that *two adjacent groups of two squares can be regarded as one group of four squares,* with Y equal to the unchanged variable if it is 1, or to its complement if its value is 0. Note that the unused X is 0. In (b) the 1s are in a group with C = 0, and hence Y is C′. In (c) the group has C = 1, and Y is C.

Finally, suppose all eight minterms of a three-variable map give 1s. The four 1s of Fig. C-18b represent C′, and those of (c) give C. Thus, Y is C′ + C, which is 1. The result was evident from the beginning, but the procedure illustrates that *two adjacent groups of four squares with 1s can be regarded as one group of eight squares.* For the opposite case in which all cells contain 0s, Y is 0.

In developing the minimal expression from a Karnaugh map, one should combine adjacent squares with 1s into groups of two, or if possible, four or perhaps eight. *The larger the groups, the simpler is the logic expression. Overlapping groups are permissible.* In a three-variable map, only groups of one, two, four, or eight squares are allowed.

To determine the minimal *product-of-sums* form for Y, the algorithm is as follows:

1. Develop the Karnaugh map for Y′.
2. From the map, find the minimal sum-of-products form for Y′.
3. Complement both sides of the expression and apply DeMorgan's laws (1) and (2).

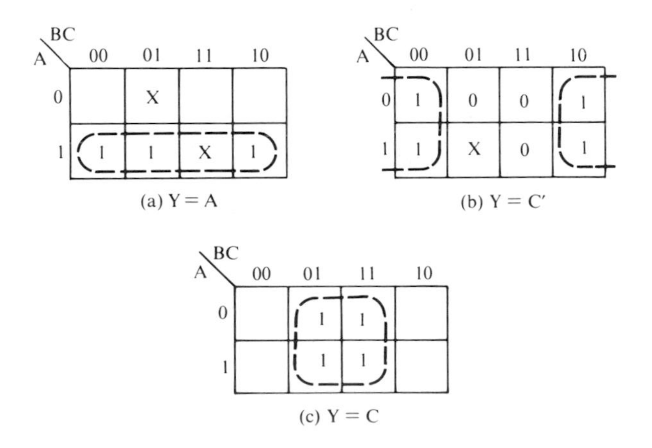

FIGURE C-18. Three-variable Karnaugh maps.

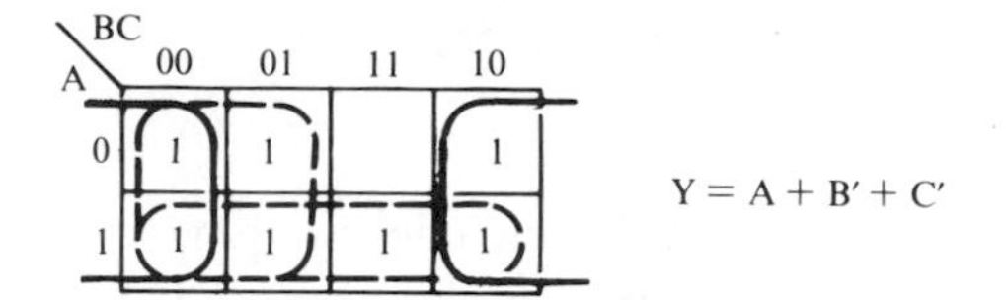

FIGURE C-19. Map and minimal expression for Example 1.

Because the 0s in the map for Y correspond to the 1s in the map for Y′, a valid method is to group the 0s in the map for Y, deducing Y′ as a sum of products directly from the 0s. Let us consider two examples.

EXAMPLE 1

For Y = A + A′BC′ + A′B′, determine the minimal expression.

SOLUTION

The Karnaugh map and the minimal expression are those of Fig. C-19. There are three "product" terms, each with only one variable, and three literals. The four 1s of the bottom row give A, those at the left side give B′, and the group consisting of the two 1s at the left and the two 1s at the right give C′. The sum (A + B′ + C′) is Y. A three-input OR gate can implement the logic, provided both true variables and their complements are available. The equivalent expression (A′BC)′ can be implemented by a three-input NAND gate. ■

EXAMPLE 2

For Y = AB + AC′ + BC, determine the simplest NOR-NOR network for implementing the logic.

SOLUTION

The map for Y is that of Fig. C-20. Grouping the 0s as indicated gives Y′ = A′C′ + B′C. By inverting and applying DeMorgan's laws (1) and (2), we find the minimal expression for Y as a product of sums. The result is (A + C)(B + C′). There are two sum terms and four literals. The simplest two-level NOR network consists of three 2-input NOR gates. At the first level the inputs to one gate are A and C, and those to the other are B and C′. ■

Four-Variable Maps

Mapping of logic expressions with four variables is similar. Variables A and B are placed at the left side, with values 00, 01, 11, and 10 from top to bottom, and C and D are arranged along the top boundary. From the left-side square of a row to

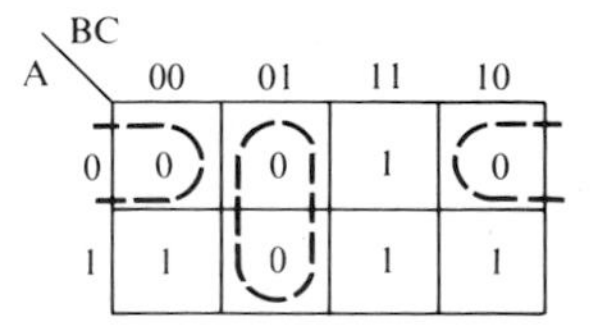

FIGURE C-20. Karnaugh map for Example 2.

the right-side square is a single variable change; there is also a single variable change from the bottom square of a column to the top square. Consequently, *the left and right columns are considered to be adjacent,* as before, and *so are the bottom and top rows.*

Each square with a 1 corresponds to a product term of four variables, and two adjacent squares correspond to a product term of three variables. A group of four 1s represents a product term of two variables, whereas a group of eight represents a single variable. If all 16 cells contain 1s, the logic expression is Y = 1, and for all 0s, Y is 0. The two examples that follow illustrate the use of four-variable maps.

EXAMPLE 3

Find the minimal sum-of-products form for

$$Y = \overline{C}D(\overline{A + B}) + AB(\overline{C + D}) + BC(A\overline{D} + \overline{A}D) + A\overline{B}\overline{C}D$$

SOLUTION

The Karnaugh map and the minimal expression are given in Fig. C-21. The 1s in cells 12 and 14 of the left and right columns correspond to ABD′. Those in cells 1 and 9 of the top and bottom rows give B′C′D, and the remaining 1 represents minterm A′BCD. Assuming that the true variables and their complements are available, the logic can be implemented with an AND-OR or a NAND-NAND network. At the first level are three gates with three, three, and four inputs. ■

EXAMPLE 4

For Y = A′B′D + A′CD + AD′, design a two-level NOR network implementing the logic. Complements are available as inputs.

SOLUTION

The Karnaugh map for Y is that of Fig. C-22. Grouping the 0s as indicated, we find Y′ to be the sum of AD, A′D′, and BC′D. Application of DeMorgan's laws to the sum-of-products form of Y′ gives

$$Y = (A + D)(\overline{A} + \overline{D})(\overline{B} + C + \overline{D})$$

The first NOR level of the network has three gates with two, two, and three inputs. ■

In Fig. C-22, the 0 in cell 5 can be grouped with the one to its left instead of the one below it. The choice is optional. We have examined Karnaugh maps for

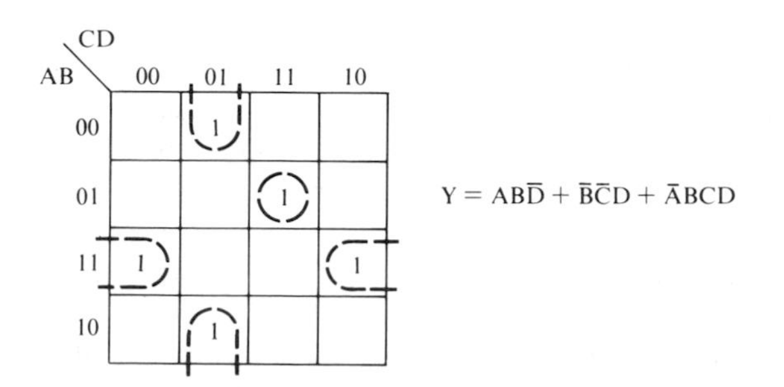

FIGURE C-21. Map and minimal expression for Example 3.

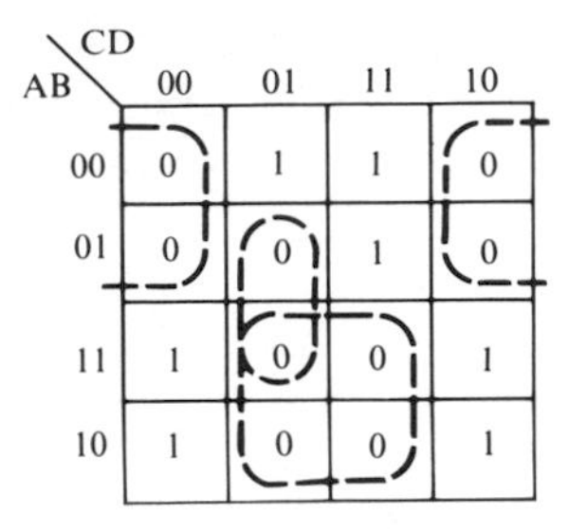

FIGURE C-22. Four-variable Karnaugh map for Example 4.

logic expressions with one to four variables. The technique can be extended to the simplification of logic functions with five and six variables, but other minimization methods are usually preferred for functions with more than four variables.

Exercise 1. For $Y = [A'B'(AB)']' + \{[A' + B'][AB + (AB)']\}'$, sketch the Karnaugh map and from it deduce the minimal logic expression for Y. Implement the logic with a quad two-input NAND IC, and repeat with a quad 2-input-NOR IC. Complements of the variables are not available as inputs.

Exercise 2. $Y = A'B'C' + A'B'C + A'BC' + AB'C' + AB'C + ABC$. Sketch the Karnaugh map and find the minimal sum-of-products and product-of-sums forms for Y. Implement the logic with a triple three-input NAND chip, and repeat with a triple three-input NOR chip. Complements of the variables are available as inputs.

Exercise 3. A four-variable Karnaugh map in the form of Fig. C-22 has 1s in cells 0, 2, 8, 10, and 13, don't-care X's in cells 5, 7, and 15, and 0s in the remaining ones. Cell identification follows the method indicated in Fig. C-17b. Find the minimal sum-of-products and product-of-sums forms for Y.

Exercise 4. $Y = A'B'C' + A'B'C + A'BC' + A'BC + AB'C$. From the map, find the minimal sum-of-products and the products-of-sums forms for Y. Implement the logic with a two-level NAND network, and repeat with a two-level OR-AND network. Complements of the variables are available as inputs.

C-4

FLIP-FLOPS

A *flip-flop* is a binary sequential logic element with two stable states. These are the 1 state, or *set* state, and the 0 state, or *reset* state. The element is a static memory cell, and the stored bit is present on an output line, usually labeled Q. Many flip-flops have a second output that provides the complement Q′ of the stored bit Q. Input signals, or a time sequence of input signals, determine the bit that is stored. Most flip-flops have either one or two input lines. There are four general types of clocked flip-flops, and the main characteristics of each are examined here.

When no dynamic indicator appears on the symbol, only the level of the clock is important. For such flip-flops, the inputs can change the stored bit whenever the clock has its indicated 1 state relative to the symbol. These are usually called *clocked latches;* they are considered to have *direct coupling* between input and output. With an active clock, the response is instantaneous, except for the small propagation delay. Unclocked flip-flops are known simply as *latches*. The name *flip-flop* is usually understood to denote one having a dynamic clock input, unless clearly stated otherwise.

The RS Flip-Flop

The *RS flip-flop* has set and reset inputs S and R. With input word SR having the normal latch state 00, the element stores bit Q, which may be either 1 or 0. Although the bit on the output line can be used by another circuit, it remains latched in the flip-flop as long as SR is 00 and power is on.

The cell is set by changing the set input S from 0 to 1 and holding it there for at least one active clock transition. Then S is returned to 0, and the logical 1 is latched. To reset, or clear, the cell, input R is changed to 1, held there for an active clock, and returned to 0.

There is a problem if word SR is 11. Input S denotes set, and R specifies reset. After the active clock transition, the state may be either 1 or 0. Because indeterminate conditions such as this one are usually intolerable in digital systems, allowing S and R to be 1 simultaneously is normally avoided.

Figure C-23 shows the symbol and *characteristic table* of the RS flip-flop. The qualifier FF of the symbol is optional. The dynamic input, usually called the *clock CLK input,* allows a change in the stored bit only on a negative transition. The RS inputs are *synchronous inputs,* because they affect the outputs only on the active clock transition. Flip-flops often have *asynchronous clear/preset inputs,* but the design can be such that they are also synchronous.

The characteristic table of the figure relates the output q after an active clock transition to the inputs and the prior output Q. We will often use a lowercase letter corresponding to the uppercase letter of a variable *to denote its state after a clock transition*. Don't-care states are indicated in the table with X's.

The JK Flip-Flop

The J input to a flip-flop is analogous to the S input, and the K input is analogous to the R input, but modified to eliminate the undefined state of the characteristic table of Fig. C-23. When word JK is 11, the stored bit of the element changes state on each active clock transition. With JK held at 11 and a periodic clock applied to the dynamic input, bit Q toggles from 0 to 1 to 0 to 1, and so on. The

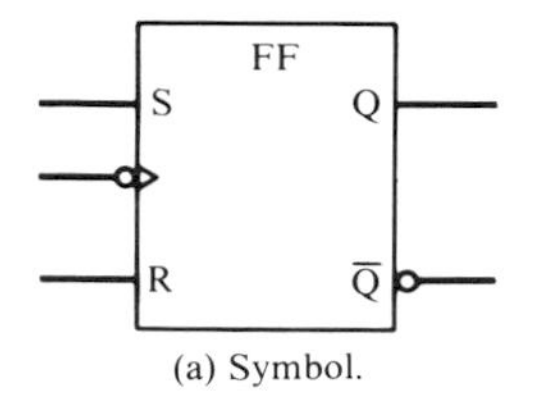

(a) Symbol.

Before CLK			After CLK	
S	R	Q	q	Condition
0	0	0	0	latch
0	0	1	1	latch
0	1	X	0	reset
1	0	X	1	set
1	1	X	?	undefined

(b) Characteristic table.

FIGURE C-23. RS flip-flop symbol and characteristic table.

characteristic table for the *JK flip-flop* is that of Fig. C-23 with JK replacing SR, except for the 11 condition.

Figure C-24 shows the JK flip-flop symbol and a circuit that converts an RS flip-flop into JK. On the symbol are optional asynchronous $\overline{\text{Preset}}$ and $\overline{\text{Clear}}$ controls, both of which are active low. When $\overline{\text{Preset}}$ is low, output Q is set, regardless of the values of J, K, and the clock. The flip-flop is reset by a 0 on input $\overline{\text{Clear}}$. If both asynchronous inputs are low, the two outputs are equal, but they change to 0 and 1 when $\overline{\text{Preset}}$ and $\overline{\text{Clear}}$ return to their inactive level. The synchronous JK inputs can change the stored bit only on a negative clock transition.

In the circuit at the right side of Fig. C-24, the AND gate outputs are

$$S = JQ' \qquad R = KQ \tag{13}$$

With JK = 00, word SR = 00, independent of Q. This is the latch state.

Now suppose JK = 01, making S equal to 0 and R equal to Q. If Q is 1, R is 1 by (13), and the output is reset on the next active clock transition. On the other hand, for Q = 0 word SR is 00, and the 0 is latched. In either case, the stored bit is 0 after the clock pulse. For JK = 10, a similar analysis shows that Q is 1 after a clock pulse, independent of its prior value.

Finally, with JK = 11, set S equals Q′ and reset R is Q. For Q = 0 before a clock pulse, word SR is 10, which sets Q when the active clock occurs. This action changes SR to 01, which resets Q on the next pulse. The process continues, with Q being complemented on each negative clock transition. This procedure is called *toggling*. Although S and R switch states whenever the clock goes low, their new values cannot affect Q until the following active clock transition because of small gate propagation delays.

The D Flip-Flop

The *D flip-flop* is usually used for temporary storage of a bit of data. There is a single data input D, and the bit on this synchronous input is transferred to output Q on an active clock transition. The symbol, the characteristic table, and a JK flip-flop converted to a D type are shown in Fig. C-25. Asynchronous Preset and Clear inputs are active in state 1, disabling both the clock and D inputs. With Preset and Clear inactive, a change in the stored bit can occur only on a positive clock transition.

The JK flip-flop of Fig. C-25 becomes a D flip-flop by the addition of the inverter, which makes K = J′. Then J is used for the data input. The same procedure converts an RS flip-flop to the D type. Available are JK′ flip-flops that

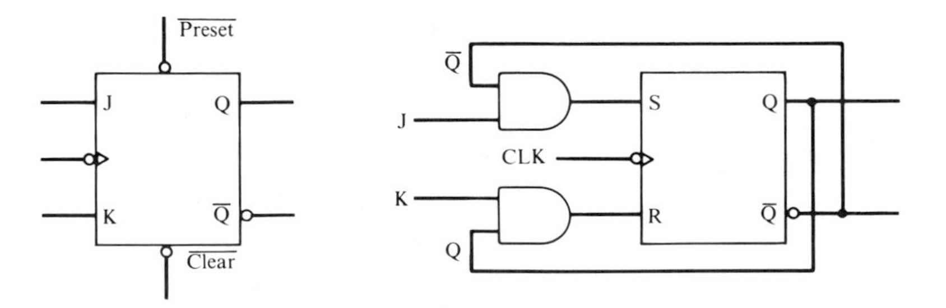

FIGURE C-24. JK flip-flop symbol and circuit.

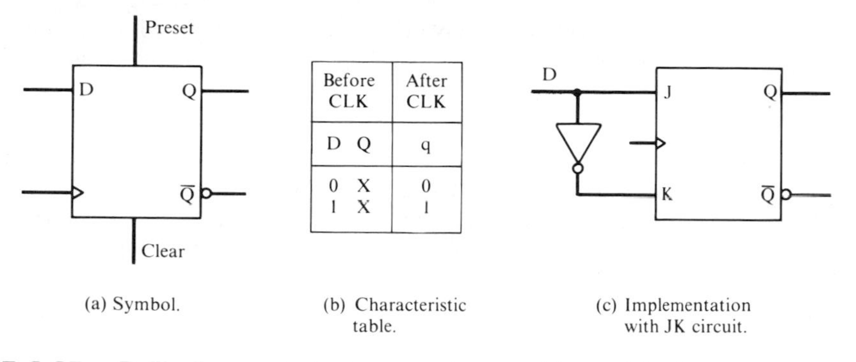

Before CLK		After CLK
D	Q	q
0	X	0
1	X	1

(a) Symbol. (b) Characteristic table. (c) Implementation with JK circuit.

FIGURE C-25. D flip-flop symbol, characteristic table, and implementation with a JK flip-flop.

can be used as D flip-flops simply by connecting together terminals J and K′. Input K′ passes through an internal inverter, and the output of this inverter is the K input of the JK flip-flop. It is also possible to convert a D flip-flop to a JK or RS type by adding appropriate combinational logic.

Exercise 1. Two 2-input AND gates have inputs AQ′ and B′Q, respectively, with B′ obtained from input B and an inverter. The AND gate outputs pass through a 2-input OR gate to the D input of a flip-flop. The flip-flop outputs are Q and Q′. Sketch the circuit and deduce the characteristic table, with columns for A, B, Q, and next state q. Identify the circuit.

The T Flip-Flop

A *T flip-flop* is simply a JK type with terminals J and K connected together to give one input T. The symbol and characteristic table are shown in Fig. C-26. It is also called a *toggle flip-flop*. When the synchronous input T is 0, the flip-flop is in the latch state, with no change in Q. However, with T = 1 the flip-flop toggles. In this case, output Q is complemented on each active clock transition. T flip-flops are not available on chips, because they are so easily made from JK circuits. When the T flip-flop is toggling, the frequency of the output pulse train is the clock frequency divided by 2. This *divide-by-2* property is utilized in many counting circuits.

Exercise 2. (a) Show combinational circuits and connections that convert an RS flip-flop to each of the following: JK, D, and T. (b) Including Preset and Clear

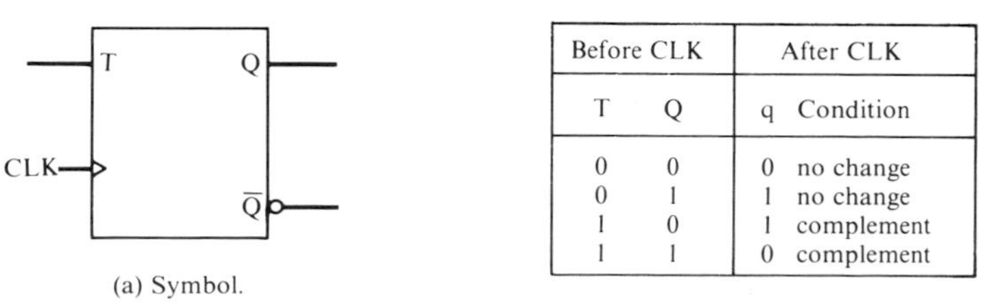

Before CLK		After CLK	
T	Q	q	Condition
0	0	0	no change
0	1	1	no change
1	0	1	complement
1	1	0	complement

(a) Symbol. (b) Characteristic table.

FIGURE C-26. T flip-flop symbol and characteristic table.

inputs that are low level active, give the characteristic table of a JK′ flip-flop. (c) Show the circuits and connections that convert a JK′ flip-flop to the D type and the T type.

Exercise 3. A positive edge-triggered T flip-flop has logic levels of 0 and 5 V. The clock frequency is 1 MHz, with a duty cycle of 20%. (a) If T is 1 at all times, determine the frequency and duty cycle of the output waveform. (b) Assume T is initially 0. At the instant of a negative clock transition, T becomes 1, and precisely three clock periods later, T returns to 0. On identical time scales, sketch and dimension the waveforms of the clock C, input T, and outputs Q and Q′.

Input Coupling

Based on input-output coupling, there are three major divisions of flip-flops. These are the *latch,* the *master-slave flip-flop,* and the *edge-triggered flip-flop.*

When enabled, latches have direct coupling, with the stored bit responding immediately to changes in the inputs. They have no dynamic indicators on their symbols. Latches can be clocked using the enable input, provided one is available, but many latches have no such control. *An even number of inverters connected in a loop constitutes a latch,* but such an arrangement has no provision for changing the stored bits at the outputs of the inverters. The RS NOR latch of Fig. C-27 consists of two inverters in a loop when R and S are both 0. The use of NOR gates instead of inverters *allows R and S to control the stored values.*

The truth table of Fig. C-27 is deduced directly from the logic. When word SR is 00, the *latch state* exists. The Q outputs are 0s when SR is 11, but this condition does not remain when SR returns to the normal latch state.

Both master-slave and edge-triggered flip-flops have dynamic inputs, usually connected to the system clock, and *coupling is not direct.* Their symbols are identical, and their performance is the same in most situations. The choice is frequently determined by availability. Whereas MOSFET flip-flops are normally the master-slave type, both types are common in TTL technology.

The master-slave flip-flop consists of a *master latch* and a *slave latch* in tandem. Let us assume that the active clock transition is low to high, so that the symbol

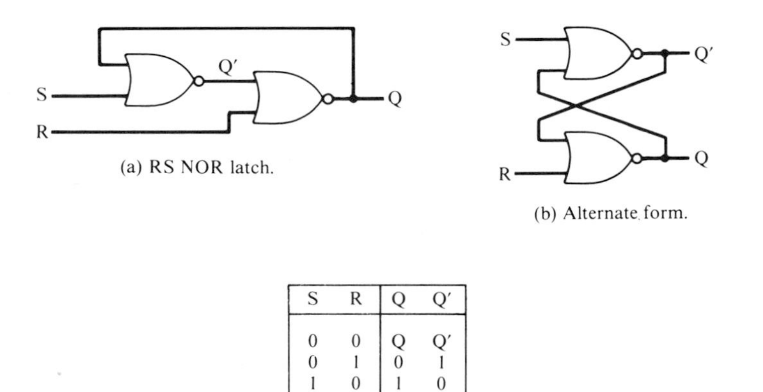

(a) RS NOR latch.

(b) Alternate form.

S	R	Q	Q′
0	0	Q	Q′
0	1	0	1
1	0	1	0
1	1	0	0

(c) Truth table.

FIGURE C-27. RS NOR latch and truth table.

has no negation indicator on its dynamic input. Then the master latch is designed to receive a bit of new data during the time when the clock is low, with the slave disabled. This bit is not available to the output at this time. When the active low-to-high clock transition occurs, the inputs to the master are disabled, and the stored bit of the master is transferred to the slave. The bit of the slave is the stored bit Q that is always present on the output line. Thus, *the output can change state only on the active clock transition,* when the bit of the master is fed to the slave.

Master-slave flip-flops are said to be *pulse triggered,* or *level triggered.* New data can enter the flip-flop whenever the clock is low, but the output is unaffected until the low-to-high transition. This condition is known as negative pulse triggering. When the clock is high, bits at the input terminals have no effect, because these inputs are disabled during this time interval. The circuit is designed so that there is *no direct coupling* at any time between input and output. *New data can be loaded into the master latch while the present bit of the slave is being transferred to the master latch of another flip-flop.*

The roles of low and high clock states are reversed when the active clock transition is high to low, as indicated by a negation indicator at the dynamic input. In this case, the flip-flop is said to be triggered on a positive pulse. A new bit is stored in the master latch during the high clock and transferred to the output when the clock goes low. An example is the TTL 7473 SSI chip, which has two independent master-slave JK flip-flops in a 14-pin DIP, each with an asynchronous clear control. Triggering occurs on a positive pulse.

In contrast to master-slave flip-flops, *edge triggered flip-flops are unaffected by data at the inputs when the clock is either high or low.* The output is established by the inputs present during the initial phase of the active clock transition. To illustrate, consider a positive edge-triggered D flip-flop with output Q′ connected to input D. With a stored bit Q of 1, input D is 0, and on the positive clock transition the stored bit changes to 1. Input D is now 0, but it has no effect until the clock once again goes low to high. The output toggles. The result is the same if the device is replaced with a master-slave flip-flop.

At no instant is there direct coupling between input and output. During the active clock transition of an edge-triggered flip-flop, input data are accepted only at the beginning of the transition; any change in output occurs near the end of the transition. Because of this timing, *the stored bit can be transmitted to another circuit while the flip-flop is simultaneously receiving a new bit.* This is an important property of both the the edge-triggered and master-slave flip-flops. It does not apply to latches. In fact, an enabled D latch is transparent.

Edge triggering may be either positive or negative, and both are readily available. An inverter at the clock input changes positive edge triggering to negative, and vice versa. The TTL 74LS364 MSI chip in a 20-pin DIP has eight positive edge-triggered D flip-flops with three-state outputs. A similar chip is the TTL 74LS363, which contains eight transparent D-type latches, also with three-state outputs.

Flip-Flop Excitation Tables

Characteristic tables of flip-flops specify the next state when the inputs and the present state are known. In the design process, the transition from the present state to the next state is determined, and the designer *must find the flip-flop inputs* that provide the desired transition. *Flip-flop excitation tables* give input values for implementation of any specified state change.

Q	q	S	R
L	L	L	X
L	H	H	L
H	L	L	H
H	H	X	L

(a) SR.

Q	q	J	K
L	L	L	X
L	H	H	X
H	L	X	H
H	H	X	L

(b) JK.

Q	q	D
L	L	L
L	H	H
H	L	L
H	H	H

(c) D.

Q	q	T
L	L	L
L	H	H
H	L	H
H	H	L

(d) T.

FIGURE C-28. Flip-Flop excitation tables with Q = present state and q = next state.

Figure C-28 shows the mixed logic excitation tables for the four basic types of flip-flops. The columns on the left side of each table show the present output Q and the next output q after a clock pulse. The columns on the right side give the inputs required for each specified transition. All the data are obtained from the flip-flop characteristic tables. Note the don't-care states. Each of these states can be L or H as convenient. As done here, the next state of a flip-flop will henceforth be represented by the lowercase form of the uppercase letter used to designate the present state. The tables in Fig. C-28 will be referred to several times in the design procedures described in the next section.

Exercise 4. A cascade of eight D flip-flops forms an 8-bit register with the LSB on the right connected through an inverter to the MSB on the left. Outputs Q′ are not connected. Initially, the stored word is 10010011. Sketch the sequential circuit. What is the initial state of the input terminal of the MSB flip-flop? Give the stored word after each of eight clock pulses from the single clock.

C-5

SEQUENTIAL CIRCUIT DESIGN

With a foundation in combinational circuit design, along with a basic knowledge of flip-flop operation, we are now ready to investigate the design of clocked sequential circuits. Although positive logic is used, the methods are also applicable to mixed logic, provided states 0 and 1 are changed to L and H, respectively.

The circuits considered here are *synchronous finite-state machines*. Each machine consists of *a combinational network, along with flip-flops connected in a feedback loop*. Inputs to the combinational network are the flip-flop outputs and the external inputs from clocked gates or registers, and outputs may include some or all of the flip-flop states. The output binary word of the flip-flops is the *machine state,* and associated with each state is a set of output variables. Changes in the flip-flop states, the external input variables, and the outputs are assumed to occur only on negative clock transitions.

The combinational circuit outputs that are connected to the flip-flops determine the next machine state. These outputs are obtained from the present state, the external inputs, and the combinational logic. Thus, *the output states and the next machine state depend on the inputs and the present machine state*. With N denoting the number of flip-flops, the number of machine states cannot exceed 2^N. It follows that a system requiring M machine states must have a number of flip-flops equal to or greater than log M, with base 2. Circuits are often designed to sequence through selected states, with other combinations of flip-flop outputs simply not encountered.

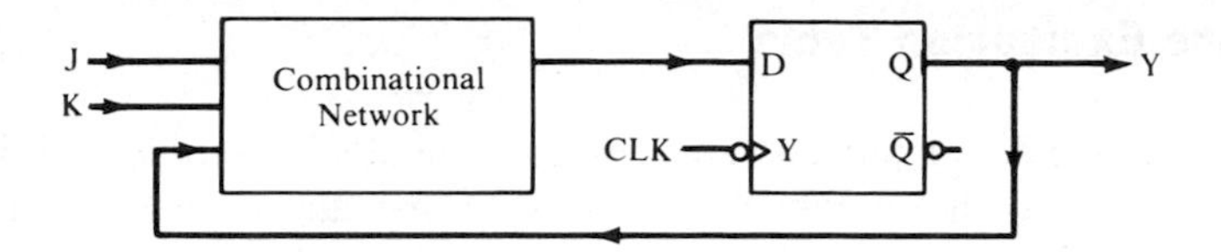

FIGURE C-29. JK flip-flop from the D type.

The State Diagram

An elementary design example is presented here for the purpose of illustrating some basic concepts of design procedure. The specific problem is *to design a JK-type flip-flop from a D-type flip-flop and combinational circuitry*. The sequential network is shown in Fig. C-29. Output Y, which equals output Q of the D flip-flop, must be related to inputs J and K, according to the JK flip-flop characteristic table. With only a single output, the circuit has two states, which are state 0 and state 1.

Let us examine the *state diagram* of the circuit, shown in Fig. C-30. The two states are identified by the two circles. One is labeled "state 0" and the other "state 1". They represent the possible flip-flop outputs. Leaving each circle are four directed lines, each identified with specific JK inputs.

Suppose the present state of the flip-flop is 0. If word JK is 00 or 01, the state must remain unchanged in accordance with the JK characteristic table. Because *a directed line must leave the present state and terminate on the next state,* the directed lines corresponding to JK words 00 and 01 leave and terminate on state 0. This action signifies *no state change* on the active clock transition. State 0 is the present state Q and also the next state q.

However, if JK is 10 or 11 with Q still cleared, the JK characteristic table specifies a transition to state 1, which becomes the next state q, and the flip-flop is set on the active clock transition. Accordingly, the corresponding lines are directed from state 0 to state 1. Arrows leaving state 1 are determined from the JK characteristic table in a similar manner.

The procedure followed here for the development of a state diagram is general. From the specified problem, the designer must determine the required number of machine states, and circles are drawn to represent them. *Within each circle, the state is identified by one or more appropriate letters or by binary bits,* as done here. Normally, the number and type of flip-flops must be selected, but in this case, a single D type was specified. Possible transitions between states are determined from the problem. These transitions are represented by directed lines, and *each line shows values of external synchronous inputs that effect the transition.* Asynchronous inputs are not usually shown on state diagrams.

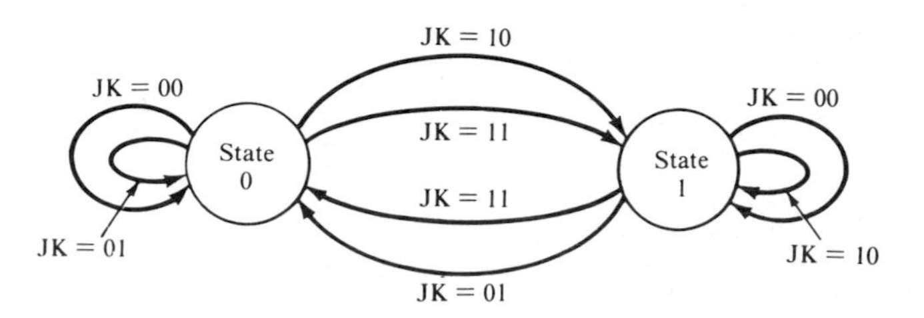

FIGURE C-30. State diagram of the JK flip-flop.

The State-Excitation Table

The next step in the design process is the development of a *state table,* or *transition table,* which presents the information of the state diagram in tabular form. It is shown in Fig. C-31, excluding the flip-flop (FF) input column at the right side, which is not part of the transition table. On the left are columns for the external inputs and all possible combinations of the bits of the present state, and to the right are columns for the next state and the present output. *All data of the state table are obtained directly from the state diagram and from the knowledge that the present output Y equals the present state Q,* as indicated on the circuit of Fig. C-29.

For example, the first data row of Fig. C-31 gives JK inputs of 0 and 0, with the present state Q being 0. From the state diagram of Fig. C-30, the next state q is 0. For the last data row with JK inputs of 11 and Q = 1, the next state q is found from the state diagram to be 0. In both cases, the present output Y is Q.

Expansion of the state table into a *state excitation table,* or simply *excitation table,* is accomplished by adding the required flip-flop input conditions. From the present state Q and the next state q of a row, the required input D can be found *with the aid of the flip-flop excitation table* of Fig. C-28. However, for the D flip-flop, D always equals the next output q. Thus, column D of Fig. C-31 is a duplication of column q.

Circuit Design

The combinational circuit is designed from data of the excitation table. The inputs are J, K, and Q, and the output is D. These columns of Fig. C-31 constitute the truth table. By rearranging the data into a Karnaugh map, the minimal sum-of-products expression for D is determined, and D is also equal to the next state q. We find that

$$q = D = J\overline{Q} + \overline{K}Q \qquad (14)$$

The combinational circuit can be implemented with a two-level NAND-NAND or AND-OR network, with the complement of K obtained from an inverter and the complement of Q obtained from the D flip-flop. CMOS JK master-slave IC flip-flops are frequently made from D-type flip-flops converted into JK by the logic of (14).

Inputs		Present State	Next State	Present Output	FF Input
J	K	Q	q	Y	D
0	0	0	0	0	0
0	0	1	1	1	1
0	1	0	0	0	0
0	1	1	0	1	0
1	0	0	1	0	1
1	0	1	1	1	1
1	1	0	1	0	1
1	1	1	0	1	0

|←——Transition Table——→|

FIGURE C-31. Excitation table for the JK flip-flop design.

The next state q of the sequential circuit that has been designed is related to the inputs and the present state Q by (14). This is a Boolean expression, *with time included,* that describes the behavior of the circuit. It also, of course, completely describes the JK characteristic table. An equation that relates the output of a flip-flop to the external inputs and the present states of the flip-flops of the circuit is called a *state equation,* or *application equation;* (14) is an example. State equations can be derived directly from state tables.

EXAMPLE 1

An N-bit binary up-counter is a circuit that counts pulses up to 2^N . It then resets to 0 and counts again. A down-counter counts in the reverse direction. Design a 2-bit synchronous up-down counter that counts up when input F is 0 and down when F is 1.

SOLUTION

Two outputs are required, with output states 00, 01, 10, and 11 corresponding to respective decimal counts of 0, 1, 2, and 3. The four-state circuit is called a *modulo-four* counter, with *modulo* referring to the number of distinct states that are sequenced by a counter before repeating. At least two flip-flops are required to generate the four states. Although any type can be chosen, let us select two negative edge-triggered JK flip-flops. This type is the most flexible, having the most don't-care states in its excitation table in Fig. C-28. Flip-flop outputs are A, B, and their complements, and the state of the synchronous sequential machine is AB.

Figure C-32 shows the state diagram. The output word of the counter is AB, and hence the present outputs are A and B. The bits indicated on the directed lines are those of external input F.

The state table is easily deduced from the state diagram. This table is shown in Fig. C-33, along with the flip-flop input conditions. Inputs JA and KA of the state excitation table refer to the JK flip-flop with output A, and these input conditions are found from columns A and a, using the JK excitation table of Fig. C-28b. Inputs JB and KB are determined similarly.

From Karnaugh maps, we find that both JA and KA can be expressed as XOR of F and B. It is evident from the excitation table of Fig. C-33 that inputs JB and KB can each be set to 1. The circuit with two flip-flops and one XOR gate is shown in Fig. C-34. A second input can be added for clearing the counter synchronously, if desired.

Analysis of a given sequential network is the inverse of the design procedure.

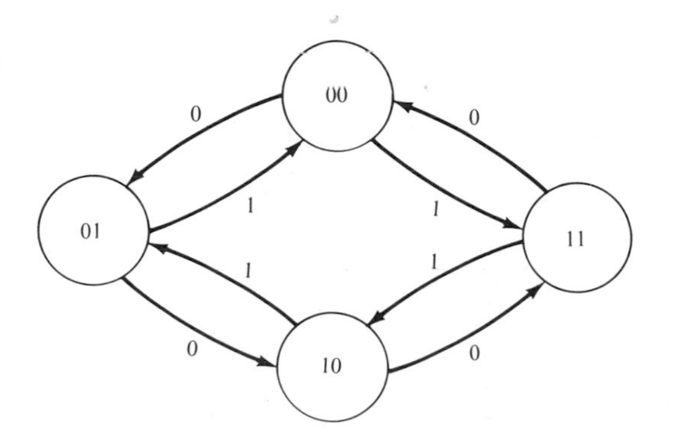

FIGURE C-32. State diagram of a 2-bit up-down counter.

Input	Present State		Next State		FF Inputs			
F	A	B	a	b	JA	KA	JB	KB
0	0	0	0	1	0	X	1	X
0	0	1	1	0	1	X	X	1
0	1	0	1	1	X	0	1	X
0	1	1	0	0	X	1	X	1
1	0	0	1	1	1	X	1	X
1	0	1	0	0	0	X	X	1
1	1	0	0	1	X	1	1	X
1	1	1	1	0	X	0	X	1

FIGURE C-33. Excitation table for the up-down counter.

First, the state excitation table is found from the network, and then the state diagram is drawn. The diagram illustrates the sequence of state transitions for specified inputs. ■

EXAMPLE 2

Analyze the synchronous sequential network of Fig. C-35.

SOLUTION

From the network we deduce that

$$JA = F \qquad JB = FA \qquad (15)$$
$$KA = \overline{F} + B \qquad KB = A \qquad Y = B$$

The excitation table is that of Fig. C-36. The three columns on the left give the eight possible combinations of F, A, and B. Output Y and the flip-flop (FF) inputs are found from (15). Then the next state ab is determined from the present state AB and the FF inputs, using the flip-flop excitation table of Fig. C-28b.

The state diagram of Fig. C-37 is deduced directly from the state excitation table. Each directed line shows the input F and the present output Y, before the transition, in the form F/Y. Output variables are often included on state diagrams in this manner. *The state diagram completely describes the operation of the given circuit.*

It should be noted that the circuit has no way of reaching state 01 from any of the other states, although it could possibly be in this state initially. If the circuit is initialized in one of the other states, then 01 becomes a don't-care state. ■

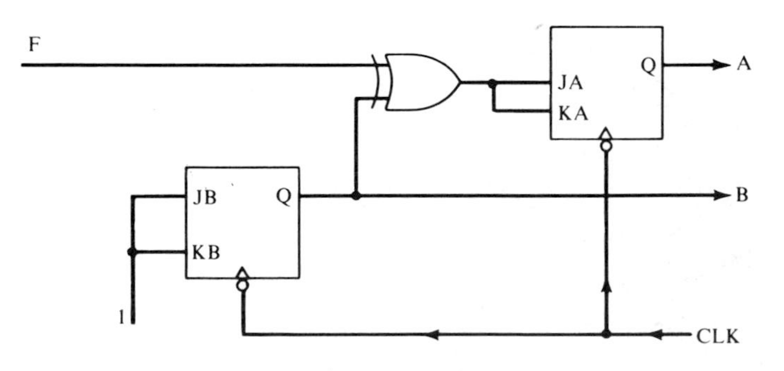

FIGURE C-34. Modulo-four up-down counter.

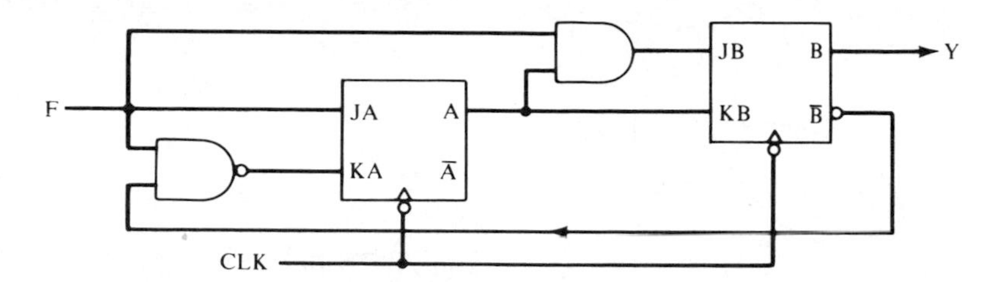

FIGURE C-35. Clocked sequential circuit for Example 2.

Exercise 1. Design a 2-bit synchronous counter that counts up when input word EF is 00 and down when it is 01. Whenever EF is 1X, the count returns to state 00 and remains there. Sketch the state diagram, showing bits EF on the directed lines, and give the excitation table in the form of that in Fig. C-33, with columns for EF. Determine the minimal sum-of-products expressions for the flip-flops. Sketch the circuit, using two JK flip-flops, an XOR gate, a two-input AND gate, a two-input OR gate, and an inverter.

Exercise 2. Using a D-type master-slave flip-flop and an XOR gate, design a T flip-flop. Show the state diagram, the state excitation table, and the Karnaugh map for input D.

Exercise 3. Using four negative edge-triggered T flip-flops and a combinational circuit based on the minimal sum-of-products expressions for the flip-flop inputs, design a binary counter that has the sequence 04285 and repeat. Others are don't-care states. Let ABCD denote both the machine state and the output word. Analyze the designed circuit, showing the state diagram that includes states treated as don't-cares.

Exercise 4. Find the state equations for a and b for the up-down counter of Fig. C-34 and for the sequential network of Fig. C-35. Give equations in minimal sum-of-products form.

Parallel-Load Register

Registers are sequential circuits and are easily designed by the methods that have been presented. Usually, the state excitation table can be developed directly from the flip-flop specifications, as illustrated by the example that follows.

Input	Present State		Next State		Present Output	FF Inputs			
F	A	B	a	b	Y	JA	KA	JB	KB
0	0	0	0	0	0	0	1	0	0
0	0	1	0	1	1	0	1	0	0
0	1	0	0	0	0	0	1	0	1
0	1	1	0	0	1	0	1	0	1
1	0	0	1	0	0	1	0	0	0
1	0	1	1	1	1	1	1	0	0
1	1	0	1	1	0	1	0	1	1
1	1	1	0	0	1	1	1	1	1

FIGURE C-36. Excitation table for Example 2.

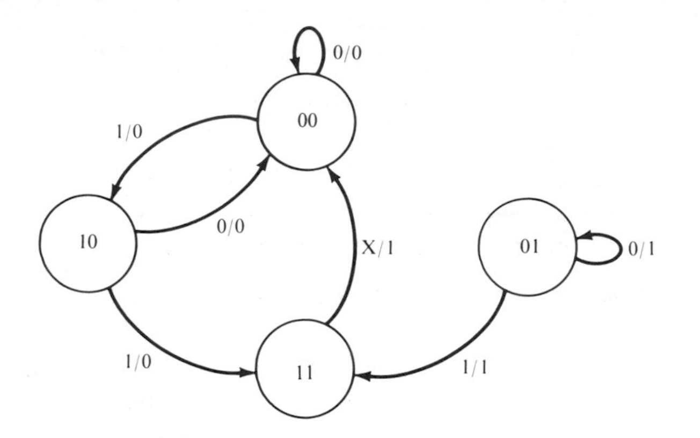

FIGURE C-37. State diagram for Example 2.

The problem is to design a parallel-load two-bit register with the function table of Fig. C-38. Input CLR is asynchronous, but loading is synchronous.

We proceed with the design. According to the specifications of the table, when CLR is 1, the register is cleared, regardless of other inputs and the present state. With both CLR and load L inactive, the next state ab equals the present state. The bits are latched, and any new input bits have no effect. Finally, for L high level with CLR inactive, the next state after a clock pulse becomes the input word EF. Because bits E and F enter simultaneously, the register has parallel loading. Output bits of the flip-flops are present on the output lines at all times, but new data enter only on an active clock transition with CLR inactive and load L active.

Either master-slave or edge-triggered JK, RS, or D flip-flops can be used. Unsuitable are T flip-flops and clocked latches. A T flip-flop does not allow a new bit to be entered independently of its present state. A clocked latch can change state whenever the clock is at its active level, and in synchronous systems it is important that state changes occur only on active transitions of the master clock.

Let us select two master-slave negative clock transition D flip-flops with asynchronous clear inputs. The clear control will be applied directly to these inputs. A state diagram can be drawn with the four states 00, 01, 10, and 11. With load L zero, each arrow leaving a state closes on the same state, but with L = 1, transitions from a state to each of the others are possible, depending on the input word EF. A diagram is unnecessary because the function table is actually an excitation table, with the flip-flop inputs being bits a and b of the next state.

If we had selected RS or JK flip-flops, additional columns for their inputs would be added to the table, with data for the columns determined from the present state, the next state, and the flip-flop excitation tables of Fig. C-28. An input bit E is transferred to the output of an RS flip-flop when S = E and R = E′, and for a JK flip-flop, J and K replace S and R.

CLR	Load L	External Inputs		Present State		Next State	
				A	B	a	b
1	X	X	X	X	X	0	0
0	0	X	X	A	B	A	B
0	1	E	F	X	X	E	F

FIGURE C-38. Function table of a 2-bit register.

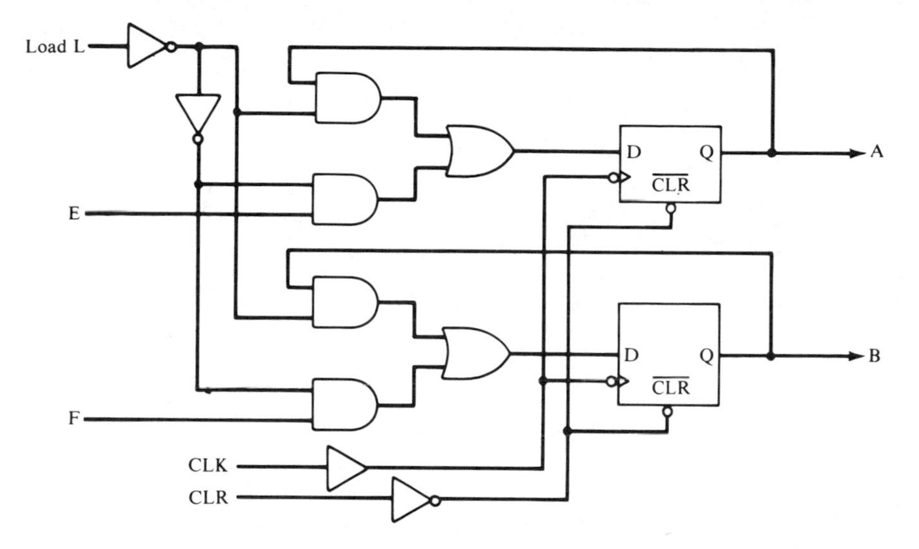

FIGURE C-39. Two-bit parallel-load D register.

From the table, the state equations are determined to be

$$a = \bar{L}A + LE \qquad (16)$$
$$b = \bar{L}B + LF$$

The present state is AB, and EF is the external input word. Flop-flop input DA is the next state a, and DB is b. The combinational circuit is designed from (16) and is shown in Fig. C-39. When load L is 0, the present state is maintained by outputs fed back into AND gate inputs. Buffers at control inputs reduce loading effects. The bit capacity of the register is easily expanded by adding additional flip-flops and logic gates, with each bit having the same circuit as those of the figure.

Exercise 5. Design a 2-bit parallel-load register with increment capability. Both load L and increment I are synchronous inputs. The register input word is EF, and the present output state is AB. Loading occurs on the active clock transition whenever L is 1, regardless of the state of I. When L is 0 and I is 1, the 2-bit stored word is incremented, and when both L and I are 0, the word is latched. Using negative edge-triggered JK flip-flops, find from a state excitation table and Karnaugh maps the flip-flop inputs in terms of L, I, E, and F. Sketch the complete circuit.

Exercise 6. Design a 4-bit parallel-load register using negative edge-triggered RS flip-flops with low-level clear inputs. The function table is that of Fig. C-38 expanded to 4 bits. The input word is EFGH and the present state is ABCD. Show the excitation table and sketch the circuit, using eight two-input AND gates, inverters, and buffered control and clock inputs. Note that the next state is E when S is E and R is E′.

Glossary

Accumulator. A storage register of the central processor unit often used for temporary storage of the result of an operation; available for general use as desired.

Acoustic coupler. A modem that uses sound wave coupling for interfacing between a computer and telephone lines.

Acronym. A word formed from the first letter of several words, such as *ROM* for *read-only memory*.

Active mode. The state of a binary signal that enables or activates a device or an operation.

Adapter card. A group of integrated circuits mounted on a circuit board that is inserted into a slot of the main system unit (motherboard), which provides the electrical connections.

Address. A binary word, with each bit combination identifying a particular location in memory.

Algorithm. An ordered, step-by-step procedure with a finite set of well-defined instructions that specify the solution to a problem or task.

Analog signal. One that can vary continuously with time. In contrast to binary digital signals, it is not restricted to two discrete states.

Arithmetic-logic unit (ALU). A digital circuit capable of performing various arithmetic and logic operations, always present in a central processor unit.

Arithmetic operator. A symbol that indicates an arithmetic operation, such as addition or multiplication.

Array. An ordered sequence of related data types assigned a single name, with the relative locations of individual elements within the array identified by index values.

ASCII. American Standard Code for Information Interchange, widely used for representing alphabetic characters, decimal digits, punctuation and special symbols, and control codes.

Assembly language. A low-level language using suggestive names, or mnemonics, to represent machine language instructions, with a one-to-one correspondence. Conversion into machine language is done with a program called an *assembler*.

Asynchronous circuit. One with an operation that is independent of the system clock.

AUTOEXEC batch file. A batch file that is automatically executed when power to the computer is applied or when a power-on reset is implemented.

BASIC. Beginner's All-purpose Symbolic Instruction Code, a widely used high-level programming language.

Batch file. A file with a single name identifying one or more programs that are executed in sequence when the file is called.

Baud. The number of times a communications signal can change per second, with each modulation change representing a change in at least one bit of a group having one or more bits as specified.

Binary coded decimal (BCD). A binary code for the decimal digits.

Binary system. A number system with only the digits 0 and 1.

Bit. Binary digit 0 or 1.

Boolean algebra. An algebra that operates on binary variables so as to facilitate the design of digital circuits; based on the logic functions NOT, AND, and OR.

Boot. To load the operating system of a computer into memory. In the PC, the boot record is the program used to boot the system.

Breakpoint. A trap within a program that temporarily stops execution so that the results can be examined.

Buffer. An intermediate storage region of memory that is used in input/output operations to hold a record of data. Also, a combinational circuit that provides increased current drive, allowing for larger loads.

Bus. A group of conductors that transmit related bits simultaneously from one point to another.

Byte. An 8-bit binary word. A kilobyte (KB) is 1024 bytes.

Cathode ray tube (CRT). The display screen.

Cell. An element that stores a single bit.

Central processor unit (CPU). The central part of a computer, containing the arithmetic-logic unit, registers, and timing and control circuits, but excluding memory and input/output devices.

Chip. An integrated-circuit package.

Clock. A pulse generator that supplies a periodic train of pulses used to control the timing of selected computer operations.

Cluster. A basic unit of diskette space assigned a 12-bit identification number. For double-sided diskettes, a cluster consists of two 512-byte sectors. For a single-sided diskette, a cluster is a sector.

COM file. A machine language program that can be derived from an EXE file that satisfies certain criteria. It can be executed simply by entering its filename, and unlike an EXE file, there is no header.

Combinational circuit. A digital circuit with one or more outputs, each of which has a binary value that is fixed by those of the inputs, with no dependence on prior history. There is no memory.

Command processor. The program COMMAND.COM, which provides many functions of the MS DOS, including all of the internal command processors, the batch file processor, and a loader for external commands.

Communications. Data transfer by means of serial bit streams, frequently over telephone lines. Telecommunication systems provide communications at a distance.

Compiler. A program that translates a high-level language into object code, which is in machine language.

Complement. To change the 0 or 1 state of each bit of a binary word to the opposite state.

Concatenate. To join together or connect character strings, modules, or data sets.

Console. The keyboard or the display screen.

Counter. A register that is either incremented or decremented on each active clock transition.

Cursor. The mark on the screen, usually a blinking underline, that indicates the position at which the next character is to be typed, deleted, or inserted.

Daisy chain. A connection of devices chained in order of priority so that signals received by the chain from a processor or polling unit are passed from each device to the next lower one in priority until a device requiring service is encountered.

Data communications equipment (DCE). Equipment designed for connection to the RS232C interface similar to the way a modem is normally connected. Data are transmitted on pin 3 and received on pin 2.

Data terminal equipment (DTE). Equipment designed for connection to the RS232C interface similar to the way a computer is normally connected. Data are transmitted on pin 2 and received on pin 3.

Debugging. The process of finding and removing errors, or bugs, from a program.

Decoder. A combinational circuit with binary bits on input lines that select and activate one of several output lines.

DeMorgan's laws. Two Boolean identities used to simplify logic expressions and diagrams.

Demultiplexer. A combinational circuit that transfers data from a single input line to an output selected from those available.

Direct memory access (DMA). A method for transfer of data directly between a peripheral device and memory without processor intervention.

Directive. An instruction to a program that operates on another program, such as an assembler directive.

Diskette. A thin, flexible disk, typically 5.25 inches in diameter, coated with an oxide that can store digital data serially in the form of magnetic flux patterns. Unlike the inflexible, hard, or fixed disk, a diskette can easily be removed from its drive by the operator. It is also called a *floppy disk*. One can store over 360,000 characters on its two sides.

Disk operating system (DOS). A program on a diskette that is loaded into memory and used to control the operation of the disk system and its files.

Don't-care state. A binary state having no effect on a digital operation, regardless of its value of 0 or 1.

Dumb terminal. A terminal that communicates with a computer, with operation restricted to simple transmission from a keyboard and reception to a display. Special features such as automatic dialing and answering are not available.

Dynamic circuit. A circuit with charges representing 0 and 1 bits stored on leaky capacitances, with the charges refreshed periodically to prevent loss of data.

Dynamic indicator. A small triangle placed at a terminal of a clock input indicating that the signal is effective only during a low-to-high transition, or during a high-to-low transition if a negation indicator is also present.

Edge-triggered input. An input pin that is activated only when its signal makes a transition from one state to the other.

Editor. A program used for editing text (ASCII) files.

Emulated disk. A portion of memory reserved for use as an artificial disk and treated as an additional disk.

Encode. To represent an entity by means of a code.

End-of-file (EOF) mark. A byte used to denote the end of a file. In the PC this mark is hexadecimal 1A. When a file is displayed or printed, the EOF mark ends the operation.

Equivalence (EQV). The Exclusive-NOR (XNOR) two-input logic function, which gives an output of 1 only for inputs of 1 and 1 or 0 and 0.

Exclusive-OR (XOR). A two-input logic function that gives 1 if one and only one input is 1. The Exclusive-NOR (XNOR) function is the complement of XOR.

EXE file. A diskette machine language file generated by either an assembler or a compiler and executed simply by entering its filename. It includes a header with data that enables the loader to assign addresses that depend on the location of the program in memory.

Extended code. A 2-byte code consisting of a high-byte scan code or pseudo scan code and a low byte of 0. It is used to identify special key and key combinations that have no defined ASCII codes.

File. A collection of related records of data treated as a unit and kept in nonvolatile storage such as a disk or a filing cabinet.

File allocation table (FAT). A table and its duplicate kept on every diskette, which links the clusters of each diskette file in the proper order and identifies the clusters that are unused and available.

Filespec. A file identifier consisting of the drive letter followed by a colon, a filename of up to eight characters, and an optional extension consisting of a period followed by up to three characters, such as A: XY.COM.

Firmware. Operating system programs in read-only memory.

Fixed disk. A hard disk referred to as fixed because it is not removable from its drive, as is a diskette.

Fixed-point number. A number containing a decimal point.

Flag bit. A bit of a flip-flop or a memory location used to indicate either a status or control condition, such as a 0 result or the enabling of interrupts.

Flip-flop. A sequential circuit element that stores a single binary bit, with the value depending on the inputs and their past history. Usually the state of the flip-flop can change only on an active clock transition. Four types are RS, JK, D, and T, and a special case is the latch.

Floating point number. A real number in scientific notation, consisting of a number followed by an exponent representing a power of 10.

Floating state. An effective open circuit of a three-state pin, also called the *high-impedance state*.

Flowchart. A graphic representation of an algorithm, using distinctively-shaped interconnected blocks.

Frequency. The number of repetitions per second of a periodic signal, such as a pulse train from a clock.

Full-duplex communications. A two-channel system that allows for transmission and reception simultaneously.

Garbage. Words with bits having no significance whatsoever.

Gate. A hardware circuit that manipulates binary input data, generating combinational logic functions.

Global character. A special character that represents any character or any sequence of characters.

Half-duplex communications. A one-channel system that allows both transmission and reception, but not simultaneously.

Handshake signals. Request and acknowledge signals used to control asynchronous data transfer.

Hard copy. A printed copy.

Hard disk. A nonflexible, nonremovable magnetic disk, also called a *fixed disk,* which has a large storage capacity.

Hertz (Hz). Unit of frequency representing one repetition per second. A kilohertz (kHz) is 1000 Hz and a megahertz (MHz) is 1 million Hz.

Hexadecimal number. A radix-16 number with 16 digits, which are 0–9, A, B, C, D, E, and F.

High-impedance state. An effective open circuit of a three-state pin, also called the *floating* state.

High-level language. A language independent of any particular machine, with statements that can be compiled into sequences of machine code.

Home. The upper-left-hand character position of the display screen.

Implication (IMP). Logical operation (NOT A) OR B.

Inclusive OR. The OR operation, which gives 0 only if all inputs are 0.

Index. A number representing an element of an array. Also, an offset address from a base address.

Integer. A signed number with neither a decimal point nor an exponent and usually restricted in range.

Intelligent device. A device containing a microprocessor that provides special features that may be nonessential.

Interface. A connection between software modules or a circuit that joins together hardware modules.

Interpreter. A program that translates a single statement of a high-level language into a sequence of machine instructions, which are immediately executed. Then the next instruction is translated and executed, and so on. The procedure is repeated each time the program is run.

Interrupt. A branch to a location in a table of vector addresses, each of which directs the program to a service routine. After execution of the routine, the program normally returns to the instruction following the last one executed prior to the interrupt. An interrupt can be initiated by a software instruction present in the program, by an internal event, or by hardware that supplies a signal to a processor interrupt input pin.

Inverter. A logic circuit that changes the binary state of the input variable

Karnaugh map. A truth table rearranged into a form that provides a direct way to reduce a Boolean logic function to the simplest possible form.

Kilobyte (KB). Denotes 1024 bytes (K = 1024).

Latch. A flip-flop with direct coupling from input to output when enabled, with the output equal to the input at every instant. In contrast, other types of flip-flops never have direct coupling. When disabled, the stored bit is not affected by the input lines and is said to be latched.

Level-triggered input. An input pin activation by a signal in its active state, in contrast to edge triggering.

LIFO stack. A last-in first-out stack, designed so that a word taken out (read) is the last one that was put in.

Line editor. An editor program that edits individual lines, but not individual words, of a text (ASCII) file.

Line printer. A printer that stores and processes the characters of entire lines, one line at a time.

Linker. A program that accepts one or more machine language object (OBJ) files, links them together into a single unit, and generates an executable (EXE) file with a header, which contains data used by the loader to fix addresses that depend on the program's location.

Listing file. A printable text file generated by an assembler or compiler, which contains the source program and other information.

Loader. A program used to load another program into memory, inserting into the program any addresses that depend on program location. For example, in the PC the bootstrap loader of ROM loads the boot record from a diskette into memory, which in turn loads the operating system programs.

Logical address. An address consisting of a segment address followed by an offset address. The physical address is the sum of the segment address multiplied by 16 and the offset. For example, hexadecimal F000:1234 is physical address F1234.

Logical expression. A combination of constants, variables, and relational operators that gives a true or false condition.

Logical operator. A Boolean operator such as NOT and AND.

Logical program line (BASIC). A line with 1 to 255 characters.

Machine. A digital system, consisting of both hardware and software, designed to implement a specified function.

Machine language. Instructions to a computer in binary code.

Macro instruction. A user-defined assembly language instruction sequence identified by a single name.

Mainframe. A large computing system designed to serve multiple users concurrently and efficiently.

Mark state. The logical 1 state of a bit on a communications line, having a voltage between -3 and -15 V.

Memory-mapped input/output (I/O). I/O with logical addresses and memory reference instructions that generate MEMR and MEMW control signals. Device I/O is treated the same as memory I/O.

Microcomputer. A computer using a microprocessor as its central processor unit, designed for individual use.

Minicomputer. A computer with greater computing power than a microcomputer but less than that of a mainframe.

Minterm. Each combination of variables of a truth table.

Mixed logic. A logic system based on voltage-level states high (H) and low (L), with 0 and 1 states denoting inactive and active states, respectively, rather than voltage levels.

Mnemonic. A name used to represent a digital signal, spelled so as to indicate its function, such as CLK.

Modem. A data set that provides the interface between circuits of the communications adapter and the serial data communications line.

Module. A functional unit of code or a discrete unit of a program that can be linked with other units. Also, a related group of interconnected integrated circuits.

Monitor. The display unit. Also, a coordinated collection of programs that continuously supervise, control, and verify operations of a computer system, usually stored in read-only memory, on a disk, or both.

Motherboard. The board that contains the microprocessor, its supporting chips, and slots for adapter cards.

MS DOS. A widely used operating system that is reasonably compatible with the PC DOS system of the IBM PC.

Multiplexer. A data selector combinational circuit that selects one of several input lines and supplies its bit to the single output.

Negation indicator. A small circle at an input or output terminal of a logic circuit denoting inversion of the normal state (not used with mixed logic).

Negative logic. Logic that uses 0 for the high voltage level and 1 for the low level.

Nibble. A 4-bit word.

Nonvolatile memory. Memory that retains stored data without the need for electrical power, such as a disk.

Null string. A defined string having no characters.

Object program. The output of an assembler or a compiler, containing machine code that is converted into executable code by the linker and loader.

Octal number. One with radix 8, having digits 1 to 7.

Operand. A sequence of bits to be operated on in some manner.

Operating system. A collection of programs that govern the operation of the system. They include routines for control of input and output devices, along with text editors, program debuggers, module linkers, and program loaders.

OR function. A function with an output of 1 if and only if at least one input has the 1 state.

Overflow. A condition that exists when significant bits of the result of an operation are lost because of inadequate space in the storage location.

Packed words. Binary words stored so that the smallest possible number of memory locations are used.

Paragraph. A 16-byte region of the address space with a starting address divisible by 16.

Parallel transfer. Data transfer of a binary word with all bits moved simultaneously.

Parity. An error-checking method that consists of appending a 0 or 1 bit to a binary word so as to make the total number of 1s even or odd, as specified. A change in parity after storage or transmission indicates an error.

Peripheral. An input or output device.

Physical address. The address placed on the address bus.

Physical line. A line of 1 to 80 characters, which fits on one line of the display.

Pipelining. Organization of a computer so that parts with different functions operate simultaneously, thereby increasing throughput.

Piping. The chaining of DOS commands with automatic redirection of standard input and output, so that the output from one command is directed to the input of the next. Chained commands are separated by the symbol |.

Pointer register. One used to hold an address of an operand.

Polarity indicator. A half-arrowhead at an input or output terminal of a logic circuit, denoting inversion of the normal state (used only with mixed logic).

Polling. Checking a flag bit at periodic intervals.

Port. A pin or a group of pins used for transfer of data between the processor and any peripheral unit.

Positive logic. Logic that uses 0 for the low voltage level and 1 for the high level.

Preset. To set a storage cell to 1.

Primitive instruction. A string instruction that is automatically repeated a specified number of times.

Program. A set of instructions that are executed in sequence, containing one or more modules.

Program segment prefix (PSP). A 256-byte segment set up by the command processor immediately ahead of each program loaded into memory, containing data useful to the program.

Pseudo code. An algorithm using statements that can be readily translated into a program in either a high-level or a low-level language.

Pseudo scan code. A keyboard scan code assigned to certain key combinations.

Radix. The base of a number system, which is 2 for binary, 8 for octal, 10 for decimal, and 16 for hexadecimal.

Random access. Access to records of a file, with each access being random, independent of any prior access.

Random access memory (RAM). Read-write volatile memory.

Real number. Either a fixed point or floating point number.

Record. A data set having a fixed number of components, possibly of different types (such as NAME and AGE).

Redirection of standard I/O. Specification of a device other than the keyboard for standard input or a device other than the display for standard output.

Refresh. To restore the charges representing 0 and 1 bits of dynamic circuits. Because these charges leak to ground, they must be restored periodically to prevent loss of data. Typically, each cell of a dynamic memory is refreshed once every 2 milliseconds.

Register. A group of binary storage cells with associated circuits for loading and manipulating their bits.

Relational operator. An operator used for comparison of expressions, giving a true or false result.

Reset. To clear the bit of a cell to 0.

Scan code. A byte that identifies a key of the keyboard. The code is sent to a buffer when a key is pressed.

Scratchpad register. A general-purpose register often used to store intermediate results of arithmetic and logic operations.

Segment address. An address of one of the four segment registers of the 8088, which gives a base address when multiplied by 16. The base address plus a specified offset gives a physical address.

Sequential access. Access to records of a file, with each access being to the next record in sequence.

Sequential circuit. A digital circuit in which the output word has more than one state for at least one combination of states of the inputs, which implies dependence on past history.

Serial data. Data with bits transmitted serially, one at a time, as along communications lines.

Set. To preset a storage cell to 1.

Shift key. A key normally used in combination with another one (INS, CAPS, NUM, SCROLL, ALT, CTRL, LEFT, RIGHT).

Sign bit. The most significant bit of a byte or word operation, with no significance for unsigned numbers. For signed numbers, a 1 denotes a negative number.

Simulated disk. An emulated disk, or RAM disk, that treats a reserved area of memory as a diskette.

Sink current. The current into a terminal.

Soft key. A key of the keyboard not dedicated to a specific function but available for special use as desired.

Software. Programs that direct the operation of a machine.

Source current. The current out of a terminal.

Source program. One written directly by the programmer.

Space state. The logical 0 state of a bit on a communications line, with a voltage between 3 and 15 V.

Stack. A region of memory with an associated pointer, which is used for temporary storage of return addresses and other data.

Standard input/output. I/O with 8-bit peripheral addresses, using IN and OUT instructions that generate IOR and IOW control signals.

State. For a binary word, each combination of its bits represents a state, which can be assigned a name.

Status flag. A bit of a flip-flop or memory that provides status information, such as the parity flag.

String. A sequence of one or more characters.

Structured program. One designed from the top down using the basic structures of sequence, choice, and repetition, with GOTO statements generally avoided.

Subroutine. A procedure that is called by the main program or another subroutine.

Subscript. The index number of an element of an array.

Synchronous input signal. One that exercises control only when the clock makes an active transition.

Synchronous system. One in which continuous time is divided into discrete intervals by means of a clock, with outputs of the digital circuits allowed to change only on active clock transitions. Circuits switch in unison.

System unit. The name of the unit containing the main computer board with its microprocessor and I/O slots.

Telecommunications. Communications at a distance, with serial data transfer.

Teletypewriter (TTY). An I/O device, called a *teletype*, with a keyboard, printer, paper reader, and paper punch, that transmits and receives serial data.

Text file. A printable file consisting of ASCII characters organized into lines ending with carriage-return line-feed codes and terminating with an end-of-file mark.

Three-state terminal. One with states 0, 1, and floating, or high impedance, or tristate, which is an effective open circuit.

Time multiplexing. Use of the same lines for different purposes at different times, such as lines used first for address bits and then for data bits.

Toggle key. A shift key with two distinct functions, including INS, CAPS, NUM, and SCROLL.

Top-down design. A method in which programming starts at the highest level of decision and control, proceeding downward through lower-level modules that implement specific functions.

Transceiver. A bidirectional integrated-circuit data buffer with an input that controls the transfer direction.

2s complement of a number. The number complemented and added to 1, with any output carry discarded. A positive number is the 2s complement of the corresponding negative number, and vice versa.

Typematic key. A key that repeats as long as it is depressed.

USART. A communications controller chip, the universal synchronous/asynchronous receiver/transmitter. Similar is the UART, with asynchronous operation only.

Variable. A quantity with a value that can be changed as often as desired, such as a named memory location.

Vector address. An address of a memory location that is stored in memory, which points to a program.

Virtual storage. Addressable space that appears to the user as real memory, although actual data may reside on magnetic disks or other storage devices.

Volatile memory. Memory that loses all data when power is removed, such as semiconductor RAMs.

Wait state. A clock state during which the processor idles, sometimes inserted into a machine cycle for delay.

Word. A string of 1 or more binary bits, often used to denote a bit length equal to the width of a register.

Word processor. A program that edits individual words and characters of a file, including insertions and deletions, along with numerous other functions such as block moves, file search, file linkage, file deletion, print control, page numbering, title insertion, margin alignment, and underlining.

XOR function. A function with an output that is 1 if and only if one and only one of two inputs is 1.

Index

B

C

D

E

F

G

H

L

M

N

O

P

Q

R

S